The Calculus Collection
A Resource for AP* and Beyond

About the Cover:
El puente del Cristo de la Expiración, popularly known as the Cachorro or Chapina bridge, spans the río Guadalquivir in Seville, Spain. The awnings over the walkways on the bridge have the approximate shape of a hyperbolic paraboloid. Canoeing (piragüismo in Spanish) is popular on the river.
Photo by Roger B. Nelsen.

*AP is a registered trademark of the College Board, which was not involved in the production of this publication.

© 2010 by
The Mathematical Association of America (Incorporated)
Library of Congress Catalog Card Number 2009937591
ISBN 978-0-88385-761-8
Printed in the United States of America
Current Printing (last digit)
10 9 8 7 6 5 4 3 2 1

The Calculus Collection
A Resource for AP* and Beyond

Editors

Caren L. Diefenderfer
Hollins University

Roger B. Nelsen
Lewis & Clark College

*Published and Distributed by the
Mathematical Association of America*

Committee on Books
Paul Zorn, *Chair*

Classroom Resource Materials Editorial Board
Gerald M. Bryce, *Editor*

William C. Bauldry
Diane L. Herrmann
Loren D. Pitt
Wayne Roberts
Susan G. Staples
Philip D. Straffin
Holly S. Zullo
Michael Bardzell

CLASSROOM RESOURCE MATERIALS

Classroom Resource Materials is intended to provide supplementary classroom material for students—laboratory exercises, projects, historical information, textbooks with unusual approaches for presenting mathematical ideas, career information, etc.

101 Careers in Mathematics, 2nd edition edited by Andrew Sterrett

Archimedes: What Did He Do Besides Cry Eureka?, Sherman Stein

The Calculus Collection: A Resource for AP* and Beyond, edited by Caren Diefenderfer and Roger B. Nelsen

Calculus Mysteries and Thrillers, R. Grant Woods

Conjecture and Proof, Miklós Laczkovich

Creative Mathematics, H. S. Wall

Environmental Mathematics in the Classroom, edited by B. A. Fusaro and P. C. Kenschaft

Exploratory Examples for Real Analysis, Joanne E. Snow and Kirk E. Weller

Geometry From Africa: Mathematical and Educational Explorations, Paulus Gerdes

Historical Modules for the Teaching and Learning of Mathematics (CD), edited by Victor Katz and Karen Dee Michalowicz

Identification Numbers and Check Digit Schemes, Joseph Kirtland

Interdisciplinary Lively Application Projects, edited by Chris Arney

Inverse Problems: Activities for Undergraduates, Charles W. Groetsch

Laboratory Experiences in Group Theory, Ellen Maycock Parker

Learn from the Masters, Frank Swetz, John Fauvel, Otto Bekken, Bengt Johansson, and Victor Katz

Math Made Visual: Creating Images for Understanding Mathematics, Claudi Alsina and Roger B. Nelsen

Ordinary Differential Equations: A Brief Eclectic Tour, David A. Sànchez

Oval Track and Other Permutation Puzzles, John O. Kiltinen

A Primer of Abstract Mathematics, Robert B. Ash

Proofs Without Words, Roger B. Nelsen

Proofs Without Words II, Roger B. Nelsen

She Does Math!, edited by Marla Parker

Solve This: Math Activities for Students and Clubs, James S. Tanton

Student Manual for Mathematics for Business Decisions Part 1: Probability and Simulation, David Williamson, Marilou Mendel, Julie Tarr, and Deborah Yoklic

Student Manual for Mathematics for Business Decisions Part 2: Calculus and Optimization, David Williamson, Marilou Mendel, Julie Tarr, and Deborah Yoklic

Teaching Statistics Using Baseball, Jim Albert

Visual Group Theory, Nathan C. Carter

Writing Projects for Mathematics Courses: Crushed Clowns, Cars, and Coffee to Go, Annalisa Crannell, Gavin LaRose, Thomas Ratliff, Elyn Rykken

MAA Service Center
P.O. Box 91112
Washington, DC 20090-1112
1-800-331-1MAA   FAX: 1-301-206-9789

Contents

Preface xvii
Introduction xix
   David M. Bressoud

Part 0. General and Historical Articles   **1**

Touring the Calculus Gallery 2
   William Dunham
   American Mathematical Monthly, (January 2005), pp. 1–19.

Calculus: A Modern Perspective 21
   Jeff Knisley
   American Mathematical Monthly, (October 1997), pp. 724–727.

Two Historical Applications of Calculus 25
   Alexander J. Hahn
   College Mathematics Journal, (March 1998), pp. 93–103.

Ideas of Calculus in Islam and India 36
   Victor J. Katz
   Mathematics Magazine, (June 1995), pp. 163–174.

A Tale of Two CD's 48
   Dan Kennedy
   American Mathematical Monthly, (August 1994), pp. 603–608.

The Changing Face of Calculus: First Semester Calculus as a High School Course 54
   David M. Bressoud
   FOCUS, (August–September 2004), pp. 6–8.

Things I Have Learned at the AP Calculus Reading 59
   Dan Kennedy
   College Mathematics Journal, (November 1999), pp. 346–355.

Book Review: *Calculus with Analytic Geometry* 69
   Underwood Dudley
   American Mathematical Monthly, (November 1988), pp. 888–892.

The All-Purpose Calculus Problem 74
   Dan Kennedy
   Math Horizons, (Spring 1994), p. 5.

Part 1. Functions, Graphs and Limits — 75

Graphs of Rational Functions for Computer Assisted Calculus — 76
Stan Byrd and Terry Walters
College Mathematics Journal, (September 1991), pp. 332–334.

Computer-Aided Delusions — 80
Richard L. Hall
College Mathematics Journal, (September 1993), pp. 366–369.

An Overlooked Calculus Question — 84
Eugene Couch
College Mathematics Journal, (November 2002), pp. 399–400.

Introduction to Limits, or Why Can't We Just Trust the Table? — 86
Allen J. Schwenk
College Mathematics Journal, (January 1997), p. 51.

A Circular Argument — 87
Fred Richman
College Mathematics Journal, (March 1993), pp. 160–162.

A Geometric Proof of $\lim_{d \to 0^+} (-d \ln d) = 0$ — 91
John H. Mathews
College Mathematics Journal, (May 1992), pp. 209–210.

Part 2. Derivatives — 93

The Changing Concept of Change: The Derivative from Fermat to Weierstrass — 94
Judith V. Grabiner
Mathematics Magazine, (September 1983), pp. 195–206.

Derivatives Without Limits — 106
Harry Sedinger
College Mathematics Journal, (January 1980), pp. 54–55.

Rethinking Rigor in Calculus: The Role of the Mean Vale Theorem — 107
Thomas W. Tucker
American Mathematical Monthly, (March 1997), pp. 231–240.

Rolle over Lagrange—Another Shot at the Mean Value Theorem — 118
Robert S. Smith
College Mathematics Journal, (November 1986), pp. 403–406.

An Elementary Proof of a Theorem in Calculus — 122
Donald E. Richmond
American Mathematical Monthly, (October 1985), pp. 589–590.

A Simple Auxiliary Function for the Mean Value Theorem — 123
Herb Silverman
College Mathematics Journal, (September 1989), p. 323.

A Note on the Derivative of a Composite Function — 124
V. N. Murty

College Mathematics Journal, (January 1980), p. 50.

Do Dogs Know Calculus? **125**
Timothy J. Pennings
College Mathematics Journal, (May 2003), pp. 178–182.

Do Dogs Know Related Rates Rather than Optimization? **130**
Pierre Perruchet and Jorge Gallego
College Mathematics Journal, (January 2006), pp. 16–19.

Do Dogs Know Bifurcations? **133**
Roland Minton and Timothy J. Pennings
College Mathematics Journal, (November 2007), pp. 356–361.

The Lengthening Shadow: The Story of Related Rates **139**
Bill Austin, Don Barry and David Berman
Mathematics Magazine, (February 2000), pp. 3–12.

The Falling Ladder Paradox **149**
Paul Sholten and Andrew Simonson
College Mathematics Journal, (January 1986), pp. 49–54.

Solving the Ladder Problem on the Back of an Envelope **154**
Dan Kalman
Mathematics Magazine, (June 2007), pp. 163–182.

How Not to Land at Lake Tahoe! **175**
Richard Barshinger
American Mathematical Monthly, (May 1992), pp. 453–455.

The Best Shape for a Tin Can **178**
P. L. Roe
College Mathematics Journal, (May 1993), pp. 233–236.

To Build a Better Box **182**
Kay Dundas
College Mathematics Journal, (January 1984), pp. 30–36.

The Curious 1/3 .... **189**
James E. Duemmel
College Mathematics Journal, (May 1993), pp. 236–237.

Hanging a Bird Feeder: Food for Thuoght **191**
John W. Dawson, Jr.
College Mathematics Journal, (March 1990), pp. 129–130.

Honey, Where Should We Sit? **193**
John A. Frohliger and Brian Hahn
Mathematics Magazine, (December 2005), pp. 379–385.

A Dozen Minima for a Parabola **200**
Leon M. Hall
College Mathematics Journal, (March 2003), pp. 139–141.

Maximizing the Area of a Quadrilateral **203**
Thomas Peter
College Mathematics Journal, (September 2003), pp. 315–316.

A Generalization of the Minimum Area Problem 205
Russell A. Gordon
College Mathematics Journal, (January 2003), pp. 21–23.

Constrained Optimization and Implicit Differentiation 208
Gary W. DeYoung
College Mathematics Journal, (March 2003), pp. 148–152.

For Every Answer There are Two Questions 213
A. M. Fink and Juan A. Gatica
Mathematics Magazine, (June 1992), pp. 182–185.

Old Calculus Chestnuts: Roast, or Light a Fire? 217
Margaret Cibes
College Mathematics Journal, (May 1993), pp. 241–243.

Cable-laying and Intuition 220
Yael Roitberg and Joseph Roitberg
College Mathematics Journal, (January 2001), pp. 52–54.

Descartes Tangent Lines 223
William Barnier and James Jantosciak
College Mathematics Journal, (January 2007), pp. 47–49.

Can We Use the First Derivative to Determine Inflection Points? 226
Duane Kouba
College Mathematics Journal, (January 1995), pp. 31–34.

Differentiate Early, Differentiate Often! 230
Robert Dawson
College Mathematics Journal, (November 2005), pp. 404–407.

A Calculus Exercise for the Sums of Integer Powers 235
Joseph Wiener
Mathematics Magazine, (October 1992), pp. 249–251.

L'Hôpital's Rule Via Integration 238
Donald Hartig
American Mathematical Monthly, (February 1991), pp. 156–157.

Indeterminate Forms Revisited 240
R. P. Boas
Mathematics Magazine, (June 1990), pp. 155–159.

The Indeterminate Form 0^0 245
Louis M. Rotando and Henry Korn
College Mathematics Journal, (January 1977), pp. 41–42.

On the Indeterminate Form 0^0 247
Leonard J. Lipkin
College Mathematics Journal, (January 2003), pp. 55–56.

Variations on a Theme of Newton 249
Robert M. Corless
Mathematics Magazine, (June 1990), pp. 155–159.

A Useful Notation for Rules of Differentiation 257

Robert B. Gardner
College Mathematics Journal, (September 1993), pp. 351–352.

Wavefronts, Box Diagrams, and the Product Rule: A Discovery Approach . . **259**
John W. Dawson, Jr.
College Mathematics Journal, (March 1980), pp. 102–106.

$(x^n)' = nx^{n-1}$: Six Proofs **264**
Russel Jay Hendel
College Mathematics Journal, (September 1990), pp. 312–313.

Sines and Cosines of the Times **266**
Victor J. Katz
Math Horizons, (April 1995), pp. 5–6.

The Spider's Spacewalk Derivation of sin′ and cos′ **268**
Tim Hesterberg
College Mathematics Journal, (March 1995), pp. 144–145.

Differentiability of Exponential Functions **270**
Philip M. Anselone and John W. Lee
College Mathematics Journal, (November 2005), pp. 388–393.

A Discover-*e* **276**
Helen Skala
College Mathematics Journal, (March 1997), pp. 128–129.

An Exponential Rule **278**
G. E. Bilodeau
College Mathematics Journal, (September 1993), pp. 350–351.

The Derivative of Arctan*x* **279**
Norman Schaumberger
College Mathematics Journal, (September 1982), pp. 274–276.

The Derivative of the Inverse Sine **281**
Craig Johnson
College Mathematics Journal, (September 1998), p. 313.

Graphs and Derivatives of the Inverse Trig Functions **282**
Daniel A. Moran
College Mathematics Journal, (November 1991), p. 417.

Logarithmic Differentiation: Two Wrongs Make a Right **283**
Noah Samuel Brannen and Ben Ford
College Mathematics Journal, (November 2004), pp. 388–390.

A Comparison of Two Elementary Approximation Methods **286**
Harvey Diamond and Louise Raphael
Mathematics Magazine, (December 1994), pp. 359–365.

Part 3. Integrals   293

How Should We Introduce Integration **294**
David M. Bressoud
College Mathematics Journal, (September 1992), pp. 296–298.

Evaluating Integrals Using Self-Similarity **297**
  Robert S. Strichartz
  American Mathematical Monthly, (April 2000), pp. 316–326.

Self-Integrating Polynomials **308**
  Jeffrey A. Graham
  College Mathematics Journal, (September 2005), pp. 318–320.

Symmetry and Integration **311**
  Roger Nelsen
  College Mathematics Journal, (January 1995), pp. 39–41.

Sums and Differences vs Integrals and Derivatives **314**
  Gilbert Strang
  College Mathematics Journal, (January 1990), pp. 20–27.

How Do You Slice the Bread? **322**
  James Colin Hill, Gail Nord, Eric Malm and John Nord
  College Mathematics Journal, (September 2005), pp. 323–326.

Disks and Shells Revisited **326**
  Walter Carlip
  American Mathematical Monthly, (February 1991), pp. 154–156.

Disks, Shells, and Integrals of Inverse Functions **328**
  Eric Key
  College Mathematics Journal, (March 1994), pp. 136–138.

Characterizing Power Functions by Volumes of Revolution **331**
  Bettina Richmond and Tom Richmond
  College Mathematics Journal, (January 1998), pp. 40–41.

Gabriel's Wedding Cake **334**
  Julian F. Fleron
  College Mathematics Journal, (January 1999), pp. 35–38.

Can You Paint a Can of Paint? **338**
  Robert M. Gethner
  College Mathematics Journal, (November 2005), pp. 400–402

A Paradoxical Paint Pail **341**
  Mark Lynch
  College Mathematics Journal, (November 2005), pp. 402–404.

Dipsticks for Cylindrical Storage Tanks—Exact and Approximate **343**
  Palm Littleton and David A. Sanchez
  College Mathematics Journal, (November 2001), pp. 352–358.

Finding Curves with Computable Arc Length **350**
  John Ferdinands
  College Mathematics Journal, (May 2007), pp. 221–222.

Arc Length and Pythagorean Triples **352**
  Courtney Moen
  College Mathematics Journal, (May 2007), pp. 222–223.

Maximizing the Arclength in the Cannonball Problem **354**

Ze-Li Dou and Susan G. Staples
College Mathematics Journal, (January 1999), pp. 44–45.

An Example Demonstrating the Fundamental Theorem of Calculus 356
Bob Palais
College Mathematics Journal, (September 1998), pp. 311–313.

Barrow's Fundamental Theorem 358
Jack Wagner
College Mathematics Journal, (January 2001), pp. 58–59.

The Point-slope Formula leads to the Fundamental Theorem of Calculus .. 360
Anthony J. Macula
College Mathematics Journal, (March 1985), pp. 135–139.

A Generalization of the Mean Value Theorem for Integrals 365
M. Sayrafiezadeh
College Mathematics Journal, (November 2002), pp. 408–409.

Proof Without Words: Look Ma, No Substitution! 367
Marc Chamberland
Mathematics Magazine, (February 2001), p. 55.

Integration by Parts 368
V. N. Murty
College Mathematics Journal, (March 1980), pp. 90–93.

Tabular Integration by Parts 372
David Horowitz
College Mathematics Journal, (September 1990), pp. 307–311.

More on Tabular Integration by Parts 377
Leonard Gillman
College Mathematics Journal, (November 1991), pp. 407–410.

A Quotient Rule Integration by Parts Formula 381
Jennifer Switkes
College Mathematics Journal, (January 2005), pp. 58–60.

Partial Fraction Decomposition by Division 384
Sidney H. Kung
College Mathematics Journal, (March 2006), pp. 132–134.

Partial Fractions by Substitution 387
David A. Rose
College Mathematics Journal, (March 2007), pp. 145–147.

Proof Without Words: A Partial Fraction Decomposition 390
Steven J. Kifowit
College Mathematics Journal, (March 2005), p. 122.

Four Crotchets on Elementary Integration 391
Leroy F. Meyers
College Mathematics Journal, (November 1991), pp. 410–413.

An Application of Geography to Mathematics: History of the Integral of the Secant 395

V. Frederick Rickey and Philip M. Tuchinsky
Mathematics Magazine, (May 1980), pp. 162–166.

How to Avoid the Inverse Secant (and Even the Secant Itself) **400**
S. A. Fulling
College Mathematics Journal, (November 2005), pp. 381–387.

The Integral of $x^{1/2}$, etc. **407**
John H. Mathews
College Mathematics Journal, (March 1994), pp. 142–144.

A Direct Proof of the Integral Formula for Arctangent **410**
Arnold J. Insel
College Mathematics Journal, (May 1989), pp. 235–237.

Riemann Sums and the Exponential Function **413**
Sheldon P. Gordon
College Mathematics Journal, (January 1994), pp. 39–40.

Proofs Without Words Under the Magic Curve **414**
Füsan Akman
College Mathematics Journal, (January 2002), pp. 42–46.

Mathematics Without Words: Integrating the Natural Logarithm **419**
Roger Nelsen
College Mathematics Journal, (November 2001), p. 368.

Integrals of Products of Sine and Cosine with Different Arguments **420**
Sherrie J. Nicol
College Mathematics Journal, (March 1993), pp. 158–160.

Moments on a Rose Petal **422**
Douglass L. Grant
College Mathematics Journal, (May 1990), pp. 225–227.

A Calculation of $\int_0^\infty e^{-x^2}\,dx$ **425**
Alberto L. Delgado
College Mathematics Journal, (September 2003), pp. 321–323.

Calculus in the Operating Room **428**
Pearl Toy and Stan Wagon
American Mathematical Monthly, (February 1995), p. 101.

Physical Demonstrations in the Calculus Classroom **430**
Tom Farmer and Fred Gass
College Mathematics Journal, (March 1992), pp. 146–148.

Who Needs the Sine Anyway? **433**
Carlos C. Huerta
College Mathematics Journal, (January 1992), pp. 43–44.

Finding Bounds for Definite Integrals **435**
W. Vance Underhill
College Mathematics Journal, (November 1984), pp. 426–429.

Estimating Definite Integrals **439**

Norton Starr
College Mathematics Journal, (January 2005), pp. 60–63.

Proof Without Words: The Trapezoidal Rule (for Increasing Functions) ... 443
Jesús Urías
Mathematics Magazine, (June 1995), p. 192.

Behold! The Midpoint Rule is Better Than the Trapezoidal Rule for Concave Functions 444
Frank Buck
College Mathematics Journal, (January 1985), p. 56.

An Elementary Proof of Error Estimates for the Trapezoidal Rule 445
D. Cruz-Uribe and C. J. Neugebauer
Mathematics Magazine, (October 2003), pp. 305–306.

Pictures Suggest How to Improve Elementary Numeric Integration 449
Keith Kendig
College Mathematics Journal, (January 1999), pp. 45–50.

Part 4. Polynomial Approximations and Series   455

The Geometric Series in Calculus 456
George E. Andrews
American Mathematical Monthly, (January 1998), pp. 36–40.

A Visual Approach to Geometric Series 462
Beata Randrianantoanina
College Mathematics Journal, (January 2004), pp. 43–47.

The Telescoping Series in Perspective 467
Marc Frantz
Mathematics Magazine, (October 1998), pp. 313–314.

Proof Without Words 469
Richard Hammack and David Lyons
College Mathematics Journal, (January 2005), p. 72.

The Bernoullis and the Harmonic Series 470
William Dunham
College Mathematics Journal, (January 1987), pp. 18–23.

On Rearrangements of the Alternating Harmonic Series 476
Fon Brown, L. O. Cannon, Joe Elich and David G. Wright
College Mathematics Journal, (March 1985), pp. 135–138.

An Improved Remainder Estimate for Use with the Integral Test 479
Roger Nelsen
College Mathematics Journal, (November 2003), pp. 397–399.

A Differentiation Test for Absolute Convergence 482
Yaser S. Abu-Mostafa
Mathematics Magazine, (September 1984), pp. 228–231.

Math Bite: Equality of Limits in Ratio and Root Tests 486
Prem N. Bajaj
Mathematics Magazine, (October 1998), p. 299.

Another Proof of the Formula $\sum_0^\infty (1/n!)$ **487**
Norman Schaumberger
College Mathematics Journal, (January 1994), pp. 38–39.

Taylor Polynomial Approximations in Polar Coordinates **489**
Sheldon P. Gordon
College Mathematics Journal, (September 1993), pp. 325–330.

Proof Without Words: The Taylor Polynomials of sin θ **495**
John Quintanilla
College Mathematics Journal, (January 2007), pp. 58–59.

Appendixes
    I. Topic Outline for AP Calculus Courses **497**
    II. Suggested Uses for the Articles in a First-year Calculus Course **501**

Author Index **505**

About the Editors **507**

Preface

In June 2005, approximately 700 high school and college calculus teachers met on the campus of Colorado State University for the annual weeklong grading of the Advanced Placement Program (AP) calculus exams. One evening during that week, our colleague Susan Kornstein invited a small group to meet and discuss the possibility of collecting and disseminating published books and articles that would be interesting and useful to AP Calculus teachers. After some discussion, we decided to concentrate on a collection of journal articles focusing on single-variable calculus that were originally published for college teachers. This book is a direct result of that 2005 discussion.

With the help of an Advisory Panel of experienced AP teachers, we have selected 123 articles, each of which was published originally in one of the MAA periodicals: *The American Mathematical Monthly*; *The College Mathematics Journal*; *Mathematics Magazine*; *Math Horizons*; and MAA *FOCUS*. In this book, the articles appear in the approximate order of the topics in the outline for the AP calculus course, which we have included as Appendix I. We have also included our suggestions for the possible use of the articles in a first-year calculus course as Appendix II.

The most joyful part of this project was reading the comments that the Advisory Panel submitted. We were happily surprised by how quickly they completed their reviews and at their enthusiasm regarding many of the articles. Their comments encouraged us to hope that this book will serve several purposes. Here we state our goals and give reviewer quotes to demonstrate that our reviewers found the articles to meet these criteria.

- To generate enthusiasm and a passion for exploring topics in calculus. *"This is an excellent article. It will inform, inspire and challenge students and encourage some chuckles in the process."*

- To emphasize real life applications of calculus. *"This nice article takes a classic optimization problem and shows the complexity of applying it to more realistic situations and illustrates the value of calculus in finding an ideal solution".*

- To show that Calculus topics connect with other branches of mathematics and with other disciplines. *"Anyone who has a passion for mathematics, history and geography will love this article."*

- To highlight articles that go beyond what appears in your textbook. *"This is an article that would be beneficial in pushing students and teachers to strengthen their content knowledge."*

- To support and reinforce what does appear in your textbook. *"I like this alternate approach to a common limit problem."*

- To stimulate discovery learning, promote class discussions, and encourage students to generate questions. *"Interesting discussions could occur in comparing advantages and disadvantages of calculus and non-calculus methods and acceptable error in real life situations."*

- To provide ideas for group and individual projects. *"I really like this article. My mind is racing with all the things I could do with it. I might use it after the exam. I think I might have my students collect some boxes of different shapes and sizes and design their own max/min problem with it. Maybe, I would make it a competition of some sort."*

We extend our deepest thanks to the members of our Advisory Panel. All six are veteran AP calculus teachers and have years of experience grading the AP calculus exams. Without their expert help in selecting the articles, the book simply would not exist.

Tom Becvar, Saint Louis University High School, St. Louis, MO
Rhea Caldwell, Providence Day School, Charlotte, NC
Bonney Daves, Christian Academy of Knoxville, Knoxville, TN
Ruth Dover, Illinois Mathematics & Science Academy, Aurora, IL
Dan Lotesto, Riverside University High School, Milwaukee, WI
John Mahoney, Benjamin Banneker Academic High School, Washington, DC

Thanks too to Jerry Bryce and the members of the Editorial Board of the Classroom Resource Materials book series for their support and helpful suggestions, and to David Bressoud, President of the MAA, for writing the book's Introduction. We would also like to thank Elaine Pedreira, Assistant Director of Publications; Beverly Ruedi, Electronic Publications Manager; Rebecca Elmo, Production and Marketing Assistant; and Don Albers, Editorial Director for MAA Books for their expertise in preparing this book for publication. Finally, thanks to our many friends in the Advanced Placement Calculus community who encouraged us to prepare this collection.

Caren Diefenderfer and Roger Nelsen

Introduction

The Mathematical Association of America has twice previously issued a calculus reader, collecting articles on calculus from its journals going all the way back to 1894, the first year of publication of the *Monthly*. The first of these was *Selected Papers in Calculus*, published in 1969 and reprinted as Part I of *A Century of Calculus*. Part II, published in 1992, covered the years 1969 to 1991. In some sense, this is the third volume in that series, bringing us up through 2007. In fact, this collection slightly overlaps with Part II and, as the full title of this volume reveals, it is also something quite different, a collection chosen for its usefulness to those who teach calculus in high school.

The earlier volumes include expositions of many sophisticated and subtle points at the level of advanced analysis. The focus here is on engaging students who are meeting the core ideas of calculus for the first time. It is filled with insights, alternate explanations of difficult ideas, and suggestions for how to take a standard problem and open it up to the rich mathematical explorations available when you encourage students to dig a little deeper. Some of the articles reflect an enthusiasm for bringing calculators and computers in the classroom, illustrating how they can be used to push students toward deeper understanding. Some consciously address themes from the calculus reform movement of the past two decades. But most of the articles are just interesting and timeless explorations of the mathematics encountered in a first course in calculus.

This resource is useful for anyone who teaches calculus, but the fact that the articles have been selected primarily by and chosen for those teaching AP Calculus reflects the fact that today most students have their first encounter with calculus while in high school.

Over the past quarter century, the number of students taking an AP Calculus exam has grown from 35,000 in 1983 to 294,000 in 2008. The total number of high school students in the United States studying calculus now exceeds half a million each year, greater than the total number who study mainstream Calculus I in all 2- and 4-year colleges. The pressures to make high school calculus available in every school and to channel students toward this course are enormous, resulting in consistent growth in the range of 6% to 8% per year. Yet calculus is and always will be college-level mathematics.

The mission of the MAA is "to advance the mathematical sciences, especially at the collegiate level." This means collegiate-level mathematics wherever it occurs. The publication of this volume is one of the ways in which the MAA recognizes its responsibility to support high school teachers of mathematics. MAA has run workshops on organizing Math Circles for Teachers and on preparing students for success in the American Invitational Mathematics Examination (AIME). It also runs the American Mathematics Competitions (AMC) and the USA Mathematical Olympiad (USAMO).

The special role of high school teachers within the MAA is reflected by SIGMAA TAHSM (Special Interest Group of the MAA, Teaching Advanced High School Mathematics). This group of high school and college teachers works to identify and publicize the many MAA publications and other resources that are useful to those teaching advanced mathematics in our high schools. It organizes special sessions, short courses, and workshops that meet the needs of these teachers. And it seeks to further dialog and collaboration between the college and high school communities.

Yet what the reader should take from this collection is not an impression of what the MAA can do for high school teachers, but rather how much the communities of high school and college teachers have in common. Most of these articles were written by college teachers with college teachers as the intended audience. What I hope will come through clearly is that this makes them no less relevant to those who teach calculus in our high schools. We are all teachers of mathematics, and we can all learn from each other.

David M. Bressoud, President of the MAA

Part 0. General and Historical Articles

Touring the Calculus Gallery

William Dunham

1. INTRODUCTION. The premise of this article—what our friends in theatre would call its "conceit"—is that we are about to visit a gallery devoted to the history of calculus. Admittedly no such institution exists, but it is easily imagined. One need only think of an art museum, albeit one whose masterpieces are not canvases but theorems and whose masters are not Courbet and Cezanne but Leibniz and Lebesgue.

Our stroll through the Calculus Gallery must be brief, and we can stop only occasionally to examine particular works in detail. Even so, the visit should provide a glimpse of the development of calculus/analysis from its appearance in the late seventeenth century, through its expansion in the eighteenth, to its "Classical Period" in the first two-thirds of the nineteenth, and on to the mature subject of today.

Like any museum, we enter through a grand hall whose walls are inscribed with noble sentiments. One, from John von Neumann, catches our eye [**16**, p. 3]:

> The calculus was the first achievement of modern mathematics, and it is difficult to overestimate its importance.

In response to such an encomium, we have but one choice: pay the admission and take a look around.

2. THE NEWTON ROOM. Calculus has many antecedents. Some reach back to classical times with the work of Eudoxus or Archimedes, and certainly seventeenth-century Europe saw important contributions from the likes of Fermat, Pascal, and Barrow. But it was Isaac Newton (1642–1727) who first cobbled the assorted ideas into a unified subject, and so it is in the gallery's Newton Room that we start.

Some dates may be helpful. Newton's discoveries began in the mid-1660s while he was a student at Trinity College, Cambridge. His first attempt to put these thoughts on paper resulted in the "October, 1666 tract," which he subsequently refined and expanded as the *De analysi* of 1669 and further polished into the *Methodus fluxionum et serierum infinitarum* of 1671. These writings treated maxima and minima, infinite series, tangents, and areas—still the basic topics of elementary calculus.

The excerpt shown in Figure 1 is taken from the *De analysi* [**22**, p. 20]. Here Newton considered the arc of a circle of radius 1 centered at the origin. Denoting the length of circular arch αD by z, he sought "the Base from the Length of the Curve," that is, an expression for the length of segment AB in terms of z. Because z is the arclength on the unit circle, it is also the (radian) measure of $\angle \alpha AD$. Thus $x = \overline{AB} = \sin z$.

Through a series of manipulations too convoluted to include here, he showed that

$$\sin z = x = z - \frac{1}{6}z^3 + \frac{1}{120}z^5 - \frac{1}{5040}z^7 + \frac{1}{362880}z^9 - \cdots,$$

Touring the Calculus Gallery

and for good measure also provided the cosine series, easily visible in section 46 of the excerpt:

$$\cos z = A\beta = \sqrt{1 - x^2} = 1 - \frac{1}{2}z^2 + \frac{1}{24}z^6 + \frac{1}{403200}z^8 - \frac{1}{3628800}z^{10} + \cdots.$$

Newton

Mathematics historian Derek Whiteside observed that these famous series "here appear for the first time in a European manuscript" [**23**, p. 237]. (His qualification recognized the discovery of equivalent formulas by Indian mathematicians more than a century before, albeit in documents unknown to Westerners during Newton's time. See [**4**].)

It is reasonable to presume that Newton's triumphs swept the world in 1669. Of course, it did not happen that way. In spite of their mathematical brilliance, his results remained unpublished for decades. For reasons as much psychological as scientific, Newton chose not to share these discoveries in print, and so the glory of first publication would fall to another.

Before continuing this tale, we should say a word about Newton's logical justification of his methods. In particular, we address his description of the "ultimate ratio of vanishing quantities"—what we now call the derivative. Perceiving this as a quotient

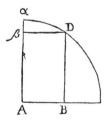

To find the Base from the Length of the Curve given.

45. If from the Arch αD given the Sine AB was required; I extract the Root of the Equation found above, viz. $z = x + \frac{1}{6}x^3 + \frac{3}{40}x^5 + \frac{5}{112}x^7$ (it being supposed that $AB = x$, $\alpha D = z$, and $A\alpha = 1$) by which I find $x = z - \frac{1}{6}z^3 + \frac{1}{120}z^5 - \frac{1}{5040}z^7 + \frac{1}{362880}z^9$ &c.

46. And moreover if the Cosine $A\beta$ were required from that Arch given, make $A\beta$ ($= \sqrt{1 - xx}$) $= 1 - \frac{1}{2}z^2 + \frac{1}{24}z^4 - \frac{1}{720}z^6 + \frac{1}{40320}z^8 - \frac{1}{3628800}z^{10}$, &c.

Figure 1. Newton's series for sine and cosine (1669).

whose numerator and denominator were shrinking to zero, Newton characterized the *ultimate* ratio as "the ratio of the quantities not before they vanish, nor afterwards, but with which they vanish" [**17**, p. 300].

This is highly problematic. We might agree that the derivative is not the ratio of quantities *before* they vanish, but what are we to make of the ratio *after* they have vanished? One is tempted to ask, "How long after?" A minute? A week? It seems that Newton wanted to capture the ratio at that precise instant when—poof!—numerator and denominator simultaneously disappear into nothingness. His ideas, although conveying a useful imagery, would surely have to be revisited.

But such criticism is insignificant compared to Newton's towering achievements. He later described his days of youthful discovery as those when he was "in the prime of my age for invention and minded Mathematicks and Philosophy more than at any time since" [**21**, p. 143]. For that period of invention, mathematicians will be forever grateful.

3. THE LEIBNIZ ROOM. As observed, it was not Newton who taught the world calculus. That distinction belongs to his celebrated contemporary, Gottfried Wilhelm Leibniz (1646–1716). The multitalented Leibniz had reached adulthood with only modest mathematical training, a situation he recalled in these words [**7**, p. 11]:

> When I arrived in Paris in the year 1672, I was self-taught as regards geometry, and indeed had little knowledge of the subject, for which I had not the patience to read through the long series of proofs.

Leibniz

With guidance from Christian Huygens (1629–1695), Leibniz sought to remedy this defect. Those daunting proofs notwithstanding, he wrote that

> [I]t seemed to me, I do not know by what rash confidence in my own ability, that I might become the equal of these if I so desired [**7**, p. 12].

Because few subjects could withstand his blinding intellect, Leibniz moved from novice to master in short order. Within a few years, he had created the calculus.

Figure 2 is a diagram that Leibniz used in an argument from 1674 [**7**, p. 42]. It is included here for two reasons. First, with all its curves and triangles, it illustrates the geometrical flavor of the subject in those formative days. Second, although its purpose may be far from evident, it lay behind one of Leibniz's great early discoveries, for from this tangle of lines he deduced that $1 - \frac{1}{3} + \frac{1}{5} - \frac{1}{7} + \frac{1}{9} - \cdots = \frac{\pi}{4}$.

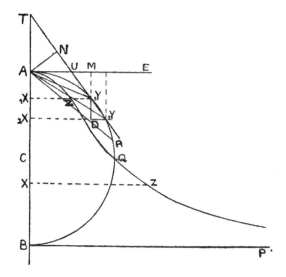

Figure 2. Leibniz excerpt (1674).

This result, the so-called Leibniz series, is worthy of a place of honor in our museum. The left side of the equation displays a trivial pattern; the right side, half of half of π, is anything but obvious. It was believed that here "for the first time ... the area of a circle was exactly equal to a series of rational quantities" [**7**, p. 47]. We might take issue with this use of "exactly," but it is hard not to echo the reaction of Huygens who, in [**7**, p. 46],

> praised [the proof] very highly, and when he returned the dissertation said, in the letter that accompanied it, that it would be a discovery always to be remembered among mathematicians. . . .

Probably no artifact in the Calculus Gallery holds a more venerated place than the document shown in Figure 3 [**17**, p. 273]. This was the first page of the first publication on calculus. In the October 1684 issue of the *Acta eruditorum*, Leibniz presented a "new method" for finding maxima, minima, and tangents and, in the last line of the title, promised "a remarkable type of calculus for this." The name stuck.

His article was not easy reading. A look at the second paragraph reveals Leibniz's presentation of such rules as:

- $d(ax) = a\,dx$
- $d(z - y + vv + x) = dz - dy + d(vv) + dx$
- $d(xv) = x\,dv + v\,dx$

Sprinting through formulas at breakneck speed, he introduced a calculus that may not have been lively but most certainly was lean.

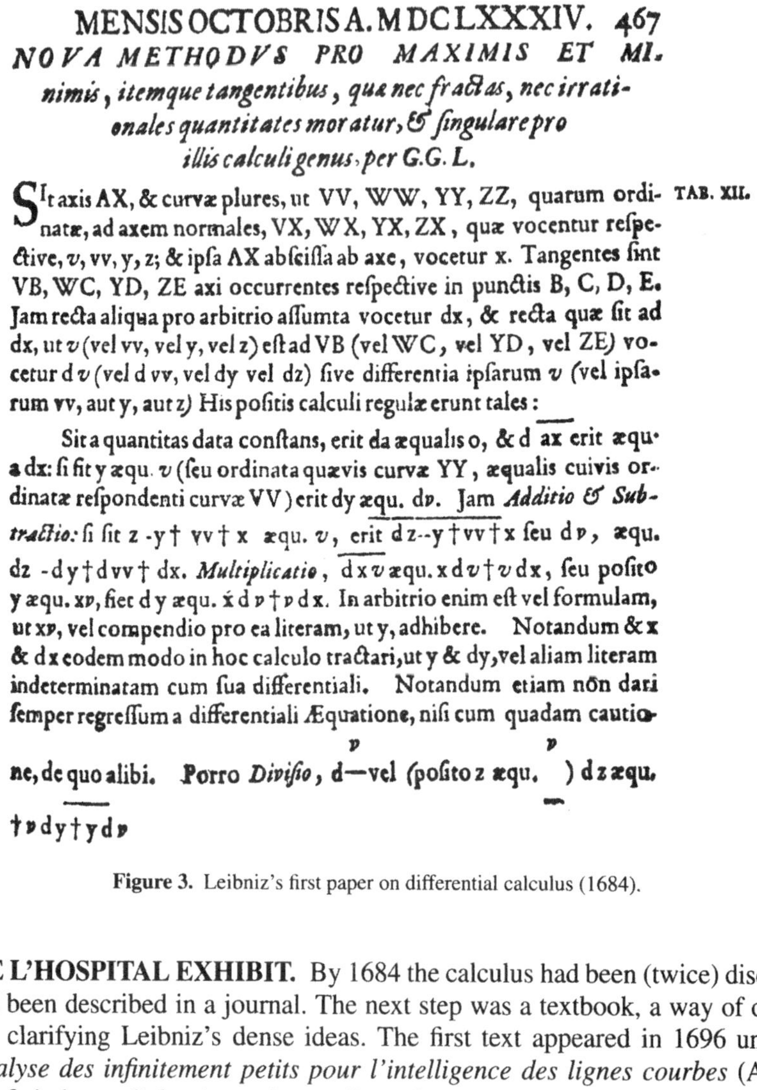

MENSIS OCTOBRIS A. MDCLXXXIV. 467

NOVA METHODVS PRO MAXIMIS ET MInimis, itemque tangentibus, quæ nec fractas, nec irrationales quantitates moratur, & singulare pro illis calculi genus, per G.G. L.

TAB. XII.

Sit axis AX, & curvæ plures, ut VV, WW, YY, ZZ, quarum ordinatæ, ad axem normales, VX, WX, YX, ZX, quæ vocentur respective, *v*, vv, y, z; & ipsa AX abscissa ab axe, vocetur x. Tangentes sint VB, WC, YD, ZE axi occurrentes respective in punctis B, C, D, E. Jam recta aliqua pro arbitrio assumta vocetur dx, & recta quæ sit ad dx, ut *v* (vel vv, vel y, vel z) est ad VB (vel WC, vel YD, vel ZE) vocetur d*v* (vel d vv, vel dy vel dz) sive differentia ipsarum *v* (vel ipsarum vv, aut y, aut z) His positis calculi regulæ erunt tales:

Sit a quantitas data constans, erit da æqualis o, & d ax erit æqu. a dx: si sit y æqu. *v* (seu ordinata quævis curvæ YY, æqualis cuivis ordinatæ respondenti curvæ VV) erit dy æqu. d*v*. Jam *Additio & Subtractio*: si sit z -y † vv † x æqu. *v*, erit d z--y † vv † x seu d*v*, æqu. dz -dy † dvv † dx. *Multiplicatio*, dx*v* æqu. x d*v* † *v* dx, seu posito y æqu. x*v*, fiet d y æqu. x d*v* † *v* d x. In arbitrio enim est vel formulam, ut x*v*, vel compendio pro ea literam, ut y, adhibere. Notandum & x & dx eodem modo in hoc calculo tractari, ut y & dy, vel aliam literam indeterminatam cum sua differentiali. Notandum etiam non dari semper regressum a differentiali Æquatione, nisi cum quadam cautione, de quo alibi. Porro *Divisio*, d $\frac{v}{y}$ vel (posito z æqu. $\frac{v}{y}$) d z æqu. $\frac{\dagger v\,dy \dagger y\,dv}{yy}$

Figure 3. Leibniz's first paper on differential calculus (1684).

4. THE L'HOSPITAL EXHIBIT. By 1684 the calculus had been (twice) discovered and had been described in a journal. The next step was a textbook, a way of organizing and clarifying Leibniz's dense ideas. The first text appeared in 1696 under the title *Analyse des infinitement petits pour l'intelligence des lignes courbes* (Analysis of the infinitely small for the understanding of curved lines). Its author was the Marquis de l'Hospital (1661–1704). Not a mathematician of the highest rank, l'Hospital acquired much of his material through a financial arrangement with Johann Bernoulli (1667–1748) in which the latter, for a fee, provided him with lectures that were to become this book. It is important to note that l'Hospital was candid about the source of his work. Referring to Leibniz and the Bernoullis, he wrote [**17**, p. 312]:

> I have made free use of their discoveries so that I frankly return to them whatever they please to claim as their own.

He began with some definitions, chief among which was that of the differential. According to l'Hospital [**15**, p. 2],

The infinitely small part by which a variable quantity is continually increased or decreased is called the differential (*Différence*) of that quantity.

In modern notation, if $y = y(x)$ and if x is increased by its infinitely small differential dx, then the corresponding, infinitely small change in y is $dy = y(x + dx) - y(x)$ and so

$$y(x + dx) = y + dy. \tag{1}$$

The notion of infinite smallness does not appeal to modern sensibilities any more than does Newton's idea of vanishing quantities. But l'Hospital used it to derive key results of differential calculus, among them the product rule.

Theorem. *The differential of xy is $y\,dx + x\,dy$.*

Proof. As x becomes $x + dx$, we see that y becomes $y + dy$. It follows that the product xy becomes $(x + dx)(y + dy)$, and so the differential of xy is

$$d(xy) = (x + dx)(y + dy) - xy = [xy + y\,dx + x\,dy + dx\,dy] - xy$$
$$= y\,dx + x\,dy + dx\,dy.$$

L'Hospital then jettisoned the $dx\,dy$ term because it "is an infinitely small quantity with respect to both terms $x\,dy$ and $y\,dx$." He was left with the fact that "the differential of the product of two quantities is equal to the product of the differential of the first by the second quantity plus the product of the differential of the second by the first" [**15**, p. 4]. ■

Skeptical readers may wish to debate what exactly happened to that $dx\,dy$, but l'Hospital was in no mood to quibble. He moved quickly to his next topic, the quotient rule.

Corollary. *The differential of x/y is*

$$\frac{y\,dx - x\,dy}{yy}.$$

(In those days, the square of a variable was written as the product of the variable by itself.)

Proof. Introduce $z = x/y$, so that $x = yz$. By the product rule, we know that $dx = y\,dz + z\,dy$ and thus

$$d\left(\frac{x}{y}\right) = dz = \frac{dx - z\,dy}{y} = \frac{dx - (x/y)\,dy}{y} = \frac{y\,dx - x\,dy}{yy}. \quad\blacksquare$$

From there, his text ranged across topics now standard in differential calculus. Section 2 was devoted to tangent lines, section 3 to maxima and minima, and section 4 to inflection points. Some of his problems are perfectly suitable for a modern course, such as, "Among all cones that can be inscribed in a sphere, find that which has the greatest convex surface" [**15**, pp. 45–46]. In his solution, l'Hospital set the appropriate differential equal to zero and found that the height of the largest cone is two-thirds the diameter of the sphere. This 300-year-old max/min problem has lost none of its luster.

Before moving on, we note that the book contained the first published account of "l'Hospital's rule" along with its first published example, namely, to find the value of the quotient

$$\frac{\sqrt{2a^3x - x^4} - a\sqrt[3]{aax}}{a - \sqrt[4]{ax^3}}$$

"when $x = a$" [**15**, p. 146]. Modern readers may be surprised to learn that the rule was framed without explicit mention of limits, a concept that still lay in the far distant future. L'Hospital's answer to this (surprisingly complicated) initial problem was $16a/9$, which is not only correct but also indicative of how thoroughly calculus outpaces intuition.

It is well known that the rule had been discovered by Johann Bernoulli and included in the materials sent, for compensation, to l'Hospital. To the suggestion that we thus call it "Bernoulli's rule," the math historian Dirk Struik once scoffed, "Let the good Marquis keep his elegant rule. He paid for it" [**18**, p. 260].

5. EULER HALL. With that, we hurry off to the next stop: a hall devoted to the preeminent mathematician of the eighteenth century, Leonhard Euler (1707–1783).

Euler

Truth to tell, Euler was a mathematical force of nature. His collected works contain over 8,000 pages of analysis—that's not eighty pages, nor eight hundred, but a colossal eight *thousand*. Among these are contributions to differential equations and the calculus of variations, to integrals and infinite series, and much more. Euler composed three textbooks whose influence on the subject was profound: his *Introductio in analysin infinitorum* of 1748, his *Institutiones calculi differentialis* of 1755, and the three-volume *Institutiones calculi integralis* of 1768. With such an output, it is little wonder that Euler was described as "Analysis Incarnate."

Touring the Calculus Gallery

Here we make do with a few morsels from this analytic feast. The first may have cast the longest shadow: Euler enshrined the *function* as the core idea of analysis. In the *Introductio*, he examined those functions that have been in the toolkit of analysts ever since—the logarithmic and exponential functions, trigonometric and inverse trigonometric functions, and so on. Although his 1748 definition of a function as an "analytic expression composed in any way whatsoever of the variable quantity and numbers or constant quantities" is too restrictive for modern tastes, it was a marked improvement over the ill-defined curves of his predecessors [**9**, p. 3]. As Euler matured, so did his notion of function, ending up closer to the modern idea of a correspondence not necessarily tied to a particular formula or "analytic expression." It is fair to say that we now study functions in analysis because of him.

Our second Eulerian exhibit is his evaluation of

$$\int_0^1 \frac{\sin(\ln x)}{\ln x}\,dx,$$

the sort of calculus challenge he loved [**10**, vol. 18, p. 4]. Ever ready to introduce an infinite series, he applied Newton's expansion for the sine by writing

$$\frac{\sin(\ln x)}{\ln x} = \frac{(\ln x) - (\ln x)^3/3! + (\ln x)^5/5! - (\ln x)^7/7! + \cdots}{\ln x}$$

$$= 1 - \frac{(\ln x)^2}{3!} + \frac{(\ln x)^4}{5!} - \frac{(\ln x)^6}{7!} + \cdots.$$

This he integrated termwise to get

$$\int_0^1 \frac{\sin(\ln x)}{\ln x}\,dx = \int_0^1 dx - \frac{1}{3!}\int_0^1 (\ln x)^2\,dx + \frac{1}{5!}\int_0^1 (\ln x)^4\,dx$$

$$- \frac{1}{7!}\int_0^1 (\ln x)^6\,dx + \cdots,$$

where the integral of a sum was replaced by the sum of the integrals without missing a beat. Euler then noted that $\int_0^1 (\ln x)^n\,dx = n!$ for even n, so

$$\int_0^1 \frac{\sin(\ln x)}{\ln x}\,dx = 1 - \frac{1}{3!}[2!] + \frac{1}{5!}[4!] - \frac{1}{7!}[6!] + \cdots = 1 - \frac{1}{3} + \frac{1}{5} - \frac{1}{7} + \cdots = \frac{\pi}{4}.$$

Here the Leibniz series made an unexpected appearance. This splendid derivation combined Newtonian and Leibnizian insights with a strong dose of Eulerian genius—quite a pedigree.

This is just the tip of the integration iceberg. Elsewhere in his career (see [**10**, vol. 19, p. 227; vol. 18, p. 8; and vol. 17, p. 407]), Euler evaluated such integrals as

$$\int_0^\infty \frac{\sin x}{x}\,dx, \quad \int_0^1 \frac{\sin(4\ln x)\cos(7\ln x)}{\ln x}\,dx, \quad \int_0^1 \frac{(\ln x)^5}{1+x}\,dx,$$

and found them to equal, respectively,

$$\frac{\pi}{2}, \quad \frac{1}{2}\arctan\left(\frac{4}{17}\right), \quad -\frac{31\pi^6}{252}.$$

No one could integrate like Uncle Leonhard!

Euler also considered foundational questions. We recall that mathematicians had had a much easier time applying calculus than explaining it. Faced with Newton's vanishing quantities and Leibniz's infinitely small ones, Euler thought he could see his way through the logical thicket. "An infinitely small quantity," he wrote, "is nothing but a vanishing quantity, and so it is really equal to 0." He continued, "There is ... not such a great mystery lurking in this idea as some commonly think and thereby render suspect the calculus of the infinitely small" [**8**, p. 51].

We can watch him apply these ideas in his treatment of l'Hospital's rule [**10**, vol. 10, p. 565]. Beginning with functions P and Q for which $P(a) = Q(a) = 0$, he sought the value of $y = P/Q$ when $x = a$. Because dx was "really equal to zero," there could be no harm in replacing a with $a + dx$. Euler thus reasoned that (recall (1))

$$y(a) = y(a+dx) = \frac{P(a+dx)}{Q(a+dx)} = \frac{P(a)+dP}{Q(a)+dQ} = \frac{0+dP}{0+dQ} = \frac{dP}{dQ}.$$

Voila! Among his examples of the rule was to "find the value of the expression

$$\frac{x^x - x}{1 - x + \ln(x)}$$

if we place $x = 1$." His answer, obtained with the help of logarithmic differentiation, was -2, a problem that remains a good one to this day.

6. THE BERKELEY ROOM. In spite of Euler's facile explanation, the mathematical community was unconvinced that the last word had been spoken on foundational questions. This was due in no small part to the critique of a nonmathematician, the philosopher and Bishop of Cloyne, George Berkeley (1685–1763).

Berkeley had read justifications of the calculus and remained skeptical. He famously described Newton's vanishing magnitudes as "the ghosts of departed quantities" and was no more charitable to Leibniz, whose infinitely small magnitudes were "above my capacity" to understand [**2**, p. 89 and p. 68, respectively].

To illustrate his concern, Berkeley reconsidered the standard proof that the derivative of x^n is nx^{n-1} [**2**, p. 72]. Mathematicians of the day began by augmenting x with a small but nonzero quantity o and looking at the ratio

$$\frac{(x+o)^n - x^n}{o}.$$

Expanding the numerator as a binomial series, they found this to equal

$$\frac{x^n + nx^{n-1}o + \frac{n(n-1)}{2}x^{n-2}o^2 + \cdots + o^n - x^n}{o} = nx^{n-1} + \frac{n(n-1)}{2}x^{n-2}o + \cdots + o^{n-1}.$$

Berkeley

At this point in the argument, the substitution $o = 0$ produced the derivative, nx^{n-1}.

Berkeley was having none of it. As he observed, we initially require that $o \neq 0$, "without which I should not have been able to have made so much as a single step." Indeed, if we allow o to be zero in

$$\frac{(x+o)^n - x^n}{o},$$

not only do we have an illegitimate denominator, but x^n would never have been *augmented* in the first place. As Berkeley suggested, the whole proof would stop dead in its tracks. But we then change our minds and decide that o is zero after all. Berkeley noted that "the former supposition that the increments were something... is destroyed, and yet a consequence of that supposition, i.e., an expression got by virtue thereof, is retained."

He acknowledged the correctness of the answer, but a procedure yielding correct answers from a string of mistakes was not to be trusted. "Error," he wrote, "may bring forth truth, though it cannot bring forth science" [**2**, p. 77]. And he offered this caustic rejoinder [**2**, p. 76]:

> I say that, in every other Science, Men prove their Conclusions by their Principles, and not their Principles by their Conclusions.

Clearly, foundational issues could not be long ignored in the face of such criticism. Someone had to resolve this matter, and that someone was Augustin-Louis Cauchy (1789–1857).

7. THE CAUCHY WING. Cauchy's analytic contributions run broad and deep. Many can be found in his seminal texts, *Cours d'analyse* of 1821 and *Résumé des leçons données a l'école royale polytechnique, sur le calcul infinitésimal* of 1823. Surveying this work, math historian Carl Boyer wrote that "Cauchy did more than anyone else to impress upon the subject the character which it bears at the present time" [**3**, p. 271].

For starters, Cauchy believed that a rigorous treatment of calculus required neither vanishing quantities nor infinitely small ones. For him, the heart of the matter was the *limit*, and he offered this definition [**6**, vol. 4, p. 13]:

> When the values successively attributed to a variable approach indefinitely to a fixed value, in a manner so as to end by differing from it by as little as one wishes, this last is called the limit of all the others.

Cauchy

A modern reader may be surprised by the purely verbal nature of the definition and troubled by terms like "approach" and "to end by." Yet the insight here is solid. Cauchy perceived that, for limits, the key was to get as close as one wishes—i.e., within any preassigned target—which is the essence of the modern idea.

Equally important, Cauchy put limits front and center, building his calculus upon this foundation. Gone were proofs by picture; gone were appeals to intuition; gone were quasi-metaphysical digressions. "It would be a serious error," he wrote, "to think that one can find certainty only in geometrical demonstrations or in the testimony of the senses" [**12**, p. 947].

We shall linger a while in the Cauchy Wing in order to consider two important exhibits: his proofs of the intermediate and mean value theorems.

Intermediate Value Theorem. *If f is continuous on $[x_0, X]$ with $f(x_0) < 0$ and $f(X) > 0$, then $f(a) = 0$ for some a between x_0 and X.*

Proof. Cauchy introduced $h = X - x_0$ as the width of the interval in question. Choosing a whole number $m > 1$, he noted that the values

$$f(x_0), f(x_0 + h/m), f(x_0 + 2h/m), \ldots, f(x_0 + [m-1]h/m), f(X)$$

begin with a negative and end with a positive, so somewhere must appear two consecutive values of opposite sign. That is, for some n we have $f(x_0 + nh/m) \leq 0$ and $f(x_0 + [n+1]h/m) \geq 0$. Cauchy let $x_1 = x_0 + nh/m$ and $X_1 = x_0 + [n+1]h/m$ so that $x_0 \leq x_1 \leq X_1 \leq X$ and the length of subinterval $[x_1, X_1]$ is h/m.

He next divided $[x_1, X_1]$ into m equal pieces of length h/m^2, chose two consecutive points of opposite sign, and continued, thereby generating sequences

$$x_0 \leq x_1 \leq x_2 \leq \cdots \leq X_2 \leq X_1 \leq X$$

where $f(x_k) \leq 0$ and $f(X_k) \geq 0$. It was clear to him that both the nondecreasing sequence $\{x_k\}$ and the nonincreasing sequence $\{X_k\}$ converged and, because $X_k - x_k = h/m^k \to 0$, that they shared a common limit. Letting $a = \lim_{k\to\infty} x_k = \lim_{k\to\infty} X_k$, he saw that a belonged to $[x_0, X]$, and from the continuity of f he deduced that

$$f(a) = f\left[\lim_{k\to\infty} x_k\right] = \lim_{k\to\infty} f(x_k) \leq 0, \qquad f(a) = f\left[\lim_{k\to\infty} X_k\right] = \lim_{k\to\infty} f(X_k) \geq 0.$$

Thus, as Cauchy put it, "$f(a)$ cannot differ from zero" [**6**, vol. 3, pp. 378–380]. ■

With a small modification to tidy up the completeness property, this proof is transferable to the modern classroom. It is reminiscent of the bisection method for approximating roots of equations, but, as Judith Grabiner wrote in her treatise on Cauchy's mathematics [**11**, p. 69]:

> [T]hough the mechanics ... are simple, the conception of the proof is revolutionary. Cauchy transformed the approximation technique into something entirely different: a proof of the existence of a limit.

Cauchy now turned to the mean value theorem, beginning with a preliminary result [**6**, vol. 4, pp. 44–46].

Lemma. *If f is continuous with A the smallest and B the largest value of f' on $[x_0, X]$, then*

$$A \leq \frac{f(X) - f(x_0)}{X - x_0} \leq B.$$

(He did not say how he knew the derivative *had* smallest and largest values.)

Proof. Cauchy began with a momentous idea. He introduced δ and ε as "very small numbers" chosen such that, for any x in $[x_0, X]$ and for any $i < \delta$, we have

$$f'(x) - \varepsilon < \frac{f(x+i) - f(x)}{i} < f'(x) + \varepsilon. \qquad (2)$$

On the one hand, we celebrate the appearance of the derivative in terms of its ε-δ definition—certainly a modern touch. On the other, we recognize here the assumption of uniformity for, given an ε, Cauchy believed that "one δ fits all." This was a misconception both subtle and pernicious.

Be that as it may, he let $x_0 < x_1 < x_2 < \cdots < x_{n-1} < X$ subdivide the interval, where the distance between any two points of subdivision was less than δ. Recalling the roles of A and B and applying (2) repeatedly, he generated the inequalities:

$$A - \varepsilon \leq f'(x_0) - \varepsilon < \frac{f(x_1) - f(x_0)}{x_1 - x_0} < f'(x_0) + \varepsilon \leq B + \varepsilon,$$

$$A - \varepsilon \leq f'(x_1) - \varepsilon < \frac{f(x_2) - f(x_1)}{x_2 - x_1} < f'(x_1) + \varepsilon \leq B + \varepsilon,$$

$$A - \varepsilon \leq f'(x_2) - \varepsilon < \frac{f(x_3) - f(x_2)}{x_3 - x_2} < f'(x_2) + \varepsilon \leq B + \varepsilon,$$

$$A - \varepsilon \leq f'(x_{n-1}) - \varepsilon < \frac{f(X) - f(x_{n-1})}{X - x_{n-1}} < f'(x_{n-1}) + \varepsilon \leq B + \varepsilon.$$

At this point Cauchy wrote that "If one divides the sum of the numerators by the sum of the denominators, one obtains a mean fraction that is ... contained between $A - \varepsilon$ and $B + \varepsilon$." In symbols, he was saying that

$$A - \varepsilon < \frac{f(x_1) - f(x_0) + f(x_2) - f(x_1) + \cdots + f(X) - f(x_{n-1})}{x_1 - x_0 + x_2 - x_1 + \cdots + X - x_{n-1}} < B + \varepsilon,$$

which telescopes to

$$A - \varepsilon < \frac{f(X) - f(x_0)}{X - x_0} < B + \varepsilon.$$

And, "as this holds however small the number ε," Cauchy concluded that

$$A \leq \frac{f(X) - f(x_0)}{X - x_0} \leq B. \qquad \blacksquare$$

He then was ready for one of the great theorems of differential calculus.

Mean Value Theorem. *If f and f' are continuous between x_0 and X, then for some θ between 0 and 1, it is the case that*

$$\frac{f(X) - f(x_0)}{X - x_0} = f'[x_0 + \theta(X - x_0)].$$

Before giving his proof, we make two observations. First, because $0 < \theta < 1$, the number $x_0 + \theta(X - x_0)$ lies strictly between x_0 and X and so plays the role of "c" in modern statements of the mean value theorem.

Second, Cauchy assumed the continuity of f'. As we shall see, he did so to ensure that the derivative possess the intermediate value property. A later theorem of Gaston Darboux (1842–1917) would establish that, even if discontinuous, f' *must* exhibit this property, so Cauchy did not need his assumption. (Nowadays, of course, we prove the mean value theorem quite differently, without mentioning the continuity of f' at all.)

Proof. The lemma guarantees that

$$\frac{f(X) - f(x_0)}{X - x_0}$$

is intermediate between A and B, the minimum and maximum values of f' on $[x_0, X]$. By the assumed continuity of the derivative, f' takes this intermediate value somewhere between x_0 and X. In other words, there is a θ between 0 and 1 with

$$f'[x_0 + \theta(X - x_0)] = \frac{f(X) - f(x_0)}{X - x_0}. \qquad \blacksquare$$

Thanks to Cauchy, the calculus had come a long way.

8. TRANSITION TO THE MODERN WING. The following half century saw three developments that carried the subject into ever more sophisticated realms. We mention each briefly as we proceed to the next major exhibit.

1. *Mathematicians refined the fundamental ideas of analysis.* By the time of Karl Weierstrass (1815–1897), the limit definition had been distilled to:

Definition. $\lim_{x \to a} f(x) = L$ if and only if, for each $\varepsilon > 0$, there exists a $\delta > 0$ so that $|f(x) - L| < \varepsilon$ whenever $0 < |x - a| < \delta$.

This string of well-chosen inequalities gave an unambiguous and purely static meaning to limits. In the hands of the next generation of mathematicians, it allowed for new standards of analytic rigor.

There were other matters in need of attention. It was Weierstrass who emphasized the distinction between continuity and uniform continuity and between pointwise and uniform convergence. And Georg Friedrich Bernhard Riemann (1826–1866) developed his integral, in the process forever separating the ideas of continuity and integrability. So natural and powerful was his approach that the Riemann integral seemed incapable of improvement.

2. *Pathological counterexamples proliferated.* These were sufficiently bizarre as to defy belief, yet their existence forced mathematicians to think more deeply about critical ideas. For instance, Weierstrass described the seemingly impossible: an everywhere continuous, nowhere differentiable function [**20**, vol. 2, pp. 71–74]. His original example, rarely found in modern texts, was $f(x) = \sum_{k=0}^{\infty} b^k \cos(\pi a^k x)$, where

- a is an odd integer greater than one
- $0 < b < 1$
- $ab > 1 + \frac{3\pi}{2}$

(These side conditions, as unexpected as they were essential to Weierstrass's reasoning, are sure signs of an analyst at work.) Any partial sum of the series yields a function everywhere continuous and everywhere differentiable, yet when we pass to the limit, the result is everywhere continuous but *nowhere* differentiable. So much for intuition.

Another provocative counterexample came from Vito Volterra (1860–1940). In 1881 he described a function that was differentiable with a bounded derivative, yet whose derivative was so wildly discontinuous that its (Riemann) integral did not exist [**19**, pp. 16–48]. His example thus destroyed an "unqualified" fundamental theorem of calculus. That is, if f is differentiable with f' bounded on $[a, b]$, then we would hope that

$$\int_a^b f'(x)\, dx = f(x)\Big|_a^b.$$

Volterra's function invalidated this equation, not because the left side failed to equal the right side but because the left side failed even to *exist* as a Riemann integral.

These and other pathological counterexamples raised significant questions, among which were:

- How discontinuous can an integrable function be?
- How discontinuous can a derivative be?
- How, if at all, can the fundamental theorem of calculus be repaired?

We shall have more to say about each of these before our visit to the Calculus Gallery is over.

3. *Set theory arrived as an ally of the analyst.* The innovator here was Georg Cantor (1845–1918). Highlighting his initial paper on this subject from 1874 was a proof that a sequence of real numbers cannot exhaust an open interval [**5**, pp. 115–118]. In modern parlance, this established the nondenumerability of the continuum, and one might naturally assume that Cantor used diagonalization as his weapon of choice. That line of attack, however, dates to 1891. His original argument invoked the completeness property of the reals and thus lay squarely in the analytic domain. Employing sets in the service of analysis would prove to be a fruitful undertaking, as Weierstrassian rigor fused with Cantorian set theory to create the subject we know today.

9. THE BAIRE EXHIBIT. One of the first to move in this direction was René Baire (1874–1932). Baire, who admired Cantor and had studied with Volterra, took their lessons to heart. His 1899 dissertation *Sur les fonctions de variables réelles* contained what has come to be known as the Baire category theorem, a featured exhibit of the Calculus Gallery.

Baire began by defining a set P of real numbers to be *nowhere dense* if every open interval (α, β) has an open subinterval (a, b) containing no point of P. Examples include all finite sets as well as the set $\{1/n : n \text{ is a whole number}\}$. He then wrote [**1**, p. 65]:

> If there exists a denumerable infinity of nowhere dense sets $P_1, P_2, P_3, \ldots$ such that every point of F belongs to at least one of them, ... I shall say that F is of the *first category*.

Baire

In coining so nondescriptive a term—"first category," after all, conjures up absolutely nothing in the imagination—Baire earned a reputation for colorless terminology. This was cemented when he gave all other sets the moniker "second category."

If his terms were bland, his mathematics was not. In that 1899 thesis, Baire proved the result that carries his name:

Theorem. *A first-category set cannot exhaust an open interval.*

Proof. Begin with an open interval (α, β) and let $F = P_1 \cup P_2 \cup P_3 \cup \cdots$ be a set of the first category, where each P_k is nowhere dense. Because P_1 is nowhere dense, there is an open subinterval of (α, β) containing no points of P_1. By shrinking the subinterval if necessary, we can find points a_1 and b_1 with $a_1 < b_1$ such that and $[a_1, b_1] \subseteq (\alpha, \beta)$ and $[a_1, b_1] \cap P_1 = \emptyset$.

But (a_1, b_1) is open and P_2 is nowhere dense, so there exists a_2 and b_2 with $a_2 < b_2$ such that $[a_2, b_2] \subseteq (a_1, b_1)$ and $[a_2, b_2] \cap P_2 = \emptyset$. Repeating the argument, Baire generated descending intervals

$$[a_1, b_1] \supseteq [a_2, b_2] \supseteq \cdots \supseteq [a_k, b_k] \supseteq \cdots,$$

where $[a_k, b_k] \cap P_k = \emptyset$ for each k. By completeness, there is a point c common to all these closed intervals. Thus $c \in (\alpha, \beta)$ but $c \notin P_1 \cup P_2 \cup P_3 \cup \cdots = F$. In short, a first category set is insufficient to fill an open interval. ∎

Two remarks are in order. First, the debt to Cantor is evident. As we noted, Cantor had proved that a denumerable set cannot exhaust an open interval, and Baire extended this principle to any first category set (of which denumerable sets are special cases). Second, the theorem's conclusion could be recast to say that the complement of a first category set is dense, for points of this complement must show up in any open interval.

With these powerful ideas, Baire was able to answer one of our earlier questions: How discontinuous can a derivative be? By proving that the discontinuity points of a derivative form a set of the first category, he concluded:

Corollary. *If f is differentiable on an interval I, then f' must be continuous on a dense subset of I.*

Thus, derivatives can be discontinuous, but not so discontinuous that there is an open interval free of their continuity points.

René Baire was among the new wave of analysts, active at the turn of the twentieth century, who viewed the subject through the lens of set theory. He asserted (see [**1**, p. 121]) that

> [A]ny problem relative to the theory of functions leads to certain questions relative to the theory of sets and, insofar as these latter questions are or can be addressed, it is possible to resolve, more or less completely, the given problem.

10. THE LEBESGUE ROOM. With that pronouncement as background, we enter the gallery's final room, one devoted to Baire's great contemporary, Henri Lebesgue (1875–1941). Lebesgue's analytic reputation was established with his 1904 classic *Leçons sur l'intégration*, a work featuring a bewildering array of seminal ideas.

He began with a survey of the Riemann integral, reviewing its history and noting those shortcomings described earlier. In response, Lebesgue defined a set to have *measure zero* if, in his words, it "can be enclosed in a finite or *a denumerable infinitude* of intervals whose total length is as small as we wish" [**13**, p. 28]. Then, in an ingenious argument using the Heine-Borel theorem, he proved that a bounded function on

Lebesgue

[a, b] is Riemann integrable over [a, b] if and only if its set of discontinuities on [a, b] has measure zero. This resolved the question as to how discontinuous a Riemann integrable function can be: in terms of measure, the answer is "not very." With this result, Henri Lebesgue could lay claim to having understood the *Riemann* integral more fully than anyone. In light of what was to come, this held a certain irony.

Lebesgue next developed a theory of measure for subsets of the real line—not just for those of measure zero. From there, in a manner used to this day, he explored the notion of a *measurable function* and then moved to his most famous creation: the Lebesgue integral.

Obviously, he had studied Riemann well. Lebesgue acknowledged the debt by noting that his new integral [**14**, p. 136]

> is analogous to that of Riemann; but whereas he divided into small subintervals the interval of variation of x, it is the interval of variation of $f(x)$ that we have subdivided.

In short, partition the range not the domain, and that will make all the difference.

Lebesgue carried the theory far enough to convince him of the superiority of his integral over Riemann's. For instance, in what we now call the bounded convergence theorem, he proved:

Theorem. *If measurable functions f_k ($k = 1, 2, \ldots$) are uniformly bounded on $[a, b]$ and have pointwise limit f on this interval, then*

$$\lim_{k \to \infty} \left[\int_a^b f_k(x)\, dx \right] = \int_a^b f(x)\, dx = \int_a^b \left[\lim_{k \to \infty} f_k(x) \right] dx.$$

This spectacular result says that, under minimal restrictions, the limit of the (Lebesgue) integrals is the (Lebesgue) integral of the limit [**13**, p. 114]. The theorem fails for Riemann's integral.

But there was more. Lebesgue established that, with *his* integral, the fundamental theorem of calculus could be salvaged as follows [**13**, p. 120]:

Theorem. *If f is differentiable on $[a, b]$ with bounded derivative, then*

$$\int_a^b f'(x)\, dx = f(x) \Big|_a^b.$$

This repaired the damage inflicted by Volterra's counterexample. In that sense it restored the fundamental theorem of calculus to its "natural" state, so long as we adopt Lebesgue's version of the integral. One hopes that Newton and Leibniz, from their simpler vantage points, would have been pleased that their creation had returned to a kind of analytic nirvana.

And that, perhaps, is a fitting way to end our visit to the Calculus Gallery. As is customary, we pop into the Museum Shop with its array of scarves, jewelry, and other noteworthy items, chief among which are reproductions of the original texts. Anyone interested in primary sources is urged to consult the works of Newton, Euler, Cauchy, or Lebesgue. Their achievements seem magnified when encountered in original form.

For those interested in secondary sources, the shop carries three books that come highly recommended: Ivor Grattan-Guinness's *The Development of the Foundations of Mathematical Analysis from Euler to Riemann* (MIT Press, 1970), Judith Grabiner's *The Origins of Cauchy's Rigorous Calculus* (MIT Press, 1981), and Thomas Hawkins's classic *Lebesgue's Theory of Integration* (Chelsea, 1975). Each is an excellent read.

With a bag full of new purchases, we are ready to go. Approaching the exit, we notice one more inscription on the rotunda walls. As before, it is due to von Neumann, and we leave the Calculus Gallery with his words in our minds [**16**, p. 3]:

> I think it [the calculus] defines more unequivocally than anything else the inception of modern mathematics, and the system of mathematical analysis, which is its logical development, still constitutes the greatest technical advance in exact thinking.

ACKNOWLEDGMENT. This article is based upon "Selections from the Calculus Museum," an invited MAA address delivered at the 2004 Joint Meetings in Phoenix.

REFERENCES

1. R. Baire, *Sur les fonctions des variables réelles*, Imprimerie Bernardoni de C. Rebeschini & Co., Milan, 1899.
2. G. Berkeley, *The Works of George Berkeley*, vol. 4, Nelson & Sons, London, 1951.
3. C. Boyer, *The Concepts of the Calculus*, Hafner, New York, 1949.
4. D. Bressoud, Was calculus invented in India? *College Math. J.* **33** (2002) 2–13.
5. G. Cantor, *Gesammelte Abhandlungen*, Georg Olms, Hildesheim, 1962.
6. A.-L. Cauchy, *Oeuvres*, ser. 2, vols. 3 and 4, Gauthier-Villars, Paris, 1882 and 1899.
7. J. M. Child, trans., *The Early Mathematical Manuscripts of Leibniz*, Open Court, Chicago, 1920.
8. L. Euler, *Foundations of Differential Calculus* (trans. J. Blanton), Springer-Verlag, New York, 2000.
9. ———, *Introduction to Analysis of the Infinite*, Book I (trans. J. Blanton), Springer-Verlag, New York, 1988.
10. ———, *Opera omnia*, ser. 1, vols. 10, 17, 18, and 19, Societatis Scientiarum Naturalium Helveticae, Leipzig, Berlin, and Zurich, 1913, 1915, 1920, and 1932.
11. J. Grabiner, *The Origins of Cauchy's Rigorous Calculus*, MIT Press, Cambridge, 1981.
12. M. Kline, *Mathematical Thought from Ancient to Modern Times*, Oxford University Press, New York, 1972.
13. H. Lebesgue, *Leçons sur l'integration et la recherché des fonctions primitives*, Gauthier-Villars, Paris, 1904.
14. ———, *Leçons sur l'integration et la recherché des fonctions primitives*, reprint of 2nd ed. (1928), American Mathematical Society, Providence, 2000.
15. M. de L'Hospital, *Analyse des infiniment petits*, reprint of original by ACL-editions, Paris, 1988.
16. J. v. Neumann, *Collected Works*, vol. 1, Pergamon Press, New York, 1961.
17. D. Struik, ed., *A Source Book in Mathematics, 1200–1800*, Harvard University Press, Cambridge, 1969.
18. ———, The origin of L'Hôpital's rule, *Math. Teacher* **56** (1963) 257–260.
19. V. Volterra, *Opere Matematiche*, vol. 1, Accademia Nazionale dei Lincei, Rome, 1954.
20. K. Weierstrass, *Mathematische Werke*, vol. 2, Mayer and Müller, Berlin, 1895.
21. R. Westfall, *Never at Rest*, Cambridge University Press, Cambridge, 1980.
22. D. Whiteside, ed., *The Mathematical Works of Isaac Newton*, vol. 1, Johnson Reprint Corp., New York, 1964.
23. ———, *The Mathematical Papers of Isaac Newton*, vol. 2, Cambridge University Press, Cambridge, 1968.

Calculus: A Modern Perspective

Jeff Knisley

1. INTRODUCTION. An American history curriculum that ended with the Civil War would be no more acceptable than a philosophy curriculum that ended with Kant. Yet an acceptable history of mathematics curriculum gives little more than a cursory nod to the mathematics of the twentieth century. This is not to say that the two hundred years following Newton and Leibniz do not deserve seven chapters in a history of mathematics textbook, but rather that the one hundred years leading up to the present deserve more than one [1].

Unfortunately, our entire undergraduate curriculum has the same focus as our history of mathematics course—the two hundred years following Newton and Leibniz. Our graduates are more prepared for the period in which the steam engine replaced the horse than they are for the period in which compact disks replaced the vinyl LP. It is no wonder that a reform movement emerged early in this century, as evidenced by several articles in [2], nor is it surprising that the reform movement is stronger than ever today. Tragically, many mathematicians have responded to reform as in [3], where it was lamented that "Mathematics is losing its soul. Its priests are pawning it off to a different god." Such a call to arms only reinforces the popular image of mathematicians as the last practitioners of some ancient art.

We know better. We know that mathematics is growing and thriving, fruitful and strong. However, the new growth in mathematics is all but absent from our undergraduate curriculum. Indeed, our traditional calculus course is packed with intellectual deadwood—contrived applications, outdated examples, and obsolete definitions. It is time we allowed some new growth in a curriculum that has been antiquated for most of this century [7].

2. THE NEW GROWTH. Much of modern mathematics is derived from modern trends in calculus. Many of the ideas in differential geometry, statistics, and numerical analysis are descended from the study of calculus in the present century. Correspondingly, any new ideas introduced into the calculus curriculum should be cultivated from these modern trends. Technology allows us to incorporate linear regression, Markov processes, and probability distributions into an introductory calculus course, and when we do so, our students sense that they are studying ideas relevant to the world in which they live.

Similarly, students appreciate applications that have a modern perspective, and such applications need not be outside of mathematics [4]. Applications of the derivative can be enhanced with the study of cubic splines and Bezier curves. The study of Newton's method can be generalized to the study of fixed point theorems in general. The study of spectral theory begins with boundary value problems such as

$$y'' = -\alpha^2 y \qquad (1)$$

$$y(0) = y(\pi) = 0 \qquad (2)$$

and students genuinely enjoy being shown that (1)–(2) has nontrivial solutions only for integral values of α.

Finally, allowing such new growth can greatly simplify our efforts at instruction. Complex numbers make partial fractions and trigonometric substitutions much more accessible. A simple matrix exponential is a great illustration of the power series concept. And stating the laws of exponents as axioms for an abelian group shows a student that these are more than rules for manipulating superscripts.

3. THE DEAD WOOD. Allowing new growth into the calculus curriculum means something old must go, but such additions mean more than simple pruning. Indeed, patching the new into the old destroys the continuity and coherence of the original structure. Our present curriculum bears witness to this fact. The applications of the integral contribute little to the remainder of the course, and the customary chapter on analytic geometry seems misplaced. Sequences are introduced in the context of convergence of series, and thus it should be no surprise that students get the two confused.

Rather than further fragment our curriculum, we need to transform the calculus so that it is a coherent mix of timeless concepts and new ideas. I believe such a transformation must address the following:

Approximation and limits. Our current calculus course relies on several seemingly unrelated notions of limit. The $\epsilon - \delta$ definition is introduced en route to the definition of the tangent line. Infinite limits are defined using ϵ and N sufficiently large. Newton's method is left to intuition. The definite integral is defined using the norm of a partition that goes to zero. Numerical integration introduces the idea of bounded error. Limits of sequences are defined with the monotone convergence theorem. And after spending section after section using limits of sequences to develop convergence tests, the idea of a converging Taylor series is developed from the remainder formula for Taylor polynomials. No attempt is ever made to connect all these ideas of limit into a coherent concept.

We need to introduce and define the limit so that all our applications of approximation and convergence are derived from a single concept. This concept will likely have to be a principle rather than a definition. It may be more appropriate to explore approximation with a graphing calculator than with a formal system of definitions and theorems.

Intuition and Rigor. We prove that the Mean Value Theorem is a consequence of the extreme value theorem, but we do not prove the extreme value theorem itself.

Instead, we argue that the extreme value theorem is intuitively obvious. The reason for such an intuitive appeal is that the proof of the extreme value theorem depends on the Heine Borel Theorem, and the Heine Borel Theorem depends on the topology of the real line. Thus, the proof of the Mean Value Theorem is not a proof at all. We might as well argue that the Mean Value Theorem is intuitively obvious and skip its proof altogether.

But we should not skip the Mean Value Theorem altogether. The point is that an introductory calculus course is not about theorems but is rather about definitions. If the difference operator

$$\Delta f(x) = f(x+h) - f(x)$$

could be used easily, then we would not even need the concept of a limit. However, our computationally attractive chain and product rules come solely from the definition

$$f'(x) = \lim_{h \to 0} \frac{f(x+h) - f(x)}{h}.$$

Good theorems are the stuff of graduate courses. Good definitions are the stuff of introductory calculus.

In my mind, this all but eliminates the traditional Riemann definition of the integral. It requires too much time and machinery—Riemann sums, the norm of a partition, the arbitrary choice of a point in a subinterval—and it is machinery that will not be used again. In contrast, Lesbegue's definition of the integral is intuitively simple and can be stated without a lot of machinery. The integral of a simple function is a picture. It can be computed using a table. The integral is a measure of simple function approximation. Simple functions can be used later in developing the integral test for convergence of a positive term series.

Technology and Modern Science. The computer was developed by mathematicians like Von Neumann for mathematicians like us, so it is absurd that mathematicians would not enthusiastically embrace their own creation. Technology should be a tool we welcome with excitement—no more rigging problems so the algebra comes out right. If a problem requires the solution to

$$6x^5 + 3x^2 + 1 = 0,$$

then the student can pull out the trusty graphing calculator, estimate the roots graphically and use a root finder to polish off the answer to the desired number of decimals. We should forget extremum problems with functions like

$$f(x) = x^6 + x^3 + x,$$

and instead to ask them to find the extrema of functions like

$$f(x) = \int_0^x \frac{6t^5 + 3t^2 + 1}{\sqrt{t^4 + 1}} \, dt.$$

As a result, calculus can be presented to the student as the true foundation of modern mathematics and science instead of as a hodge-podge of problems restricted to angle multiples of 30° and 45°. Indeed, modern science is more than the study of classical mathematics. It runs parallel to modern mathematics and often intertwines with it. Technology means that regression and curve-fitting can be studied in a calculus course and then immediately applied in a chemistry course. Our debate should be about how we should use technology, not if we should use technology.

4. CONCLUSION. The present dilemma is both fortunate and obvious. Calculus *is* modern, but our calculus curriculum is not. Calculus will continue to prosper with time and technology. We can take comfort in knowing that mathematics will continue to enlighten and enliven the minds of generations to come.

But poetry does not have to be taught by poets, and likewise, mathematics need not be taught by mathematicians. Our traditional course barely even tests the abilities of software tools such as Maple and Mathematica, and the days when every student carries a laptop are not that far off. Already our first semester calculus course is little more than a supplement to the high school curriculum. Without a modern perspective and an enthusiasm for the technology we mathematicians created for ourselves, our relevance to society will continue to dwindle. It will be only a matter of time before we mathematical horsemen are replaced by the intellectual equivalent of the steam engine.

REFERENCES

1. Eves, H., *An Introduction to the History of Mathematics*, 5th edition, Saunders College Publishing, Philadelphia, 1983.
2. Ewing, J., *A Century of Mathematics—Through the Eyes of the Monthly*, The Mathematical Association of America, Washington, 1994.
3. Greenman, C., On Articles in the October Issue, *Notices Amer. Math. Soc.* 43 (1996), 5-6.
4. Kleinfeld, M., Calculus: Reformed or Deformed, *Amer. Math. Monthly* 103 (1996), 230-232.
5. Krantz, S., Math for Sale, *Notices Amer. Math. Soc.* 42 (1995), 1116.
6. Royden, H. L., *Real Analysis*, 2nd edition, MacMillan Publishing Company, Inc., New York, 1968.
7. Woods, R., How Can Interest in Calculus be Increased?, *Amer. Math. Monthly* 36 (1929), 28-32.

Two Historical Applications of Calculus

Alexander J. Hahn

Calculus has been energized throughout its development by the demands of basic science and engineering. In turn, mathematics has enlightened these fields and has provided a clarifying point of view. I will present two applications of elementary calculus that illustrate this point. Both have interesting historical roots in the seventeenth century. One is a statics problem from the calculus book of the Marquis De L'Hospital and the other is an analysis of a page in Galileo's notebooks where he records an experiment with balls rolling down an inclined plane.

Let's begin with De L'Hospital (this is the spelling that the gentleman himself used). It appears that the marquis desired to be among the first of his French countrymen to understand Leibniz's powerful calculus and to become a part of these sensational developments of mathematics. To reach this goal efficiently he took the young Johann Bernoulli (one of Leibniz's star pupils, in his mid-twenties at the time) into his employ, first as his teacher and then as his consultant and collaborator. It became Bernoulli's job to clarify subtle points, to solve (or assist in solving) problems, and to proofread and translate the marquis' manuscripts. For these services Bernoulli received payment that he thought very generous, judging by the gratitude expressed in his letters to the marquis. It is likely therefore that the marquis thought in good conscience that he had bought the right to deal as he saw fit with everything he learned from Bernoulli. In 1696 he saw no difficulty in using this information as the basis for his own new book, the *Analyse des Infiniment Petits*. This was the world's very first elementary calculus textbook. Isaac Newton's *Principia* had appeared earlier, in 1687, but it was by no means a basic text. De L'Hospital's volume developed Leibniz's "analysis of the infinitely small" with elegance and clarity, and it soon became well known.

The correspondence [1] between Bernoulli and De L'Hospital makes it clear that most of the ideas in the *Analyse* were due to Leibniz, Newton, and Johann and Jakob Bernoulli, and that none were due to the marquis. For instance, *L'Hospital's rule* is due to Johann Bernoulli. As the marquis showed no interest in publicizing the role played by his collaborator, the *Analyse* in effect became his own brilliant contribution. (De L'Hospital's standard of research ethics would certainly not be acceptable today.) One of the most interesting problems in the *Analyse* is a pulley problem. As we will see in a moment, this problem quickly translates to finding the maximum value of an algebraic function on a closed interval. The computational difficulties that the solution presents are at just the right level to provide a good workout for beginning students.

Let's now turn to Galileo. Until recently there were historians of science who held that Galileo's experiments were primarily thought experiments—carefully conceived speculations about hypothetical experiments that Galileo never really carried out. This reflected the point of view and also the influence of the prominent historian and philosopher Alexandre Koyré (1892–1964). It is now apparent that this assessment was incorrect. In the early 1970s the historian Stillman Drake delved into Galileo's working papers and found explicit data about results of experiments with inclined planes [2]. One page from the year 1608—certainly one of the most important pages in the history of science—considers a ball that rolls down an inclined plane and across a short horizontal expanse of table, before falling freely to the ground. The numerical data recorded on the page suggest that Galileo has run several trials.

After students work through the development of a mathematical model for the motion of the ball, they will see that Galileo's data match the predictions of the model very well. The differences can be explained by the fact that the model is based on an idealization that ignores some of the complications that would be present in an actual experiment. The mathematical analysis of the rolling ball was, most certainly, beyond Galileo's mathematical expertise, so it provides strong evidence that his data came from experiments that he actually carried out.

Many other examples in my textbook [3] show how calculus has illuminated and provided essential structure to many important enterprises in engineering, science, and economics. These enterprises range in time from the seventeenth to the twentieth centuries, and in topics from the analysis of nuclear clocks and what they tell us, to the dynamics of a bullet in the barrel of a rifle, the design of a suspension bridge, the workings of the solar system, and consequences of the big bang. The text is designed for the beginning college student, presents all the relevant mathematics in complete detail in today's notation, and includes the surrounding engineering, science, and economics. I hope that my book will also serve as a resource for teachers who wish to insert such connections into the flow of their calculus courses.

De L'Hospital's Pulley Problem

At a point C on the horizontal ceiling of a room, attach a cord of length r that has a pulley affixed at its other end, the point F. At a second point B on the ceiling, a distance d from point C, attach a cord of length l, pass it through the pulley at F and connect a weight W at the other end, at D. Release the weight and allow the system to achieve its equilibrium position, as shown in Figure 1.

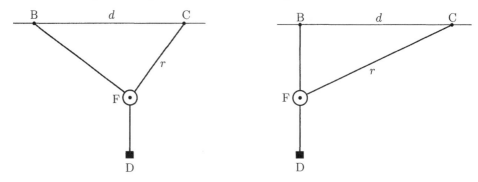

Figure 1. Typical equilibrium configurations. *Left,* $r < d$; *right,* $r \geq d$.

Two Historical Applications of Calculus 27

De L'Hospital supposes that the attached weight is heavy enough that the weights of the cords and the pulley can be ignored, and he asks: What is the geometry of the equilibrium configuration? More precisely, what are the dimensions of the triangle BCF, in terms of the constants r, d, l, and W? We will see that this seemingly elementary problem is surprisingly challenging.

First, note that if $r \geq d$ (so the first cord is long enough to reach B), then the weight at D will hang directly below B, since the weightless pulley and cord connecting it to C will not deflect it. Such a situation is shown on the right in Figure 1. We will therefore assume from now on that $r < d$.

Let E be the intersection of the extension of the segment DF with BC, and let x denote the distance EC, as in Figure 2. Applying the Pythagorean theorem yields

$$\text{BF} = \sqrt{\text{BE}^2 + \text{EF}^2}$$
$$= \sqrt{(d-x)^2 + r^2 - x^2}.$$

Thus FD $= l - \sqrt{(d-x)^2 + r^2 - x^2}$. The system is in equilibrium, which means that the weight at D must be at the lowest point that the geometry of the situation will allow. Put another way, the distance ED must be as large as possible. Observe that

$$\text{ED} = \text{EF} + \text{FD}$$
$$= \left(r^2 - x^2\right)^{1/2} + l - \left[(d-x)^2 + r^2 - x^2\right]^{1/2}.$$

If De L'Hospital can find the value of x for which ED is a maximum, in terms of the constants r, d, l, and W, his problem will be solved. It therefore remains only to find the points at which the function

$$f(x) = \left(r^2 - x^2\right)^{1/2} + l - \left[(d-x)^2 + r^2 - x^2\right]^{1/2}$$

attains its maximum value on $0 \leq x \leq r$.

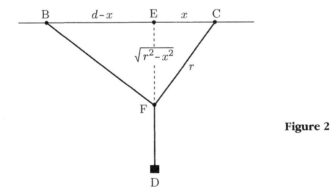

Figure 2

If $x = 0$ (so E = C), we have the configuration shown on the left in Figure 3 (page 96). This configuration cannot be in equilibrium, since the tension in the cord BF due to the weight surely pulls the cord CF to the side. Similarly, if $x = r$ (so E = F), we would have the configuration on the right in Figure 3, and this too is impossible because the weight certainly pulls down on the pulley at F. It follows that the maximum value of $f(x)$ must occur at some interior point of the interval $[0, r]$.

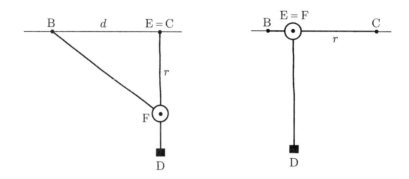

Figure 3

We compute the derivative and find where on this interval it either fails to exist or is zero. By the chain rule,

$$f'(x) = \frac{1}{2}\left(r^2 - x^2\right)^{-1/2}(-2x) - \frac{1}{2}\left[(d-x)^2 + r^2 - x^2\right]^{-1/2}\left[-2(d-x) - 2x\right]$$

$$= \frac{d}{\left[(d-x)^2 + r^2 - x^2\right]^{1/2}} - \frac{x}{\left(r^2 - x^2\right)^{1/2}}$$

$$= \frac{d\left(r^2 - x^2\right)^{1/2} - x\left[(d-x)^2 + r^2 - x^2\right]^{1/2}}{\left(r^2 - x^2\right)^{1/2}\left[(d-x)^2 + r^2 - x^2\right]^{1/2}}.$$

The denominator in this expression is never zero for $0 < x < r$, so the only critical points of f in this open interval are points where the numerator of $f'(x)$ is zero; that is, points where

$$d\left(r^2 - x^2\right)^{1/2} = x\left[(d-x)^2 + r^2 - x^2\right]^{1/2}.$$

Squaring and simplifying, we get

$$2dx^3 - 2d^2x^2 - r^2x^2 + r^2d^2 = 0.$$

One root of this polynomial is $x = d$, but by assumption $d > r$, so this root is outside the relevant interval $(0, r)$. Dividing by $x - d$ we see that the other roots satisfy the quadratic equation

$$2dx^2 - r^2x - dr^2 = 0.$$

The quadratic formula then tells us that f' has two other zeros, namely $x = (r/4d)\left(r \pm \sqrt{r^2 + 8d^2}\right)$. Now $r - \sqrt{r^2 + 8d^2} < 0$, so this root is outside the interval $(0, r)$. But it is easy to verify that $x = (r/4d)\left(r + \sqrt{r^2 + 8d^2}\right) < r$, so we conclude that the maximum value of $f(x)$ on $[0, r]$ occurs at this point.

Thus, the equilibrium configuration of L'Hospital's pulley and cords occurs when

$$x = \frac{r}{4d}\left(r + \sqrt{r^2 + 8d^2}\right).$$

Note that neither the weight W nor the length l of the cord holding this weight appears in the expression for x. Therefore, the equilibrium configuration is the same regardless of the exact amount of the weight W; this was not obvious when our analysis began.

Students who have learned to solve statics problems by balancing forces are invited to try this approach. It involves the separation of the forces acting on the pulley into their horizontal and vertical components, and the realization that the tension in the cord to which the weight is attached is constant. Those who persevere will discover that the equilibrium equations eventually lead to the same cubic polynomial that resulted in the solution presented above.

An Experiment of Galileo

In 1608, Galileo took an inclined plane with a shallow groove running along it, placed it on top of a table, and let a bronze ball of about 2 centimeters in diameter roll repeatedly down the groove. The ball started from rest at various heights, rolled down the plane, and rolled briefly along the horizontal table, before falling on the ground (also horizontal) along a parabolic path. Galileo carefully measured the initial height from which the ball was released and the corresponding distance from the base of the table to the ball's point of impact on the ground, and he recorded these findings in his notebook. Figure 4 depicts one of Galileo's diagrams, with his unit of length, the *punto* (≈ 0.94 millimeters), converted to meters. The table is 0.778 m high. Our goal is to develop a mathematical model for the motion of the ball and to compare Galileo's observations with the predictions of the model.

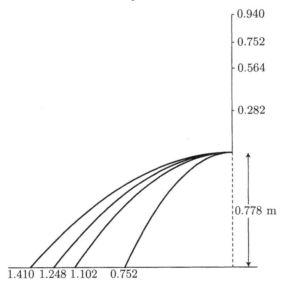

Figure 4

Denote the mass of the ball by m. The gravitational force on the ball is the weight of the ball; it acts downward and has magnitude mg, where g is the constant acceleration experienced by falling objects near the surface of the earth. If the angle of elevation of the inclined plane is β, the gravitational force can be decomposed into a component perpendicular to the plane, and a component $mg \sin \beta$ parallel to the

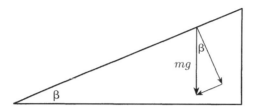

Figure 5

plane, as indicated in Figure 5. Since the motion of the ball is parallel to the inclined plane, only the component $mg \sin \beta$ acts to accelerate the ball. The component of the gravitational force on the ball perpendicular to the plane is exactly balanced by an equal and opposite force exerted by the plane on the ball.

Consider a typical trial of the experiment. Galileo sets the ball into the groove at a (vertical) height of h above the table and releases the ball at time $t = 0$. Consider the ball in typical position and let t be the corresponding elapsed time. Note that the ball does not slide down the plane, as it would in a frictionless situation; instead, it rolls. The rotation is produced by friction (i.e., by the contact between the ball and the plane). Denote the magnitude of the frictional force acting along the plane and against the motion of the ball by $f(t)$. Notice that at any time t the total force on the ball parallel to the inclined plane is $mg \sin \beta - f(t)$.

Let's introduce a coordinate axis along the inclined plane, taking as origin the point at which the ball is released, with the positive direction pointing downward along the plane. Denote the coordinate of the point of contact of the ball with the plane by $s(t)$. So $s(t)$ is the distance that the ball has moved during the time interval $[0, t]$. Notice that $s(0) = 0$. According to Newton's law of motion, the total force $mg \sin \beta - f(t)$ on the ball is equal to its mass m times the acceleration $a(t) = s''(t)$. Therefore,

$$ma(t) = mg \sin \beta - f(t). \tag{1}$$

To understand the motion we must assess the impact of the frictional force $f(t)$. When the ball has rolled a distance s, the angle θ through which it has turned is $s = r\theta$, where r is the radius of the ball. Differentiating this equation twice with respect to t gives $a(t) = r\alpha(t)$, where $\alpha(t) = \theta''(t)$ is the *angular acceleration* of the ball. There is a connection between the force $f(t)$ and $\alpha(t)$. It is required to complete our analysis and follows next.

Torques and moments of inertia. The angular acceleration α of a rigid body about an axis in response to a force of magnitude f applied at a distance r from the axis and acting in a direction perpendicular to the axis is given by $r \cdot f = I\alpha$, where the constant I, called the *moment of inertia* of the body about that axis, depends on how the mass of the body is distributed with respect to the axis. The product $r \cdot f$ is called the *torque* produced by the force f about the given axis. Just as Newton's law of motion says that the mass of an object measures its resistance to being accelerated by a force, the equation

$$r \cdot f = I\alpha \tag{2}$$

says that the moment of inertia of a rigid body about an axis describes its resistance to angular acceleration by a torque about that axis.

For a point mass of mass m joined to an axis by a rigid rod of negligible mass, as in Figure 6, a force f acting on the point and perpendicular to the axis produces a linear acceleration a given by $f = ma$. Since the resulting motion is circular, the angular acceleration α satisfies $a = r\alpha$, as we have seen. Substituting and multiplying by r gives $r \cdot f = (mr^2)\alpha$. Thus, according to the defining equation (2), the moment of inertia of a point mass m at one end of a weightless rod of length r, about an axis through the other end and perpendicular to the rod, is given by $I = mr^2$.

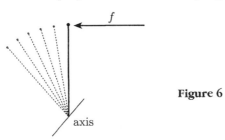

Figure 6

The moment of inertia is additive. In other words, if a body is composed of several parts whose moments of inertia about some axis are say $I_1, \ldots, I_n$, then the moment of inertia of the whole about this axis is the sum $I_1 + \cdots + I_n$. It should be clear from the interpretation of the moment of inertia as a measure of resistance to angular acceleration, that the resistance of a composite body is the sum of the resistances of its constituent parts.

As an example of the usefulness of this additivity, consider a thin homogeneous ring of radius r and total mass m. What is its moment of inertia about a central axis perpendicular? If we think of the ring as composed of many point masses, say n of them, each with mass m/n and joined by a weightless spoke to the central axis, then since each of these point masses has moment of inertia $(m/n)r^2$, the moment of inertia of the entire ring is $n \cdot (m/n)r^2 = mr^2$.

The moment of inertia of a uniform disk of radius r, about an axis through its center and perpendicular to the plane of the disk, can now be found by using the additivity property in the form of an integral. We think of the disk as a composite of concentric rings of radius x and infinitesimal thickness dx, where $0 \leq x \leq r$. A ring of radius x has circumference $2\pi x$, so the area of the ring with radius x and thickness dx is (approximately) $2\pi x\, dx$. If the mass of the whole disk is m, then the density of the material forming the disk is $m/(\pi r^2)$ mass units per unit area. So the mass of the ring of radius x is

$$\frac{m}{\pi r^2}(2\pi x\, dx) = \frac{2m}{r^2}(x\, dx).$$

The moment of inertia of this ring about the central axis is the product of its mass by the square of its radius x, or $(2m/r^2)x^3\, dx$. The moment of inertia of the whole disk is therefore

$$\frac{2m}{r^2}\int_0^r x^3\, dx = \frac{1}{2}mr^2. \tag{3}$$

A final use of the additivity property will compute the moment of inertia of a uniform ball, which is what we need for our analysis of Galileo's experiment. Think of a ball of mass m and radius r as sliced into very thin uniform disks centered

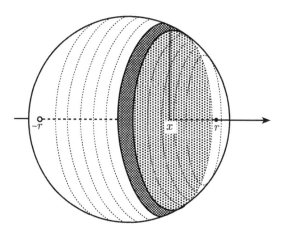

Figure 7

on the interval $[-r, r]$ of the x-axis, as in Figure 7. The disk centered at x has radius $\sqrt{r^2 - x^2}$, so if its thickness is dx then its volume will be approximately $\pi \left(r^2 - x^2\right) dx$. If the mass of the whole ball is m, then the density of the material forming the ball is $m/(4/3\pi r^3)$ mass units per unit volume. So the mass of the disk at x is

$$\frac{3m}{4\pi r^3} \cdot \pi\left(r^2 - x^2\right) dx = \frac{3m}{4r^3}\left(r^2 - x^2\right) dx.$$

The moment of inertia of this disk about the central axis is half the product of the mass by the square of its radius, according to (3), which gives $(3m/8r^3)\left(r^2 - x^2\right)^2 dx$. The moment of inertia of the whole ball is, therefore,

$$\frac{3m}{8r^3} \int_{-r}^{r} \left(r^4 - 2r^2 x^2 + x^4\right) dx = \frac{3m}{8r^3} \frac{16}{15} r^5 = \frac{2}{5} mr^2.$$

Returning to the defining equation (2), we now have $r \cdot f = \frac{2}{5} mr^2 \alpha$, or $f = \frac{2}{5} mr\alpha$. Substituting $r\alpha = a$ gives the key equation $f = \frac{2}{5} ma$. For $f = f(t)$, a function of time, this becomes

$$f(t) = \frac{2}{5} ma(t). \qquad (4)$$

A mathematical model for Galileo's experiment. Substitution of (4) into the basic equation (1) for a ball rolling down Galileo's inclined plane yields $ma(t) = mg \sin \beta - \frac{2}{5} ma(t)$, or

$$a(t) = \frac{5g}{7} \sin \beta. \qquad (5)$$

The rest is routine! Integrating (5), using the fact that the ball starts from rest, gives

$$v(t) = \left(\frac{5g}{7} \sin \beta\right) t,$$

and another integration, using the initial condition $s(0) = 0$, gives

$$s(t) = \left(\frac{5g}{14}\sin\beta\right)t^2.$$

We wish to determine the speed of the ball when it reaches the bottom of the inclined plane. This occurs when the ball has rolled the length of the hypotenuse of the right triangle in Figure 8. If t_b denotes the time when the ball reaches the bottom of the inclined plane, then $s(t_b) = h/\sin\beta$. Comparing this with our general formula for $s(t)$ gives

$$\left(\frac{5g}{14}\sin\beta\right)t_b^2 = \frac{h}{\sin\beta},$$

and hence

$$t_b = \sqrt{\frac{14h}{5g}\frac{1}{\sin\beta}}.$$

Substituting this into the velocity formula shows that the ball reaches the bottom of the inclined plane with velocity

$$v(t_b) = \left(\frac{5g}{7}\sin\beta\right)\sqrt{\frac{14h}{5g}\frac{1}{\sin\beta}} = \sqrt{\frac{10}{7}gh}.$$

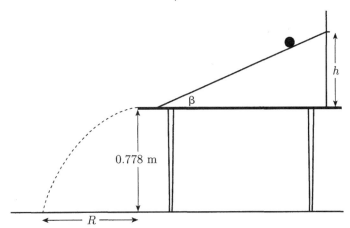

Figure 8

Let's step back from our analysis for a moment. If Galileo releases a ball from rest at height h above the table, we can predict its speed when it arrives at the bottom of the inclined plane. The table serves to stop the downward motion of the ball, deflecting it so that it leaves the table *horizontally* with speed $\sqrt{(10/7)gh}$.

It remains to take a minute to derive the range formula for the case of a projectile with horizontal initial velocity. Take a coordinate system with origin on the horizontal ground directly below the point of release with the y-axis pointing upward and the x-axis pointing along the floor in the direction of the initial velocity vector. The only force acting on the projectile is the downward gravitational force mg. Therefore, the motion (Newton's law) is given by

$$x''(t) = 0, \quad x'(0) = v_0, \quad x(0) = 0,$$
$$y''(t) = -g, \quad y'(0) = 0, \quad y(0) = y_0.$$

Integrating each equation twice gives

$$x'(t) = v_0, \quad x(t) = v_0 t$$
$$y'(t) = -gt, \quad y(t) = -\frac{1}{2}gt^2 + y_0.$$

The projectile hits the floor when $y(t) = 0$, in other words, when $t = \sqrt{(2y_0)/g}$, and at that time $x = v_0\sqrt{(2y_0)/g}$. For Galileo's ball we know $v_0 = \sqrt{(10/7)gh}$, so the range is given by

$$R = \sqrt{\frac{10}{7}gh}\sqrt{\frac{2y_0}{g}} = \sqrt{\frac{20}{7}y_0 h}.$$

Taking $y_0 = 0.778$ meters, the height of Galileo's table, gives the desired formula for the distance R from the base of the table at which the ball strikes the ground in terms of its starting height h above the table:

$$R = 1.491\sqrt{h}.$$

Table 1 shows the heights h used in Galileo's experiment, together with his observed values of R and the theoretical values predicted by our mathematical model. All of Galileo's measurements fall a little short of the distances predicted by the theory. But this is to be expected! After all, the ball is not perfectly round and its mass is not perfectly uniformly distributed. The groove in the plane will have bumps. As the ball rolls down the groove its axis of rotation will not maintain a perfectly horizontal position. Given these (and other) retarding influences, the ball will arrive at the bottom of the inclined plane with slightly lower speed than the model predicts. The impact of the moving ball with the table will also absorb a small amount of the ball's kinetic energy, thus reducing its speed slightly more. All of these minor effects will tend to reduce the distance of the ball's parabolic fall. So the experimental values of this distance should be slightly lower, but *consistently* lower, than the values predicted by the theory. This is exactly the pattern exhibited by Table 1.

Table 1. Comparison of Data: Model vs. Galileo

h	Calculated value of R	Galileo's value of R	Difference (in inches)
0.282	0.792	0.752	1.6
0.564	1.120	1.102	0.7
0.752	1.293	1.248	1.8
0.940	1.446	1.410	1.4

We have seen how calculus and some basic physics make a convincing case that the data in Galileo's notebook are a record of a genuine experiment and not the result of speculations. If it were indeed true that Galileo arrived at his numerical data via a thought experiment, then the only plausible conclusion would be that he was in possession of a mathematical model that allowed him to compute these data. Since it was more than a hundred years later that Euler and others worked out the basic theory of moments of inertia and rotational motion, we must assume that Galileo had no such quantitative model at hand.

Sit back and picture Galileo's rolling ball. It seems rather lifeless and uninteresting without the illuminating information that this analysis has provided. The basic physical principles in combination with the language and computational power of mathematics have certainly deepened our insight into Galileo's experiment. This is also true for De L'Hospital's contraption, but to a lesser extent. His brute force calculation is not as instructive as the vector solution to the problem.

References

1. Johann Bernoulli, *Der Briefwechsel von Johann Bernoulli*, Band I, Birkhäuser, Basel, 1955.
2. Stillman Drake, *Galileo: Pioneer Scientist*, University of Toronto Press, 1990.
3. Alexander J. Hahn, *Basic Calculus: From Archimedes to Newton to Its Role in Science*, Springer-Verlag, 1998.

Ideas of Calculus in Islam and India

Victor J. Katz

Isaac Newton created his version of the calculus during the years from about 1665 to 1670. One of Newton's central ideas was that of a power series, an idea he believed he had invented out of the analogy with the infinite decimal expansions of arithmetic [9, Vol. III, p. 33]. Newton, of course, was aware of earlier work done in solving the area problem, one of the central ideas of what was to be the calculus, and he knew well that the area under the curve $y = x^n$ between $x = 0$ and $x = b$ was given by $b^{n+1}/(n+1)$. (This rule had been developed by several mathematicians in the 1630s, including Bonaventura Cavalieri, Gilles Persone de Roberval, and Pierre de Fermat.) By developing power series to represent various functions, Newton was able to use this basic rule to find the areas under a wide variety of curves. Conversely, the use of the area formula enabled him to develop power series. For example, Newton developed the power series for $y = \arcsin x$, in effect by defining it in terms of an area and using the area formula. He then produced the power series for the sine by solving the equation $y = \arcsin x$ for $x = \sin y$ by inversion of the series. What Newton did not know, however, was that both the area formula—which he believed had been developed some 35 years earlier—and the power series for the sine had been *known for hundreds of years* elsewhere in the world. In particular, the area formula had been developed in Egypt around the year A.D. 1000 and the power series for the sine, as well as for the cosine and the arctangent, had been developed in India, probably in the fourteenth century. It is the development of these two ideas that will be discussed in this article.

Before going back to eleventh-century Egypt, however, we will first review the argument used both by Fermat and Roberval in working out their version of the area formula in 1636. In a letter to Fermat in October of that year, Roberval wrote that he had been able to find the area under curves of the form $y = x^k$ by using a formula—whose history in the Islamic world we will trace—for the sums of powers of the natural numbers: "The sum of the square numbers is always greater than the third part of the cube which has for its root the root of the greatest square, and the same sum of the squares with the greatest square removed is less than the third part of the same cube; the sum of the cubes is greater than the fourth part of the fourth power and with the greatest cube removed, less than the fourth part, etc." [5, p. 221]. In other words, finding the area of the desired region depends on the formula

$$\sum_{i=1}^{n-1} i^k < \frac{n^{k+1}}{k+1} < \sum_{i=1}^{n} i^k.$$

Ideas of Calculus in Islam and India

Fermat wrote back that he already knew this result and, like Roberval, had used it to determine the area under the graph of $y = x^k$ over the interval $[0, x_0]$. Both men saw that if the base interval was divided into n equal subintervals, each of length x_0/n, and if over each subinterval a rectangle whose height is the y-coordinate of the right endpoint was erected (see FIGURE 1), then the sum of the areas of these N circumscribed rectangles is

$$\frac{x_0^k}{n^k}\frac{x_0}{n} + \frac{(2x_0)^k}{n^k}\frac{x_0}{n} + \cdots + \frac{(nx_0)^k}{n^k}\frac{x_0}{n} = \frac{x_0^{k+1}}{n^{k+1}}(1^k + 2^k + \cdots + n^k).$$

Similarly, they could calculate the sum of the areas of the inscribed rectangles, those whose height is the y-coordinate of the left endpoint of the corresponding subinterval. In fact, if A is the area under the curve between 0 and x_0, then

$$\frac{x_0^{k+1}}{n^{k+1}}\left(1^k + 2^k + \cdots + (n-1)^k\right) < A < \frac{x_0^{k+1}}{n^{k+1}}(1^k + 2^k + \cdots + n^k).$$

The difference between the outer expressions of this inequality is simply the area of the rightmost circumscribed rectangle. Because x_0 and $y_0 = x_0^k$ are fixed, Fermat knew that the difference could be made less than any assigned value simply by taking n sufficiently large. It follows from the inequality cited by Roberval that both the area A and the value $x_0^{k+1}/(k+1) = x_0 y_0/(k+1)$ are squeezed between two values whose difference approaches 0. Thus Fermat and Roberval found that the desired area was $x_0 y_0/(k+1)$.

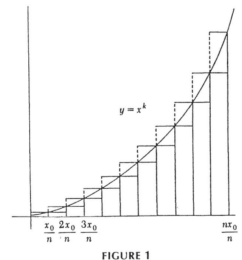

FIGURE 1

The obvious question is how either of these two men discovered formulas for the sums of powers. But at present, there is no answer to this question. There is nothing extant on this formula in the works of Roberval other than the letter cited, and all we have from Fermat on this topic, in letters to Marin Mersenne and Roberval, is a general statement in terms of triangular numbers, pyramidal numbers, and the other numbers that occur as columns of Pascal's triangle. (We note that Fermat's work was done some twenty years before Pascal published his material on the arithmetical triangle; the triangle had, however, been published in many versions in China, the Middle East, North Africa, and Europe over the previous 600 years. See [4], pp. 191–192; 241–242; 324–325.) Here is what Fermat says: "The last side multiplied by the next greater makes twice the triangle. The last side multiplied by the triangle of the next greater side makes three times the pyramid. The last side multiplied by the pyramid of the next greater side makes four times the triangulotriangle. And so on by the same progression in infinitum" [5, p. 230]. Fermat's statement can be written using the modern notation for binomial coefficients as

$$n\binom{n+k}{k} = (k+1)\binom{n+k}{k+1}.$$

We can derive from this formula for each k in turn, beginning with $k = 1$, an explicit formula for the sum of the kth powers by using the properties of the Pascal triangle. For example, if $k = 2$, we have

$$n\binom{n+2}{2} = 3\binom{n+2}{3} = 3\sum_{j=2}^{n+1}\binom{j}{2}$$

$$= 3\sum_{j=2}^{n+1} \frac{j(j-1)}{2} = 3\sum_{i=1}^{n} \frac{i(i+1)}{2} = 3\sum_{i=1}^{n} \frac{i^2+i}{2}.$$

Therefore,

$$2\frac{n}{3}\frac{(n+2)(n+1)}{2} - \sum_{i=1}^{n} i = \sum_{i=1}^{n} i^2$$

and

$$\sum_{i=1}^{n} i^2 = \frac{n^3 + 3n^2 + 2n}{3} - \frac{n^2+n}{2} = \frac{2n^3 + 3n^2 + n}{6} = \frac{n^3}{3} + \frac{n^2}{2} + \frac{n}{6}.$$

In general, the sum formula is of the form

$$\sum_{i=1}^{n} i^k = \frac{n^{k+1}}{k+1} + \frac{n^k}{2} + p(n),$$

where $p(n)$ is a polynomial in n of degree less than k, and Roberval's inequality can be proved for each k. We do not know if Fermat's derivation was like that above, however, because he only states a sum formula explicitly for the case $k = 4$ and gives no other indication of his procedure.

Sums of Integer Powers in Eleventh-Century Egypt

The formulas for the sums of the kth powers, however, at least through $k = 4$, as well as a version of Roberval's inequality, were developed some 650 years before the mid-seventeenth century by Abu Ali al-Hasan ibn al-Hasan ibn al-Haytham (965–1039), known in Europe as Alhazen. The formulas for the sums of the squares and cubes were stated even earlier. The one for squares was stated by Archimedes around 250 B.C. in connection with his quadrature of the parabola, while the one for cubes, although it was probably known to the Greeks, was first explicitly written down by Aryabhata in India around 500 [2, pp. 37–38]. The formula for the squares is not difficult to discover, and the one for cubes is virtually obvious, given some experimentation. By contrast, the formula for the sum of the fourth powers is not obvious. If one can discover a method for determining this formula, one can discover a method for determining the formula for the sum of any integral powers. Ibn al-Haytham showed in fact how to develop the formula for the kth powers from $k = 1$ to $k = 4$; all his proofs were similar in nature and easily generalizable to the discovery and proof of formulas for the sum of any given powers of the integers. That he did not state any such generalization is probably due to his needing only the formulas for the second and fourth powers to solve the problem in which he was interested: computing the volume of a certain paraboloid.

Before discussing ibn al-Haytham's work, it is good to briefly describe the world of Islamic science. (See [1] for more details.) During the ninth century, the Caliph al-Ma'mun established a research institute, the House of Wisdom, in Baghdad and invited scholars from all parts of the caliphate to participate in the development of a scientific tradition in Islam. These scientists included not only Moslem Arabs, but also Christians, Jews, and Zoroastrians, among others. Their goals were, first, to translate into Arabic the best mathematical and scientific works from Greece and India, and, second, by building on this base, to create new mathematical and scientific ideas. Although the House of Wisdom disappeared after about two centuries, many of the rulers of the Islamic states continued to support scientists in their quest for knowledge, because they felt that the research would be of value in practical applications.

Thus it was that ibn al-Haytham, born in Basra, now in Iraq, was called to Egypt by the Caliph al-Hakim to work on a Nile control project. Although the project never came to fruition, ibn al-Haytham did produce in Egypt his most important scientific work, the *Optics* in seven books. The *Optics* was translated into Latin in the early thirteenth century and was studied and commented on in Europe for several centuries thereafter. Ibn al-Haytham's fame as a mathematician from the medieval period to the present chiefly rests on his treatment of "Alhazen's problem," the problem of finding the point or points on some reflecting surface at which the light from one of two points outside that surface is reflected to the other. In the fifth book of the *Optics* he set out his solutions to this problem for a variety of surfaces, spherical, cylindrical, and conical, concave and convex. His results, based on six separately proved lemmas on geometrical constructions, show that he was in full command of both the elementary and advanced geometry of the Greeks.

The central idea in ibn al-Haytham's proof of the sum formulas was the derivation of the equation

$$(n+1)\sum_{i=1}^{n} i^k = \sum_{i=1}^{n} i^{k+1} + \sum_{p=1}^{n}\left(\sum_{i=1}^{p} i^k\right). \qquad (*)$$

Naturally, he did not state this result in general form. He only stated it for particular integers, namely $n = 4$ and $k = 1, 2, 3$, but his proof for each of those k is by induction on n and is immediately generalizable to any value of k. (See [7] for details.) We consider his proof for $k = 3$ and $n = 4$:

$$(4+1)(1^3 + 2^3 + 3^3 + 4^3) = 4(1^3 + 2^3 + 3^3 + 4^3) + 1^3 + 2^3 + 3^3 + 4^3$$
$$= 4 \cdot 4^3 + 4(1^3 + 2^3 + 3^3) + 1^3 + 2^3 + 3^3 + 4^3$$
$$= 4^4 + (3+1)(1^3 + 2^3 + 3^3) + 1^3 + 2^3 + 3^3 + 4^3.$$

Because equation $(*)$ is assumed true for $n = 3$,

$$(3+1)(1^3 + 2^3 + 3^3) = 1^4 + 2^4 + 3^4 + (1^3 + 2^3 + 3^3) + (1^3 + 2^3) + 1^3.$$

Equation $(*)$ is therefore proved for $n = 4$. One can easily formulate ibn al-Haytham's argument in modern terminology to give a proof for any k by induction on n.

Ibn al-Haytham now uses his result to derive formulas for the sums of integral powers. First, he proves the sum formulas for squares and cubes:

$$\sum_{i=1}^{n} i^2 = \left(\frac{n}{3} + \frac{1}{3}\right) n \left(n + \frac{1}{2}\right) = \frac{n^3}{3} + \frac{n^2}{2} + \frac{n}{6}$$

$$\sum_{i=1}^{n} i^3 = \left(\frac{n}{4} + \frac{1}{4}\right) n(n+1) n = \frac{n^4}{4} + \frac{n^3}{2} + \frac{n^2}{4}.$$

We will not deal with these proofs here, but only with the derivation of the analogous result for the fourth powers. Although ibn al-Haytham himself derives this result only for $n = 4$, he asserts it for arbitrary n. We will therefore use modern techniques modeled on ibn al-Haytham's method to derive it for that case. We begin by using the formulas for the sums of squares and cubes to rewrite equation (∗) in the form

$$(n+1) \sum_{i=1}^{n} i^3 = \sum_{i=1}^{n} i^4 + \sum_{p=1}^{n} \left(\frac{p^4}{4} + \frac{p^3}{2} + \frac{p^2}{4}\right)$$

$$= \sum_{i=1}^{n} i^4 + \frac{1}{4} \sum_{i=1}^{n} i^4 + \frac{1}{2} \sum_{i=1}^{n} i^3 + \frac{1}{4} \sum_{i=1}^{n} i^2.$$

It then follows that

$$(n+1) \sum_{i=1}^{n} i^3 = \frac{5}{4} \sum_{i=1}^{n} i^4 + \frac{1}{2} \sum_{i=1}^{n} i^3 + \frac{1}{4} \sum_{i=1}^{n} i^2$$

$$\frac{5}{4} \sum_{i=1}^{n} i^4 = \left(n + 1 - \frac{1}{2}\right) \sum_{i=1}^{n} i^3 - \frac{1}{4} \sum_{i=1}^{n} i^2$$

$$\sum_{i=1}^{n} i^4 = \frac{4}{5}\left(n + \frac{1}{2}\right) \sum_{i=1}^{n} i^3 - \frac{1}{5} \sum_{i=1}^{n} i^2$$

$$= \frac{4}{5}\left(n + \frac{1}{2}\right)\left(\frac{n}{4} + \frac{1}{4}\right) n(n+1)n - \frac{1}{5}\left(\frac{n}{3} + \frac{1}{3}\right) n \left(n + \frac{1}{2}\right)$$

$$= \left(\frac{n}{5} + \frac{1}{5}\right)\left(n + \frac{1}{2}\right) n(n+1)n - \left(\frac{n}{5} + \frac{1}{5}\right)\left(n + \frac{1}{2}\right) n \cdot \frac{1}{3}.$$

Ibn al-Haytham stated his result verbally in a form we translate into modern notation as

$$\sum_{i=1}^{n} i^4 = \left(\frac{n}{5} + \frac{1}{5}\right) n \left(n + \frac{1}{2}\right)\left[(n+1)n - \frac{1}{3}\right].$$

The result can also be written as a polynomial:

$$\sum_{i=1}^{n} i^4 = \frac{n^5}{5} + \frac{n^4}{2} + \frac{n^3}{3} - \frac{n}{30}.$$

It is clear that this formula can be used as Fermat and Roberval used Roberval's inequality to show that

$$\lim_{n \to \infty} \frac{\sum_{i=1}^{n} i^4}{n^5} = \frac{1}{5}.$$

Ibn al-Haytham used his result on sums of integral powers to perform what we would call an integration. In particular, he applied his result to determine the volume of the solid formed by rotating the parabola $x = ky^2$ around the line $x = kb^2$, perpendicular to the axis of the parabola, and showed that this volume is 8/15 of the volume of the cylinder of radius kb^2 and height b. (See FIGURE 2.) His formal argument was a typical Greek-style exhaustion argument using a double *reductio ad absurdum*, but in essence his method involved slicing the cylinder and paraboloid into n disks, each of thickness $h = b/n$, and then adding up the disks. The ith disk in the paraboloid has radius $kb^2 - k(ih)^2$ and therefore has volume $\pi h(kh^2 n^2 - ki^2 h^2)^2 = \pi k^2 h^5 (n^2 - i^2)^2$. The total volume of the paraboloid is therefore approximated by

$$\pi k^2 h^5 \sum_{i=1}^{n-1} (n^2 - i^2)^2 = \pi k^2 h^5 \sum_{i=1}^{n-1} (n^4 - 2n^2 i^2 + i^4).$$

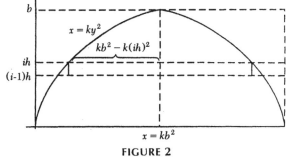

FIGURE 2

But since ibn al-Haytham knew the formulas for the sums of integral squares and fourth powers, he could calculate that

$$\sum_{i=1}^{n-1} (n^4 - 2n^2 i^2 + i^4) = \frac{8}{15}(n-1)n^4 + \frac{1}{30}n = \frac{8}{15}n \cdot n^4 - \frac{1}{2}n^4 - \frac{1}{30}n$$

and therefore that

$$\frac{8}{15}(n-1)n^4 < \sum_{i=1}^{n-1}(n^2-i^2)^2 < \frac{8}{15}n \cdot n^4.$$

But the volume of a typical slice of the circumscribing cylinder is $\pi h(kb)^2)^2 = \pi k^2 h^5 n^4$, and therefore the total volume of the cylinder is $\pi k^2 h^5 n \cdot n^4$, while the volume of the cylinder less its "top slice" is $\pi k^2 h^5 (n-1)n^4$. The inequality then shows that the volume of the paraboloid is bounded between 8/15 of the cylinder less its top slice and 8/15 of the entire cylinder. Because the top slice can be made as small as desired by taking n sufficiently large, it follows that the volume of the paraboloid is exactly 8/15 of the volume of the cylinder as asserted.

Ibn al-Haytham's formula for the sum of fourth powers shows up in other places in the Islamic world over the next few centuries. It appears in the work of Abu-l-Hasan ibn Haydur (d. 1413), who lived in what is now Morocco, and in the work of Abu Abdallah ibn Ghazi (1437–1514), who also lived in Morocco. (See [3] for details.) Furthermore, one also finds the formula in *The Calculator's Key* of Ghiyath al-Din Jamshid al-Kashi (d. 1429), a mathematician and astronomer whose most productive years were spent in Samarkand, now in Uzbekistan, in the court of Ulugh Beg. We do not know, however, how these mathematicians learned of the formula or for what purpose they used it.

Trigonometric Series in Sixteenth-Century India

The sum formulas for integral powers surface in sixteenth-century India and they are used to develop the power series for the sine, cosine, and arctangent. These power series appear in Sanskrit verse in the *Tantrasangraha-vyakhya* (of about 1530), a commentary on a work by Kerala Gargya Nilakantha (1445–1545) of some 30 years earlier. Unlike the situation for many results of Indian mathematics, however, a detailed derivation of these power series exists, in the *Yuktibhasa*, a work in Malayalam, the language of Kerala, the southwestern region of India. This latter work was written by Jyesthadeva (1500–1610), who credits these series to Madhava, an Indian mathematician of the fourteenth century.

Even though we do not know for sure whether Madhava was the first discoverer of the series, it is clear that the series were known in India long before the time of Newton. But why were the Indians interested in these matters? India had a long tradition of astronomical research, dating back to at least the middle of the first millennium B.C. The Indians had also absorbed Greek astronomical work and its associated mathematics during and after the conquest of northern India by Alexander the Great in 327 B.C. Hence the Indians became familiar with Greek trigonometry, based on the chord function, and then gradually improved it by introducing our sine, cosine, and tangent. Islamic mathematicians learned trigonometry from India, introduced their own improvements, and, after the conquest of northern India by a Moslem army in the twelfth century, brought the improved version back to India. (See [4] for more details.)

The interaction of astronomy with trigonometry brings an increasing demand for accuracy. Thus Indian astronomers wanted an accurate value for π (which comes from knowing the arctangent power series) and also accurate values for the sine and cosine (which comes from their power series) so they could use these values in determining planetary positions. Because a recent article [8] in this MAGAZINE discussed the arctangent power series, we will here consider only the sine and cosine series.

The statement of the Indian rule for determining these series is as follows: "Obtain the results of repeatedly multiplying the arc [s] by itself and then dividing by $2, 3, 4, \ldots$ multiplied by the radius [ρ]. Write down, below the radius (in a column) the even results [i.e. results corresponding to $n = 2, 4, 6$ in $s^n / n! \rho^{n-1}$], and below the radius (in another column) the odd results [corresponding to $n = 3, 5, 7, \ldots$ in $s^n / n! \rho^{n-1}$]. After writing down a number of terms in each column, subtract the last term of either column from the one next above it, the remainder from the term next above, and so on, until the last subtraction is made from the radius in the first column and from the arc in the second. The two final remainders are respectively the cosine and the sine, to a certain degree of approximation." [6, p. 3] These words can easily be translated into the formulas:

$$x = \cos s = \rho - \frac{s^2}{2!\rho} + \frac{s^4}{4!\rho^3} - \cdots + (-1)^n \frac{s^{2n}}{(2n)!\rho^{2n-1}} + \cdots$$

$$y = \sin s = s - \frac{s^3}{3!\rho^2} + \frac{s^5}{5!\rho^4} - \cdots + (-1)^n \frac{s^{2n+1}}{(2n+1)!\rho^{2n}} + \cdots$$

(These formulas reduce to the standard power series when ρ is taken to be 1.)

Ideas of Calculus in Islam and India

The Indian derivations of these results begin with the obvious approximations to the cosine and sine for small arcs and then use a "pull yourself up by our own bootstraps" approach to improve the approximation step by step. The derivations also make use of the notion of differences, a notion used in other aspects of Indian mathematics as well. In our discussion of the Indian method, we will use modern notation to enable the reader to follow these sixteenth-century Indian ideas.

We first consider the circle of FIGURE 3 with a small arc $\alpha = \widehat{AC} \approx AC$. From the similarity of triangles AGC and OEB, we get

$$\frac{x_1 - x_2}{\alpha} = \frac{y}{\rho} \quad \text{and} \quad \frac{y_2 - y_1}{\alpha} = \frac{x}{\rho}$$

$$\text{or} \quad \frac{\alpha}{\rho} = \frac{x_1 - x_2}{y} = \frac{y_2 - y_1}{x}.$$

In modern terms, if $\angle BOF = \theta$ and $\angle BOC = \angle AOB = d\theta$, these equations amount to

$$\sin(\theta + d\theta) - \sin(\theta - d\theta) = \frac{y_2 - y_1}{\rho} = \frac{\alpha x}{\rho^2} = \frac{2\rho \, d\theta}{\rho} \cos\theta = 2\cos\theta \, d\theta$$

and

$$\cos(\theta + d\theta) - \cos(\theta - d\theta) = \frac{x_2 - x_1}{\rho} = -\frac{\alpha y}{\rho^2} = -\frac{2\rho \, d\theta}{\rho} \sin\theta = -2\sin\theta \, d\theta.$$

Now, suppose we have a small arc s divided into n equal subarcs, with $\alpha = s/n$. For simplicity we take $\rho = 1$, although the Indian mathematicians did not. By applying the previous results repeatedly, we get the following sets of differences for the y's (FIGURE 4) (where $y_n = y = \sin s$):

$$\Delta_n y = y_n - y_{n-1} = \alpha x_n$$
$$\Delta_{n-1} y = y_{n-1} - y_{n-2} = \alpha x_{n-1}$$
$$\cdots$$
$$\Delta_2 y = y_2 - y_1 = \alpha x_2$$
$$\Delta_1 y = y_1 - y_0 = y_1 = \alpha x_1.$$

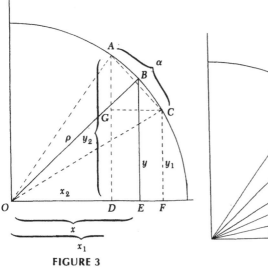

FIGURE 3

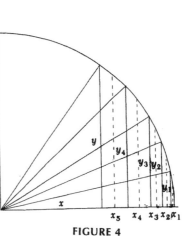

FIGURE 4

Similarly, the differences for the x's can be written
$$\Delta_{n-1} x = x_n - x_{n-1} = -\alpha y_{n-1}$$
$$\cdots$$
$$\Delta_2 x = x_3 - x_2 = -\alpha y_2$$
$$\Delta_1 x = x_2 - x_1 = -\alpha y_1.$$

We next consider the second differences on the y's:
$$\Delta_2 y - \Delta_1 y = y_2 - y_1 - y_1 + y_0 = \alpha(x_2 - x_1) = -\alpha^2 y_1.$$
In other words, the second difference of the sines is proportional to the negative of the sine. But since $\Delta_1 y = y_1$, we can write this result as
$$\Delta_2 y = y_1 - \alpha^2 y_1.$$
Similarly, since
$$\Delta_3 y - \Delta_2 y = y_3 - y_2 - y_2 + y_1 = \alpha(x_3 - x_2) = -\alpha^2 y_2,$$
it follows that
$$\Delta_3 y = \Delta_2 y - \alpha^2 y_2 = y_1 - \alpha^2 y_1 - \alpha^2 y_2,$$
and, in general, that
$$\Delta_n y = y_1 - \alpha^2 y_1 - \alpha^2 y_2 - \cdots - \alpha^2 y_{n-1}.$$
But the sine equals the sum of its differences:
$$y = y_n = \Delta_1 y + \Delta_2 y + \cdots + \Delta_n y$$
$$= n y_1 - [y_1 + (y_1 + y_2) + (y_1 + y_2 + y_3) + \cdots + (y_1 + y_2 + \cdots + y_{n-1})]\alpha^2.$$
Also, $s/n \approx y_1 \approx \alpha$, or $ny_1 \approx s$. Naturally, the larger the value of n, the better each of these approximations is. Therefore,
$$y \approx s - \lim_{n \to \infty} \left(\frac{s}{n}\right)^2 [y_1 + (y_1 + y_2) + \cdots + (y_1 + y_2 + \cdots + y_{n-1})].$$
Next we add the differences of the x's. We get
$$x_n - x_1 = -\alpha(y_1 + y_2 + \cdots + y_{n-1}).$$
But $x_n \approx x = \cos s$ and $x_1 \approx 1$. It then follows that
$$x \approx 1 - \lim_{n \to \infty} \left(\frac{s}{n}\right)(y_1 + y_2 + \cdots + y_{n-1}).$$

To continue the calculation, the Indian mathematicians needed to approximate each y_i and use these approximations to get approximations for $x = \cos s$ and $y = \sin s$. Each new approximation in turn is placed back in the expressions for x and y and leads to a better approximation. Note first that if y is small, y_i can be approximated by is/n. It follows that
$$x \approx 1 - \lim_{n \to \infty} \left(\frac{s}{n}\right)\left[\frac{s}{n} + \frac{2s}{n} + \cdots + \frac{(n-1)s}{n}\right]$$
$$= 1 - \lim_{n \to \infty} \left(\frac{s}{n}\right)^2 [1 + 2 + \cdots + (n-1)]$$

$$= 1 - \lim_{n \to \infty} \frac{s^2}{n^2} \left[\frac{(n-1)n}{2} \right]$$

$$= 1 - \frac{s^2}{2}.$$

Similarly,

$$y \approx s - \lim_{n \to \infty} \left(\frac{s}{n}\right)^2 \left[\frac{s}{n} + \left(\frac{s}{n} + \frac{2s}{n}\right) + \cdots + \left(\frac{s}{n} + \frac{2s}{n} + \cdots + \frac{(n-1)s}{n}\right) \right]$$

$$= s - \lim_{n \to \infty} \frac{s^3}{n^3} \left[1 + (1+2) + (1+2+3) + \cdots + (1+2+\cdots+(n-1)) \right]$$

$$= s - \lim_{n \to \infty} \frac{s^3}{n^3} \left[n(1+2+\cdots+(n-1)) - (1^2+2^2+\cdots+(n-1)^2) \right]$$

$$= s - s^3 \lim_{n \to \infty} \left[\frac{\sum_{i=1}^{n-1} i}{n^2} - \frac{\sum_{i=1}^{n-1} i^2}{n^3} \right]$$

$$= s - s^3 \left(\frac{1}{2} - \frac{1}{3} \right)$$

$$= s - \frac{s^3}{6},$$

and there is a new approximation for y and therefore for each y_i. Note that in the transition from the second to the third lines of this calculation the Indians used ibn al-Haytham's equation (∗) for the case $k = 1$. Although the Indian mathematicians did not refer to ibn al-Haytham or any other predecessor, they did explicitly sketch a proof of this result in the general case and used it to show that, for any k, the sum of the kth powers of the first n integers is approximately equal to $n^{k+1}/(k+1)$. This result was used in the penultimate line of the above calculation in the cases $k = 1$ and $k = 2$ and in the derivation of the power series for the arctangent as discussed in [8].

To improve the approximation for sine and cosine, we now assume that $y_i \approx (is/n) - (is)^3/(6n^3)$ in the expression for $x = \cos s$ and use the sum formula in the case $k = 3$ to get

$$x \approx 1 - \lim_{n \to \infty} \frac{s}{n} \left[\frac{s}{n} - \frac{s^3}{6n^3} + \frac{2s}{n} - \frac{(2s)^3}{6n^3} + \cdots + \frac{(n-1)s}{n} - \frac{((n-1)s)^3}{6n^3} \right]$$

$$= 1 - \frac{s^2}{2} + \lim_{n \to \infty} \frac{s^4}{6n^4} \left[1^3 + 2^3 + \cdots + (n-1)^3 \right]$$

$$= 1 - \frac{s^2}{2} + \frac{s^4}{6} \lim_{n \to \infty} \frac{\sum_{i=1}^{n-1} i^3}{n^4}$$

$$= 1 - \frac{s^2}{2} + \frac{s^4}{6} \cdot \frac{1}{4}$$

$$= 1 - \frac{s^2}{2} + \frac{s^4}{24}.$$

Similarly, ibn al-Haytham's formula for the case $j = 3$ and the sum formula for the cases $j = 3$ and $j = 4$ lead to a new approximation for $y = \sin s$:

$$y \approx s - \frac{s^3}{6} + \lim_{n \to \infty} \left(\frac{s}{n}\right)^2 \left[\frac{s^3}{6n^3} + \left(\frac{s^3}{6n^3} + \frac{(2s)^3}{6n^3}\right) \right.$$
$$\left. + \cdots + \left(\frac{s^3}{6n^3} + \frac{(2s)^3}{6n^3} + \cdots + \frac{((n-1)s)^3}{6n^3}\right)\right]$$

$$= s - \frac{s^3}{6} + \lim_{n \to \infty} \frac{s^5}{6n^5}\left[1^3 + (1^3 + 2^3) + \cdots + (1^3 + 2^3 + \cdots + (n-1)^3)\right]$$

$$= s - \frac{s^3}{6} + \lim_{n \to \infty} \frac{s^5}{6n^5}\left[n(1^3 + 2^3 + \cdots + (n-1)^3) - (1^4 + 2^4 + \cdots + (n-1)^4)\right]$$

$$= s - \frac{s^3}{6} + \frac{s^5}{6} \lim_{n \to \infty} \left[\frac{\sum_{i=1}^{n-1} i^3}{n^4} - \frac{\sum_{i=1}^{n-1} i^4}{n^5}\right]$$

$$= s - \frac{s^3}{6} + \frac{s^5}{6}\left(\frac{1}{4} - \frac{1}{5}\right)$$

$$= s - \frac{s^3}{6} + \frac{s^5}{120}.$$

Because Jyesthadeva considers each new term in these polynomials as a correction to the previous value, he understood that the more terms taken, the more closely the polynomials approach the true values for the sine and cosine. The polynomial approximations can thus be continued as far as necessary to achieve any desired approximation. The Indian authors had therefore discovered the sine and cosine power series!

Conclusion

How close did Islamic and Indian scholars come to inventing the calculus? Islamic scholars nearly developed a general formula for finding integrals of polynomials by A.D. 1000—and evidently could find such a formula for any polynomial in which they were interested. But, it appears, they were not interested in any polynomial of degree higher than four, at least in any of the material which has so far come down to us. Indian scholars, on the other hand, were by 1600 able to use ibn al-Haytham's sum formula for arbitrary integral powers in calculating power series for the functions in which they were interested. By the same time, they also knew how to calculate the differentials of these functions. So some of the basic ideas of calculus were known in Egypt and India many centuries before Newton. It does not appear, however, that either Islamic or Indian mathematicians saw the necessity of connecting some of the disparate ideas that we include under the name calculus. There were apparently only specific cases in which these ideas were needed.

There is no danger, therefore, that we will have to rewrite the history texts to remove the statement that Newton and Leibniz invented the calculus. They were certainly the ones who were able to combine many differing ideas under the two unifying themes of the derivative and the integral, show the connection between them, and turn the calculus into the great problem-solving tool we have today. But what we do not know is whether the immediate predecessors of Newton and Leibniz, including in particular Fermat and Roberval, learned of some of the ideas of the Islamic or Indian mathematicians through sources of which we are not now aware.

The entire question of the transmission of mathematical knowledge from one culture to another is a matter of current research and debate. In particular, with more medieval Arabic manuscripts being discovered and translated into European languages, the route of some mathematical ideas can be better traced from Iraq and Iran into Egypt, then to Morocco and on into Spain. (See [3] for more details.) Medieval Spain was one of the meeting points between the older Islamic and Jewish cultures and the emerging Latin-Christian culture of Europe. Many Arabic works were translated there into Latin in the twelfth century, sometimes by Jewish scholars who also wrote works in Hebrew. But although there is no record, for example, of ibn al-Haytham's work on sums of integral powers being translated at that time, certain ideas he used do appear in both Hebrew and Latin works of the thirteenth century. And since the central ideas of his work occur in the Indian material, there seems a good chance that transmission to India did occur. Answers to the questions of transmission will require much more work in manuscript collections in Spain and the Maghreb, work that is currently being done by scholars at the Centre National de Recherche Scientifique in Paris. Perhaps in a decade or two, we will have evidence that some of the central ideas of calculus did reach Europe from Africa or Asia.

Acknowledgement. I want to thank the referee for thoughtful comments and suggestions on the original version of this paper.

REFERENCES

1. J. L. Berggren, *Episodes in the Mathematics of Medieval Islam*, Springer-Verlag, New York, 1986.
2. Walter E. Clark, *The Aryabhatiya of Aryabhata* University of Chicago Press, Chicago, 1930.
3. Ahmed Djebbar, *Enseignement et Recherche Mathématiques dans le Maghreb des XIIIe − XIVe Siecles* (Publications Mathématiques D'Orsay No. 81-02) Université de Paris-Sud, Orsay, France, 1981.
4. Victor Katz, *A History of Mathematics: An Introduction*, Harper Collins Publishers, New York, 1993.
5. Michael Mahoney, *The Mathematical Career of Pierre de Fermat 1601–65*, Princeton University Press, Princeton, NJ, 1973.
6. C. T. Rajagopal and A. Venkataraman, The sine and cosine power-series in Hindu mathematics, *J. of the Royal Asiatic Society of Bengal—Science* 15 (1949), 1–13.
7. Roshdi Rashed, Ibn al-Haytham et la measure du paraboloide, *J. for the History of Arabic Science* 5 (1981), 262–291.
8. Ranjan Roy, The discovery of the series formula for π by Leibniz, Gregory and Nilakantha, this MAGAZINE 63 (1990), 291–306.
9. Derek Whiteside, *The Mathematical Papers of Isaac Newton*, Cambridge University Press, Cambridge, 1967–1981.

A Tale of Two CD's

Dan Kennedy

These are indeed exciting times in the world of Mathematics. Stirred by the NCTM *Standards*, the winds of change are blowing through every level of the K-12 curriculum. At the same time, in colleges and universities the finest minds of our profession are turning their attention to the forging of a Calculus for a New Century. Earlier this year, we even saw the unexpected verification of Fermat's Last Theorem, something which surely none of us thought we would see in this lifetime. There is so much to talk about when mathematicians get together these days, yet, because we are teachers, we have so little time for talking. That is why I have chosen to write this article about none of these things.

Instead, I would like to write about records.

I have collected records since I was in grade school. By the time I was in graduate school, I owned more than two thousand 45's, three hundred albums, and miscellaneous 78's and EP's. I shared my hobby with friends, including some memorable years as music director and station manager of the college radio station at Holy Cross. I was one of those guys who could, upon hearing a golden oldie on the radio, quote the title, artist, and year of the song, and quite often the label and songwriter as well. My interest in the music of the moment naturally declined about the time disco music became popular, but by that time I had accumulated enough vinyl classics to keep myself and my party guests entertained forever. For example, one of my favorite ways to pass an evening with friends is to stage a "nostalgia playoff" between two guests, playing alternately the hit songs from their respective high school graduation years, until the quality of one year's hits is clearly unable to keep up with the quality of the other's. (In case you are interested, no guest has ever gone up against a graduate of 1957 without conceding after the 15th round or so.)

As a student of the recording industry, I would always sit up and take notice when some new product emerged which the prophets predicted would change the way people listened to music. The first big pretender to the vinyl throne was the 8-track tape. "The 8-track tape," they predicted, "will redefine the recording industry." It required no threading onto a spool, it did not scratch or shatter, it did not collect dust, it required no needle, it produced high fidelity sound for multiple speakers, and *you could play it in your car*! I was momentarily impressed, but I continued to buy records, and so did apparently a lot of other people. Today if you want to buy an 8-track tape you have to go to an antique show.

Then came the cassette tape. "The cassette tape," they predicted, "will refine the recording industry." It was smaller than the clumsy 8-track, but it had all its same advantages, including that of being playable in your automobile. Moreover, you could actually stick a few into your glove compartment. You could also buy a "portable tape player" which would play your cassettes on an arbitrary street corner at an arbitrary volume level. I myself continued to buy records, although I did eventually buy a cassette recorder so that I could tape my records at home for playing later in my car. Although record stores eventually began selling albums on cassette, it was usually from a shelf toward the back of the store. Vinyl was still king.

Next came the laser disc. "The laser disc," they predicted, "will redefine the recording industry." The laser disc had the music encoded digitally, promising virtually perfect fidelity forever. It also would not scratch, smudge, or collect dust, and in place of the old diamond needle, which everyone always suspected would be fatal to plastic records eventually, there was a neat, powerful laser beam to lift the music off of the disc as cleanly as Scottie might beam up Captain Kirk. This was dazzling technology indeed; unfortunately, you could not play a laser disc in your car. For one thing, each disc was the size of a medium pizza; for another thing, you had to *sell* your car in order to buy a laser disc *player*, which cost several thousand dollars. Needless to say Hugh Hefner bought one for every room of his mansion, while the rest of us just kept buying records.

Then along came the compact disc. "The compact disc," they predicted, "will redefine the recording industry." And in an incredibly short period of time, it did.

Walk into a record store today and it probably won't even be *called* a record store. The bins that once stored the vinyl now store row upon row of CD's. Oh, the biggest stores will still carry a few records, but you often have to walk past the cassette section in order to find them. If you want a *real* adventure, try replacing your old phonograph needle! The recording industry has been completely taken over by the compact disc.

Why did the CD succeed where the other technologies had failed? Simply put, it was such a perfect idea that nobody who loved music could resist it. The sound was virtually perfect; the discs themselves were rugged; the players were affordable; and these things were, as their name implied, *compact*—you could fit dozens of them into a shoebox and carry them to the home of your friend. There, you would almost assuredly find another CD player. Everyone could share in the miracle.

Of course, for a while I held out. After all, I had this enormous investment in records, not to mention the means for playing them. But I would go the homes of my friends and hear the crystalline strains of CD music, and I would be jealous of that incredible *sound*. Finally, I realized that it was not the *records* that I liked; it was the *music*, and the music could be heard better on CD's. I bought myself a player and began collecting compact discs. Most of my first compact disc purchases were actually albums that I already owned on vinyl, but I bought them so that I could rediscover them on a new level. Now I hardly ever play my records unless I am hosting a graduation year playoff. I still own them, but they are doing something that records do unfortunately well: They are gathering dust.

I have lived long enough to see the very essence of my lifelong hobby redefined.

But records are only my hobby; my profession is teaching mathematics. I suppose I became interested in mathematics at about the same time I started collecting records. I was in seventh grade when the Russians launched Sputnik, thereby kicking off some interesting times for my mathematics teachers. My high school courses were taught out of paperback textbooks authored by the School Mathematics Study Group, code SMSG, whose approach, they predicted, would redefine the way we taught and learned mathematics.

That was the New Math, and it lasted long enough to develop a reputation bad enough to spawn the Back to Basics movement. "The Back to Basics Movement," they predicted, "will redefine the way we teach and learn mathematics."

Then suddenly we all got distracted by computers. Computers were doing well in the stock market, and were pretty much running most companies and the government, so there was a strong feeling that we should all find out how they work. What most people discovered was that they worked by *mathematics*, which was enough for most people, because it meant that mathematics teachers could henceforth be held responsible for explaining the remaining details to their children. Computers were installed in many schools, because, they predicted, "computers will redefine the way we teach and learn mathematics."

I was part of a three-man committee in 1978 that persuaded our Board of Trustees to sink 100 thousand dollars into a computer system that featured a Data General Nova 830 with 32K RAM, seven CRT terminals, and a teletype printer. Five years later we were back before them, hats in hand, pleading for another 100 thousand dollars to upgrade to a Hewlett-Packard 8000/30, plus ten more terminals and a sensible printer. Five years later we all but abandoned the HP and built a computer lab with shiny new Apple II's, at a cost of another 100 thousand dollars. Now we have an entire multimedia lab, fully stocked with Macintoshes, laser printers, scanners, CD-ROMs, and networking hardware, while our gleeful supplier has another 100 thousand of our school dollars. The amazing thing about this buying frenzy is that *each time* we pleaded our case with the Trustees, we assured them that *computers would redefine the way we taught and learned mathematics*. Incredibly, they fell for it every time.

But the sad truth of the matter was that we were still teaching and learning mathematics the way we had been doing it for decades. For all its marvelous capabilities, the computer was not changing the way that we taught and learned mathematics, and it was costing our school approximately 100 thousand dollars every five years to prove to ourselves that this was so.

Meanwhile, pocket scientific calculators had quietly appeared on the scene and had been welcomed in most mathematics classes. They *did* change a few things, but at such a mechanical level that people hardly noticed. Trig tables and log tables died a hasty and largely unmourned death, and now everyone had an equal chance at finding the purchase price of 4 CD's at $16.95 a piece, after a 10% discount and a 7.5% sales tax have been figured in. To most teachers, this was the sort of inconsequential drudgery that machines were *supposed* to do. Indeed, such was the inflexibility of the mathematics curriculum that these machines were welcomed precisely *because* they made so little difference in what we taught and learned. Significantly, the only place where they were at all controversial was at the elementary level, where people were nervous about children losing the ability to multiply and divide on paper. In any event, pocket calculators did not redefine the way we taught and learned mathematics.

By now you have probably figured out my little parable. SMSG and Back to Basics may have shifted things a little bit, but only in terms of decades, while the inflexibility of the mathematics curriculum must be measured on a geological scale. New points of emphasis come and go periodically, denting the curriculum monolith with the same approximate impact as that of the 8-track tape on the recording industry.

Scientific calculators were nice, and they even got into the classrooms, but they were the cassettes of our profession. Sure, everyone has one, but cassettes never replaced the records, which were still the way that serious people played their music, and scientific calculators never replaced factoring, which was what students *really* needed to know if they wanted to succeed in serious mathematics.

Computers took us to the laser disc stage, and failed to redefine the way we taught and learned mathematics for the same reasons that laser discs failed to redefine the recording industry. It had nothing to do with the capabilities of the technology, and everything to do with the mood of the market place. No machine can inspire a revolution if Hugh Hefner is the only one who owns one.

Of course, there is another chapter in this story. We are still writing that chapter as I write these words, but I firmly believe that you and I have lived long enough to see the compact disc of our profession: an instrument which is so perfectly suited to what we do, that it is in the process of redefining the way we teach and learn mathematics. We call this wonderful machine a graphing calculator, but it is in fact a computer, with ironically the same computing power in kilobytes as that first computer system we bought at my school fifteen years and 400 thousand dollars ago. With this machine my students can do far more than compute; they can conjecture, they can model, and they can make connections—the very things that I want to teach them to do. Moreover, they *own* this technology; they do not have to go over to a rich friend's home or to a special room at school to use it. Nor do I even have to tell them how or when to use it. Like the compact disc, it has become a part of their lives.

Oh sure, for a while I tried to treat this technology the way I treated the cassettes in the record store. I bought one and used it, but I never thought it would redefine my profession. After all, I had twenty years of my life invested in the math curriculum monolith, and I had become pretty successful at teaching the traditional courses in some embarrassingly traditional ways. But I was open to change, and I had read the *Standards*, so I began to chip away at my preconceptions of what and how I had to teach. The first thing I did was to let them use their graphing calculators all the time. The next thing I did was to start every class with a problem, which the students would talk out until a solution emerged that they could explain to each other. What I discovered, of course, was how useless my crisp set of lecture notes had been all these years. The students were discovering the results *without me*, and then *showing each other how to solve the problems*. That left me free to walk around and answer questions. Every so often I still tie things together or generalize, but for the most part I let the course evolve through what the students are doing, and I provide the direction by the problems I select. I'm still not sure what the heck I'm doing, but I do know this: There is more mathematics going on in my classroom these days than there ever has been before. Now there are more people doing it.

Was the graphing calculator responsible for transforming my entire approach to teaching? Well, yes and no. What the graphing calculator did was get me to question what I had been doing for twenty years. It also got me focused on how I would get the students using it, which in turn got me focused on student learning rather than my own teaching. Eventually, after sacrificing the first few sacred cows, I acquired a taste for sacred beef and the rest was easy. And believe me, that same thing is going on in mathematics classrooms all across the country, in a movement that is growing exponentially.

When I was young, all of my radical friends were in reform school. Today, all of my radical friends are in school reform. It is a crazy, free-wheeling time, not unlike the political scene in eastern Europe, and it's really pretty exciting—once you overcome the initial sensation of being totally lost in a Brave New World. Let's face it, though: For years we mathematicians were in a rather unique position in the world of academia, being smugly certain that we could *all* teach the *exact* same curriculum in the *exact* same way because we knew *exactly* what was best for our students. The other subject area committees in the AP program fought among themselves all the time, while my colleagues and I on the Calculus committee nodded sagely in unison, reviewing yet another problem about a region being rotated about the x-axis. It was a happy, homogeneous, unrealistic world, totally incompatible with the spirit of creative discovery that had characterized the evolution of mathematics since the dawn of cognition.

Did you ever wonder what Newton would say if he could come back today and watch a traditional calculus class in action—if you could call it "action"? Do you suppose he would be flattered to see that, 300 years after his death, we were all teaching *his same results* to our students? I don't want to put words into the old Lion's mouth, but knowing that Newton once wrote to Robert Hooke, "If I have seen further it is by standing on the shoulder of giants," I dare say that he might pick up a graphing calculator, stare at it for several minutes with amazement, then say something like this: "The creators of this magic are giants. Is there nobody here who would stand on their shoulders?"

It is almost axiomatic in the academic world that the most creative minds belong to the mathematicians. Sadly, in the world of mathematics education that renowned creativity has been stifled for too long. To counter the trend, I tried a few years ago to release my own creative Muse in an evening of unbridled mathematical activity. I sat down with a fresh pad of paper and a pitcher of martinis, and before the evening was over and the martinis were gone I had produced 75 proofs of never-before-seen theorems. (Actually, I underachieved. It was an 80-proof gin.) I also created the perfect calculus problem, and here it is:

THE ALL-PURPOSE CALCULUS PROBLEM

©1993 Dan Kennedy

A particle starts at rest and moves with velocity $v(t) = \int_1^t e^{-x^2} dx$ along a 10-foot ladder, which leans against a trough with a triangular cross-section two feet wide and one foot high.

Sand is flowing out of the trough at a constant rate of two cubic feet per hour, forming a conical pile in the middle of a sandbox which has been formed by cutting a square of side x from each corner of an 8" by 15" piece of cardboard and folding up the sides.

An observer watches the particle from a lighthouse one mile off shore, peering through a window shaped like a rectangle surmounted by a semicircle.

(a) How fast is the tip of the shadow moving?
(b) Find the volume of the solid generated when the trough is rotated about the y-axis.
(c) Justify your answer.
(d) Using the information found in parts (a), (b), and (c), sketch the curve on the pair of coordinate axes given.

Okay, maybe I went too light on the vermouth. But here's the sad part: Every teacher reading this article knew those problems by heart, because we've been teaching those same basic problems for years. That, indeed, is what calculus teachers have been doing for decades and decades. We have been teaching the math literate of tomorrow with the problems of yesterday, while explaining to them all the while that they will *need this mathematics in the future*. My friends, this is the stuff of which the Emperor's New Clothes are made! While we have been spinning golden oldies on the phonograph, perhaps more accurately the victrola, the world of Mathematical Reality has gone CD.

Today, thankfully, all of that is changing. To cite just one close-to-home example of that, the College Board has given final approval to the AP Committee's recommendation that the AP examinations be made graphing-calculator-active beginning in 1995. We will also go to a new test format, splitting the multiple choice section into two 45-minute parts, one with 25 questions to be taken with *no* calculator, the other with 15 questions, some requiring a graphing calculator. The free response section will remain the same, 6 questions in 90 minutes, and will be designed with graphing calculators in mind. The *extent* to which calculators will be required will be slight at first, but may increase over time as emphases in the curriculum change.

For the next year or so, the Committee will be buried under test development details resulting from our decision. After that we will emerge to begin a careful evaluation of the calculus curriculum, from a to z. Or, if you like, from ε to δ. I recommend, however, that every teacher in the AP program get started without us, just as you started without us with graphing calculators. Don't worry about whether you are doing the same thing in your classroom as I am doing in mine; we have all worried about that for too long. Just worry about whether your students are learning calculus and enjoying it. If that's the *only* thing that our classrooms have in common, then our students are still much better off than when we had *everything else* in common at the expense of that.

In the next few years, the vibrations from all these education reform movements will reach a crescendo, and from it all will emerge a new paradigm for teaching and learning. I have seen enough to realize that it will not be confined only to mathematics; nonetheless, it is apparent to any observer that mathematics is leading the way. If Newton does come back to visit us, I hope he waits a few years so that all of this will have had a chance to develop. Then, when he asks to see the Calculus for the Twenty-first Century, we can show him something which Newton would truly appreciate: an entire Renaissance Curriculum.

And I'll bet we get to show it to him on a CD—hooked up to our graphing calculator.

The Changing Face of Calculus: First-Semester Calculus as a High School Course

David M. Bressoud

Once upon a time, calculus was the first college-level mathematics course taken by mathematically talented students. The students in first-semester calculus were mathematically motivated, generally well prepared, and they were seeing these ideas for the very first time. This is no longer true. Most of our best-prepared mathematics students arrive in college with credit for at least the first semester of calculus, many of them with credit for both semesters. Despite steady growth in majors in science and engineering, enrollment in first semester calculus has been flat or slightly declining at both two- and four-year undergraduate programs. It is the College Board's Advanced Placement Calculus Program that has been growing steadily at 7–8% per year (see figure 1).

In 2004 over 225,000 high school students took the AP calculus exam. This number is far larger than the number of students who took mainstream first-semester calculus in all four-year undergraduate programs in the Fall of 2000. By the time of the next CBMS survey in 2005–06, we can expect that more students will take an AP Calculus exam than will take mainstream Calculus I in the Fall of 2005 in all 2-year and 4-year institutions combined.[2]

First-semester calculus has become a high school topic for most of our strongest students. This has several implications:

1. We should ensure that students who take calculus in high school are prepared for the further study of mathematics.
2. We should address the particular needs of those students who arrive in college with credit for calculus.
3. We should recognize that the students who take first-semester calculus in college may need more support and be less likely to continue with further mathematics than those of a generation ago.

This article will address the implications for calculus taught in high schools. A second article, "The Changing Face of Calculus: First- and Second-Semester Calculus as College Courses," will look at the implications for how we teach calculus in colleges and universities.

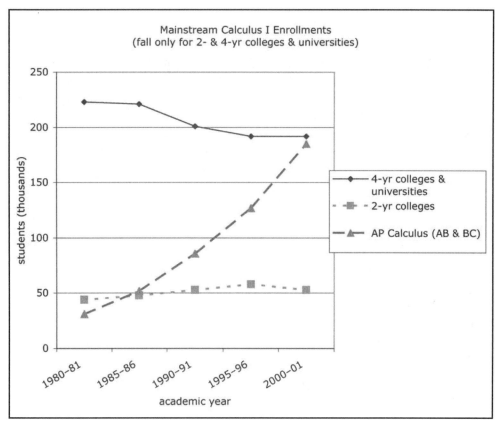

Note 1: Mainstream Calculus I Enrollments. Fall two- and four-year college and university enrollments from [4][1]. AP Calculus enrollments from The College Board (most recent years available at [1]).

Recommendations for High School Calculus

The pressure to take calculus in high school is understandable. Competition for admission to the best colleges and universities is fierce. It has helped to create strong growth in AP programs across the board. Many mathematicians deplore this movement of calculus from the college to the high school curriculum, but the pressures are too strong to stop or even substantially slow it. What we can hope is to shape it.

With this in mind, the presidents of the MAA and NCTM issued a joint statement in 1986 [3] with two strong recommendations which I paraphrase here:

1. In spite of the pressures to take calculus while still in high school, students should never short-change their mathematical preparation in subjects such as algebra, geometry, or trigonometry. Solid mathematical preparation is far more important than exposure to calculus.
2. When calculus is taught in high school it should be a college-level course. This means that the goal of the course should be to give students the same breadth of topics and mastery of calculus obtained by students taking such a

course in college. It means that the course should be taught with the expectation that students who perform satisfactorily will be able to place into the succeeding college calculus course.

I believe that these recommendations need to be repeated and re-emphasized. One of the inevitable weaknesses of the AP program is that student enrollment in an AP class appears on the transcript that is reviewed for college admission, but the test that evaluates whether or not the student has learned this material at a college level is not administered until after college acceptances have been sent out. This is why many students enroll in AP courses but do not take the examinations. Many schools are under pressure to offer a course that is nominally an AP Calculus course, even if they expect few students will be able to pass the AP exam. These recommendations are intended to back up the teachers who are trying to resist rushing students into calculus before they are properly prepared.

It is particularly important that the calculus taught in high school should be a substantive course that prepares students for further work in mathematics. A weak overview of calculus does little to reinforce student knowledge of algebra, geometry, or trigonometry. In fact, it may encourage slighting these subjects in order to get into the calculus course that will improve the appearance of one's transcript. On the other hand, a solid calculus course should require and help develop a level of mastery of these core subjects that is essential for any further work in mathematics.

Finally, these recommendations recognize that the students who take calculus in high school are among our best students. They must be prepared for college-level mathematics. Once they are ready for and are studying calculus, they should be learning it in a course that is comparable to what they would see in a mainstream college course.

The Responsibilities of Mathematicians

How calculus is taught is important. As I argued in 1992 [2], calculus is not only essential for building mathematical models of the world around us and thus informing disciplines such as physics, economics, and biology, its creation/discovery was the defining moment in the birth of modern mathematics. It has shaped our modern conception of and expectations for mathematics. Calculus should not be the only pillar supporting the undergraduate curriculum in mathematics. Discrete mathematics, geometry, and data analysis have equally important if very different roles to play. But calculus must remain one of those pillars. To ensure that it remains so, mathematicians must be concerned about how it is taught both in colleges and in high schools.

Calculus can be and is being taught well in high schools, but as the number of high school calculus courses expands, so does the number of high school teachers who must teach these courses without much more preparation than the undergraduate course they themselves took, often many years before. At many high schools, only one person teaches calculus, and so peer support may be lacking. The purpose of the AP Calculus examinations is to provide a common standard against which to measure students from all of these classes, but it can only accomplish so much. Ultimately, the way to ensure that what is taught in high school calculus really is a college-level course is through the preparation and support of the teachers who will lead these classes.

The College Board runs many workshops for AP Calculus teachers. NCTM meetings include well-attended sessions that address their needs. The MAA is beginning to realize its own potential in this area.[3] But there still remain far too few university-level mathematicians who are willing to assist in the task of preparing and supporting high school teachers. At the very least, all mathematicians have a responsibility to be aware of the AP Calculus program: its course expectations and the nature of its examinations. Every department should encourage at least one individual to attend the annual AP Reading (the grading of the free response questions), to work with local AP Calculus teachers, or to help prepare and support those who will teach calculus in high school.

Calculus II as a High School Class

The same pressures that are pushing Calculus I into the high school curriculum are doing the same for Calculus II. Traditionally, it was a very elite group of students who took BC Calculus, covering the entire two-semester college syllabus. That group of students also grew by 6–8% per year until the mid-1990s. Over the period 1995–98, the rate of growth of BC calculus accelerated to 10–11% per year, a rate that has held up since then.[4] In 2004, the number of students taking the BC Calculus exam exceeded 50,000. It will likely exceed 60,000 by 2005–06, the year of the next CBMS survey.

In 2002, 23% of the students who took BC Calculus did so before their senior year [7]. These high school students are not necessarily well served by taking classes in linear algebra, several variable calculus, or differential equations at a local college. Picking up additional college credits is far less useful for them than deepening and broadening the mathematics they already think they know. These students need to be challenged, but they also need to be prepared for and enticed into a deep study of further mathematics in the company of their peers.

There are many local programs that recognize this. In Minnesota, we have the University of Minnesota Talented Youth Math Program (UMTYMP). At the North Carolina School of Science and Mathematics, the mathematics department is developing courses that return to calculus, using several variables, differential equations, and modeling to explore its topics in greater depth. But not enough students have access to these kinds of programs. There is a need for a substantial national effort to create materials that can be used with these students and to help teachers learn how to use them.

The movement of calculus into the high schools is not necessarily bad, but it does require the efforts of the mathematical community—individuals, departments, and professional associations—to prepare and support those who will teach it and to resist the pressures that would weaken it.

Acknowledgement: Thanks to Ben Klein, Johnny Lott, Bernie Madison, Bob Megginson, Carol Miller, and Dan Teague for helpful comments.

Bibliography

[1] AP Central, AP Research and Data, http://apcentral.collegeboard.com/program/research/.

[2] Bressoud, David M., Why do we teach calculus?, *Amer. Math. Monthly*, vol. 99, no. 7 (1992), 615–617.

[3] Dossey, John A. and Lynn A. Steen, *Calculus in the Secondary School*, joint letter of the MAA and NCTM Presidents, 1986.

[4] Lutzer, David J., James W. Maxwell, and Stephen B. Rodi, *Statistical Abstract of Undergraduate Programs in the Mathematical Sciences in the United States: Fall 2000 CBMS Survey*, American Mathematical Society, Providence, RI, 2002.

[5] Small, Don, Report of the CUPM Panel on Calculus Articulation: Problems in Transition from High School Calculus to College Calculus, *Amer. Math. Monthly*, vol. 94, no. 8 (1987), 776–785.

[6] Snyder, Thomas D. and Charlene M. Hoffman, *Digest of Education Statistics 2002*, NCES 2003-060, National Center for Education Statistics, U.S. Department of Education.

[7] 2002 AP Yearbook, The College Board, New York.

Notes

1. For 1980 and 1985, the CBMS Survey only reports total numbers of students taking all calculus classes in the fall. The numbers of students taking Calculus I were estimated by taking 55% of this total for 4-year institutions and 60% for 2-year institutions (the approximate percentages in the years 1990, 1995, and 2000).

2. The total number of high school students taking calculus each year is unknown. Numbers range from 300,000, the NCES figure for 2000 ([6], table 141), to 500,000 or more. The larger number is based on the College Board estimate that 60% of AP Calculus students take the examination and the observation that many students take a high school calculus course that is not an AP course. This includes students in the International Baccalaureate program and students in joint programs between high schools and community colleges.

3. Daniel Teague, Benjamin Klein, and I are in the process of establishing a SIGMAA for high school teachers that will focus on support for teaching college-level mathematics courses in high schools.

4. The increased rate was almost certainly helped by the fact that an AB subscore was made available for the BC exam beginning in 1998.

Things I Have Learned at the AP Calculus Reading

Dan Kennedy

It almost goes without saying among those who have participated in the process that the benefits of the Advanced Placement reading are many and varied. There is, for example, the practical benefit of getting 161,000 exams graded in a smooth and professional manner. For AP readers, there are also the social benefits of seeing old friends and making new acquaintances with teaching colleagues from across the country. There are the nutritional benefits of storing up the valuable reserves of fat and cholesterol that their bodies need to make it through the long, hot summer. But one of the most important professional benefits for AP readers is the opportunity for calculus teachers to learn more about calculus, and that is the aspect of the reading that I would like to reflect upon in this paper.

Only the most naive of God's creatures—for example, our students—would assume that their teachers know everything about the subjects that they teach. We who teach calculus have learned to be particularly humble in that respect, encountering fairly regularly questions that at least cause us to tell our students, "Say, that's a good question, but let's not take class time on that right now; I'll explain it to you tomorrow." This buys us enough time, we hope, to solve the problem in private, where we can consult the solution manual.

We have all learned a lot of calculus that way. But those of us who have read AP calculus papers over the years have also learned that the simplest of problems can have remarkable subtleties buried just beneath the surface. Such subtleties might go undetected under normal circumstances, but under the abnormal circumstances of thousands of panicky students thrashing around with unfamiliar mathematical tools, they are soon exposed and demand our scrutiny. It is in these unexpected moments that I have learned some fascinating calculus facts over the years, and I would like to share a few of them with you in this paper. I hope that some of them will give you the same thrill of discovery that they gave me.

1. Radical Lies That We Tell Students in Algebra

This was AB-1 in 1988:

1. Let f be the function given by $f(x) = \sqrt{x^4 - 16x^2}$.
   a) Find the domain of f.
   b) Describe the symmetry, if any, of the graph of f.
   c) Find $f'(x)$.
   d) Find the slope of the line *normal* to the graph of f at $x = 5$.

This is a pretty harmless-looking problem, but imagine your best algebra student looking at that radical and thinking, "Uh-oh. Better simplify that thing before I go on." We have almost programmed our students to think that way. So, the student writes

$$\sqrt{x^4 - 16x^2} = |x|\sqrt{x^2 - 16}.$$

(Notice the absolute value. I said that we were imagining your *best* algebra student.)

Now, how does the student answer part (a)? If she uses the simplified expression, she sets $x^2 - 16 \geq 0$ and arrives at a domain of $(-\infty, -4] \cup [4, \infty)$, losing the domain point $\{0\}$ (and, of course, a point from her AB-1 score). If she uses the unsimplified expression, she sets $x^4 - 16x^2 \geq 0$ and arrives at the correct domain of $(-\infty, -4] \cup \{0\} \cup [4, \infty)$. (Actually, many AB students that year divided by x^2 and arrived at the wrong domain anyway, but your best algebra student would surely not have done *that*.)

I will leave it to the reader's imagination to consider what happened to the absolute value crowd when they moved on to part (c).

There are many examples of expressions that *pick up* domain values when simplified, such as $\frac{x^2 - 4}{x - 2}$ or $\frac{\sin x}{\tan x}$, but this problem provides a rare example of an expression that *loses* a domain value when simplified. The dozens of AB students who lost that point in 1988 were probably not as charmed by this subtlety as we were, but it made for some nice conversation among teachers that year. We all resolved to be a little more careful to think about equations like

$$\sqrt{x^4 - 16x^2} = |x|\sqrt{x^2 - 16}$$

as *identities*. Identities have domains of validity, and this one is invalid at $x = 0$.

2. Proving Differentiability at a Point

There have been several "split-function" questions over the years, such as BC-4 in 1992:

4. Let f be a function defined by $f(x) = \begin{cases} 2x - x^2 & \text{for } x \leq 1, \\ x^2 + kx + p & \text{for } x > 1. \end{cases}$
   (a) For what values of k and p will f be continuous and differentiable at $x = 1$?
   (b) For the values of k and p found in part (a), on what intervals is f increasing?
   (c) Using the values of k and p found in part (a), find all points of inflection of the graph of f. Support your conclusion.

To solve part (a), the typical good student will paste together the two sides of the function at $x = 1$ to make it continuous

$$2(1) - 1^2 = 1^2 + k(1) + p \Rightarrow k + p = 0$$

and then paste together the derivatives of the two sides at $x = 1$ to make it differentiable

$$2 - 2(1) = 2(1) + k \Rightarrow k = -2.$$

The two conditions taken together imply that $k = -2$ and $p = 2$.

Things I Have Learned at the AP Calculus Reading

This is all well and good, but it leads many students (and once led me) to conclude that this is how to show that a function is differentiable at a point: establish continuity and show that the derivative is the same coming in from the left as it is coming in from the right. Luckily for everyone doing BC-4, that approach *will* successfully establish differentiability *if* it works. But it might *not* work, as was shown in 1982 on BC-7:

7. Let f be the function defined by $f(x) = \begin{cases} x^2 \sin\left(\dfrac{1}{x}\right) & \text{for } x \neq 0, \\ 0 & \text{for } x = 0. \end{cases}$

   (a) Using the definition of the derivative, prove that f is differentiable at $x = 0$.
   (b) Find $f'(x)$ for $x \neq 0$.
   (c) Show that f' is not continuous at $x = 0$.

We solve the problem as follows:

(a) $f'(0) = \lim\limits_{h \to 0} \dfrac{f(0+h) - f(0)}{h} = \lim\limits_{h \to 0} \dfrac{h^2 \sin(1/h) - 0}{h} = \lim\limits_{h \to 0} h \sin\left(\dfrac{1}{h}\right) = 0.$

(b) $f'(x) = 2x \sin\left(\dfrac{1}{x}\right) + x^2 \cos\left(\dfrac{1}{x}\right) \cdot \left(-\dfrac{1}{x^2}\right) = 2x \sin\left(\dfrac{1}{x}\right) - \cos\left(\dfrac{1}{x}\right).$

(c) Because the function $\cos\left(\dfrac{1}{x}\right)$ diverges by oscillation on both sides of 0, it follows that $\lim\limits_{x \to 0} f'(x) = \lim\limits_{x \to 0} \left(2x \sin\left(\dfrac{1}{x}\right) - \cos\left(\dfrac{1}{x}\right)\right)$ does not exist. Therefore, f' is not continuous at $x = 0$.

The implication of this problem is subtle, but profound. Notice that $f'(x)$ *does* exist at $x = 0$, but that this fact could not be discovered by looking at $f'(x)$ to the left and right of $x = 0$.

This same problem also taught me the true meaning of right- and left-hand derivatives. Notice that f is differentiable at 0, so both its right- and left-hand derivatives exist there. They are the two limits $\lim\limits_{h \to 0^-} \dfrac{h^2 \sin(1/h) - 0}{h}$ and $\lim\limits_{h \to 0^+} \dfrac{h^2 \sin(1/h) - 0}{h}$, both of which equal 0. They are *not* the same as $\lim\limits_{x \to 0^-} f'(x)$ and $\lim\limits_{x \to 0^+} f'(x)$, both of which fail to exist.

Students could have made perfect scores on these problems and never realized half of what I just said, but because I graded their papers at the reading those same problems afforded me the chance to learn those nuggets of calculus from my colleagues.

I later learned other interesting things about this particular split function, which is shown in Figure 1. As we have just seen, this function is both continuous and differentiable at $x = 0$. The graph of

$$g(x) = \begin{cases} x \sin\left(\dfrac{1}{x}\right) & \text{for } x \neq 0, \\ 0 & \text{for } x = 0 \end{cases}$$

is also shown in Figure 1. This function is continuous but it is an easy exercise to show that it is *not* differentiable at $x = 0$. You can almost see why, too: you could "zoom in" endlessly on the right-hand graph at the origin and the graph would get no straighter than it is now, whereas if you were to zoom in on the left-hand graph

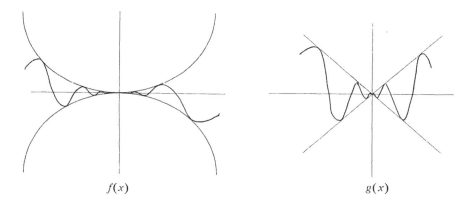

$f(x)$ $\qquad$ $g(x)$

Figure 1

the curve would get flatter and flatter at the origin. If you find that your eye cannot quite distinguish why both $\lim_{x \to 0^-} f'(x)$ and $\lim_{x \to 0^+} f'(x)$ fail to exist in the left-hand graph, welcome to the club. This is why there will always be the need for algebraic limits in first-year calculus!

3. The First Derivative Test for Points of Inflection

There have been so many questions on AP exams over the years asking students to justify points of inflection that it is impossible to say when this issue first arose, so I will put it in the context of the following hypothetical question:

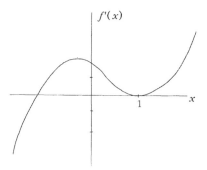

The expected response is that there is a point of inflection because there is a turning point of the graph of $y = f'(x)$ at $x = 1$ (or some equivalent statement about f'' changing sign), but what would you do with a student who says the following:

There is a point of inflection at $x = 1$ because $f'(1) = 0$ while $f'(x)$ is positive on either side of $x = 1$.

This student has noted the "shelf point" at $x = 1$ and is arguing that it must be a point of inflection, as in the familiar graph of $y = x^3$. There is no mention of the sign of the second derivative; indeed, the whole argument hinges on the sign of the first derivative. Does this student get credit for a valid justification?

Since this response actually occurred one year, the table leaders had to decide whether this was a valid theorem. The smart money being on "no" from the outset, most of their efforts were concentrated on finding a counterexample. My notes do not record who came up with one first, but later conversations have variously attributed it to Robert Ellis, Tom Tucker, or Bruce Peterson. (Bruce, chair of the Committee in those days, actually published a paper on the topic.) This is the function that I found in my notes from the reading:

$$f(x) = \begin{cases} 12x^5\sin(1/x) + x^3 & \text{for } x \neq 0, \\ 0 & \text{for } x = 0. \end{cases}$$

We can show that $f'(0) = 0$ and that $f'(x)$ is positive on both sides of $x = 0$, and yet the graph of f does not have a point of inflection at the origin. The salient feature of the function is that its graph changes concavity infinitely often in every neighborhood of zero. Here is a sketch of the proof:

1. (Proof that $f'(0) = 0$):

$$f'(0) = \lim_{h \to 0} \frac{12h^5 \sin(1/h) + h^3 - 0}{h}$$

$$= \lim_{h \to 0} 12h^4\sin(1/h) + h^2 = 0 \cdot (\text{bounded}) + 0 = 0.$$

2. (Proof that $f'(x) > 0$ in some neighborhood of 0):

$$f'(x) = 60x^4\sin(1/x) + 12x^5\cos(1/x) \cdot (-1/x^2) + 3x^2$$

$$= x^2(60x^2\sin(1/x) - 12x\cos(1/x) + 3)$$

$$\downarrow \quad \Downarrow \quad \downarrow \quad \Downarrow \quad \downarrow$$

$$0 \quad (\text{bounded}) \quad 0 \quad (\text{bounded}) \quad 3$$

So for small x on either side of 0, $f'(x)$ has the same sign as $3x^2$, namely positive.

3. (Proof that f'' changes sign infinitely often in any neighborhood of $x = 0$):

$$f''(x) = 240x^3\sin(1/x) + 60x^4\cos(1/x) \cdot (-1/x^2) - 36x^2\cos(1/x)$$

$$+ 12x^3\sin(1/x) \cdot (-1/x^2) + 6x$$

$$= 240x^3\sin(1/x) - 96x^2\cos(1/x) + 6x - 12x\sin(1/x)$$

$$= x\left[(240x^2\sin(1/x) - 96x\cos(1/x)) + 6(1 - 2\sin(1/x))\right]$$

$$\downarrow \quad \Downarrow \quad \downarrow \quad \Downarrow \quad \Downarrow$$

$$0 \quad (\text{bounded}) \quad 0 \quad (\text{bounded}) \quad (\text{oscillating between } \pm 2)$$

So for small x on either side of 0, $f''(x)$ has the same sign as $18x$ infinitely often and the same sign as $-6x$ infinitely often. There can be no well-defined change in concavity at $x = 0$, and so the graph does not have a point of inflection there.

A simpler function with the same property (brought to my attention by Ray Cannon) is $f(x) = x^5 + x^6\sin(1/x)$. It requires a little more trigonometry to show that f'' changes sign infinitely often in every neighborhood of 0, but the interested reader can verify that it does.

Incidentally, as you might have suspected by now, the student's argument is actually valid if it is known that the function only has a finite number of inflection points near the critical point in question. (The proof, omitted by the student, is a nice exercise.)

BC-2 in 1996 was about Maclaurin series:

2. The Maclaurin series for $f(x)$ is given by $1 + \dfrac{x}{2!} + \dfrac{x^2}{3!} + \dfrac{x^3}{4!} + \cdots + \dfrac{x^n}{(n+1)!} + \cdots$
   (a) Find $f'(0)$ and $f^{(17)}(0)$.
   (b) For what values of x does the given series converge? Show your reasoning.
   (c) Let $g(x) = xf(x)$. Write the Maclaurin series for $g(x)$, showing the first three nonzero terms and the general term.
   (d) Write $g(x)$ in terms of a familiar function without using series. Then, write $f(x)$ in terms of the same familiar function.

I had no problem recognizing this as a "series manipulation" problem when I took the exam in the privacy of my own room. The series in part (c) was clearly the series for $e^x - 1$, and so my answer in part (d) was:

$$g(x) = e^x - 1;$$
$$f(x) = \frac{e^x - 1}{x}.$$

When I arrived at the reading I discovered that I had only scored 8 out of 9 on BC-2. I am not sure whether the Committee had intended to catch me on a technicality, but I had certainly taken their bait, including the hook, line, and sinker. It was somewhat consoling to note, after all the exams had been graded, that I was not the only sucker in the Committee's boat: flopping around beside me were more than 99% of the BC population. (There were only 5 perfect solutions among the 21,020 BC exams that year.) The problem, you see, is that $f(x) = \dfrac{e^x - 1}{x}$ is not even *defined* at $x = 0$, let alone infinitely differentiable, and is therefore not eligible to have a Maclaurin series. The correct f should be

$$f(x) = \begin{cases} \dfrac{e^x - 1}{x} & \text{for } x \neq 0, \\ 1 & \text{for } x = 0. \end{cases}$$

For additional perspective let me hearken back to 1993, when the following problem appeared as BC-5:

5. Let f be the function given by $f(x) = e^{\frac{x}{2}}$.
   (a) Write the first four nonzero terms and the general term for the Taylor series expansion of $f(x)$.
   (b) Use the result from part (a) to write the first three nonzero terms and the general term of the series expansion about $x = 0$ for $g(x) = \dfrac{e^{\frac{x}{2}} - 1}{x}$.
   (c) For the function g in part (b), find $g'(2)$ and use it to show that $\sum_{n=1}^{\infty} \dfrac{n}{4(n+1)!} = \dfrac{1}{4}$.

Notice in part (b) the clever avoidance of the names "Taylor" and "Maclaurin." This allows the student to construct a power series of the form $a_0 + a_1 x + a_2 x^2 + \cdots + a_n x^n + \cdots$ without being concerned about whether or not $g(0)$, $g'(0)$, $g''(0)$, etc., exist—which they do not.

Since learning my lesson in 1996 I have enjoyed looking through various textbooks to see how many include exercises in their "series manipulation" sections that look like this:

Use the Maclaurin series for $\cos x$ to construct a Maclaurin series for $\dfrac{\cos x - 1}{x}$.

Each one I find means that there is one more sucker in the boat.

5. The Differential Equation with Too Many Solutions

This was BC-6 in 1993:

6. Let f be a function that is differentiable throughout its domain and that has the following properties.
   (i) $f(x+y) = \dfrac{f(x)+f(y)}{1-f(x)f(y)}$ for all real numbers x, y, and $x+y$ in the domain of f.
   (ii) $\lim_{h \to 0} f(h) = 0$
   (iii) $\lim_{h \to 0} \dfrac{f(h)}{h} = 1$

   (a) Show that $f(0) = 0$.
   (b) Use the definition of the derivative to show that $f'(x) = 1 + [f(x)]^2$. Indicate clearly where properties (i), (ii), and (iii) are used.
   (c) Find $f(x)$ by solving the differential equation in part (b).

Not only was I chair of the Committee when this problem appeared, I actually wrote the thing. It then went through the usual rigorous scrutiny and polishing, wherein quite a few competent people solved it multiple times. All this quality control notwithstanding, it was only at the reading itself that someone (the Chief Reader) discovered that there was a lot more involved in pinning f down than anyone had theretofore realized (including the author of the problem).

The solution to the differential equation is pretty straightforward: You separate variables and find that $f(x) = \tan(x+C)$. One is then supposed to use the initial condition from part (a) to conclude that $\tan(C) = 0$, from which it follows that $f(x) = \tan(x + k\pi) = \tan x$. It looks simple enough, but it is not that simple. Can you see why?

Here is the rub: the initial condition is fine for "pinning down" the function at $x = 0$, but the function we are talking about is the *tangent* function, which has asymptotes. So we can narrow our solution down to $y = \tan x$ on the interval $(-\pi/2, \pi/2)$, but *outside* that interval we could jump to a different, shifted tan curve without affecting $f(0)$ a bit! Fortunately for the integrity of the exam, condition (i) precludes this from happening. (This is another nice exercise.) Still, nobody expected the BC students to catch this subtlety, let alone prove their way out of it, so the grading standard was written without mentioning it. It was those of us who discussed the subtlety at the grading who gained some wonderful insights into differential equations.

6. Methods of Solution That Should Not Work

Few of us will ever forget the "cola problem" on the 1996 exam:

3. The rate of consumption of cola in the United States is given by $S(t) = Ce^{kt}$, where S is measured in billions of gallons per year and t is measured in years

from the beginning of 1980.
(a) The consumption rate doubles every 5 years and the consumption rate at the beginning of 1980 was 6 billion gallons per year. Find C and k.
(b) Find the average rate of consumption of cola over the 10-year period beginning January 1, 1983. Indicate units of measure.
(c) Use the trapezoidal rule with four equal subdivisions to estimate $\int_5^7 S(t)\,dt$.
(d) Using correct units, explain the meaning of $\int_5^7 S(t)\,dt$ in terms of cola consumption.

There were numerous lessons we learned from this problem, including the influence of units of measurement on the constants C and k. The most unexpected lesson, however, came (as usual) from a student solution.

Here is the way you expect students to find the average value in part (b):

$$\text{Average value} = \frac{1}{b-a}\int_a^b Ce^{kt}\,dt = \left(\frac{1}{b-a}\right)\left(\frac{C}{k}\right)(e^{kb}-e^{ka}).$$

A few students, however, proceeded quite differently. They found the "special c" value guaranteed by the Mean Value Theorem for derivatives (which I will call p here to avoid confusion with C), then plugged it into the function S:

$$S'(p) = kCe^{kp} = \frac{Ce^{kb}-Ce^{ka}}{b-a}$$

$$\Rightarrow S(p) = Ce^{kp} = \left(\frac{1}{k}\right)\frac{Ce^{kb}-Ce^{ka}}{b-a}.$$

Notice that the answers are the same! This is no coincidence arising from the numbers involved, either; that is why I used all those general constants. It is actually *true*, for an exponential function, that the average value of the function occurs at the point found by the MVT for derivatives! Is this a general theorem?

Well, of course not. Consider the graph of $y = \sin x$ on the interval $[0, \pi]$, for example (Figure 2). The MVT value occurs where the derivative is zero, and the y-coordinate there is the maximum, not the average, value.

It is not even true for monotonic functions in general (as the curious reader is invited to check), but since it is true for exponential functions, the students got full credit!

1. Given the graph of $y = f'(x)$ at the right, explain why the graph of $y = f(x)$ has a point of inflection at $x = 1$.

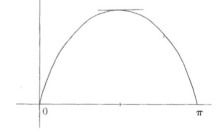

Figure 2

And that wasn't the only surprise on the 1996 test. Check out AB-2:

2. Let R be the region in the first quadrant under the graph of $y = 1/\sqrt{x}$ for $4 \le x \le 9$.
   (a) Find the area of R.
   (b) If the line $x = k$ divides the region R into two regions of equal area, what is the value of k?
   (c) Find the volume of the solid whose base is the region R and whose cross sections cut by planes perpendicular to the x-axis are squares.

Here is a reasonable approach to solving part (b):

$$\int_a^k \frac{1}{\sqrt{x}} \, dx = \int_k^b \frac{1}{\sqrt{x}} \, dx$$

$$2\sqrt{x}\Big|_a^k = 2\sqrt{x}\Big|_k^b$$

$$2\sqrt{k} - 2\sqrt{a} = 2\sqrt{b} - 2\sqrt{k}$$

$$4\sqrt{k} = 2\sqrt{a} + 2\sqrt{b}$$

$$\sqrt{k} = \frac{\sqrt{a} + \sqrt{b}}{2}.$$

For our interval $[4, 9]$ we conclude that $\sqrt{k} = \frac{5}{2}$, and so $k = \frac{25}{4}$.

Again, I left all the letters in the solution so that I could contrast it with another method of solution that was actually used by some students. These creative individuals chose to find k by finding the x in $[4, 9]$ at which the average value of the function occurs. (This is the number whose existence is guaranteed by the Mean Value Theorem for Integrals.) Here's the work:

$$\frac{1}{\sqrt{k}} = \frac{1}{b-a} \int_a^b \frac{1}{\sqrt{x}} \, dx$$

$$\frac{1}{\sqrt{k}} = \frac{1}{b-a} \left(2\sqrt{x}\Big|_a^b \right)$$

$$\frac{1}{\sqrt{k}} = \frac{2(\sqrt{b} - \sqrt{a})}{b-a}$$

$$\frac{1}{\sqrt{k}} = \frac{2}{\sqrt{b} + \sqrt{a}}$$

$$\sqrt{k} = \frac{\sqrt{a} + \sqrt{b}}{2}$$

Needless to say, these students got the same answer. In fact, they also got the same credit, since we can see that their algorithm is valid for this function. But is this a general theorem?

Once again, the answer is, "Of course not." Indeed, the same picture of $y = \sin x$ on $[0, \pi]$ that debunked the previous "theorem" will serve to debunk this one, for virtually the same reason: the area splits evenly where the function value is the maximum, not the average, value of f on the interval.

So, in these two problems did the students who used the "methods that should not work" actually know what they were doing? In all likelihood, no. Did they really deserve full credit? According to our rubrics, yes. Notice that these students *showed their work* and demonstrated that they were using what was, like it or not, a *valid algorithm for solving the given problem.* This is quite a different story from those frequently-seen papers that have correct answers with little or no work. In those cases we take off credit precisely because we *cannot tell* whether the students are using a valid algorithm or not. We can take off credit for mathematics that is not there, but as the current Chief Reader, Bernie Madison, has often pointed out, We can't take off credit for correct mathematics.

On that philosophical quote I will end my brief historical tour of the Lessons I have Learned from AP Readings Past. If you have never had the experience of grading exams all day every day for a week, I urge you to become involved as an AP reader. Then, for your continued happiness and professional development, I wish that each of you might enjoy learning experiences similar to these examples at every reading you attend. Meanwhile, I hope that this paper might have given you a few in absentia.

Book Review: *Calculus with Analytic Geometry*

Underwood Dudley

This is about calculus books, and there are seven important conclusions.

Let us go back to the beginning and look at the first calculus book, *Analyse des Infiniment Petits, pour L'intelligence des Lignes Courbes*, published in Paris in 1696 and written by Guillaume François Antoine L'Hospital, Marquis de Sainte-Mesme, Comte d'Entrement, Seigneur d'Orques, etc. To almost everyone, L'Hospital is just a name attached to L'Hospital's Rule and almost no one knows anything about him. His memory deserves better. He displayed mathematical talent early, solving a problem about cycloids at fifteen, and he was a lifelong lover and supporter of mathematics who, unfortunately, died young, at the age of 43. Also, L'Hospital was no mean mathematician. He published several papers in the journals of the day, solving various nontrivial problems. I know that I could not have found, as he did, the shape of a curve such that a body sliding down it exerts a normal force on it always equal to the weight of the body. Further, as Abraham Robinson has written,

> According to the testimony of his contemporaries, L'Hospital possessed a very attractive personality, being, among other things, modest and generous, two qualities which were not widespread among the mathematicians of his time.

His book was a huge success. There was a second edition in 1715, and there were commentaries written on it. I have the 1781 edition, with additions made by another author. Not many textbooks last almost 100 years. Birkhoff and MacLane is not yet 50 years old. L'Hospital's book is about differentials and their applications to curves and the style is exclusively geometrical. There are not many equations, but there are an awful lot of letters and pictures, just as I remember in my tenth grade geometry text. Mathematics was geometry then, and mathematicians were geometers.

There are many things not in the book. There are no sines or cosines, no exponentials or logarithms, only algebraic functions and algebraic curves. There are also no derivatives, only differentials. Here is L'Hospital's proof of L'Hospital's Rule, using differentials. In Figure 1, points on the graph of $f(x)/g(x)$ are found by dividing lengths of abscissas, except at a. But if you dx past a, the point on the graph will be df/dg. Since dx is infinitesimal, that ratio gives you the point at a. Is that not nice?

It took some time for calculus to become generally taught in colleges. Eventually it made it, and calculus textbooks began to appear in the nineteenth century. I have a copy of one, *Elements of the Differential and Integral Calculus*, by Elias Loomis,

Ll. D., Professor of Natural Philosophy at Yale College. His calculus was first published in 1851, and my copy of it is the 1878 edition. It sold in excess of 25,000 copies, so it must reflect accurately the style and content of calculus teaching of the time. Just as with L'Hospital, the differential was the important idea. Loomis derived the formula for the differential of x^n with no use of the binomial expansion, $(d(xy)/xy = dx/x + dy/y,\ d(x^n)/x^n = dx/x + \cdots + dx/x$, add and simplify) and his proof of L'Hospital's Rule was short, simple, and clear, and also one which

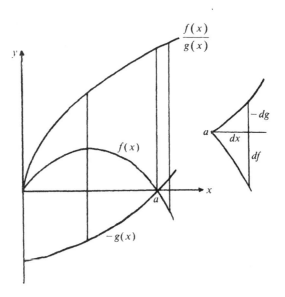

FIG. 1.

does not appear in modern texts because it fails for certain pathological examples. Also, Loomis put *all* of his formulas in words, italicized words. After deriving the formula for the differential of a power of a function, he wrote

> The differential of a function affected with any exponent whatever is the continued product of the exponent, the function itself with its exponent diminished by unity, and the differential of the function.

It was a good idea. It is also probably a good idea to do as L'Hospital and Loomis did and talk about differentials instead of derivatives whenever possible. Little bits of things are easier to understand than rates of change. It is a still better idea to strive for clarity and let students see what is really going on, which is what Loomis did, rather than putting rigor first. But nowadays authors cannot do that. They must protect themselves against some colleague snootily writing to the publisher, "Evidently Professor Blank is unaware that his so-called proof of L'Hospital's Rule is faulty, as the following well-known example shows. I could not possibly adopt a text with such a serious error." It is a shame, and probably inevitable that calculus books are written for calculus teachers, but I have nevertheless concluded

CONCLUSION #1: CALCULUS BOOKS SHOULD BE WRITTEN FOR STUDENTS.

It would be worth a try. *Calculus Made Easy* by Silvanus P. Thompson was quite successful in its time, which ran for quite a while. The second edition appeared in 1914, and my copy was printed in 1935. It is still in print. The book has a motto:

What one fool can do, another can.

and a prologue:

> Considering how many fools can calculate, it is surprising that it should be thought either a difficult or a tedious task for any other fool to learn how to master the same tricks. ...
>
> Being myself a remarkably stupid fellow, I have had to unteach myself the difficulties, and now beg to present to my fellow fools the parts that are not hard. Master these thoroughly, and the rest will follow. What one fool can do, another can.

Chapter 1, whose title is "To Deliver You From The Preliminary Terrors" forthrightly says that dx means "a little bit of x." Thompson did not include L'Hospital's Rule.

Both Loomis and Thompson are like L'Hospital when it comes to giving applications and examples: they are all almost entirely geometrical. Loomis's applications of maxima and minima are all about inscribing and circumscribing things, and so are all of Thompson's except one. In fact, all three books are full of geometry. Thompson concluded with arc length and curvature, Loomis had involutes and evolutes, cusps and multiple points, and lots of curve sketching. Did you know that the asymptote to $y^3 = x^3 + x^2$ is $y = x + 1/3$? I didn't until I read Loomis. It is nice to know. There must have been some reason why calculus books for more than 200 years taught so much geometry. Mathematics may no longer be synonymous with geometry, but we have discarded, wrongly I think, the wisdom of the ages, and I have concluded

CONCLUSION #2: CALCULUS BOOKS NEED MORE GEOMETRY.

Before writing this essay, I examined 85 separate and distinct calculus books. I looked at all of their prefaces, all of their applications of maxima and minima, and all of their treatments of L'Hospital's Rule. By the way, I found five different spellings of L'Hospital. There were the two you would expect, and Lhospital, as L'Hospital sometimes spelled his name. In addition, one author, not wanting to take chances, had it L'Hôspital, and one thought it was Le Hospital. Why are there so many calculus books, and why do they keep appearing? One could be cynical and say that the authors are all motivated by greed. But I do not think so. I think that authors write new calculus books because they have observed that students do not learn much from the old calculus books. Therefore, prospective authors think, "if I write a text and do things properly, students will be able to learn." They are wrong, all of them. The reason for that is

CONCLUSION #3: CALCULUS IS HARD.

Too hard, I think, to teach to college freshmen in the United States in the 1980s, but that is another topic.

If you plot the books' numbers of pages against their year of publication, you have a chart in which an ominous increasing trend is clear. The 1000-page barrier, first pierced in 1960, has been broken more and more often as time goes on. New highs on the calculus-page index are made almost yearly. Where will it all end? We can get an indication. The magic of modern statistics packages produces the least-squares line: Pages = 2.94 (Year) − 5180, showing that in the middle of the next millennium, the average calculus book will have 2,270 pages and the longest one, just published, will have 3,783 pages exclusive of index.

Why do we need 1000 pages to do what L'Hospital did in 234, Loomis in 309, Thompson in 301, and the text I learned calculus from, used exclusively for four whole semesters, 14 semester-hours in all, in 416? There are several reasons. One, of course, is the large number of reviewers of prospective texts. No more can an editor make up his mind about the merits of a text, it has to go out to fifteen different people for opinions. And if one of them writes that the author has left out the $\tan(x/2)$ substitution in the section on techniques of integration, how can he or she do that, we won't be able to integrate $3/(4 + 5\sin 6x)$, how can anyone claim to know calculus who can't do that; isn't the easiest response to include the $\tan(x/2)$ substitution? Of course it is, in it goes, and in goes everything else that is in every other 1000-page text. It is impossible to escape

CONCLUSION #4: CALCULUS BOOKS ARE TOO LONG.

Another reason for the length is the current mania for Applications. If you go to *Books in Print* and look in the subject index under "Calculus" what you see is

The Usefulness of Calculus for the Behavioral, Life, and Managerial Sciences
Essentials of Calculus for Business, Economics, Life Sciences, and Social Sciences

and many, many similar titles. Now authors have to explain, with examples, what marginal revenue is, and consumer surplus, and what tracheae are whereas in the old days, all their readers knew what a cone was. A third reason is the supposed need to be rigorous. Now we see statements of L'Hospital's Rule that take up half a page and proofs of it that go on for three pages. My 416-page calculus book never even mentioned L'Hospital's Rule, and I never felt the lack. Its author never proved that the derivative of x^e was ex^{e-1}, but I was willing to believe it. Trying to include everything and trying to prove everything makes for long books. Everything gets longer. Prefaces used to be short, a page or less. Now they are five and six pages, hard sells for the incredible virtues of the text that follows, full of thanks to reviewers, to five or six editors, to wives, to students, even to cats.

Let me return to "applications." There aren't many, you know. In the 85 calculus books I examined, almost all of them had the Norman window problem—the rectangle surmounted by a semicircle, fixed perimeter, maximize the area. The semicircle always "surmounts." This is the sole surviving use of "surmounted" in the English language, except for the silo, a cylinder surmounted by a hemisphere. Only one author had the courage to say that the window was a semicircle on top of a rectangle. All the books had the box made by cutting the corners out of a flat sheet, all have the ladder sliding down the wall, all had the conical tank with changing height of water, all had the tin can with fixed surface area and maximum volume, all had the V-shaped trough, all had the field to fence, with or without a river flowing (in a dead straight line) along one side, all had the wire—usually wire, but sometimes string—cut into two pieces to be formed into a circle and a square, though some daring authors made circles and equilateral triangles. There are only finitely many calculus problems, and their number is *very* finite.

"Applications" are so phony. Ladders do not slide down walls with the base moving away from the wall at a constant rate. Authors know the applications are phony. One book has the base of the ladder sliding away from the wall at a rate of 2 feet per *minute*. At that rate, you could finish up your painting with time to spare and easily step off the ladder when it was a foot from the ground. Another author has the old run-and-swim problem—you know, minimize the time to get somewhere on the other side of the river—with the person able to run 25 feet per second and swim 20 feet per second. That's not bad for running (it's a 3:31.2 mile), but it is super swimming, 100 yards in 15 seconds, a new world's record by far. There are no conical reservoirs outside of calculus books. Real reservoirs are cylindrical, or perhaps rectangular. The reason for this is found in the texts: in the problems, the conical reservoirs usually have a leak at the bottom. Tin cans are not made to minimize surface area. I could give any number of examples of absurd applications in which businessmen "observe" the price of their product decreasing at the rate of \$1 per month, or where the S. D. S. (remember them?) "find" that staging x demonstrations costs $\$250x^3$. Why will authors not be honest and say that these artificial problems provide valuable practice in translating from English into mathematics and that is all they are for? Surely they cannot disagree with

CONCLUSION #5: FIRST-SEMESTER CALCULUS HAS *NO* APPLICATIONS.

Before getting to my next conclusion, here is my favorite "application."

A cow has 90 feet of fence to make a rectangular pasture. She has the use of a cliff for one side. She decides to leave a 10 foot gap in the fence in case the grass should get greener on the other side. Find ... .

Hardly any authors dare to do that. Calculus books are Serious. The text from which that problem came was titled *Calculus Without Analytic Geometry* and it is no surprise that it did not catch on.

The existence of all those calculus books with "Applications" in their titles implies a market for them. There must be students out there who are being forced to undergo a semester of calculus before they can complete their major in botany and take over the family flower shop. I cannot believe that any more than a tiny fraction of them will ever see a derivative again, or need one. Calculus is a splendid screen for screening out dummies, but it also screens out perfectly intelligent people who find it difficult to deal with quantities. I don't know about you, but I long ago concluded

CONCLUSION #6: NOT EVERYONE NEEDS TO LEARN CALCULUS.

The book by Simmons is a fine one. It was written with care and intelligence. It has good problems, and the historical material is almost a course in the history of mathematics. It is nicely printed, well bound, and expensive. Future historians of mathematics will look back on it and say, "Yes, that is an excellent example of a late twentieth-century calculus book." This leads to my last conclusion

CONCLUSION #7: THAT'S *ENOUGH* ABOUT CALCULUS BOOKS.

The All-Purpose Calculus Problem

Dan Kennedy

Here's a calculus problem to end all calculus problems. (And you thought your professor assigned you hard ones!) See how many familiar themes you can find embedded in this problem.

A particle starts at rest and moves with velocity

$$v(t) = \int_1^t e^{-x^2}\,dx$$

along a 10-foot ladder, which leans against a trough with a triangular cross-section two feet wide and one foot high. Sand is flowing out of the trough at a constant rate of two cubic feet per hour, forming a conical pile in the middle of a sandbox which has been formed by cutting a square of side x from each corner of an 8" by 15" piece of cardboard and folding up the sides. An observer watches the particle from a lighthouse one mile offshore, peering through a window shaped like a rectangle surmounted by a semicircle.

(a) How fast is the tip of the shadow moving?
(b) Find the volume of the solid generated when the trough is rotated about the y-axis.
(c) Justify your answer.
(d) Using the information found in parts (a), (b), and (c) sketch the curve on a pair of coordinating axes.

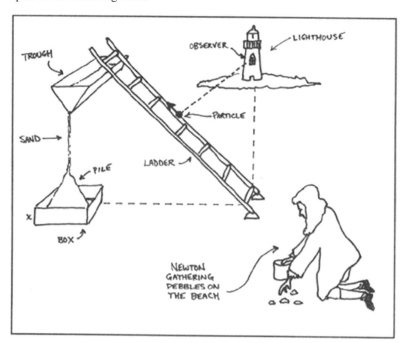

Part 1. Functions, Graphs and Limits

Graphs of Rational Functions for Computer Assisted Calculus

Stan Byrd and Terry Walters

In this capsule, we suggest some calculus problems whose solutions involve pencil-and-paper techniques and some form of computer assistance. These are problems that can be used as calculus laboratory projects. We expect the computer to act as a strong and convenient number-cruncher, but we expect the student to supply the conceptual framework. For some of the problems below, finding the proper scaling so that one can see the extrema is a bit difficult, but we feel that a student will profit from this trial-and-error experience. The main computational difficulty of these problems is approximating all the roots of a polynomial, so your computer package should have a reliable polynomial root finder.

Assuming a, b, c, and d are positive real numbers, we determine the important properties of the graphs of the family of rational functions,

$$f(x) = \frac{ax^2+b}{(x+b)^2} + \frac{cx^2+d}{(x+d)^2}. \tag{1}$$

(We encountered this family in a problem in the *SIAM Review* [3], where f is described as a mean-squared-error function for a class of regression models, and the author asks for conditions under which the minimum of f on $[0,\infty)$ is unique.) As we will show in the following discussion, this family of functions is a good place to make the transition from usual textbook rational function graphs to graphs that should be analyzed with the aid of a computer. (Students who hope to push buttons and get sufficient information will be disappointed.)

In the following list we suggest some problems, following each with remarks about its solutions. Problems 1 and 2 should be solved via pencil and paper analysis, while Problems 3 and 4 should be solved with the aid of one or two of the many available computer programs. (We have used *Derive* and *Mathematica*, but these powerful programs are not necessary.)

Problem 1. Graph the function

$$f(x) = \frac{ax^2+b}{(x+b)^2}, \tag{2}$$

finding the asymptotes, monotonicity intervals, concavity intervals, extrema, and inflection points. (The assumption that a and b are positive insures that each graph has the same basic shape.)

We prefer that students graph some specific instances of these functions using hand calculations and then graph the general case, labeling important points on the graph with expressions involving a and b. The derivatives are messy, but we have prepared our students by doing similar calculations before looking at these problems. They should especially use hand calculations to find the minimum at $x = 1/a$, the unique inflection point at $x = 3/(2a) + b/2$ and the horizontal asymptote at $y = a$, since it would be possible to overlook these features on a computer-generated plot.

Problem 2. Use the results from problem 1 to make a rough sketch of the graph of (1) for $(a, b, c, d) = (2, 3, 3, 1)$ and those values of x *not* between $x = m = \min\{1/a, 1/c\}$ and $x = n = \max\{3/(2a) + b/2, 3/(2c) + d/2\}$. Explain why it is difficult to sketch this function on the omitted domain (m, n).

As in problem 1, pencil and paper analysis is sufficient to work problem 2. By thinking of the function in (1) as the sum of two terms (each in the form of (2)), and realizing that both f' and f'' are the sum of the derivatives of the two terms, one has no difficulty sketching this function outside the interval (m, n). Although functions in the form of (1) clearly must have at least one minimum on (m, n), a complete analytical determination of the roots of the derivatives is difficult.

Problem 3. Use a function plotter to plot some examples of (2) such as $(a, b) = (2, 3)$ and $(a, b) = (3, 1)$.

A student using a function plotter will notice the scaling problem that often occurs when a computer plots a function (see [2]). If one scales the y-axis to see more of the vertical asymptote at $x = -b$, then one will have trouble seeing the unique minimum at $x = 1/a$ and vice versa.

Problem 4. Use a function plotter program to plot (1) for $(a, b, c, d) = (2, 3, 3, 1)$. Use the zoom technique and a root finding program to find the approximate location of the extrema in the interval (m, n) and the minimum in the interval between $-b$ and $-d$. Repeat this problem for $(a, b, c, d) = (2, 1, 2, 2)$ and $(a, b, c, d) = (0.02, 0.7, 10.0, 0.01)$.

Students will find that a scaling of the axis to show the gross features of the graph will not show the extrema in the interval (m, n). Still, they can use the zoom technique and a numerical root extraction routine to find very good approximations to the x coordinates of the extrema. (Using a computer algebra system to extract the roots of the fourth degree polynomial in the first derivative's numerator would be slow and its results difficult to interpret.) After working the first two examples of problem 4, students might guess that the graphs of functions of the form (1) all have the same shape. The last example illustrates that this guess is incorrect. For it, students should find four extrema and three inflection points. Figure 1 and Figure 2 are plots of this example with a different scaling on each plot.

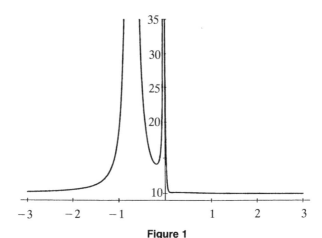

Figure 1
Plot of (1) with $(a, b, c, d) = (0.02, 0.7, 10.0, 0.01)$ showing a minimum and two vertical asymptotes.

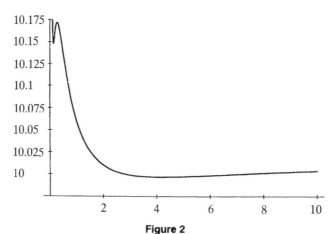

Figure 2
Plot of (1) with $(a, b, c, d) = (0.02, 0.7, 10.0, 0.01)$ showing three extrema.

The instructor might wish to discuss the implications of the facts that the graph of (1) must have positive concavity for x less than m and that the degree of the numerator of the derivative of (1) is only four.

We conclude by noting the existence of a computational test to determine if the functions (1) (with b not equal to d) have one or two minima in the interval (m, n) (see [3]). One can determine if the roots of a fourth degree polynomial (such as the numerator of the derivative of (1)) are real, complex, or repeated by computing the discriminant of the polynomial directly from its coefficients and applying results from the theory of equations (see [1]). Using a few facts about any function in the form of (1) with the positivity assumptions on a, b, c, and d, it can be shown that if the discriminant is zero or negative then the graph of function (1) has one minimum in the interval (m, n), while if the discriminant is positive then the graph has two minima and a maximum in (m, n).

References

1. W. S. Burnside, *The Theory of Equations*, Dover Publications, New York, 1960.
2. F. Demana and B. Waits, Pitfalls in graphical computation, or why a single graph isn't enough, *The College Mathematics Journal*, 19(1988) 177–183.
3. T.-S. Lee, On the uniqueness of a minimum problem, *SIAM Review*, 28(1986) 395.

Computer-Aided Delusions

Richard L. Hall

The so-called real numbers used in computing are actually a small subset of the rational numbers. This simple fact can easily lead to some interesting effects. The example we study in this note illustrates a pitfall that may be encountered when studying the graph of a function with the aid of a computer. Demana and Waits [1] discuss some interesting examples in which computer generated graphs overstep important features of a function or suggest erroneous asymptotic values. The example we discuss here illustrates a case where a computer-generated graph appears to indicate incorrect slope behavior at a point.

We consider the following well-known continuous function used in the teaching of calculus to illustrate logarithmic differentiation:

$$f(x) = \begin{cases} |x|^x, & x \neq 0 \\ 1, & x = 0. \end{cases} \quad (1)$$

Suppose that we were interested in the shape of f and, in particular, suppose that we were curious about the slope S_0 of $f(x)$ at $x = 0$. Figures 1 and 2 indicate a

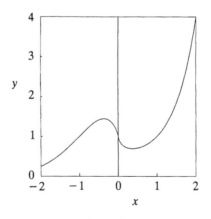

Figure 1

Computer-Aided Delusions

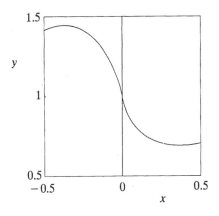

Figure 2

slope of about $S_0 \approx -4.5$, as estimated by laying a ruler along the graph and measuring its slope. Since the graphs are very nice and they were, after all, generated by a *computer*, this rough estimate might be considered satisfactory.

However, imagine we were to look a little more closely at the problem, and plot the graphs shown in Figures 3 and 4, which show magnified portions of the graph of f near $x = 0$. Figure 3 indicates a slope of about $S_0 \approx -9$, and Figure 4 indicates a slope of about $S_0 \approx -18$. Similar graphs covering x intervals with half-widths of respectively 10^{-10} and 10^{-20} indicate slopes of about -27 and -50. This really is turning out to be a slippery slope.

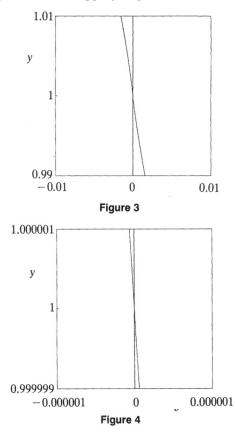

Figure 3

Figure 4

Since the function f is elementary, we can analyze the situation exactly. Using logarithmic differentiation on (1) we find

$$f'(x) = f(x)(\ln|x| + 1). \qquad (2)$$

Consequently, since $f(x)$ has limit 1 as x tends to 0, we see that the slope is unbounded in this limit; evidently $f'(x)$ tends to $-\infty$. The approach to large (negative) values is *very* slow because of the nature of the log function. Some approximate values of $f'(x)$ are shown in Table 1. In Figure 5 we exhibit graphs of

Table 1

n	$f'(10^{-n})$	n	$f'(10^{-n})$
1	-1.03	10	-22.03
2	-3.44	20	-45.05
3	-5.87	30	-68.08
4	-8.20	40	-91.10
5	-10.51	50	-114.13
6	-12.82	60	-137.16
7	-15.12	70	-160.18
8	-17.42	80	-183.21
9	-19.72	90	-206.23
10	-22.03	100	-229.26

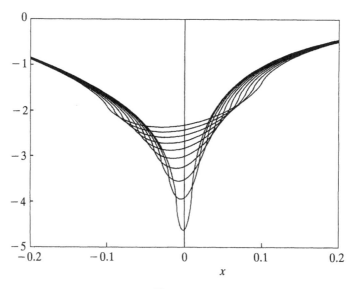

Figure 5

a symmetrical approximation

$$g(x, h) = \frac{1}{2h}[f(x+h) - f(x-h)]$$

for the slope of f for values of h diminishing from $h = 0.1$ down to $h = 0.01$ in steps of size 0.01. As h decreases, the minimum of $g(x, h)$ with respect to x decreases without bound.

Of course, alternatively, we could use a computer to investigate the slope S_0 without graphics; we could simply compute values of $f(x)$ near $x = 0$. If none of the points we consider is *very* close to 0, then it is clear in principle that the same misleading evidence concerning S_0 could arise from a purely numerical investigation.

The pitfall we address is not to do with accuracy nor with computer arithmetic. The danger arises because the graphical presentation of the data, with straight-line segments joining adjacent computed points, may easily *suggest* to us that we have explored the situation in enough detail when in fact we have not. The graphs of f shown in Figures 1 and 2 look extremely convincing. In fact, they are just the kind of computer-generated graphs we should be proud to include in a research report. We usually turn to the computer, not for classroom examples like (1) but when the going gets tough, when even computing *values* of the function f is not easy. This, of course, is precisely the situation where we do *not* have analytical expressions like (2) available to us for checking our results. The lesson is a very general one. We should always be very careful to explore the territory round about any result that we first find with the aid of a computer. Perhaps we should remember one of Alan Turing's exhortations made towards the beginning of the present era of computing with the aid of electronic machines, that *one should be fair to computers*.

Acknowledgment. The author gratefully acknowledges partial financial support of this work by the Natural Sciences and Engineering Research Council of Canada.

Reference

1. F. Demana and B. K. Waits, Pitfalls in graphical computation, or Why a single graph isn't enough, *College Mathematics Journal* 19 (1988) 177–183.

An Overlooked Calculus Question

Eugene Couch

For $a > 1$, do the graphs of a^x and $\log_a x$ ever intersect? And if so, for what bases a and at what points x? The answers must be well known in some realms and times, but not, it seems, in the present North American calculus culture. Several of the most widely used first-year calculus texts, including ones that have gone through several editions, present non-intersecting graphs of a^x and $\log_a x$ on the same set of axes to characterize the case $a > 1$. It is difficult to avoid the conclusion that these texts say the two graphs never intersect in this case. But clearly they must intersect for some values of $a > 1$ since for a sufficiently close to 1, the graph of a^x stays close to 1 for x as large as we like. So it must be intersected by $\ln x$, for example, and thus by $\log_a x$ because $\log_a x$ lies above $\ln x$ for $x > 1$ when $a < e$.

The answers to our questions may be obtained quickly by first-year calculus methods. Observe that $a^x > x$ implies $a^x > x > \log_a x$, and $a^x < x$ implies $a^x < x < \log_a x$. So if the graphs of a^x and $\log_a x$ intersect at a point x, then $a^x = x$. Furthermore, if $a^x = x$, then $\log_a x = x$ and the graphs of a^x and $\log_a x$ intersect at x. Thus, the graphs of a^x and $\log_a x$ intersect at x if and only if $a^x = x$, which is equivalent to

$$a = x^{1/x}.$$

Using standard graphing procedures, we sketch the graph of $x^{1/x}$ and observe which horizontal line $y = a$ intersects this graph. As Figure 1 illustrates, a^x and $\log_a x$ intersect exactly twice for every a in the interval $1 < a < e^{1/e}$, exactly once at $x = e$ for $a = e^{1/e}$, and not at all for $a > e^{1/e}$. The number $e^{1/e} = 1.4446\ldots$ is the least upper bound of the set of values of the base a for which the two graphs intersect. Thus the correct graphs of a^x and $\log_a x$ for $a > 1$ are displayed in Figures 2a–2c.

Since complete characterization of the case $a > 1$ requires more than just Figure 2c, calculus courses would be somewhat enriched if our questions about logarithmic

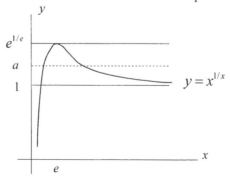

Figure 1.

An Overlooked Calculus Question

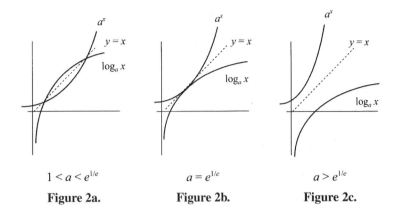

Figure 2a. $1 < a < e^{1/e}$

Figure 2b. $a = e^{1/e}$

Figure 2c. $a > e^{1/e}$

and exponential functions were discussed in texts, or correct graphs were given with pointers to one or more exercises for justification.

Introduction to Limits, or Why Can't We Just Trust the Table?

Allen J. Schwenk

In a first-semester calculus class, or in any class where limits have just been introduced, students tend to think that they can reliably predict the correct limit for $\lim_{x \to a} f(x)$ by selecting a few well-chosen values for x near a. Moreover, they are certain the professor is just being picky by insisting on developing theorems and algebraic methods for "proving" limits.

I find that the following example produces spirited debate about the "correct" limiting value. Ultimately, it convinces the class that *strange things can happen*.

First, I partition the class into nine groups and assign each group a different digit d from $d = 1$ to $d = 9$. Each group is then asked to predict the limit

$$\lim_{x \to 0} \sin \frac{\pi}{x}$$

by completing the following table, where d stands for their assigned digit.

x	$f(x) = \sin \frac{\pi}{x}$
0.d	
0.0d	
0.00d	
0.000d	
0.0000d	
0.00000d	

The curious fact is that eight of the groups will generally conclude that this function has a limit, while the ninth group will think there is no limit. But among the eight groups observing a limit, five get 0, and one each gets the values

$$0.866025403784\ldots$$

$$-0.866025403784\ldots$$

$$-0.342020143\ldots.$$

Of course the knowledgeable reader understands why this function has no limit, but the selection process has been rigged to lead most students to various faulty answers.

I leave it to you and to your class:

- to deduce what these three curious decimals represent,
- to discover which d gives which "limit," and
- to explain why!

A Circular Argument

Fred Richman

Sketch of the circle. The first interesting limit that the student of calculus is exposed to is often

$$\lim_{x \to 0} \frac{\sin x}{x} = 1. \tag{$*$}$$

This limit has received some attention recently, see [5], [6], [9] and [13]. It is not usually recognized that the standard proof is circular, as was suggested in [13] and denied in [9]. Archimedes proved a variant of ($*$) in order to show that the area of a circle is equal to the area of a right triangle whose perpendicular sides are the radius and the circumference of the circle. By the definition of π, the circumference of a circle is $2\pi r$, so his theorem establishes the area formula πr^2. Despite this dependence of the area formula on ($*$), the area formula is the basis for the argument used in most calculus texts to prove ($*$); see [3], [7], [10], [11], [12].

The usual proof. The standard argument for ($*$) hinges on the inequalities

$$\sin x < x < \tan x, \tag{$**$}$$

from which ($*$) readily follows. The usual proof of ($**$) considers an arc AB of length x on the circle of radius one with center at O, and the point B' on the extension of OB such that AB' is perpendicular to OA. The triangle OAB is contained in the circular sector OAB which is contained in the triangle OAB'. Moreover

(1) the area of the triangle OAB is $(\sin x)/2$,
(2) the area of the circular sector OAB is $x/2$,
(3) the area of the triangle OAB' is $(\tan x)/2$.

Statements (1) and (3) are clearly true; Statement (2) is true because the area of the sector is to the area π of the circle as the length x of the arc AB is to the circumference 2π of the circle.

What's wrong with this proof? The problem lies in how we know that the area of the circle is π. The answer that we learned it in elementary school is not good enough. The fact is that to prove that the area of the circle is π, we have to invoke ($*$) in some form; for example, in the form of the inequalities ($**$).

Archimedes' proof that the area of a circle is πr^2. Archimedes was perhaps the first to prove that the area of a circle of radius r is πr^2. Euclid had shown earlier [4; XII.2] that the area of a circle is *proportional to r^2*. Archimedes inscribes and circumscribes the circle with regular n-sided polygons. The length of a side of the inscribed polygon is, in our terms, $2\sin \pi/n$, the length of a side of the circumscribed polygon is $2\tan \pi/n$, and $2\pi/n$ is the length of the circular arc between adjacent points of contact of the circle with either polygon.

Archimedes argues that $2\sin \pi/n < 2\pi/n$ because the shortest distance between two points is a straight line. This principle is intuitively clear, despite the fact that Euclid felt it necessary to prove that any side of a triangle was shorter than the sum of the other two [4, I.20]. To prove that $2\pi/n < 2\tan \pi/n$ Archimedes invokes the following more complicated principle [2, p. 145]:

(∗∗∗) If two plane curves C and D with the same endpoints are concave in the same direction, and C is included between D and the straight line joining the endpoints, then the length of C is less than the length D.

This is not an implausible principle—I find it rather attractive. Still, it doesn't have the immediacy of "the shortest distance between two points is a straight line," and I am not sure that it is easier to accept than (∗) itself.

Principle (∗∗∗) is applied with C the circular arc between two adjacent points of contact of the circumscribed polygon with the circle, and D the polygonal path consisting of the two adjacent halves of the sides of the circumscribed polygon that touch the circle at these points. Thus *Archimedes proves* (∗∗) *in order to show that the area of a circle is πr^2*. He needs (∗∗) to show that the lengths of the perimeters of the polygons approximate the length of the circumference of the circle.

Related proofs. In the proof of (∗) in [8, Chap. 10, Sec. 2], Johnson and Kiokemeister tacitly assume that if A and B are points on a circle, P is a point outside the circle, and PA and PB are tangent to the circle, then the length of the arc from A to B is less than the sum of the lengths of PA and PB. So their proof is like that of Archimedes—and is not circular—but they do not justify this step.

In [9] Rose shows that (∗) is equivalent to the validity of the area formula $sr/2$ for a circular sector of radius r and arc length s. Nevertheless, he claims that the standard proof of (∗) "need not involve circular reasoning since the sector area formula can be obtained geometrically." Indeed Archimedes obtained the formula geometrically, but he first had to establish (∗∗); the standard proof uses the area formula to show (∗), completing the circle. The alternative is to assume (∗) by *defining* arc length to be the supremum of the lengths of inscribed polygonal paths.

True by definition? Ultimately the problem may be that (∗) is true by definition, as suggested by Gillman in [6]: "the theorem on the limit $(\sin \alpha)/\alpha$ in its natural setting as essentially just the definition of the circumference of a circle." The idea underlying the definition of arc length is that small chords approximate small arcs, so polygonal paths can be used to approximate curves. There is no explanatory value in proving (∗) because (∗) is presupposed in the definition of arc length. The best we can do is provide an informal motivation for (∗). Drawing a few isosceles triangles OAA' on a fixed base AA', with the angle at O equal to $2x$ getting smaller and smaller, might convince a student that the length $2rx$ of the circular arc AA', with center at O, approaches the length $2r \sin x$ of the line AA'. Getting across an intuitive feeling for the idea that small arcs are approximately linear is worthwhile in any case: that idea, after all, is the basis for the notion of a derivative.

An ingenious solution. Apostol [1, page 102] circumvents circularity by using area rather than arc length to define the radian measure of an angle. He does not mention the problem of circularity explicitly; instead he comments that he has on hand a general notion of area via the definite integral, but not yet a general notion of arc length. The measure of an angle is defined to be twice the area of the circular sector it subtends, divided by the square of the radius. Then (∗∗) follows immediately from the inclusion of the areas because $x/2$ is *defined* to be the area of the sector. Later, when arc length is defined as the limit of the lengths of polygonal approximations, it can be shown that this definition of the measure of an angle agrees with the standard one that uses arc length.

Although Apostol's solution is elegant, the use of a nonstandard definition of angle measure is a serious drawback in a calculus course. Nevertheless, a case can be made that plane area is a more accessible concept than arc length, and so provides a better way to measure angles. Area is readily bounded by polygons from above and below; arc length, absent a principle like (∗∗∗), is bounded by polygonal lines only from below. If a circle can be put between two polygons, then we can be confident that the area of the circle—whatever it might be—lies between the areas of those two polygons. But why couldn't the circumference of a circle be greater than the supremum of the perimeters of its inscribed polygons?

Recap. The usual proof of (∗) uses the area formula to prove (∗∗). The classical proof of the area formula uses (∗∗), which is established by appeal to the unfamiliar, and not quite obvious principle (∗∗∗). The modern alternative to postulating (∗∗∗) is to define arc length as the supremum of the lengths of polygonal approximations, but this amounts to postulating (∗).

There are various ways out.

- Continue being circular. After all, a proof is just a completely convincing argument. The students accept the area formula, so why not use it?
- Use Archimedes' (∗∗∗), explicitly or tacitly. Maybe it's obvious—Johnson and Kiokemeister evidently thought that it was.
- Postulate a suitable form of (∗). In the form of the definition of arc length, this is the current view of a logical development. Moreover Gillman says that deriving the area formula from this "is what so pleased the students."
- Define angle measure using area, as Apostol does, thus postponing the whole question of arc length.

References

1. T. M. Apostol, *Calculus*, Blaisdell, Waltham, Mass., 1967.
2. E. J. Dijksterhuis, *Archimedes*, Princeton University Press, Princeton, NJ, 1987.
3. C. H. Edwards, and D. E. Penney, *Calculus and Analytic Geometry*, third edition, Prentice Hall, Englewood Cliffs, NJ, 1990.

4. Euclid, *The Elements*, Translated by T. L. Heath, Dover, New York, 1956.
5. W. B. Gearhart, and H. S. Shultz, The function sin x/x, *CMJ* 21 (1990) 90–99.
6. L. Gillman, π and the limit of $(\sin \alpha)/\alpha$, *American Mathematical Monthly* 98 (1991) 345–348.
7. R. A. Hunt, *Calculus with Analytic Geometry*, Harper & Row, New York, 1988.
8. R. E. Johnson, and F. L. Kiokemeister, *Calculus with Analytic Geometry*, third edition, Allyn and Bacon, 1969.
9. D. A. Rose, The differentiability of sin x, *CMJ* 22 (1991) 139–142.
10. S. K. Stein, *Calculus and Analytic Geometry*, fourth edition, McGraw-Hill, New York, 1987.
11. E. W. Swokowski, *Calculus*, fifth edition, PWS-KENT, Boston, 1991.
12. G. B. Thomas, and R. L. Finney, *Calculus and Analytic Geometry*, sixth edition, Addison-Wesley, Reading, MA 1984.
13. P. Ungar, Reviews, *American Mathematical Monthly* 93 (1986) 221–230.

A Geometric Proof of $\lim_{d \to 0^+} (-d \ln d) = 0$

John H. Mathews

Almost every calculus book contains the example $\lim_{d \to 0^+}(-d \ln d) = 0$, and the computation of this limit is usually done by applying L'Hôpital's rule. However, it can be viewed geometrically as follows: the portion of the unit square lying to the right of the vertical line $x = d$ and under the hyperbola $xy = d$ has area $-d \ln d$, (see Figure 1). As $d \to 0$ the area of this region will shrink to zero (see animation in Figures 2–5). The area of the region in Figure 1 is:

$$\int_d^1 \frac{d}{x}\, dx = d \ln x \Big|_{x=d}^{x=1} = -d \ln d.$$

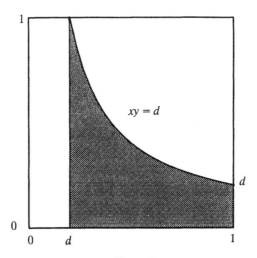

Figure 1
The region whose area is $-d \ln d$.

The sequence of graphical animations corresponding to $d = 0.1$, 0.05, 0.01 and 0.005 are shown in Figures 2–5.

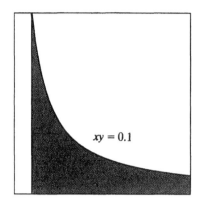

Figure 2
area = $-0.1 \ln(0.1)$.

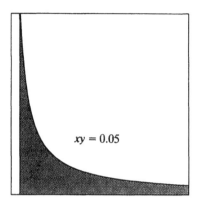

Figure 3
area = $-0.05 \ln(0.05)$.

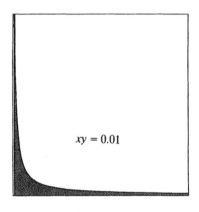

Figure 4
area = $-0.01 \ln(0.01)$.

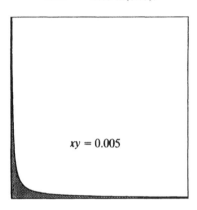

Figure 5
area = $-0.005 \ln(0.005)$.

Part 2. Derivatives

The Changing Concept of Change: The Derivative from Fermat to Weierstrass

Judith V. Grabiner

Some years ago while teaching the history of mathematics, I asked my students to read a discussion of maxima and minima by the seventeenth-century mathematician, Pierre Fermat. To start the discussion, I asked them, "Would you please define a relative maximum?" They told me it was a place where the derivative was zero. "If that's so," I asked, "then what is the definition of a relative minimum?" They told me, *that's* a place where the derivative is zero. "Well, in that case," I asked, "what is the difference between a maximum and a minimum?" They replied that in the case of a maximum, the second derivative is negative.

What can we learn from this apparent victory of calculus over common sense?

I used to think that this story showed that these students did not understand the calculus, but I have come to think the opposite: they understood it very well. The students' answers are a tribute to the power of the calculus in general, and the power of the concept of derivative in particular. Once one has been initiated into the calculus, it is hard to remember what it was like *not* to know what a derivative is and how to use it, and to realize that people like Fermat once had to cope with finding maxima and minima without knowing about derivatives at all.

Historically speaking, there were four steps in the development of today's concept of the derivative, which I list here in chronological order. The derivative was first *used*; it was then *discovered*; it was then *explored and developed*; and it was finally *defined*. That is, examples of what we now recognize as derivatives first were used on an ad hoc basis in solving particular problems; then the general concept lying behind these uses was identified (as part of the invention of the calculus); then many properties of the derivative were explained and developed in applications both to mathematics and to physics; and finally, a rigorous definition was given and the concept of derivative was embedded in a rigorous theory. I will describe the steps, and give one detailed mathematical example from each. We will then reflect on what it all means—for the teacher, for the historian, and for the mathematician.

The seventeenth-century background

Our story begins shortly after European mathematicians had become familiar once more with Greek mathematics, learned Islamic algebra, synthesized the two traditions, and struck out on their own. François Vieta invented symbolic algebra in 1591; Descartes and Fermat independently

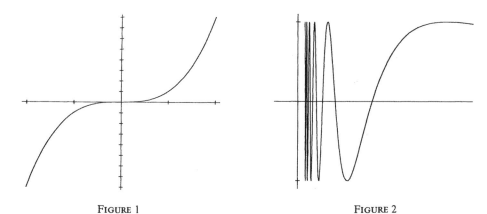

FIGURE 1   FIGURE 2

invented analytic geometry in the 1630's. Analytic geometry meant, first, that curves could be represented by equations; conversely, it meant also that every equation determined a curve. The Greeks and Muslims had studied curves, but not that many—principally the circle and the conic sections plus a few more defined as loci. Many problems had been solved for these, including finding their tangents and areas. But since any equation could now produce a new curve, students of the geometry of curves in the early seventeenth century were suddenly confronted with an explosion of curves to consider. With these new curves, the old Greek methods of synthetic geometry were no longer sufficient. The Greeks, of course, had known how to find the tangents to circles, conic sections, and some more sophisticated curves such as the spiral of Archimedes, using the methods of synthetic geometry. But how could one describe the properties of the tangent at an arbitrary point on a curve defined by a ninety-sixth degree polynomial? The Greeks had defined a tangent as a line which touches a curve without cutting it, and usually expected it to have only one point in common with the curve. How then was the tangent to be defined at the point $(0,0)$ for a curve like $y = x^3$ (FIGURE 1), or to a point on a curve with many turning points (FIGURE 2)?

The same new curves presented new problems to the student of areas and arc lengths. The Greeks had also studied a few cases of what they called "isoperimetric" problems. For example, they asked: of all plane figures with the same perimeter, which one has the greatest area? The circle, of course, but the Greeks had no general method for solving all such problems. Seventeenth-century mathematicians hoped that the new symbolic algebra might somehow help solve all problems of maxima and minima.

Thus, though a major part of the agenda for seventeenth-century mathematicians—tangents, areas, extrema—came from the Greeks, the subject matter had been vastly extended, and the solutions would come from using the new tools: symbolic algebra and analytic geometry.

Finding maxima, minima, and tangents

We turn to the first of our four steps in the history of the derivative: its *use*, and also illustrate some of the general statements we have made. We shall look at Pierre Fermat's method of finding maxima and minima, which dates from the 1630's [8]. Fermat illustrated his method first in solving a simple problem, whose solution was well known: *Given a line, to divide it into two parts so that the product of the parts will be a maximum.* Let the length of the line be designated B and the first part A (FIGURE 3). Then the second part is $B - A$ and the product of the two parts is

$$A(B-A) = AB - A^2. \tag{1}$$

Fermat had read in the writings of the Greek mathematician Pappus of Alexandria that a problem which has, in general, two solutions will have only one solution in the case of a maximum. This remark led him to his method of finding maxima and minima. Suppose in the problem just stated there is a second solution. For this solution, let the first part of the line be designated as $A + E$; the second part is then $B - (A + E) = B - A - E$. Multiplying the two parts together, we obtain

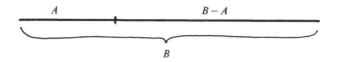

FIGURE 3

for the product

$$BA + BE - A^2 - AE - EA - E^2 = AB - A^2 - 2AE + BE - E^2. \qquad (2)$$

Following Pappus' principle for the maximum, instead of two solutions, there is only one. So we set the two products (1) and (2) "sort of" equal; that is, we formulate what Fermat called the pseudo-equality:

$$AB - A^2 = AB - A^2 - 2AE + BE - E^2.$$

Simplifying, we obtain

$$2AE + E^2 = BE$$

and

$$2A + E = B.$$

Now Fermat said, with no justification and no ceremony, "suppress E." Thus he obtained

$$A = B/2,$$

which indeed gives the maximum sought. He concluded, "We can hardly expect a more general method." And, of course, he was right.

Notice that Fermat did not call E infinitely small, or vanishing, or a limit; he did not explain why he could first divide by E (treating it as nonzero) and then throw it out (treating it as zero). Furthermore, he did not explain what he was doing as a special case of a more general concept, be it derivative, rate of change, or even slope of tangent. He did not even understand the relationship between his maximum-minimum method and the way one found tangents; in fact he followed his treatment of maxima and minima by saying that the same method—that is, adding E, doing the algebra, then suppressing E—could be used to find tangents [**8**, p. 223].

Though the considerations that led Fermat to his method may seem surprising to us, he did devise a method of finding extrema that worked, and it gave results that were far from trivial. For instance, Fermat applied his method to optics. Assuming that a ray of light which goes from one medium to another always takes the quickest path (what we now call the Fermat least-time principle), he used his method to compute the path taking minimal time. Thus he showed that his least-time principle yields Snell's law of refraction [**7**] [**12**, pp. 387–390].

Though Fermat did not publish his method of maxima and minima, it became well known through correspondence and was widely used. After mathematicians had become familiar with a variety of examples, a pattern emerged from the solutions by Fermat's method to maximum-minimum problems. In 1659, Johann Hudde gave a general verbal formulation of this pattern [**3**, p. 186], which, in modern notation, states that, *given a polynomial of the form*

$$y = \sum_{k=0}^{n} a_k x^k,$$

there is a maximum or minimum when

$$\sum_{k=1}^{n} k a_k x^{k-1} = 0.$$

Of even greater interest than the problem of extrema in the seventeenth century was the finding of tangents. Here the tangent was usually thought of as a secant for which the two points came closer and closer together until they coincided. Precisely what it meant for a secant to "become" a

The Changing Concept of Change

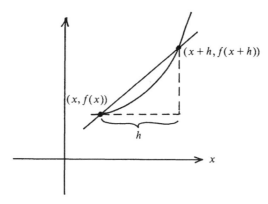

FIGURE 4

tangent was never completely explained. Nevertheless, methods based on this approach worked. Given the equation of a curve

$$y = f(x),$$

Fermat, Descartes, John Wallis, Isaac Barrow, and many other seventeenth-century mathematicians were able to find the tangent. The method involves considering, and computing, the slope of the secant,

$$\frac{f(x+h) - f(x)}{h},$$

doing the algebra required by the formula for $f(x+h)$ in the numerator, then dividing by h. The diagram in FIGURE 4 then suggests that when the quantity h vanishes, the secant becomes the tangent, so that neglecting h in the expression for the slope of the secant gives the slope of the tangent. Again, a general pattern for the equations of slopes of tangents soon became apparent, and a rule analogous to Hudde's rule for maxima and minima was stated by several people, including René Sluse, Hudde, and Christiaan Huygens [3, pp. 185–186].

By the year 1660, both the computational and the geometric relationships between the problem of extrema and the problem of tangents were clearly understood; that is, a maximum was found by computing the slope of the tangent, according to the rule, and asking when it was zero. While in 1660 there was not yet a general concept of derivative, there was a general method for solving one type of geometric problem. However, the relationship of the tangent to other geometric concepts—area, for instance—was not understood, and there was no completely satisfactory definition of tangent. Nevertheless, there was a wealth of methods for solving problems that we now solve by using the calculus, and in retrospect, it would seem to be possible to generalize those methods. Thus in this context it is natural to ask, how did the derivative as we know it come to be?

It is sometimes said that the idea of the derivative was motivated chiefly by physics. Newton, after all, invented both the calculus and a great deal of the physics of motion. Indeed, already in the Middle Ages, physicists, following Aristotle who had made "change" the central concept in his physics, logically analyzed and classified the different ways a variable could change. In particular, something could change uniformly or nonuniformly; if nonuniformly, it could change uniformly-nonuniformly or nonuniformly-nonuniformly, etc. [3, pp. 73–74]. These medieval classifications of variation helped to lead Galileo in 1638, without benefit of calculus, to his successful treatment of uniformly accelerated motion. Motion, then, could be studied scientifically. Were such studies the origin and purpose of the calculus? The answer is no. However plausible this suggestion may sound, and however important physics was in the later development of the calculus, physical questions were in fact neither the immediate motivation nor the first application of the calculus.

Certainly they prepared people's thoughts for some of the properties of the derivative, and for the introduction into mathematics of the concept of change. But the immediate motivation for the general concept of derivative—as opposed to specific examples like speed or slope of tangent—did not come from physics. The first problems to be solved, as well as the first applications, occurred in mathematics, especially geometry (see [1, chapter 7]; see also [3; chapters 4–5], and, for Newton, [17]). The concept of derivative then developed gradually, together with the ideas of extrema, tangent, area, limit, continuity, and function, and it interacted with these ideas in some unexpected ways.

Tangents, areas, and rates of change

In the latter third of the seventeenth century, Newton and Leibniz, each independently, invented the calculus. By "inventing the calculus" I mean that they did three things. First, they took the wealth of methods that already existed for finding tangents, extrema, and areas, and they subsumed all these methods under the heading of two general concepts, the concepts which we now call **derivative** and **integral**. Second, Newton and Leibniz each worked out a notation which made it easy, almost automatic, to use these general concepts. (We still use Newton's $\dot{x}$ and we still use Leibniz's dy/dx and $\int y\,dx$.) Third, Newton and Leibniz each gave an argument to prove what we now call the Fundamental Theorem of Calculus: the derivative and the integral are mutually inverse. Newton called our "derivative" a *fluxion*—a rate of flux or change; Leibniz saw the derivative as a ratio of infinitesimal differences and called it the *differential quotient*. But whatever terms were used, the concept of derivative was now embedded in a general subject—the calculus—and its relationship to the other basic concept, which Leibniz called the integral, was now understood. Thus we have reached the stage I have called *discovery*.

Let us look at an early Newtonian version of the Fundamental Theorem [13, sections 54–5, p. 23]. This will illustrate how Newton presented the calculus in 1669, and also illustrate both the strengths and weaknesses of the understanding of the derivative in this period.

Consider with Newton a curve under which the area up to the point $D = (x, y)$ is given by z (see FIGURE 5). His argument is general: "Assume any relation betwixt x and z that you please;" he then proceeded to find y. The example he used is

$$z = \frac{n}{m+n} ax^{(m+n)/n};$$

however, it will be sufficient to use $z = x^3$ to illustrate his argument.

In the diagram in FIGURE 5, the auxiliary line bd is chosen so that $Bb = o$, where o is not zero. Newton then specified that $BK = v$ should be chosen so that area $BbHK$ = area $BbdD$. Thus ov = area $BbdD$. Now, as x increases to $x + o$, the change in the area z is given by

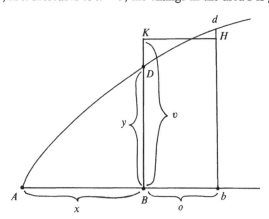

FIGURE 5

$$z(x+o) - z(x) = x^3 + 3x^2o + 3xo^2 + o^3 - x^3 = 3x^2o + 3xo^2 + o^3,$$

which, by the definition of v, is equal to ov. Now since $3x^2o + 3xo^2 + o^3 = ov$, dividing by o produces $3x^2 + 3ox + o^2 = v$. Now, said Newton, "If we suppose Bb to be diminished infinitely and to vanish, or o to be nothing, v and y in that case will be equal and the terms which are multiplied by o will vanish: so that there will remain..."

$$3x^2 = y.$$

What has he shown? Since $(z(x+o) - z(x))/o$ is the rate at which the area z changes, that rate is given by the ordinate y. Moreover, we recognize that $3x^2$ would be the slope of the tangent to the curve $z = x^3$. Newton went on to say that the argument can be reversed; thus the converse holds too. We see that derivatives are fundamentally involved in areas as well as tangents, so the concept of derivative helps us to see that these two problems are mutually inverse. Leibniz gave analogous arguments on this same point (see, e.g. [16, pp. 282–284]).

Newton and Leibniz did not, of course, have the last word on the concept of derivative. Though each man had the most useful properties of the concept, there were still many unanswered questions. In particular, what, exactly, is a differential quotient? Some disciples of Leibniz, notably Johann Bernoulli and his pupil the Marquis de l'Hospital, said a differential quotient was a ratio of infinitesimals; after all, that is the way it was calculated. But infinitesimals, as seventeenth-century mathematicians were well aware, do not obey the Archimedean axiom. Since the Archimedean axiom was the basis for the Greek theory of ratios, which was, in turn, the basis of arithmetic, algebra, and geometry for seventeenth-century mathematicians, non-Archimedean objects were viewed with some suspicion. Again, what is a fluxion? Though it can be understood intuitively as a velocity, the proofs Newton gave in his 1671 *Method of Fluxions* all involved an "indefinitely small quantity o," [14, pp. 32–33] which raises many of the same problems that the o which "vanishes" raised in the Newtonian example of 1669 we saw above. In particular, what is the status of that little o? Is it zero? If so, how can we divide by it? If it is not zero, aren't we making an error when we throw it away? These questions had already been posed in Newton's and Leibniz's time. To avoid such problems, Newton said in 1687 that quantities defined in the way that $3x^2$ was defined in our example were the *limit* of the ratio of vanishing increments. This sounds good, but Newton's understanding of the term "limit" was not ours. Newton in his *Principia* (1687) described limits as "ultimate ratios"—that is, the value of the ratio of those vanishing quantities just when they are vanishing. He said, "Those ultimate ratios with which quantities vanish are not truly the ratios of ultimate quantities, but limits towards which the ratios of quantities decreasing without limit do always converge; and to which they approach nearer than by any given difference, but never go beyond, nor in effect attain to, till the quantities are diminished in infinitum" [15, Book I, Scholium to Lemma XI, p. 39].

Notice the phrase "but never go beyond"—so a variable cannot oscillate about its limit. By "limit" Newton seems to have had in mind "bound," and mathematicians of his time often cite the particular example of the circle as the limit of inscribed polygons. Also, Newton said, "nor...attain to, till the quantities are diminished in infinitum." This raises a central issue: it was often asked whether a variable quantity ever actually reached its limit. If it did not, wasn't there an error? Newton did not help clarify this when he stated as a theorem that "Quantities and the ratios of quantities which in any finite time converge continually to equality, and before the end of that time approach nearer to each other than by any given difference, become ultimately equal" [15, Book I, Lemma I, p. 29]. What does "become ultimately equal" mean? It was not really clear in the eighteenth century, let alone the seventeenth.

In 1734, George Berkeley, Bishop of Cloyne, attacked the calculus on precisely this point. Scientists, he said, attack religion for being unreasonable; well, let them improve their own reasoning first. A quantity is either zero or not; there is nothing in between. And Berkeley characterized the mathematicians of his time as men "rather accustomed to compute, than to think" [2].

Perhaps Berkeley was right, but most mathematicians were not greatly concerned. The concepts of differential quotient and integral, concepts made more effective by Leibniz's notation and by the Fundamental Theorem, had enormous power. For eighteenth-century mathematicians, especially those on the Continent where the greatest achievements occurred, it was enough that the concepts of the calculus were understood sufficiently well to be applied to solve a large number of problems, both in mathematics and in physics. So, we come to our third stage: *exploration and development.*

Differential equations, Taylor series, and functions

Newton had stated his three laws of motion in words, and derived his physics from those laws by means of synthetic geometry [15]. Newton's second law stated: *"The change of motion [our 'momentum'] is proportional to the motive force impressed, and is made in the direction of the [straight] line in which that force is impressed"* [15, p. 13]. Once translated into the language of the calculus, this law provided physicists with an instrument of physical discovery of tremendous power—because of the power of the concept of the derivative.

To illustrate, if F is force and x distance (so $m\dot{x}$ is momentum and, for constant mass, $m\ddot{x}$ the rate of change of momentum), then Newton's second law takes the form $F = m\ddot{x}$. Hooke's law of elasticity (when an elastic body is distorted the restoring force is proportional to the distance [in the opposite direction] of the distortion) takes the algebraic form $F = -kx$. By equating these expressions for force, Euler in 1739 could easily both state and solve the differential equation $m\ddot{x} + kx = 0$ which describes the motion of a vibrating spring [10, p. 482]. It was mathematically surprising, and physically interesting, that the solution to that differential equation involves sines and cosines.

An analogous, but considerably more sophisticated problem, was the statement and solution of the partial differential equation for the vibrating string. In modern notation, this is

$$\frac{\partial^2 y}{\partial t^2} = \frac{T \partial^2 y}{\mu \partial x^2},$$

where T is the tension in the string and μ is its mass per unit length. The question of how the solutions to this partial differential equation behaved was investigated by such men as d'Alembert, Daniel Bernoulli, and Leonhard Euler, and led to extensive discussions about the nature of continuity, and to an expansion of the notion of function from formulas to more general dependence relations [10, pp. 502–514], [16, pp. 367–368]. Discussions surrounding the problem of the vibrating string illustrate the unexpected ways that discoveries in mathematics and physics can interact ([16, pp. 351–368] has good selections from the original papers). Numerous other examples could be cited, from the use of infinite-series approximations in celestial mechanics to the dynamics of rigid bodies, to show that by the mid-eighteenth century the differential equation had become the most useful mathematical tool in the history of physics.

Another useful tool was the Taylor series, developed in part to help solve differential equations. In 1715, Brook Taylor, arguing from the properties of finite differences, wrote an equation expressing what we would write as $f(x + h)$ in terms of $f(x)$ and its quotients of differences of various orders. He then let the differences get small, passed to the limit, and gave the formula that still bears his name: the Taylor series. (Actually, James Gregory and Newton had anticipated this discovery, but Taylor's work was more directly influential.) The importance of this property of derivatives was soon recognized, notably by Colin Maclaurin (who has a special case of it named after him), by Euler, and by Joseph-Louis Lagrange. In their hands, the Taylor series became a powerful tool in studying functions and in approximating the solution of equations.

But beyond this, the study of Taylor series provided new insights into the nature of the derivative. In 1755, Euler, in his study of power series, had said that for any power series,

$$a + bx + cx^2 + dx^3 + \cdots,$$

one could find x sufficiently small so that if one broke off the series after some particular term—say x^2—the x^2 term would exceed, in absolute value, the sum of the entire remainder of the series [6, section 122]. Though Euler did not prove this—he must have thought it obvious since he usually worked with series with finite coefficients—he applied it to great advantage. For instance, he could use it to analyze the nature of maxima and minima. Consider, for definiteness, the case of maxima. If $f(x)$ is a relative maximum, then by definition, for small h,

$$f(x-h) < f(x) \quad \text{and} \quad f(x+h) < f(x).$$

Taylor's theorem gives, for these inequalities,

$$f(x-h) = f(x) - h\frac{df(x)}{dx} + h^2\frac{d^2f(x)}{dx^2} - \cdots < f(x) \tag{3}$$

$$f(x+h) = f(x) + h\frac{df(x)}{dx} + h^2\frac{d^2f(x)}{dx^2} + \cdots < f(x). \tag{4}$$

Now if h is so small that $h\,df(x)/dx$ dominates the rest of the terms, the only way that both of the inequalities (3) and (4) can be satisfied is for $df(x)/dx$ to be zero. Thus the differential quotient is zero for a relative maximum. Furthermore, Euler argued, since h^2 is always positive, if $d^2f(x)/dx^2 \neq 0$, the only way both inequalities can be satisfied is for $d^2f(x)/dx^2$ to be negative. This is because the h^2 term dominates the rest of the series—unless $d^2f(x)/dx^2$ is itself zero, in which case we must go on and think about even higher-order differential quotients. This analysis, first given and demonstrated geometrically by Maclaurin, was worked out in full analytic detail by Euler [6, sections 253–254], [9, pp. 117–118]. It is typical of Euler's ability to choose computations that produce insight into fundamental concepts. It assumes, of course, that the function in question has a Taylor series, an assumption which Euler made without proof for many functions; it assumes also that the function is uniquely the sum of its Taylor series, which Euler took for granted. Nevertheless, this analysis is a beautiful example of the exploration and development of the concept of the differential quotient of first, second, and nth orders—a development which completely solves the problem of characterizing maxima and minima, a problem which goes back to the Greeks.

Lagrange and the derivative as a function

Though Euler did a good job analyzing maxima and minima, he brought little further understanding of the nature of the differential quotient. The new importance given to Taylor series meant that one had to be concerned not only about first and second differential quotients, but about differential quotients of any order.

The first person to take these questions seriously was Lagrange. In the 1770's, Lagrange was impressed with what Euler had been able to achieve by Taylor-series manipulations with differential quotients, but Lagrange soon became concerned about the logical inadequacy of all the existing justifications for the calculus. In particular, Lagrange wrote in 1797 that the Newtonian limit-concept was not clear enough to be the foundation for a branch of mathematics. Moreover, in not allowing variables to surpass their limits, Lagrange thought the limit-concept too restrictive. Instead, he said, the calculus should be reduced to algebra, a subject whose foundations in the eighteenth century were generally thought to be sound [11, pp. 15–16].

The algebra Lagrange had in mind was what he called the algebra of infinite series, because Lagrange was convinced that infinite series were part of algebra. Just as arithmetic deals with infinite decimal fractions without ceasing to be arithmetic, Lagrange thought, so algebra deals with infinite algebraic expressions without ceasing to be algebra. Lagrange believed that expanding $f(x+h)$ into a power series in h was always an algebraic process. It is obviously algebraic when one turns $1/(1-x)$ into a power series by dividing. And Euler had found, by manipulating formulas, infinite power-series expansions for functions like $\sin x, \cos x, e^x$. If functions like those have power-series expansions, perhaps everything could be reduced to algebra. Euler, in his book *Introduction to the analysis of the infinite* (*Introductio in analysin infinitorum*, 1748), had studied infinite series, infinite products, and infinite continued fractions by what he thought of as purely algebraic methods. For instance, he converted infinite series into infinite products by treating a series as a very long polynomial. Euler thought that this work was purely algebraic, and—what is crucial here—Lagrange also thought Euler's methods were purely algebraic. So Lagrange tried to make the calculus rigorous by reducing it to the algebra of infinite series.

Lagrange stated in 1797, and thought he had proved, that any function (that is, any analytic expression, finite or infinite) had a power-series expansion:

$$f(x+h) = f(x) + p(x)h + q(x)h^2 + r(x)h^3 + \cdots, \qquad (5)$$

except, possibly, for a finite number of isolated values of x. He then defined a new function, the coefficient of the linear term in h which is $p(x)$ in the expansion shown in (5)) and called it the **first derived function** of $f(x)$. Lagrange's term "derived function" (*fonction dérivée*) is the origin of our term "derivative." Lagrange introduced a new notation, $f'(x)$, for that function. He defined $f''(x)$ to be the first derived function of $f'(x)$, and so on, recursively. Finally, using these definitions, he proved that, in the expansion (5) above, $q(x) = f''(x)/2$, $r(x) = f'''(x)/6$, and so on [**11**, chapter 2].

What was new about Lagrange's definition? The concept of *function*—whether simply an algebraic expression (possibly infinite) or, more generally, any dependence relation—helps free the concept of derivative from the earlier ill-defined notions. Newton's explanation of a fluxion as a rate of change appeared to involve the concept of motion in mathematics; moreover, a fluxion seemed to be a different kind of object than the flowing quantity whose fluxion it was. For Leibniz, the differential quotient had been the quotient of vanishingly small differences; the second differential quotient, of even smaller differences. Bishop Berkeley, in his attack on the calculus, had made fun of these earlier concepts, calling vanishing increments "ghosts of departed quantities" [**2**, section 35]. But since, for Lagrange, the derivative was a function, it was now the same sort of object as the original function. The second derivative is precisely the same sort of object as the first derivative; even the nth derivative is simply another function, defined as the coefficient of h in the Taylor series for $f^{(n-1)}(x+h)$. Lagrange's notation $f'(x)$ was designed precisely to make this point.

We cannot fully accept Lagrange's definition of the derivative, since it assumes that every differentiable function is the sum of a Taylor series and thus has infinitely many derivatives. Nevertheless, that definition led Lagrange to a number of important properties of the derivative. He used his definition together with Euler's criterion for using truncated power series in approximations to give a most useful characterization of the derivative of a function [**9**, p. 116, pp. 118–121]:

$$f(x+h) = f(x) + hf'(x) + hH, \text{ where } H \text{ goes to zero with } h.$$

(I call this the *Lagrange property of the derivative*.) Lagrange interpreted the phrase "H goes to zero with h" in terms of inequalities. That is, he wrote that,

> Given D, h can be chosen so that $f(x+h) - f(x)$ lies between $h(f'(x) - D)$ and $h(f'(x) + D)$. $\qquad (6)$

Formula (6) is recognizably close to the modern delta-epsilon definition of the derivative.

Lagrange used inequality (6) to prove theorems. For instance, he proved that a function with positive derivative on an interval is increasing there, and used that theorem to derive the Lagrange remainder of the Taylor series [**9**, pp. 122–127], [**11**, pp. 78–85]. Furthermore, he said, considerations like inequality (6) are what make possible applications of the differential calculus to a whole range of problems in mechanics, in geometry, and, as we have described, the problem of maxima and minima (which Lagrange solved using the Taylor series remainder which bears his name [**11**, pp. 233–237]).

In Lagrange's 1797 work, then, the derivative is defined by its position in the Taylor series—a strange definition to us. But the derivative is also *described* as satisfying what we recognize as the appropriate delta-epsilon inequality, and Lagrange applied this inequality and its nth-order analogue, the Lagrange remainder, to solve problems about tangents, orders of contact between

The Changing Concept of Change

P. Fermat	R. Descartes	I. Newton	G.W. Leibniz
1637–38		1669	1684

Dates refer to these mathematician's major works which

curves, and extrema. Here the derivative was clearly a function, rather than a ratio or a speed.

Still, it is a lot to assume that a function has a Taylor series if one wants to define only *one* derivative. Further, Lagrange was wrong about the algebra of infinite series. As Cauchy pointed out in 1821, the algebra of finite quantities cannot automatically be extended to infinite processes. And, as Cauchy also pointed out, manipulating Taylor series is not foolproof. For instance, e^{-1/x^2} has a zero Taylor series about $x = 0$, but the function is not identically zero. For these reasons, Cauchy rejected Lagrange's definition of derivative and substituted his own.

Definitions, rigor, and proofs

Now we come to the last stage in our chronological list: *definition*. In 1823, Cauchy defined the derivative of $f(x)$ as the limit, when it exists, of the quotient of differences $(f(x+h)-f(x))/h$ as h goes to zero [4, pp. 22–23]. But Cauchy understood "limit" differently than had his predecessors. Cauchy entirely avoided the question of whether a variable ever reached its limit; he just didn't discuss it. Also, knowing an absolute value when he saw one, Cauchy followed Simon l'Huilier and S.-F. Lacroix in abandoning the restriction that variables never surpass their limits. Finally, though Cauchy, like Newton and d'Alembert before him, gave his definition of limit in words, Cauchy's understanding of limit (most of the time, at least) was algebraic. By this, I mean that when Cauchy needed a limit property in a proof, he used the algebraic inequality-characterization of limit. Cauchy's proof of the mean value theorem for derivatives illustrates this. First he proved a theorem which states: *if $f(x)$ is continuous on $[x, x+a]$, then*

$$\min_{[x, x+a]} f'(x) \leqslant \frac{f(x+a)-f(x)}{a} \leqslant \max_{[x, x+a]} f'(x). \tag{7}$$

The first step in his proof is [4, p. 44]:

> Let δ, ε be two very small numbers; the first is chosen so that for all [absolute] values of h less than δ, and for any value of x [on the given interval], the ratio $(f(x+h)-f(x))/h$ will always be greater than $f'(x) - \varepsilon$ and less than $f'(x) + \varepsilon$.

(The notation in this quote is Cauchy's, except that I have substituted h for the i he used for the increment.) Assuming the intermediate-value theorem for continuous functions, which Cauchy had proved in 1821, the mean-value theorem is an easy corollary of (7) [4, pp. 44–45], [9, pp. 168–170].

Cauchy took the inequality-characterization of the derivative from Lagrange (possibly via an 1806 paper of A.-M. Ampère [9, pp. 127–132]). But Cauchy made that characterization into a definition of derivative. Cauchy also took from Lagrange the name derivative and the notation $f'(x)$, emphasizing the functional nature of the derivative. And, as I have shown in detail elsewhere [9, chapter 5], Cauchy adapted and improved Lagrange's inequality proof-methods to prove results like the mean-value theorem, proof-methods now justified by Cauchy's definition of derivative.

L. Euler	J.-L. Lagrange	A.-L. Cauchy	K. Weierstrass
1755	1797	1823	1861

contributed to the evolution of the concept of the derivative.

But of course, with the new and more rigorous definition, Cauchy went far beyond Lagrange. For instance, using his concept of limit to define the integral as the limit of sums, Cauchy made a good first approximation to a real proof of the Fundamental Theorem of Calculus [9, pp. 171–175], [4, pp. 122–125, 151–152]. And it was Cauchy who not only raised the question, but gave the first proof, of the existence of a solution to a differential equation [9, pp. 158–159].

After Cauchy, the calculus itself was viewed differently. It was seen as a rigorous subject, with good definitions and with theorems whose proofs were based on those definitions, rather than merely as a set of powerful methods. Not only did Cauchy's new rigor establish the earlier results on a firm foundation, but it also provided a framework for a wealth of new results, some of which could not even be formulated before Cauchy's work.

Of course, Cauchy did not himself solve all the problems occasioned by his work. In particular, Cauchy's definition of the derivative suffers from one deficiency of which he was unaware. Given an ε, he chose a δ which he assumed would work for any x. That is, he assumed that the quotient of differences converged uniformly to its limit. It was not until the 1840's that G. G. Stokes, V. Seidel, K. Weierstrass, and Cauchy himself worked out the distinction between convergence and uniform convergence. After all, in order to make this distinction, one first needs a clear and algebraic understanding of what a limit is—the understanding Cauchy himself had provided.

In the 1850's, Karl Weierstrass began to lecture at the University of Berlin. In his lectures, Weierstrass made algebraic inequalities replace words in theorems in analysis, and used his own clear distinction between pointwise and uniform convergence along with Cauchy's delta-epsilon techniques to present a systematic and thoroughly rigorous treatment of the calculus. Though Weierstrass did not publish his lectures, his students—H. A. Schwartz, G. Mittag-Leffler, E. Heine, S. Pincherle, Sonya Kowalevsky, Georg Cantor, to name a few—disseminated Weierstrassian rigor to the mathematical centers of Europe. Thus although our modern delta-epsilon definition of derivative cannot be quoted from the *works* of Weierstrass, it is in fact the *work* of Weierstrass [3, pp. 284–287]. The rigorous understanding brought to the concept of the derivative by Weierstrass is signaled by his publication in 1872 of an example of an everywhere continuous, nowhere differentiable function. This is a far cry from merely acknowledging that derivatives might not always exist, and the example shows a complete mastery of the concepts of derivative, limit, and existence of limit [3, p. 285].

Historical development versus textbook exposition

The span of time from Fermat to Weierstrass is over two hundred years. How did the concept of derivative develop? Fermat implicitly used it; Newton and Liebniz discovered it; Taylor, Euler, Maclaurin developed it; Lagrange named and characterized it; and only at the end of this long period of development did Cauchy and Weierstrass define it. This is certainly a complete reversal of the usual order of textbook exposition in mathematics, where one starts with a definition, then explores some results, and only then suggests applications.

This point is important for the teacher of mathematics: the historical order of development of the derivative is the reverse of the usual order of textbook exposition. Knowing the history helps us as we teach about derivatives. We should put ourselves where mathematicians were before Fermat, and where our beginning students are now—back on the other side, before we had any concept of derivative, and also before we knew the many uses of derivatives. Seeing the historical origins of a concept helps motivate the concept, which we—along with Newton and Leibniz—want for the problems it helps to solve. Knowing the historical order also helps to motivate the rigorous definition—which we, like Cauchy and Weierstrass, want in order to justify the uses of the derivative, and to show precisely when derivatives exist and when they do not. We need to remember that the rigorous definition is often the end, rather than the beginning, of a subject.

The real historical development of mathematics—the order of discovery—reveals the creative mathematician at work, and it is creation that makes doing mathematics so exciting. The order of exposition, on the other hand, is what gives mathematics its characteristic logical structure and its incomparable deductive certainty. Unfortunately, once the classic exposition has been given, the order of discovery is often forgotten. The task of the historian is to recapture the order of discovery: not as we think it might have been, not as we think it should have been, but as it really was. And this is the purpose of the story we have just told of the derivative from Fermat to Weierstrass.

This article is based on a talk delivered at the Conference on the History of Modern Mathematics, Indiana Region of the Mathematical Association of America, Ball State University, April 1982; earlier versions were presented at the Southern California Section of the M. A. A. and at various mathematics colloquia. I thank the MATHEMATICS MAGAZINE referees for their helpful suggestions.

References

[1] Margaret Baron, Origins of the Infinitesimal Calculus, Pergamon, Oxford, 1969.
[2] George Berkeley, The Analyst, or a Discourse Addressed to an Infidel Mathematician, 1734. In A. A. Luce and T. R. Jessop, eds., The Works of George Berkeley, Nelson, London, 1951 (some excerpts appear in [16, pp. 333–338]).
[3] Carl Boyer, History of the Calculus and Its Conceptual Development, Dover, New York, 1959.
[4] A.-L. Cauchy, Résumé des leçons données à l'école royale polytechnique sur le calcul infinitésimal, Paris, 1823. In Oeuvres complètes d'Augustin Cauchy, Gauthier-Villars, Paris, 1882- , series 2, vol. 4.
[5] Pierre Dugac, Fondements d'analyse, in J. Dieudonné, Abrégé d'histoire des mathématiques, 1700–1900, 2 vols., Hermann, Paris, 1978.
[6] Leonhard Euler, Institutiones calculi differentialis, St. Petersburg, 1755. In Operia omnia, Teubner, Leipzig, Berlin, and Zurich, 1911- , series 1, vol. 10.
[7] Pierre Fermat, Analysis ad refractiones, 1661. In Oeuvres de Fermat, ed., C. Henry and P. Tannery, 4 vols., Paris, 1891–1912; Supplement, ed. C. de Waard, Paris, 1922, vol. 1, pp. 170–172.
[8] _____, Methodus ad disquirendam maximam et minimum et de tangentibus linearum curvarum, Oeuvres, vol. 1, pp. 133–136. Excerpted in English in [16, pp. 222–225].
[9] Judith V. Grabiner, The Origins of Cauchy's Rigorous Calculus, M. I. T. Press, Cambridge and London, 1981.
[10] Morris Kline, Mathematical Thought from Ancient to Modern Times, Oxford, New York, 1972.
[11] J.-L. Lagrange, Théorie des fonctions analytiques, Paris, 2nd edition, 1813. In Oeuvres de Lagrange, ed. M. Serret, Gauthier-Villars, Paris, 1867–1892, vol. 9.
[12] Michael S. Mahoney, The Mathematical Career of Pierre de Fermat, 1601–1665, Princeton University Press, Princeton, 1973.
[13] Isaac Newton, Of Analysis by Equations of an Infinite Number of Terms [1669], in D. T. Whiteside, ed., Mathematical Works of Isaac Newton, Johnson, New York and London, 1964, vol. 1, pp. 3–25.
[14] _____, Method of Fluxions [1671], in D. T. Whiteside, ed., Mathematical Works of Isaac Newton, vol. 1, pp. 29–139.
[15] _____, Mathematical Principles of Natural Philosophy, tr. A. Motte, ed. F. Cajori, University of California Press, Berkeley, 1934.
[16] D. J. Struik, Source Book in Mathematics, 1200–1800, Harvard University Press, Cambridge, MA, 1969.
[17] D. T. Whiteside, ed., The Mathematical Papers of Isaac Newton, Cambridge University Press, 1967–1982.

Derivatives Without Limits

Harry Sedinger

In an introductory calculus course, the derivative of a function $f(x)$ at $x = p$ is frequently described as the slope of the line tangent to the graph of f at $(p, f(p))$. It can be desirable to exhibit examples and applications of this before confronting the idea of limit. Moreover, for a large class of curves, students have the intuitive notion of tangency being the intersection of the curve and the line at exactly one point. Our approach uses this idea to compute the slopes of such tangent lines.

Given a function $f(x)$ and a point with coordinates $(p, f(p))$, the idea is to choose m so that the line $y = f(p) + m(x - p)$ intersects the graph of f at no point other than $(p, f(p))$. In other words, choose m so that

$$f(x) = f(p) + m(x - p) \qquad (*)$$

has exactly one solution. For a large class of functions f, equation $(*)$ is quadratic and there are techniques to insure exactly one solution.

Example 1. Let $f(x) = 6/x$ and consider $p = 3$. Then $6/x = 2 + m(x - 3)$ yields

$$mx^2 + (2 - 3m)x - 6 = 0.$$

If we set the discriminant to zero and simplify, we obtain

$$9m^2 + 12m + 4 = 0 \quad \text{and} \quad m = -\tfrac{2}{3}.$$

Example 2. For $f(x) = \sqrt{ax^2 + b}$ and given point $(p, f(p))$,

$$\sqrt{ax^2 + b} = \sqrt{ap^2 + b} + m(x - p)$$

can be simplified to

$$(x - p)\left[m^2(x - p) + 2m\sqrt{ap^2 + b} - a(x + p) \right] = 0,$$

which can be rewritten as

$$(x - p)\left[(m^2 - a)(x - p) + 2m\sqrt{ap^2 + b} - 2ap \right] = 0.$$

Thus $x = p$ is a double root if $2m\sqrt{ap^2 + b} = 2ap$; that is, if $m = ap/\sqrt{ap^2 + b}$.

The class of functions $f(x) = \sqrt{ax + b}$ also yields a quadratic in $(*)$. For those functions where $(*)$ does not reduce to a quadratic, other methods of insuring one solution may exist. In general, limits are needed.

Rethinking Rigor in Calculus: The Role of the Mean Value Theorem

Thomas W. Tucker

1. INTRODUCTION. Mathematicians have been struggling with the theoretical foundations of the calculus ever since its inception. Bishop Berkeley's attack on Newton's "ghosts of departed quantities," Euler's claim that $1 - 1 + 1 - 1 \cdots = 1/2$, Cauchy's $\varepsilon - \delta$ definition of limit, all are part of the fascinating history of this struggle (see [7]). Calculus instructors and textbooks face the same struggle, but the tack taken, although formal, is often not sensible or honest. Instead of an admission that Newton, Leibnitz, the Bernoullis, and Euler all managed quite well without any rigorous foundations, instead of the story how a rigorous calculus took mathematicians two hundred years to get right, the Mean Value Theorem is waved, like a cross in front of a vampire, to hold the difficulties at bay. The origin of the Mean Value Theorem in the structure of the real numbers is not addressed; that is much too difficult for a standard course. Maybe it is traced back to the Extreme Value Theorem, but the trail ends there. The result is that a technical existence theorem is introduced without proof and used to prove intuitively obvious statements, such as "if your speedometer reads zero, you are not going anywhere" (if $f' = 0$ on an interval, then f is constant on that interval). That's the sort of thing that gives mathematics a bad name: assuming the nonobvious to prove the obvious. And by the way, there is nothing obvious about the Mean Value Theorem without the hypothesis of continuity of the derivative. Cauchy himself was never able to prove it in that form.

I have serious reservations about the need for formal theorems and proofs in a standard calculus course. On the other hand, for those mathematicians who do feel that need, I have a suggestion for an alternative theoretical cornerstone to replace the Mean Value Theorem (MVT); I hope textbook authors adopt it. It is much easier to state, much more intuitively obvious, and much more powerful than most mathematicians realize. It is simply this:

The Increasing Function Theorem (IFT). *If $f' \geq 0$ on an interval, then f is increasing on that interval.*

Here, *increasing* means that if $c \leq d$, then $f(c) \leq f(d)$. This would usually be called nondecreasing, but that term is awkward; for example, nondecreasing and not decreasing mean different things. It seems to make more sense to use the term *strictly increasing* for the condition that if $c < d$, then $f(c) < f(d)$. A function that is increasing, but not strictly increasing, we call *weakly increasing*.

Most of the rest of this paper is concerned with the consequences of the IFT, treating it as an axiom. I will give, however, a short independent proof of the IFT, for the sake of completeness and for readers who have probably never thought of proving the IFT directly without the MVT. Of course, the IFT follows easily from the MVT. In fact, the contrapositive of the IFT is a weak form of the MVT: if $a < b$ and $f(b) < f(a)$, there is a number c, $a \leq c \leq b$, such that $f'(c) < 0$.

It is impossible to be a pioneer in territory as well-trodden as the Mean Value Theorem. Others have championed calculus without the Mean Value Theorem (see [1], [4], [6]). The first two sections of this paper follow Lax, Burstein, and Lax [9] quite closely, although unintentionally. In fact, after searching through dozens of calculus books for the Taylor remainder proof given in this paper and finally finding it in Lax-Burstein-Lax (LBL), I felt a little uncomfortable. Maybe this paper shouldn't be published and all that is needed is an announcement "Go read LBL." Then I read Grabiner [7] and found that the Taylor remainder proof given here and in LBL is actually Lagrange's original proof. I was surprised that such a simple, direct proof could have been covered over by years of second-growth jungle.

Moreover, the idea of Lagrange's proof keeps being rediscovered for special cases like $\sin x$ or $\cos x$. For example, the *Monthly* published such an article recently [2], which then generated a subsequent Editor's Note [2] citing calculus textbooks and *Monthly* articles where the idea of [2] had already been presented. None of these references noted that the same idea works for all functions; LBL is still the only book that does that, to my knowledge. And hardly anyone seems to know the idea is really Lagrange's! Under these circumstances, it appears that some dissemination is badly needed to clear up a memory lapse of generations of mathematicians. It also appears that previous calls ([4], [6]) to downplay the Mean Value Theorem have fallen on deaf ears. Perhaps the recent debates about calculus instruction have unplugged some ears and it is time to try the call again.

2. A PROOF OF THE INCREASING FUNCTION THEOREM. There is a reasonably elementary proof of the IFT that depends only on the nested interval property of the reals: if $a_n \leq a_{n+1} \leq b_{n+1} \leq b_n$ for all $n \geq 1$ and $\lim_{n \to \infty}(b_n - a_n) = 0$, then there is a number c such that $\lim_{n \to \infty} a_n = \lim_{n \to \infty} b_n = c$. The proof of the IFT given here does not require the continuity of f' and is so self-contained that it probably could be given in a standard calculus course. Although I generated this proof in response to some remarks of Peter Lax, I should have known the proof is too natural to be original. In revising this paper, I discovered Richmond's article [10], which contains essentially the same proof, and as I already knew, Ampère and Cauchy used the key observation in their own proofs.

Proof of the IFT. The proof depends on the following simple

Observation. Given a function f, define slope(a, b) to be the usual quotient $(f(b) - f(a))/(b - a)$. If slope$(a, b) = m$ and c is between a and b, then one of slope(a, c) and slope(c, b) is greater than or equal to m and one is less than or equal to m. For a proof, draw the obvious picture.

Suppose now that $f'(x) \geq 0$ on $[a, b]$ and that f is not increasing; that is, for some a_1, b_1 with $a \leq a_1 < b_1 \leq b$, we have $f(a_1) > f(b_1)$. Let $m = \text{slope}(a_1, b_1)$. Note that $m < 0$. By repeated bisection and our observation, we can find a nested sequence of intervals $[a_n, b_n]$ with $\text{slope}(a_n, b_n) \leq m$ and $\lim_{n \to \infty}(b_n - a_n) = 0$. Let $c = \lim_{n \to \infty} a_n = \lim_{n \to \infty} b_n$ (the possibility $c = a$ or $c = b$ causes no difficulty). Since $f'(c) \geq 0$ and $m < 0$, for all x sufficiently near c, $\text{slope}(x, c) > m$. Thus for all large enough n, $\text{slope}(a_n, c) > m$ and $\text{slope}(c, b_n) > m$, which contradicts our observation and the fact that, by construction, $\text{slope}(a_n, b_n) \leq m$. If $a_n = c$ or $b_n = c$, the contradiction is immediate. ∎

As we have observed, the contrapositive of the IFT is an existence statement that if f is not increasing on the interval $[a, b]$, there exists a number c between a and b where $f'(c) < 0$. The preceding proof is constructive, in that once one finds $a_1 < b_1$ with $f(a_1) > f(b_1)$, the bisection procedure effectively computes a number c such that $f'(c) < 0$.

3. IMMEDIATE CONSEQUENCES OF THE IFT. We first consider some consequences and variations of the IFT.

Theorem 1. *The following statements are consequences of the IFT. Assume f is differentiable on $[a, b]$ and $a < b$.*

a) *If $f'(x) \leq 0$ on the interval $[a, b]$, then f is decreasing on the interval $[a, b]$.*
b) *If $f'(x) = 0$ on the interval $[a, b]$, then f is constant on the interval $[a, b]$.*
c) *If $f'(x) > 0$ on the interval $[a, b]$, then f is strictly increasing on the interval $[a, b]$.*
d) *If $f'(x) \leq g'(x)$ on the interval $[a, b]$, then $f(x) - f(a) \leq g(x) - g(a)$ for all x in $[a, b]$.*
e) *If $m \leq f'(x) \leq M$ on the interval $[a, b]$, then $m(x - a) \leq f(x) - f(a) \leq M(x - a)$ for all x in $[a, b]$.*

Proof:

(a) Multiplication by -1 reverses inequalities and interchanges "increasing" and "decreasing".
(b) By the IFT and (a), it follows that f is both (weakly) increasing and (weakly) decreasing on $[a, b]$. That means f is constant.
(c) By the IFT, f is increasing. Suppose that $a \leq c < d \leq b$ and $f(c) = f(d)$. Since f is increasing on $[c, d]$ we must have $f(x) = f(c) = f(d)$ on $[c, d]$. That is, f is constant on $[c, d]$. Therefore $f'(x) = 0$ on $[c, d]$, contradicting $f'(x) > 0$ on $[a, b]$.
(d) Apply the IFT to $h(x) = g(x) - f(x)$ to conclude $g(a) - f(a) \leq g(x) - f(x)$.
(e) Apply (d) to $f(x)$ and Mx to get the right inequality and to mx and $f(x)$ to get the left inequality.

Theorem 1c could be called the Strictly Increasing Function Theorem (SIFT). Lax-Burstein-Lax [9] calls it the Criterion for Montonicity. There the IFT is derived directly from the SIFT by looking at $f(x) + mx = g(x)$, for all positive slopes m. If $f'(x) \geq 0$, then $g'(x) > 0$, so by the SIFT g is strictly increasing. Thus if $x > a$, then $f(a) + ma < f(x) + mx$. Since this inequality holds for all $m > 0$, it follows that $f(a) \leq f(x)$, that is, f is increasing. I feel, however, that this proof is a little tricky. Although the idea of perturbing a function is important throughout analysis, it comes out of the blue for a first-year calculus student. I prefer the IFT over the SIFT as a theoretical cornerstone. First, our proof that the IFT implies the SIFT is easier and more natural than a proof that the SIFT implies the IFT. More importantly, Theorem 1c, which could be called the Constant Function Theorem, follows immediately from the IFT; the only way the SIFT can get this fundamental result is via the IFT. By the way, I view the Constant Function Theorem as even more basic than the IFT. It would be nice to use it as our theoretical cornerstone, but I know of no way to use it to get the IFT.

Theorem 1d is called the Racetrack Principle by Jerry Uhl: if one car goes faster than another, it travels farther during any time interval. It is used as a theoretical cornerstone in the text [5].

Theorem 1e is perhaps the most important, especially from a historical viewpoint. If the inequalities are rewritten:

$$m \leq \frac{f(x) - f(a)}{x - a} \leq M$$

we have the Mean Value Inequality. The Mean Value Theorem follows immediately if we know that f' is continuous and that the Intermediate Value Theorem holds. That is exactly what Cauchy did [7]: he proved the Mean Value Inequality and assumed the continuity of f' and the Intermediate Value Theorem. His assumption of continuity should not be surprising since his proof of the Mean Value Inequality also assumes that the difference quotient $(f(x + h) - f(x))/h$ approaches $f'(x)$ uniformly as h approaches 0. Peter Lax has argued that, for the theoretical foundations of an introductory calculus course, one should always avoid pathology and assume uniform continuity and uniform convergence, just as Cauchy did. It is interesting to note that before Cauchy, Ampère [7] saw the importance of the Mean Value Inequality and even used it as the defining property of the derivative. One could argue in a similar vein that the Mean Value Theorem should be the defining property of the derivative; Andrew Gleason has told me that a calculus textbook by Donald Richmond around 1960 did exactly that, but I have been unable to find the book.

Finally, I should comment on the hypothesis of differentiability at the endpoints, both in the IFT and in Theorem 1. All one need assume is continuity at the endpoints, just as in the MVT. Simply observe in the proof of the IFT that the initial points a_1 and b_1 can be chosen so that $a < a_1 < b_1 < b$, since if $f(a) > f(b)$ then by continuity $f(a_1) > f(b)$ for $a_1 > a$ near enough a, and $f(a_1) > f(b_1)$ for $b_1 < b$ near enough b.

4. ERROR BOUNDS AND ERROR BEHAVIOR FOR TAYLOR POLYNOMIALS.
If Theorem 1e is rewritten

$$f(a) + m(x - a) \leq f(x) \leq f(a) + M(x - a),$$

we see a glimmering of an error bound for Taylor polynomials. The proof we are about to give is almost too transparent and simple to believe: just antidifferentiate repeatedly the inequality $f^{(n+1)}(x) \leq M$. Not only does the proof give the Lagrange form of the error bound, it also creates the Taylor polynomial itself. Moreover, as we have observed, it is Lagrange's original proof and can be found in LBL [9]. It is also the proof I wrote for the textbook of the Calculus Consortium Based at Harvard [8]. On the other hand, I have so far been unable to find it anywhere else. All the other proofs I know involve applications of Rolle's Theorem to rather elaborate auxiliary functions or repeated integration by parts or clever tricks with varying parameters. None are natural and none are likely to be discovered or appreciated by an average calculus student.

Theorem 2. (Taylor Error Bound). *Suppose that* $m \leq f^{(n+1)}(x) \leq M$ *on the interval* $[a, b]$, *where* $f^{(i)}$ *denotes the ith derivative of f. Then on* $[a, b]$

$$m\frac{(x - a)^{n+1}}{(n + 1)!} \leq f(x) - T_n(x) \leq M\frac{(x - a)^{n+1}}{(n + 1)!},$$

where $T_n(x)$ *is the degree n Taylor polynomial for f centered at* $x = a$.

Proof: To get the upper bound, we apply Theorem 1d (the Racetrack Principle) to $f^{(n)}(x)$ and Mx (since $f^{n+1} \leq M$), which gives

$$f^{(n)}(x) - f^{(n)}(a) \leq M(x - a).$$

Applying the Racetrack Principle again, we get

$$f^{(n-1)}(x) - f^{(n-1)}(a) - f^{(n)}(a)(x - a) \leq M\frac{(x - a)^2}{2},$$

and again

$$f^{(n-2)}(x) - f^{(n-2)}(a) - f^{(n-1)}(a)(x - a) - f^{(n)}(a)\frac{(x - a)^2}{2} \leq M\frac{(x - a)^3}{3!}.$$

Applying the Racetrack Principle a total of $n + 1$ times gives the upper bound. The lower bound is obtained the same way. ∎

Theorem 2 gives error bounds only for $x \geq a$. To get similar bounds for $x \leq a$, we observe that if f is increasing and $x \leq a$, then $f(x) \leq f(a)$, rather than $f(a) \leq f(x)$. Thus for $x \leq a$, each application of Theorem 1d reverses the inequalities, but since Theorem 2 sandwiches the error for $x \geq a$, reversing inequalities will simply sandwich the error again for $x \leq a$ (although which bound is the upper one depends on whether n is odd or even). The usual two-sided error bound involving absolute values then follows immediately.

It is possible for students to discover Theorem 2 for themselves. Consider the following problem. A particle is traveling along the x-axis with position $x = f(t)$ and suppose the initial position, velocity, and acceleration are all 0. If $f'''(t) \leq 5$ for $t \geq 0$, find an upper bound on the position at time $t = 2$. Since students are well-trained to antidifferentiate acceleration to get velocity and velocity to get position, it is not unnatural to see them argue as follows:

$$f'''(t) \leq 5$$
$$a = f''(t) \leq 5t + c_1, \text{ and here } c_1 = 0 \text{ since } f''(0) = 0$$

$$v = f'(t) \leq 5\frac{t^2}{2} + c_2, \quad \text{and here } c_2 = 0 \text{ since } f'(0) = 0$$

$$s = f(t) \leq 5\frac{t^3}{6} + c_3, \quad \text{and here } c_3 = 0 \text{ since } f(0) = 0.$$

Thus, we get $f(2) \leq 5 \cdot 2^3/6 = 20/3$. This is a legitimate argument as long as one can justify antidifferentiating inequalities in the same way as equalities. That is exactly the point of the Racetrack Principle!

Acceleration and velocity are not a bad way of introducing Taylor series. The usual formula students memorize from physics,

$$s = s_0 + v_0 t + \frac{1}{2}at^2,$$

is precisely the degree 2 Taylor polynomial for $s(t)$ when the constant acceleration a is interpreted as the acceleration at time 0. This fact seems worth exploiting, but I don't know any textbook that makes the connection.

Taylor's theorem is usually presented as a method of bounding the error in approximating a function by its degree n Taylor polynomial. This viewpoint is particularly appropriate in studying the error for fixed x as $n \to \infty$, as in the proof of the convergence for all values of x for the Taylor series for e^x or $\sin x$. Nevertheless, I believe that this viewpoint is overemphasized and that the true power of Taylor series is in explaining error (or convergence) behavior for fixed n as $x \to a$. Why is Simpson's Rule so much better than the Trapezoid Rule? What makes the approximation $\sin x \approx x$ so good? For numerical behavior, the important thing to know is the order of convergence for fixed n under normal circumstances and what situations might affect that order of convergence. The real point of Taylor's theorem is that the error is order $n + 1$ in $(x - a)$ with a constant depending on the $(n + 1)^{\text{st}}$ derivative.

To be more precise, we say $E(h)$ is asymptotic to Ch^n, denoted $E(h) \sim Ch^n$, if $\lim_{h \to 0} E(h)/h^n = C$. Also, we say $E(h)$ is order n with bound M if $\limsup |E(h)/h^n| \leq M$. Then Taylor's theorem can be viewed this way:

Corollary. *Let $E(h)$ be the error $f(x) - T_n(x)$ where $T_n(x)$ is the nth degree Taylor polynomial for f at $x = a$ and where $h = x - a$. If $f^{(n+1)}$ is continuous at $x = a$, then $E(h) \sim f^{(n+1)}(a) h^{n+1}/(n + 1)!$. If $|f^{(n+1)}(x)| \leq M$ in a neighborhood of $x = a$, then $E(h)$ is order $n + 1$ with bound $M/(n + 1)!$.*

5. ERROR BEHAVIOR FOR NUMERICAL INTEGRATION. Another application of the Mean Value Theorem is to explain the error behavior for various common numerical integration rules: Left Rule, Right Rule, Trapezoid Rule, Midpoint Rule, Simpson's Rule. This behavior is best described using Taylor series in Δx for the error. Numerical analysis texts sometimes do this, but calculus texts don't. Since this approach is not so well-known, I'll give a version.

The idea is to concentrate on one panel of the subdivided area. Without loss of generality, we can assume the panel is centered at the origin. Thus we wish to compute

$$I(h) = \int_{-h}^{h} f(x)\, dx, \quad \text{where } h = \Delta x/2.$$

The estimate for this single panel by the left-rectangle rule is
$$I(h) \approx L(h) = 2h(f(-h)).$$

The other estimates are given by

Left: $L(h) = 2hf(-h)$

Right: $R(h) = 2hf(h)$

Midpoint: $M(h) = 2hf(0)$

Trapezoid: $T(h) = (L(h) + R(h))/2$

Simpson: $S(h) = (2M(h) + T(h))/3$

The formula relating Simpson's Rule to the midpoint and trapezoidal rules is not as well known as it should be. Students can be led to guess the weighted mean as a better estimate, if they spend a little time looking at the error behavior of the midpoint and trapezoidal rules.

We want to compute the Taylor series centered at $a = 0$ for all these functions. For the rules, this is simply a matter of replacing $f(h)$ or $f(-h)$ by the Taylor series for f centered at $a = 0$. For $I(h)$, we observe that by the Fundamental Theorem of Calculus, $I'(h) = f(h) + f(-h)$. Thus $I''(h) = f'(h) - f'(-h)$, $I'''(h) = f''(h) + f''(-h)$, etc.

The Taylor series for $I(h)$ is therefore
$$I(h) = 2f(0)h + 2f''(0)\frac{h^3}{3!} + 2f''''(0)\frac{h^5}{5!} + \cdots.$$

The series for the rules are
$$L(h) = 2h\left[f(0) + f'(0)(-h) + f''(0)\frac{(-h)^2}{2!} + \cdots\right]$$

$$R(h) = 2h\left[f(0) + f'(0)h + f''(0)\frac{h^2}{2} + \cdots\right]$$

$$M(h) = 2h[f(0)]$$

$$T(h) = 2h\left[f(0) + f''(0)\frac{h^2}{2} + f''''(0)\frac{h^4}{4!} + \cdots\right]$$

$$S(h) = 2h\left[f(0) + f''(0)\frac{h^2}{6} + f''''(0)\frac{h^4}{3\cdot 4!} + \cdots\right].$$

The error behavior for each rule is obtained by subtracting the Taylor series for $I(h)$ from the Taylor series for the rule and looking for the first term that doesn't cancel. The errors behave asymptotically as follows:

Left Error $\sim -2f'(0)h^2$

Right Error $\sim 2f'(0)h^2$

$$\text{Midpoint Error} \sim -2f''(0)\frac{h^3}{3!}$$

$$\text{Trapezoid Error} \sim 2f''(0)\left(\frac{1}{2} - \frac{1}{6}\right)h^3 = 2f''(0)\frac{h^3}{3}$$

$$\text{Simpson Error} \sim 2f''''(0)\left(\frac{1}{3\cdot 4!} - \frac{1}{5!}\right)h^5 = 2f''''(0)\frac{h^5}{180}.$$

The error behavior of these rules for the entire interval is obtained by multiplying by the number n of subdivisions and replacing h by $\Delta x/2$ where $\Delta x = (b-a)/n$, except for Simpson's rule where $h = \Delta x$. We have to replace $f^{(k)}(0)$ by a bound M_k on $|f^{(k)}|$ for the entire interval. Using $2nh = (b-a)$, we find the absolute value of the errors have the following behavior in terms of Δx:

Left: order 1 with bound $(b-a)(1/2)M_1$

Right: order 1 with bound $(b-a)(1/2)M_1$

Midpoint: order 2 with bound $(b-a)(1/24)M_2$

Trapezoid: order 2 with bound $(b-a)(1/12)M_2$

Simpson: order 4 with bound $(b-a)(1/180)M_4$.

The typical textbook problem on numerical integration is to find the value of n that guarantees the error is within a specified tolerance. In practice, one simply keeps doubling n until the desired number of digits seems to have stabilized. Thus, error behavior, rather than error bounds, may be what we really are interested in. For example, it is useful to know that increasing n by a factor of 10 for the Left or Right rule, decreases the error by a factor of $1/10$, that is, it gives one more significant digit. Thus if it takes 1 second for a graphing calculator to compute an integral accurate to 2 digits using the Left or Right Rule, it will take 10^{10} seconds to get 12 digits of accuracy (that's 3169 years and, as my students have observed, a lot of batteries). By contrast, Simpson's Rule gets 4 extra digits for 10 times the work, and the same integral can be computed to 12 digits of accuracy in a minute or two on the same calculator (Simpson's Rule probably would get a headstart of 4 or 5 digits in the first second).

The dependence of error behavior on the higher derivatives of the integrand is also important, because it is a warning to look out for integrals whose integrand has an unbounded derivative on the interval of integration. For example, even using Simpson's Rule on $\int_0^1 \sqrt{1-x^2}\,dx$ to get an approximation for $\pi/4$ is painfully slow going. Indeed, the order of convergence is $3/2$ rather than 4.

Taylor series can be used in the same way to analyze the error behavior for numerical differentiation approximations:

$$f'(x) \approx \frac{f(x+h) - f(x)}{h}$$

$$f'(x) \approx \frac{f(x+h) - f(x-h)}{2h}$$

$$f''(x) \approx \frac{f(x+h) + f(x-h) - 2f(x)}{h^2}.$$

For example, students are often curious why some graphing calculators use the second of these approximations as a numerical derivative rather than the more familiar first approximation. Taylor series give the answer immediately: the second error for the approximation is order 2 while the first is order 1. The dependence of the error of each approximation on higher derivatives of f also has interesting effects. Try plotting the error near $x = 0$ with $h = .01$ for the second approximation to f', when f is the innocuous-looking function $f(x) = x^{8/3}$.

6. THE FUNDAMENTAL THEOREMS OF CALCULUS. The proof given in [8] for the Taylor error bound appeals to the Fundamental Theorem of Calculus to turn the inequality $f^{n+1}(x) \leq M$ into the inequality $f^{(n)}(x) - f^{(n)}(a) \leq M(x - a)$. I suspect this is the natural inclination of most mathematicians, and it shows how much under-appreciated the IFT is. No definite integrals are needed; the IFT itself is a disguised form of integration. The subtle connection between the IFT and the Fundamental Theorem of Calculus is worth discussing.

There are of course two main versions of the Fundamental Theorem of Calculus. There are also variations on what restrictions are placed on the integrand f. I will assume f is continuous. The theorems then are

First Fundamental Theorem of Calculus (FTC I). *If f is continuous on the interval $[a, b]$ and $F(x) = \int_a^x f(t)\,dt$ for x in $[a, b]$, then $F'(x) = f(x)$.*

Second Fundamental Theorem of Calculus (FTC II). *If f is continuous and $F(x) = f(x)$, then $\int_a^b f(t)\,dt = F(b) - F(a)$.*

The First Fundamental Theorem is not directly related to the IFT. The hard part of the proof is showing that continuous functions are Riemann integrable. The rest is a straightforward consequence of the integral version of the Mean Value Inequality:

$$m(b - a) \leq \int_a^b f(x)\,dx \leq M(b - a),$$

where $m \leq f(x) \leq M$ on the interval $[a, b]$. Note that unlike the Mean Value Inequality for derivatives, this inequality follows easily from the definition of the Riemann integral, so easily that it is not uncommon to view the inequality as a defining property of the definite integral (the corresponding view for the Mean Value Inequality for derivatives, Ampère notwithstanding, is much less common).

On the other hand, the Second Fundamental Theorem is closely connected to the IFT. The IFT for continuously differentiable functions follows directly from the FTC II and the fact that the integral of a nonnegative function is nonnegative. In fact, that is the way the IFT is proved in [8]. There, the FTC II, as embodied in the relation between velocity and change in position, is taken as the intuitively clear, theoretical cornerstone, and the IFT is derived from it. I suspect, however, that most students see the IFT as more "obvious" than the FTC II.

Conversely, the IFT implies the FTC II by the method used in many calculus books: simply invoke the FTC I with $x = b$ and observe that, by the IFT (the Constant Function Theorem, Theorem 1b), two antiderivatives of f differ by a constant.

The assumption of continuity in the FTC I is necessary. The assumption of continuity in the FTC II is another matter. Of course, if $F' = f$ is not continuous, the integral might not exist. For example, if $F(0) = 0$ and $F(x) = x^2 \sin(1/x^2)$ for $x \neq 0$, then F' exists everywhere but is not even Lebesgue integrable on $[0, 1]$. Suppose, however, that we assume only that $\int_a^b f(t)\, dt$ exists. Then the familiar argument using the Mean Value Theorem still works. Just represent $F(b) - F(a)$ as a telescoping sum and use the MVT on each term of the sum to turn it into a Riemann sum for $\int_a^b f(t)\, dt$. Here the IFT does not work. Just as the MVT follows from the IFT only under the assumption of continuity of the derivative, the FTC II follows from the IFT only under the assumption of continuity of the integrand.

7. CONCLUSION. Many calculus textbooks have sections where the author is writing on automatic pilot, just putting in material demanded by users. These sections have the same dreary examples; little is new, or thought over fresh from the start. This shouldn't be surprising, since writing a calculus textbook is a significant project and one can't devote the same enthusiasm and energy to all parts of the project. I have always felt that the theoretical sections of standard calculus textbooks are most prone to such a pedestrian treatment. Moreover, calculus instruction does not place much emphasis on those theoretical sections, at least when it comes to testing. For example, a study of the compendium of final exams in [11] reveals only one question (out of more than 300 on 23 exams) involving the Mean Value Theorem, and that one asked for the value of c satisfying the conclusion of the Mean Value Theorem for a quadratic function. When both textbooks and instruction appear to be just going through the motions with theory, it surprises me that some critics of new textbooks like [8] bemoan the absence of the Mean Value Theorem or a $\epsilon - \delta$ definition of limit.

I sympathize with yearnings for an occasional foray into the theoretical structure of the calculus. I just ask that it be thoughtful and sensible. Use intuitive definitions. If a theorem is to be used without proof, like the Mean Value Theorem, keep it as simple and as "obvious" as possible. Don't use tricky proofs or deus-ex-machina auxiliary functions. Don't prove things in more generality than necessary; even analysts don't usually deal with the discontinuous derivatives allowed by the Mean Value Theorem.

In this paper, I have tried to give a sensible approach to the Mean Value Theorem and its usual applications to monotonicity, Taylor error bounds, quadrature error bounds, and the Fundamental Theorems of Calculus. One standard application of the MVT I have not considered is l'Hopital's Rule; for a non-MVT approach, see [3]. LBL [9] has some other applications to concavity and the second derivative test for extrema.

In recent years, calculus content and pedagogy have been rethought completely. People have found that there is nothing sacred about related rates and the lecture method. It is time as well to rethink the theory taught in standard calculus classes. There is nothing sacred about the Mean Value Theorem.

ACKNOWLEDGMENTS. I wish to thank Andy Gleason, Peter Lax, and Jerry Uhl for numerous suggestions and corrections for this article. In particular, the proof given for the IFT was instigated by a bisection proof Lax showed me for the SIFT. He also showed me applications to l'Hopital's Rule, the Corrected Midpoint Rule for quadrature, and the definition of volumes and arclengths using antiderivatives rather than definite integrals; all of this I hope he puts into print. I am indebted to Gleason, whose meticulous reading caught a number of egregious errors and whose comments cleared up my muddy thinking at numerous points.

REFERENCES

1. L. Bers, On avoiding the mean value theorem, *Amer. Math. Monthly* 74 (1967), 583.
2. D. Bo, A simple derivation of the Maclaurin series for sine and cosine, *Amer. Math. Monthly* 97 (1990), 836. Editor's Note in the *Monthly* 98 (1991), 364.
3. R. P. Boas, Lhospital's rule without mean value theorems, *Amer. Math. Monthly* 76 (1969), 1051–1053.
4. R. P. Boas, Who needs these mean-value theorems anyway?, *Two-Year College Math J.* 12 (1981), 178–181.
5. W. Davis, H. Porta, J. Uhl, *Calculus & Mathematica: Derivatives: Measuring Growth*, Addison-Wesley, 1994.
6. J. Dieudonne, *Foundations of Modern Analysis*, Academic Press, New York, 1960.
7. J. V. Grabiner, *The Origins of Cauchy's Rigorous Calculus*, MIT Press, Cambridge, 1981.
8. D. Hughes-Hallett, A. M. Gleason, et al., *Calculus*, John Wiley & Sons, New York, 1994.
9. P. Lax, S. Burstein, and A. Lax, *Calculus with Applications and Computing, Volume 1*, Springer-Verlag, New York, 1984.
10. D. E. Richmond, An elementary proof of a theorem of calculus, *Amer. Math. Monthly* 92 (1985), 589–590.
11. L. A. Steen, editor, *Calculus for a New Century: A Pump, Not a Filter*, MAA Notes 8, Mathematical Association of America, Washington, DC, 1988.

Rolle over Lagrange—Another Shot at the Mean Value Theorem

Robert S. Smith

In a typical elementary calculus course, the status of Rolle's Theorem is somewhere between a smear of ink in the text and a "named" theorem that tradition has ordained. If this theorem of Rolle is mentioned at all, then its raison d'être is the Mean Value Theorem of Lagrange, or the former is a trivial consequence of the latter. Therefore, after lip service has been paid, this theorem is unceremoniously Rolled away until the next course offering. On the other hand, the Mean Value Theorem is debuted amidst fanfare and professorial homage.

We suggest that in an elementary calculus course, the status of these theorems should be reversed, making Rolle regal and letting Lagrange languish. Let's first recall the statements of Rolle and Lagrange, and the related statement of Cauchy.

Rolle's Theorem. *Let f be continuous on $[a,b]$ and differentiable on (a,b). If $f(a) = f(b)$, then there is a number of $c \in (a,b)$ such that $f'(c) = 0$.*

Lagrange's Mean Value Theorem. *Let f be continuous on $[a,b]$ and differentiable on (a,b). Then there is a number $c \in (a,b)$ such that $[f(b) - f(a)]/(b-a) = f'(c)$.*

Cauchy's Mean Value Theorem. *Let f and g be continuous on $[a,b]$ and differentiable on (a,b). If $g'(x) \neq 0$ for all $x \in (a,b)$, then there is a number $c \in (a,b)$ such that $[f(b) - f(a)]/[g(b) - g(a)] = f'(c)/g'(c)$.*

After Lagrange's Theorem has been unveiled, what type of supporting applications are offered? From the students' perspective, applications are usually contrived exercises that begin as follows:

Show that $f(x)$ satisfies the hypothesis of the Mean Value Theorem over
   $x \in [a,b]$, and find all numbers $c \in (a,b)$ such that $f(b) - f(a) = f'(c)(b-a)$.
On the whole, students (and many instructors) find such exercises less than inspiring. In a more enlightened environment, students may be exposed to one or two applications of the following type: Use the Mean Value Theorem to prove that $\sin t \leq t$ for $0 \leq t$.

From the instructor's perspective, applications of Lagrange's Mean Value Theorem consist of demonstrating or deriving the following:

(i) Let f be continuous on $[a, b]$ and differentiable on (a, b). If $f'(x) > 0$ (< 0) on (a, b), the f is increasing (decreasing) on $[a, b]$.
(ii) If $f'(x) = 0$ for all $x \in [a, b]$, then $f(x)$ is constant on $[a, b]$.
(iii) Let f be differentiable on an open interval (a, b) containing c. If $f''(c) > 0$ (< 0), then the graph of f is concave upward (downward) at $(c, f(c))$.
(iv) L'Hôpital's Rule.
(v) The formula for the arc length of a smooth function over an interval $[a, b]$.

While Lagrange's theorem of the mean has proven quite sufficient in demonstrating or deriving these items, it certainly is not necessary. A number of authors have coped with the above sans Lagrange. (See Bers [4] for (i); Cohen [7], Halperin [8], Powderly [9], and Richmond [10] for (ii); Boas [5] for (iv); and, Apostol [2] for (v).)

Let's consider (iii). Since $f''(c) > 0$, there must be a deleted neighborhood I of c contained in the domain of f over which $(f'(x) - f'(c))/(x - c) > 0$. Recall that a function f which is differentiable at c is concave upward at $(c, f(c))$ if there is a deleted neighborhood N of c throughout which $f(x)$ is above the tangent line to f through $(c, f(c))$. So, let $t(x)$ be the tangent line to f at $(c, f(c))$ and set $g(x) = f(x) - t(x)$. It suffices to show that $g(x) > 0$ on I.

Clearly, $g(c) = 0$ and $g'(c) = 0$. Since $g'(x) = f'(x) - t'(x) = f'(x) - f'(c)$, and $(f'(x) - f'(c))/(x - c) > 0$ for $x \in I$, we see that $g'(x) < 0$ for $x \in I \cap (-\infty, c)$ and $g'(x) > 0$ for $x \in I \cap (c, \infty)$. Hence, $g(c)$ is a local minimum, and $g(x) > 0$ on I.

From the instructor's perspective, applications of Lagrange's Mean Value Theorem consist of demonstrating or deriving the following:

(i) Let f be continuous on $[a, b]$ and differentiable on (a, b). If $f'(x) > 0$ (< 0) on (a, b), the f is increasing (decreasing) on $[a, b]$.
(ii) If $f'(x) = 0$ for all $x \in [a, b]$, then $f(x)$ is constant on $[a, b]$.
(iii) Let f be differentiable on an open interval (a, b) containing c. If $f''(c) > 0$ (< 0), then the graph of f is concave upward (downward) at $(c, f(c))$.
(iv) L'Hôpital's Rule.
(v) The formula for the arc length of a smooth function over an interval $[a, b]$.

While Lagrange's theorem of the mean has proven quite sufficient in demonstrating or deriving these items, it certainly is not necessary. A number of authors have coped with the above sans Lagrange. (See Bers [4] for (i); Cohen [7], Halperin [8], Powderly [9], and Richmond [10] for (ii); Boas [5] for (iv); and, Apostol [2] for (v).)

Let's consider (iii). Since $f''(c) > 0$, there must be a deleted neighborhood I of c contained in the domain of f over which $(f'(x) - f'(c))/(x - c) > 0$. Recall that a function f which is differentiable at c is concave upward at $(c, f(c))$ if there is a deleted neighborhood N of c throughout which $f(x)$ is above the tangent line to f through $(c, f(c))$. So, let $t(x)$ be the tangent line to f at $(c, f(c))$ and set $g(x) = f(x) - t(x)$. It suffices to show that $g(x) > 0$ on I.

Clearly, $g(c) = 0$ and $g'(c) = 0$. Since $g'(x) = f'(x) - t'(x) = f'(x) - f'(c)$, and $(f'(x) - f'(c))/(x - c) > 0$ for $x \in I$, we see that $g'(x) < 0$ for $x \in I \cap (-\infty, c)$ and $g'(x) > 0$ for $x \in I \cap (c, \infty)$. Hence, $g(c)$ is a local minimum, and $g(x) > 0$ on I.

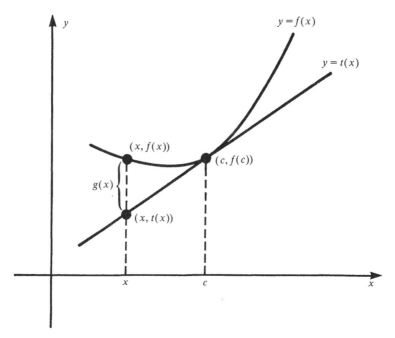

Apparently, instructors can accomplish much of what is needed in the calculus without the Mean Value Theorem. Boas [5] has even demonstrated that Cauchy's Mean Value Theorem is superfluous. Certainly, Rolle's Theorem is equivalent to, less complicated than, and more intuitive than, either mean value theorem. In view of this evidence, perhaps an alternate approach in the calculus should be considered —depose the theorem of Lagrange (and that of Cauchy, too) and enthrone the theorem of Rolle. Reinforce this new direction by proving Rolle's Theorem for your students. A variety of proofs are readily available in texts and the literature (see Abian [1] or Samelson [11]).

If abandoning Lagrange and Cauchy is too alien a notion to ponder, then the proof of Cauchy's Theorem should at least be presented as an application of Rolle's Theorem:

Let $h(x) = [f(b) - f(a)]g(x) - [g(b) - g(a)]f(x)$. Then $h(a) = h(b)$, and Rolle's Theorem applies.

Lagrange's Theorem is then an immediate corollary of Cauchy's, with $g(x) = x$.

Once those mean habits have been satisfied, instructors can proceed directly to solving equations. Students have been solving equations since the seventh grade, and they have had to do so in a variety of non-mathematics courses. Thus, instructors do not have to convince students that this type of application is important.

The discussion could begin with a word or two about the difficulty of solving polynomial equations of degree n for $n \geq 5$, and a few historical notes on the contributions of Galois and Abel. Then prove,

Theorem. *An nth degree polynomial $P(x) = a_n x^n + \cdots + a_1 x + a_0$ has at most n real zeros.*

This algebraic result is one that most students should have heard in high school but few have seen proven. Rolle's Theorem can be used to give an easy proof: If x_1 and x_2 are zeros of P, then there is a number c between x_1 and x_2 such that $P'(c) = 0$. Hence, if P has m distinct zeros, then P' has at least $m - 1$ distinct zeros. To proceed with a formal proof, use induction on n and arrive at a contradiction.

There are several instructive components to this exercise. Rolle's Theorem is put to work in an interesting application that students perceive as meaningful. An opportunity is afforded instructors to introduce (or re-introduce) the principle of mathematical induction (many instructors have already used induction to prove that

$$\frac{d}{dx}(x^n) = nx^{(n-1)}$$

for natural numbers n), and students' calculus experience will be enriched by an exposure to a proof by contradiction. In this same spirit, instructors can assign other interesting problems in which Rolle's Theorem can be used. The examples below were taken from popular calculus texts.

(a) Show that $6x^4 - 7x + 1 = 0$ has exactly two real solutions.
(b) Show that $6x^5 + 13x + 1 = 0$ has exactly one real solution.
(c) Show that $x^n + ax + b = 0$ cannot have more than two real solutions if n is even, or more than three real solutions if n is odd.
(d) Show that $x^3 + 9x^2 + 33x - 8 = 0$ has exactly one real solution.
(e) Show that $x^n + ax^2 + b = 0$ has at most three distinct real roots if n is odd, and at most four if n is even.
(f) Show that $x^2 = x \sin x + \cos x$ has exactly two solutions.

Conclusion. The above approach to Rolle's Theorem can make it a more significant part of the calculus and engender a greater appreciation of its value. However, a cautionary note is in order. Instructors who adopt this approach have been known to develop a single-mindedness that is all but consuming. While on a Rolle during a lecture, this writer once invoked the theorem in completing an extrema word problem!

Acknowledgement. I wish to thank my friend and colleague, Frederick S. Gass, for his valuable suggestions in the preparation of this article.

REFERENCES

1. Alexander Abian, "An Ultimate Proof of Rolle's Theorem," American Mathematical Monthly 86 (1979) 484–485.
2. Tom M. Apostol, *Calculus*, Volume 1, 2nd ed., Blaisdell Publishing Co., 1967.
3. Abdul K. Aziz and Joachin B. Diaz, "On Pompeiu's Proof of the Mean Value Theorem," *Contributions to Differential Equations* 1 (1963) 467–481.
4. Lipman Bers, "On Avoiding the Mean Value Theorem," American Mathematical Monthly 74 (1967) 583.
5. Ralph P. Boas, "L'Hôpital's Rule Without Mean-Value Theorems," American Mathematical Monthly 76 (1969) 1051–1053.
6. _____, "Who Needs Those Mean-Value Theorems, Anyway?," Two-Year College Mathematics Journal 12 (1981) 178–181.
7. Leon W. Cohen, "On Being Mean to the Mean Value Theorem," American Mathematical Monthly 74 (1967) 581–582.
8. Israel Halperin, "A Fundamental Theorem of the Calculus," American Mathematical Monthly 61 (1954) 122–123.
9. Mary Powderly, "A Simple Proof of a Basic Theorem of the Calculus," American Mathematical Monthly 70 (1963) 544.
10. Donald E. Richmond, "An Elementary Proof of a Theorem in Calculus," American Mathematical Monthly 92 (1985) 589–590.
11. Hans Samelson, "On Rolle's Theorem," American Mathematical Monthly 86 (1979) 486.

An Elementary Proof of a Theorem in Calculus

Donald E. Richmond

A fundamental theorem states that if $f'(x) = 0$ at every point of an interval $[a, b]$, then $f(x)$ is constant on $[a, b]$. Most proofs use the Mean Value Theorem, but several proofs have been given that are independent of this theorem (Bers [1], Halperin [3], Powderly [4]). The present note proves our fundamental theorem in an elementary way.

Let $f'(x) \equiv 0$ on $[a, b]$. If $f(x)$ is not constant on $[a, b]$ as the theorem states, then for some $u < v$ on $[a, b]$, $f(v) \neq f(u)$. Hence the chord joining $(u, f(u))$ and $(v, f(v))$ has a slope different from zero. That is

(1) $$f(v) - f(u) = C(v - u), \quad \text{or} \quad \Delta f = C \Delta x, \quad \text{with } C \neq 0.$$

Assume that $C > 0$. Bisect $[u, v]$ at w. If

$$f(v) - f(w) < C(v - w)$$

and

$$f(w) - f(u) < C(w - u),$$

then

$$f(v) - f(u) < C(v - u),$$

which contradicts (1) with $C > 0$. Hence on at least one of the intervals $[u, w]$ and $[w, v]$, $\Delta f / \Delta x \geq C$. Call this interval $[u_1, v_1]$. By repeated bisection one obtains a nested sequence of intervals $[u_n, v_n]$ over each of which $\Delta f / \Delta x \geq C$. $[u_n, v_n]$ converges to some x in $[u, v]$. Hence if $C > 0$, $\Delta f / \Delta x$ cannot approach $f'(x) = 0$ as required.

The possibility $C < 0$ is excluded by reversing the inequalities. Hence $C = 0$, contrary to the assumption that $f(v) \neq f(u)$.

The same argument shows that if $f'(x) \geq 0$ on $[a, b]$, then $f(v) \geq f(u)$, and if $f'(x) \leq 0$ on $[a, b]$, then $f(v) \leq f(u)$.

The argument may be generalized to show that if $m \leq f'(x) \leq M$ on $[a, b]$, then

(2) $$m(v - u) \leq f(v) - f(u) \leq M(v - u).$$

It is sufficient to replace $C > 0$ by $C > M$ and $C < 0$ by $C < m$. (2) has been called the "weak" form of the Mean Value Theorem [2].

References

1. L. Bers, On avoiding the mean value theorem, this MONTHLY, 74 (1967) 583.
2. R. P. Boas, Who needs those mean-value theorems, anyway?, Two-Year College Mathematics Journal, 12 (1981) 178–181.
3. I. Halperin, A fundamental theorem of the calculus, this MONTHLY, 61 (1954) 122–123.
4. M. Powderly, A simple proof of a basic theorem of the calculus, this MONTHLY, 70 (1963) 544.

A Simple Auxiliary Function for the Mean Value Theorem

Herb Silverman

Most calculus texts contain the following statement:

Rolle's Theorem. *If f is continuous on $[a, b]$ and differentiable on (a, b) with*
$$f(a) = f(b) = 0, \tag{1}$$
then $f'(c) = 0$ for some $c \in (a, b)$.

Some texts replace (1) with
$$f(a) = f(b). \tag{1'}$$

This is certainly an improvement because an additional hypothesis is justifiable only if it delivers a stronger conclusion or an easier proof. I know of no proof of Rolle's theorem that makes use of the vanishing of f at the end points.

The main (only?) purpose of Rolle's theorem is to serve as a lemma for the more general mean value theorem. Whether (1) or (1') is given in Rolle's theorem, the mean value theorem is usually proved by first introducing the auxiliary function
$$g(x) = f(x) - f(a) - \frac{f(b) - f(a)}{b - a}(x - a) \tag{2}$$
and then demonstrating that it satisfies the hypotheses and hence the conclusion of Rolle's theorem. Despite our best efforts at making the function g in (2) appear geometrically intuitive and therefore "natural," most students seem to think of g as artificial and the proof as magical. It is our intention to show that the version of Rolle's theorem with (1) replaced by (1') can lead to a more straightforward auxiliary function.

To establish the existence of a tangent line parallel to the secant line joining $(a, f(a))$ to $(b, f(b))$, it certainly makes sense to construct an auxiliary function that differs from f by a linear function whose slope is $(f(b) - f(a))/(b - a)$. But why not choose the simplest such linear function, the one passing through the origin? This is accomplished by replacing (2) with
$$g(x) = f(x) - \frac{f(b) - f(a)}{b - a} x. \tag{2'}$$

The simplicity of the auxiliary function in (2') more than offsets the computation needed to verify that it satisfies hypothesis (1').

If instead we wish our auxiliary function g (satisfying (1')) to agree with f at $x = a$ and still be a linear perturbation of f, then choose
$$g(x) = f(x) - \frac{f(b) - f(a)}{b - a}(x - a).$$

An unnecessary translation downward by $f(a)$ brings us to $g(x) - f(a)$, the traditional auxiliary function given in (2).

We leave it for the reader to make a similar simplification in the proof of the generalized mean value theorem.

A Note on the Derivative of a Composite Function

V. N. Murty

The domain of the composite function $(f \circ g)(x) = f\{g(x)\}$ is the set of all values x in the domain of g such that $g(x)$ is in the domain of f. The Chain Rule states that if g is differentiable at x_0 and f is differentiable at $g(x_0)$, then $f \circ g$ is differentiable at x_0 and

$$(f \circ g)'(x_0) = f'\{g(x_0)\} \cdot g'(x_0).$$

Most students learning calculus get the impression that $(f \circ g)'(x_0)$ does not exist unless $f'\{g(x_0)\}$ and $g'(x_0)$ exist. A number of calculus texts do not explain this point. The following simple examples should get the student back on the right track.

Example 1. Consider $f(x) = x^2$ and $g(x) = |x|$. Here $f'\{g(0)\} = f'(0) = 0$, and $g'(0)$ does not exist. Nevertheless, $(f \circ g)(x) = x^2$ and $(f \circ g)'(0) = 0$. Similarly $g'\{f(0)\} = g'(0)$ does not exist, yet $(g \circ f)(x) = x^2$ and $(g \circ f)'(0) = 0$.

Example 2. Suppose $f(x) = 3x + |x|$ and $g(x) = \frac{3}{4}x - \frac{1}{4}|x|$. Although neither $f'\{g(0)\} = f'(0)$ nor $g'\{f(0)\} = g'(0)$ exist, we have $(f \circ g)(x) = 2x$ and $(f \circ g)'(0) = 2$.

REMARK: Note that $(g \circ f)(x) = 2|x|$ has no derivative at $x = 0$.

Do Dogs Know Calculus?

Timothy J. Pennings

Most calculus students are familiar with the calculus problem of finding the optimal path from A to B. "Optimal" may mean, for example, minimizing the time of travel, and typically the available paths must transverse two different mediums, involving different rates of speed.

This problem comes to mind whenever I take my Welsh Corgi, Elvis, for an outing to Lake Michigan to play fetch with his favorite tennis ball. Standing on the water's edge (See Figure 1) at A, I throw the ball into the water to B. By the look in Elvis's eyes and his elevated excitement level, it seems clear that his objective is to retrieve it as quickly as possible rather than, say, to minimize his expenditure of energy. Thus I assume that he unconsciously attempts to find a path that minimizes the retrieval time.

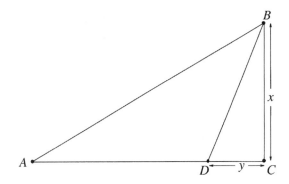

Figure 1. Paths to the ball

This being his goal, what should be his strategy? One option would be to try to minimize the time by minimizing the distance traveled. Thus he could immediately jump into the surf and swim the entire distance. On the other hand, since he runs considerably faster than he swims, another option would be to minimize the swimming distance. Thus, he could sprint down the beach to the point on shore closest to the ball, C, and then turn a right angle and swim to it. Finally, there is the option of running a portion of the way, and then plunging into the lake at D and swimming diagonally to the ball.

Depending on the relative running and swimming speeds, this last option usually turns out to minimize the time. Although this type of problem is in every calculus text, I have never seen it solved in the general form.[1] Let's do it quickly—the answer is revealing.

Let r denote the running speed, and s be the swimming speed. (Our units will be meters and seconds.) Let $T(y)$ represent the time to get to the ball given that Elvis jumps into the water at D, which is y meters from C. Let z represent the entire distance from A to C. Since time = distance/speed, we have

$$T(y) = \frac{z-y}{r} + \frac{\sqrt{x^2+y^2}}{s}. \tag{1}$$

We want to find the value of y that minimizes $T(y)$. Of course this happens where $T'(y) = 0$. Solving $T'(y) = 0$ for y, we get

$$y = \frac{x}{\sqrt{r/s+1}\,\sqrt{r/s-1}}, \tag{2}$$

where T is seen to have a minimum by using the second derivative test. Several things about the solution should be noticed. First, somewhat surprisingly, the optimal path does not depend on z, as long as z is larger than y. Second, if $r < s$, we get no solution. That makes sense; if $r < s$ then it is obviously optimal to jump into the lake and swim the entire distance. Third, note that for $r \gg s$, y is small, and for $r \approx s$, y is large, as one would reasonably expect. Finally, note that for fixed r and s, y is proportional to x.

Now, back to Elvis. I noticed when playing fetch with Elvis that he uses the third strategy of jumping into the lake at D. It also seemed that his y values were roughly proportional to the x values. Thus, I conjectured that Elvis was indeed choosing the optimal path, and decided to test it by calculating his values of r and s and then checking how closely his ratio of y to x coincided with the exact value provided by the mathematical model.

With a friend to help me, we clocked Elvis as he chased the ball a distance of 20 meters on the beach. We then timed him as he swam (pursuing me) a distance of 10 meters in the water. His times are given in Table 1.

Table 1. Running and swimming times

Running times (in seconds) for 20 meters	Swimming times (in seconds) for 10 meters
3.20	12.13
3.16	11.15
3.15	11.07
3.13	10.75
3.10	12.22

Since we wanted Elvis's greatest running speed, we averaged just the three fastest running times, giving $r = 6.40$ meters/second. Similarly, using the three fastest swim-

---

[1] Several of the standard problems found in calculus texts are much more interesting and illuminating if done in the general case. For example, if you find the longest board that can be taken around a corner from a hallway of width a to a hallway of width b, you will discover a beautiful answer of the form $(a^p + b^p)^{1/p}$.

ming times, $s = 0.910$ meters/second. Then from (2), we get the predicted relationship that

$$y = 0.144x. \tag{3}$$

To test this relationship, I took Elvis to Lake Michigan on a calm day when the waves were small. I fixed a measuring tape about 15 meters down the beach at C from where Elvis and I stood at A as I threw the ball. After throwing it, I raced after Elvis, plunging a screwdriver into the sand at the place where he entered the water at D. Then I quickly grabbed the free end of the tape measure and raced him to the ball. I was then able to get both the distance from the ball to the shore, x, and the distance y. If my throw did not land close to the line perpendicular to the shoreline and passing through C, I did not take measurements. I also omitted the couple of times when Elvis, in his haste and excitement, jumped immediately into the water and swam the entire distance. I figured that even an "A" student can have a bad day. We spent three hours getting 35 pieces of data. We stopped only when the waves grew. Elvis had no interest in stopping or slowing down. The data are in Table 2.

Table 2. Throw and fetch trials

x	y	x	y	x	y	x	y	x	y
10.5	2.0	17.0	2.1	4.7	0.9	10.9	2.2	15.3	2.3
7.2	1.0	15.6	3.9	11.6	2.2	11.2	1.3	11.8	2.2
10.3	1.8	6.6	1.0	11.5	1.8	15.0	3.8	7.5	1.4
11.7	1.5	14.0	2.6	9.2	1.7	14.5	1.9	11.5	2.1
12.2	2.3	13.4	1.5	13.5	1.8	6.0	0.9	12.7	2.3
19.2	4.2	6.5	1.0	14.2	1.9	14.5	2.0	6.6	0.8
11.4	1.3	11.8	2.4	14.2	2.5	12.5	1.5	15.3	3.3

The scatter plot of these results is given in Figure 2.

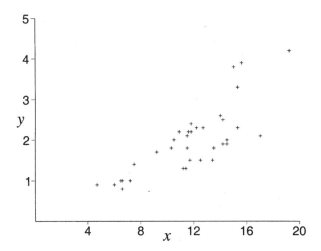

Figure 2. Scatter plot of Elvis's choices

Before looking at the scatter plot again in Figure 3, we ask the reader to imagine what line is best suggested by the data. There may be a difference of opinion here. Some may wish to take all points into account, while others may argue that the four points in the upper right lie outside of the pattern and therefore should be discounted. The plot does indeed seem to suggest that most of the points (31 of them) show a rather tight and clear pattern. Statisticians call this the "smooth", whereas the ones that don't fit the general pattern are called the "rough" or "outliers". Figure 3 shows the data points again, together with the line that is predicted from our model, $y = 0.144x$.

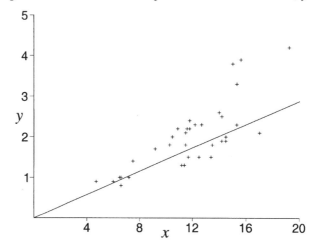

Figure 3. Scatter plot with optimal line

To my (maybe biased) eye, the agreement looks good. It seems clear that in most cases Elvis chose a path that agreed remarkably closely with the optimal path. The way to rigorously validate (and quantify) what the scatter plot suggests is to do a statistical analysis of the data. We will not do this in this paper, but it would be a natural avenue for further work. We conclude with several pertinent points.

First, we are in fact using a mathematical model. That is to say, we arrived at our theoretical figure by making many simplifying assumptions. These include

- We assumed there was a definite line between shore and lake. Because of waves, this was not the case.
- We assumed that when Elvis entered the water, he started swimming. Actually, he ran a short distance in the water. (Although given his six-inch legs, this is not too bad of an assumption!)
- We assumed the ball was stationary in the water. Actually, the waves, winds, and currents moved it a slight distance between the time Elvis plunged into the water and when he grabbed it.
- We assumed that the values of r and s are constant, independent of the distance run or swum.

Given these complicating factors as well as the error in measurements, it is possible that Elvis chose paths that were actually *better* than the calculated ideal path.

Second, we confess that although he made good choices, Elvis does not know calculus. In fact, he has trouble differentiating even simple polynomials. More seriously, although he does not do the calculations, Elvis's behavior is an example of the uncanny way in which nature (or Nature) often finds optimal solutions. Consider how soap bubbles minimize surface area, for example. It is fascinating that this optimizing ability seems to extend even to animal behavior. (It could be a consequence of natural selection, which gives a slight but consequential advantage to those animals that exhibit better judgment.)

Finally, for those intrigued by this general study, there are further experiments that are available, other than using your own favorite dog. One might do a similar experiment with a dog running in deep snow versus a cleared sidewalk. Even more interesting, one might test to determine whether the optimal path is found by six-year-old children, junior high aged pupils, or college students. For the sake of their pride, it might be best not to include professors in the study.

Do Dogs Know Related Rates Rather than Optimization?

Pierre Perruchet and Jorge Gallego

Timothy J. Pennings (this *Journal*, [2]) describes the strategy followed by his dog, Elvis, to fetch a tennis ball thrown into the water from the shore of Lake Michigan.

Let AC in Figure 1 be the water's edge. The ball is thrown from A, and falls into the water at B. Elvis, said Pennings, did not jump immediately in the water at A, a strategy that would have minimized the distance traveled (AB). Neither did he run along the beach to enter at C, which would have minimized the swimming distance (BC). Rather, he ran along the beach a part of the way, then jumped into the water at a point D, somewhere between A and C. Pennings speculated that the location of D was chosen to minimize the retrieval time. In order to test his hypothesis, he measured the running speed (r) and the swimming speed (s) of his dog (r being considerably larger than s) and computed the optimal path.

The time to get the ball is given by

$$T(y) = \frac{z-y}{r} + \frac{\sqrt{x^2 + y^2}}{s}. \tag{1}$$

The value of y providing the optimal path is the value for which $T'(y) = 0$. Solving $T''(y) = 0$ for y, Pennings obtained:

$$y = \frac{x}{\sqrt{r/s+1}\sqrt{r/s-1}} \tag{2}$$

Surprisingly, it turned out that in most cases Elvis jumped into the water at a point that agreed remarkably well with the optimal value given by the mathematical model. Pennings did *not* conclude that his dog knows calculus, but instead noted that "Elvis's

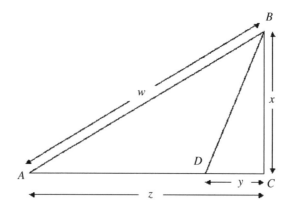

Figure 1. The problem space (adapted from Pennings)

behavior is an example of the uncanny way in which nature (or Nature) often finds optimal solutions" (p. 182).

In an effort to double the sample size, we determined that Salsa, a female Labrador, also apparently chooses the optimal path when playing fetch along a lakeside beach near Nimes in France! But we want to suggest that the dog's behavior may not be as uncanny as it might at first seem. What makes the dog's performance seem surprising is that it agrees with the result of a mathematical model minimizing *the total duration of the travel*. That is, it suggests that dogs are supposedly able to calculate optimal strategies involving knowledge of the entire route before they ever begin running. But the question is; is this ability really required?

Let us assume instead that the dogs are attempting to optimize their behavior *on a moment-to-moment basis*. For our specific concern, let us assume that a dog playing fetch chooses at each point in time the path that allows it to maximize its speed of approach to the ball. When running from A towards C at a constant speed, the ball at B appears closer and closer as the dog gets closer to C, but its speed of approach to B diminishes (reaching zero at C). At some moment of his run, his speed of approach while running on the beach becomes equal to his speed of approach when swimming directly to the ball; i.e., his swimming speed. It can be shown (a related rates problem) that if the dog jumps into the water at this moment, this strategy yields the same y value as that provided by the time of travel minimization model.

For let $W(t)$ be the distance from the dog to the ball (see Figure 1). Then

$$W(t) = \sqrt{x^2(t) + z^2(t)},$$

so

$$W'(t) = \frac{xx' + zz'}{\sqrt{x^2 + z^2}}.$$

Since $x' = 0$ and $z' = r$, we get

$$W'(t) = \frac{rz}{\sqrt{x^2 + z^2}}.$$

When $z = y$, $W'(t) = s$, so we solve

$$s = \frac{ry}{\sqrt{x^2 + z^2}}$$

for y to get

$$y = \frac{x}{\sqrt{r/s + 1}\sqrt{r/s - 1}}, \tag{3}$$

which is identical to (2).

Although this solution is identical to that proposed by Pennings, it was gained without assuming canine knowledge of the entire route, and hence can be construed as a more plausible model for dog's strategy. To perform in this way, dogs must first be able to estimate accurately their speed of approach at each moment. Second, they need to have a general awareness of their swimming speed before entering the water, since they have to jump into the water at the point when their speed of approach towards the ball while running on the beach becomes slower than their swimming speed.

Is it reasonable to postulate that dogs have this ability? To answer, we ask in turn: Would this ability have been useful for dog's ancestors, who lived in a natural environment? Obviously, the ability to detect transient changes in distance is crucial for animal species' survival, as when the ball is replaced by prey or predator. The general awareness of swimming speed is certainly a more sophisticated ability, in so far as it requires memory of relative speeds. But again, it seems to be essential for the survival of any animal to know how rapidly it can move in the various media that it may encounter when pursuing prey or escaping a predator. For animal species, such as mammals, that are destined to move in a variety of different mediums, it is reasonable to assume that there is an innate ability to learn quickly from early experiences.

In conclusion, both approaches require the use of calculus, either in solving an optimization problem or a related rates problem. However, as we showed, there is a major difference between our interpretation and the interpretation that results from taking Pennings' model as a realistic model of the dog's strategy. The ability that is required, in our view, forms part of general motion detection capabilities. As motion detection is common in most animals, it has been the focus of thorough investigations that have revealed some of its biological mechanisms (for a review, see [1]). Thus our solution provides a bridge between a specific behavioral strategy and ubiquitous biological mechanisms. Calculus then allows us to demonstrate that evolution has led to the development of biological mechanisms that are so powerful that they often lead to the optimal solution.

References

1. C.W.G. Clifford and M.R. Ibbotson, Fundamental mechanisms of visual motion detection: models, cells and functions, *Progress in Neurobiology* **68** (2003) 409–437.
2. T.J. Pennings, Do dogs know calculus? *College Math. J.* **34** (2003) 178–182.

Do Dogs Know Bifurcations?

Roland Minton and Timothy J. Pennings

Elvis burst upon the mathematical scene in May, 2003. The second author's article "Do Dogs Know Calculus?" [2] introduced his dog Elvis and Elvis's ability to solve a classic optimization problem. Peruchet and Gallego's article "Do Dogs Know Related Rates Rather Than Optimization?" [4] gave an alternative explanation of how dogs (including their own) might solve the problem. Elvis's surprising repudiation of that explanation in [3] inspired this article. Here, we explore Elvis's problem-solving ability when he must choose between two qualitatively different options. Such a situation induces a bifurcation in his optimal strategy. As a bonus, our analysis reveals a neat geometric proof of the arithmetic mean–geometric mean inequality.

In the original problem, Elvis is on the shoreline and wants to retrieve a ball thrown x meters into the water and z meters downshore, as in Figure 1. Elvis runs along the shore at speed r m/s to a point y meters upshore from the ball, then swims to the ball at speed s m/s.

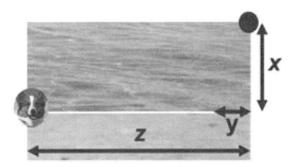

Figure 1. The original problem.

It is shown in [2] that the total time to the ball is minimized with

$$\hat{y} = \frac{x}{\sqrt{(r/s)^2 - 1}} \qquad (1)$$

if $z > \hat{y}$. This solution is remarkable because the optimal entry point is independent of the distance z. Also, the distance $\hat{y}$ is a linear function of x. In [2], Elvis's actual entry points for a large number of throws are presented. The scatter plot of these points shows a remarkable linear trend that closely matches the line of optimal entry points. Thus, it seems that Elvis is able to solve this general problem.

In [4], Perruchet and Gallego start with the function $d(t)$ giving the distance between Elvis and the ball. As he runs along the shoreline, the rate of change $d'(t)$ is negative and increasing. The position at which $|d'(t)|$ reaches the swim speed s (that is, $d'(t) = -s$) is shown to be exactly the optimal $\hat{y}$ in (1) above. This allows a different interpretation of how Elvis gets to the ball. Instead of some internal calculation of x (how far out in the water the ball is) and then y (where to enter the water), perhaps Elvis runs along the shore until he senses that he could make better progress to the ball by swimming. That is, instead of solving a global optimization problem, perhaps Elvis is solving a local related rates problem.

Fortunately, Elvis has provided more clues about his problem-solving strategy. As noted in [3], when Elvis starts *in* the water and a ball is thrown a long distance parallel to the shore, he first swims to shore, then runs along the shore, and finally swims back out to the ball. Thus, at least in this situation, Elvis is apparently viewing the task globally. However, his behavior raises three new questions.

First of all, what are the possible optimal paths? This question is easily answered. For any path, Elvis either reaches the shore or stays in the water. If he stays in the water, then swimming directly to the ball will result in the shortest time. If he reaches the shore, then the optimal path will involve swimming and running in straight lines. Thus, the optimal path will either be a straight swim to the ball (designated S), or a path (designated SRS) consisting of three straight lines. If the ball is thrown a short distance, S will be faster than SRS. The longer the throw, the more likely that SRS will be the quicker path. The second question is, what is the bifurcation point at which the optimal strategy changes from S to SRS? And third, does Elvis change his strategy at the optimal point? That is, does Elvis know bifurcations?

The swim-run-swim problem

Given the discussion above, we compare the times of the S and SRS paths. To find the optimal SRS path, suppose Elvis starts x_1 meters out in the water and races to a ball that is z meters downshore and x_2 meters out into the water, as in Figure 2. He first

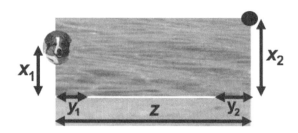

Figure 2. The SRS problem.

swims ashore with speed s m/s to a point y_1 meters downshore, then runs along the beach at speed r m/s to a point y_2 meters upshore from the ball, and finally swims out to the ball.

The total time to reach the ball is given by

$$T = \frac{\sqrt{x_1^2 + y_1^2}}{s} + \frac{z - y_1 - y_2}{r} + \frac{\sqrt{x_2^2 + y_2^2}}{s}. \tag{2}$$

If we consider T as a function of y_1 and y_2, it reaches a minimum when both partial derivatives are zero. We have

$$\frac{\partial T}{\partial y_i} = \frac{y_i}{s\sqrt{x_i^2 + y_i^2}} - \frac{1}{r}$$

for $i = 1, 2$. Setting each partial derivative equal to 0 and eliminating $\frac{1}{r}$ gives

$$\frac{y_1}{\sqrt{x_1^2 + y_1^2}} = \frac{y_2}{\sqrt{x_2^2 + y_2^2}}$$

which are the cosines of the angles in Figure 3.

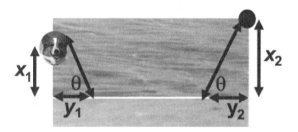

Figure 3. Equal angles.

We conclude that **angle in equals angle out**! Further, solving $\partial T/\partial y_i = 0$ for y_i gives

$$\hat{y}_i = \frac{x_i}{\sqrt{(r/s)^2 - 1}} \tag{3}$$

for $i = 1, 2$. The value for $\hat{y}_2$ coincides exactly with the solution of the original problem, if $z > \hat{y}_1 + \hat{y}_2$.

Upon reflection, these results are obvious. This is often the case when such a nice result emerges. Think of the problem in two parts. Step (i) is to go from a point in the water to a distant point on the shore. Step (ii) is to go from that point on the shore to another distant point in the water. Since the original solution (1) is independent of z, steps (i) and (ii) are independent. In fact, step (ii) is simply the original problem.

Further, the two steps are equivalent with step (i) being step (ii) covered in reverse. Therefore, the angles must be equal. Thought of in a different way, since (1) is independent of z, choose $z = \hat{y}_1 + \hat{y}_2$ so that there is no running at all. Then, analogous to light reflecting off a mirror or a billiard ball bouncing off a rail, the optimal path has angle in equal to angle out.

This suggests a possible explanation of Elvis's behavior. Perhaps Elvis uses a small set of rules. For example,

1. If the ball is close, swim directly to it. (Elvis does this.)
2. If the ball is far away, then (A) get out of the water and (B) solve the shore to ball problem.

The experience gained solving (2B) can help in (2A), since angle in equals angle out. The correct angle might "feel" right. Notice that this leaves open the question of how Elvis actually solves (2B). Such an explanation is consistent with artificial life models such as Craig Reynolds's boids [5] and with constructal theory [1].

Bifurcation points

The next step is to compare the S and SRS strategies. Substituting (3) into (2) gives the total time for the optimal SRS path, which can be written in the form

$$T_{\text{SRS}} = \frac{z}{r} + \frac{x_1 + x_2}{sr/\sqrt{r^2 - s^2}}.$$

The time to swim directly to the ball is given by

$$T_{\text{S}} = \frac{\sqrt{z^2 + (x_2 - x_1)^2}}{s}.$$

We want to find all values of z for which $T_{\text{SRS}} = T_{\text{S}}$. Squaring the equation $T_{\text{SRS}} = T_{\text{S}}$, using the quadratic formula to solve for z, and discarding the extraneous solution gives the critical value

$$\tilde{z} = \frac{x_1 + x_2 + 2\frac{r}{s}\sqrt{x_1 x_2}}{\sqrt{(r/s)^2 - 1}}. \tag{4}$$

For $z < \tilde{z}$, the fastest route is to swim directly to the ball. For $z > \tilde{z}$, the fastest route is the SRS path found above. The value $\tilde{z}$ is called a bifurcation point, since the nature of the optimal solution changes at this value.

This result leads to some interesting insights. First, there is no bifurcation point if $s > r$. If swimming is faster than running, then swimming directly to the ball is always optimal. Second, if $s < r$ and $s \approx r$, then $\tilde{z}$ is very large. For shorter distances z, the small advantage that running provides cannot compensate for having to swim the extra distance to shore.

Finally, as $\frac{r}{s} \to \infty$ in equation (4), $\tilde{z} \to 2\sqrt{x_1 x_2}$. For the physical problem with large values of $\frac{r}{s}$, the optimal strategy is to swim the shortest distance possible getting to shore and back to the ball. That is, as $\frac{r}{s} \to \infty$, the shape of the optimal SRS path will form three sides of a trapezoid, as in Figure 4.

The physical problem also helps us determine the length of the top of this trapezoid. As $\frac{r}{s} \to \infty$, the running time along the beach approaches zero, so the total SRS time equals the time to swim $x_1 + x_2$ meters. At the bifurcation point, the S and SRS times are equal, so the S path must also have length $x_1 + x_2$ meters, as in Figure 5. Check this out geometrically!

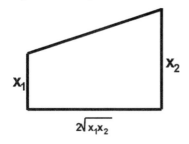
Figure 4. The SRS path.

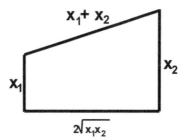
Figure 5. A mean triangle.

Figure 5 shows us that given two positive numbers, x_1 and x_2, $\sqrt{x_1 x_2} \leq \frac{1}{2}(x_1 + x_2)$. Thus, in the process of analyzing optimal retrieval strategies, we have discovered a picture proof for the relationship between the geometric mean and arithmetic mean of two numbers. The figure also reveals that equality holds only when $x_1 = x_2$. How interesting that by thinking rather intuitively about a problem in the physical world, we discover truths about the mathematical world.

In the case that a ball is thrown parallel to the shore, $x_2 = x_1$. Here, the equation $T_{\text{SRS}} = T_{\text{S}}$ simplifies and the bifurcation point is

$$\tilde{z} = 2x \sqrt{\frac{r/s + 1}{r/s - 1}}. \tag{5}$$

The bifurcation experiment

Since Elvis already revealed his willingness to bifurcate [3], the remaining question is whether he bifurcates at the correct point. To answer this, the second author and two undergraduate research students took Elvis to the same Lake Michigan beach where the first experiment [2] was done. Taking the average of several timed trials, Elvis's running speed was estimated to be 6.39 m/s and his swimming speed to be 0.73 m/s. However, once Elvis actually started chasing the ball, his running speed slowed down considerably to an average of 3.02 m/s. (This reduced speed was likely a combination of being tired from swimming to shore and just enjoying a lazy July afternoon.) Using equation (5), we see that the optimal bifurcation point is then $\tilde{z} = 2.56x$.

The second author stood 4 meters out in the water with Elvis and threw a ball various distances, but landing about 4 meters from the shore. One student measured the distance of the throw and the other recorded Elvis's choice. The results are in Table 1.

The data suggest several conclusions about Elvis's choices. First, it seems that there is a bifurcation. He consistently applies the SRS strategy for longer distances and swims directly to the ball for shorter ones. Secondly, there may or may not be a well-defined bifurcation point. If there is, it exists somewhere between 14 m and 15 m for this example, but without doing many more trials to narrow down the point and to show consistency, one cannot be sure. Lastly, if there is a well-defined bifurcation point, it is not where it should be. According to equation (5), the bifurcation should be at 10.24 m. Elvis's bifurcation distance was (a disappointing) 4 m larger. Thus it might be concluded that Elvis knows bifurcations qualitatively, but not quantitatively.

Lest we be too hard on Elvis, though, it should be remembered that dogs, like all of us, learn from experience and that bifurcations by their very nature make learning

Table 1.

Trial number	z (m)	Strategy
1	16.5	SRS
2	9.5	S
3	14.2	S
4	15.1	SRS
5	15	SRS
6	12	S
7	7.8	S
8	11.6	S
9	18.2	SRS

difficult. Bifurcations force a choice upon us, and we often do not have the opportunity to go back and try both options. But, in forcing the choice, bifurcations add interest to mathematics and richness to life. As the poet Robert Frost wrote,

> "Two roads diverged in a wood, and I—
> I took the one less traveled by,
> And that has made all the difference."

References

1. Adrian Bejan and Gil Merkx, ed., *Constructal Theory of Social Dynamics*, Springer, 2007.
2. Timothy J. Pennings, Do dogs know calculus? *College Math. J.* **34** (2003) 178–182.
3. Timothy J. Pennings, Response, *College Math. J.* **37** (2006) 19.
4. Pierre Perruchet and Jorge Gallego, Do dogs know related rates rather than optimization?, *College Math. J.* **37** (2006) 16–18.
5. Craig Reynolds, Flocks, herds and schools: a distributed behavioral model, *Computer Graphics: Proceedings of SIGGRAPH 87* (1987) 25–34.

The Lengthening Shadow: The Story of Related Rates

Bill Austin, Don Barry and David Berman

A boy is walking away from a lamppost. How fast is his shadow moving?

A ladder is resting against a wall. If the base is moved out from the wall, how fast is the top of the ladder moving down the wall?

Such "related rates problems" are old chestnuts of introductory calculus, used both to show the derivative as a rate of change and to illustrate implicit differentiation. Now that some "reform" texts [4, 14] have broken the tradition of devoting a section to related rates, it is of interest to note that these problems originated in calculus reform movements of the 19th century.

Ritchie, related rates, and calculus reform

Related rates problems as we know them date back at least to 1836, when the Rev. William Ritchie (1790–1837), professor of Natural Philosophy at London University 1832–1837, and the predecessor of J. J. Sylvester in that position, published *Principles of the Differential and Integral Calculus*. His text [21, p. 47] included such problems as:

> If a halfpenny be placed on a hot shovel, so as to expand uniformly, at what rate is its *surface* increasing when the diameter is passing the limit of 1 inch and 1/10, the diameter being supposed to increase *uniformly* at the rate of .01 of an inch per second?

This related rates problem was no mere practical application; it was central to Ritchie's reform-minded pedagogical approach to calculus. He sought to simplify the presentation of calculus so that the subject would be more accessible to the ordinary, non-university student whose background might include only "the elements of

Geometry and the principles of Algebra as far as the end of quadratic equations." [**21**, p. v] Ritchie hoped to rectify what he saw as a deplorable state of affairs:

> The Fluxionary or Differential and Integral Calculus has within these few years become almost entirely a science of symbols and mere algebraic formulae, with scarcely any illustration or practical application. Clothed as it is in a transcendental dress, the ordinary student is afraid to approach it; and even many of those whose resources allow them to repair to the Universities do not appear to derive all the advantages which might be expected from the study of this interesting branch of mathematical science.

Ritchie's own background was not that of the typical mathematics professor. He had trained for the ministry, but after leaving the church, he attended scientific lectures in Paris, and "soon acquired great skill in devising and performing experiments in natural philosophy. He became known to Sir John Herschel, and through him [Ritchie] communicated [papers] to the Royal Society" [**24**, p. 1212]. This led to his appointment as the professor of natural philosophy at London University in 1832.

To make calculus accessible, Ritchie planned to follow the "same process of thought by which we arrive at actual discovery, namely, *by proceeding step by step from the simplest particular examples till the principle unfolds itself in all its generality*." [**21**, p. vii; italics in original]

Drawing upon Newton, Ritchie takes the change in a magnitude over time as the fundamental explanatory concept from which he creates concrete, familiar examples illustrating the ideas of calculus. He begins with an intuitive introduction to limits through familiar ideas such as these: (i) the circle is the limit of inscribed regular polygons with increasing numbers of sides; (ii) $1/9$ is the limit of $1/10 + 1/100 + 1/1000 + \cdots$; (iii) $1/2x$ is the limit of $h/(2xh + h^2)$ as h approaches 0. Then—crucial to his pedagogy—he uses an expanding square to introduce both the idea of a function and the fact that a uniform increase in the independent variable may cause the dependent variable to increase at an increasing rate. Using FIGURE 1 to illustrate his approach, he writes:

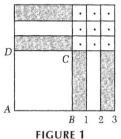

FIGURE 1
An expanding square

> Let AB be the side of a square, and let it increase uniformly by the increments $1, 2, 3$, so as to become $AB + 1$, $AB + 2$, $AB + 3$, etc., and let squares be described on the new sides, as in the annexed figure; then it is obvious that the square on the side $A1$ exceeds that on AB by the two shaded rectangles and the small white square in the corner. The square described on $A2$ has received an increase of two equal rectangles with *three* equal white squares in the corner.

The Lengthening Shadow: The Story of Related Rates

> The square on A3 has received an increase of two equal rectangles and *five* equal small squares. Hence, when the side increases uniformly the area goes on at an increasing rate [**21**, p. 11].

Ritchie continues:

> The object of the differential calculus, is to determine the *ratio* between the rate of variation of the independent variable and that of the function into which it enters [**21**, p. 11].

A problem follows:

> If the side of a square increase uniformly, at what rate does the area increase when the side becomes x? [**21**, p. 11]

His solution is to let x become $x + h$, where h is the rate at which x is increasing. Then the area becomes $x^2 + 2xh + h^2$, where $2xh + h^2$ is the rate at which the area would increase if that rate were uniform. Then he obtains this proportion [**21**, p. 12]:

$$\frac{\text{rate of increase of the side}}{\text{rate of increase of area}} = \frac{h}{2xh + h^2}.$$

Letting h tend to zero yields $1/2x$ for the ratio.

He then turns to this problem:

> If the side of a square increase uniformly at the rate of three feet per second, at what rate is the area increasing when the side becomes 10 feet? [**21**, p. 12]

Using the previous result, he observes that since 1 is to $2x$ as 3 is to $6x$, the answer is 6×10. Then he expresses the result in Newton's notation: If $\dot{x}$ denotes the rate at which a variable x varies at an instant of time and if $u = x^2$, then $\dot{x}$ is to u as 1 is to $2x$, or $\dot{u} = 2x\dot{x}$.

In his first fifty pages, Ritchie develops rules for differentiation and integration. To illustrate the product rule, he writes:

> If one side of a rectangle vary at the rate of 1 inch per second, and the other at the rate of 2 inches, at what rate is the area increasing when the first side becomes 8 inches and the last 12? [**21**, p. 28]

His problem sets ask for derivatives, differentials, integrals, and the rate of change of one variable given the rate of change of another. Some related rates problems are abstract, but on pages 45–47 Ritchie sets the stage for the future development of related rates with nine problems, most of which concern rates of change of areas and volumes. One was the halfpenny problem; here are three more [**21**, p. 47–48]:

> 25. If the side of an equilateral triangle increase uniformly at the rate of 3 feet per second, at what rate is the area increasing when the side becomes 10 feet? ...
>
> 30. A boy with a mathematical turn of mind observing an idle boy blowing small balloons with soapsuds, asked him the following pertinent question:—If the diameter of these balloons increase uniformly at the rate of 1/10 of an inch per

second, at what rate is the internal capacity increasing at the moment the diameter becomes 1 inch?...

34. A boy standing on the top of a tower, whose height is 60 feet, observed another boy running towards the foot of the tower at the rate of 6 miles an hour on the horizontal plane: at what rate is he approaching the first when he is 100 feet from the foot of the tower?

Since the next section of the book deals with such applications of the calculus as relative extrema, tangents, normals and subnormals, arc length and surface area, Ritchie clearly intended related rates problems to be fundamental, explanatory examples.

Augustus De Morgan (1806–1871) was briefly a professional colleague of Ritchie's at London University. De Morgan held the Chair of Mathematics at London University from 1828 to July of 1831, reassuming the position in October of 1836. Ritchie was appointed in January of 1832 and died in September of 1837. In *A Budget of Paradoxes*, published in 1872, De Morgan wrote [9, p. 296]:

Dr. Ritchie was a very clear-headed man. He published, in 1818, a work on arithmetic, with rational explanations. This was too early for such an improvement, and nearly the whole of his excellent work was sold as waste paper. His elementary introduction to the Differential Calculus was drawn up while he was learning the subject late in life. Books of this sort are often very effective on points of difficulty.

De Morgan, too, was concerned with mathematics education. In *On the Study and Difficulties of Mathematics* [6], published in 1831, De Morgan used concrete examples to clarify mathematical rules used by teachers and students. In his short introduction to calculus, *Elementary Illustrations of the Differential and Integral Calculus* [7, p. 1–2], published in 1832, he tried to make calculus more accessible by introducing fewer new ideas simultaneously. De Morgan's book, however, does not represent the thoroughgoing reform that Ritchie's does. De Morgan touches on fluxions, but omits related rates problems. In 1836, shortly before Ritchie's death, De Morgan began the serial publication of *The Differential and Integral Calculus*, a major work of over 700 pages whose last chapter was published in 1842. He promised to make "the theory of limits... the sole foundation of the science, without any aid from the theory of series" and stated that he was not aware "that any work exists in which this has been avowedly attempted." [8, p. 1] De Morgan was more concerned with the logical foundations of calculus than with pedagogy; no related rates problems appear in the text.

Connell, related rates, and calculus reform

Another reform text appeared shortly after Ritchie's. James Connell, LLD (1804–1846), master of the mathematics department in the High School of Glasgow from 1834 to 1846, published a calculus textbook in 1844 promising "numerous examples and familiar illustrations designed for the use of schools and private students." [5, title page] Like Ritchie, Connell complained that the differential calculus was enveloped in needless mystery for all but a select few; he, too, proposed to reform the teaching of calculus by returning to its Newtonian roots [5, p. iv]. Connell wrote that he

> ...has fallen back upon the original view taken of this subject by its great founder, and, from the single definition of a rate, has been enabled to carry it out without the slightest assistance from Limiting ratio, Infinitesimals, or any other mode which, however good in itself, would, if introduced here, only tend to mislead and bewilder the student." [5, p. v]

To introduce an instantaneous rate, Connell asks the reader to consider two observers computing the speed of an accelerating locomotive as it passes a given point. One notes its position two minutes after it passes the point, the other after one minute; they get different answers for the speed. Instead of considering observations on shorter and shorter time intervals, Connell imagines the engineer cutting off the power at the given point. The locomotive then continues (as customary, neglecting friction) at a constant speed, which both observers could compute. This gives the locomotive's rate, or differential, at that point. Connell goes on to develop the calculus in terms of rates. For example, to prove the product rule for differentials, he considers the rectangular area generated as a particle moves so that its projections along the x- and y-axes move at the rates dx and dy respectively. As with Ritchie, the product rule is taught in terms of an expanding rectangle and rates of change.

Connell illustrates a number of the simpler concepts of the differential calculus using related rates problems. Some of his problems are similar to Ritchie's, but most are novel and original and many remain in our textbooks (punctuation in original):

> 5. A stone dropped into still water produces a series of continually enlarging concentric circles; it is required to find the rate per second at which the area of one of them is enlarging, when its diameter is 12 inches, supposing the wave to be then receding from the centre at the rate of 3 inches per second. [5, p. 14]
>
> 6. One end of a ball of thread, is fastened to the top of a pole, 35 feet high; a person, carrying the ball, starts from the bottom, at the rate of 4 miles per hour, allowing the thread to unwind as he advances; at what rate is it unwinding, when the person is passing a point, 40 feet distant from the bottom of the pole; the height of the ball being 5 feet?...
>
> 12. A ladder 20 feet long reclines against a wall, the bottom of the ladder being 8 feet distant from the bottom of the wall; when in this position, a man begins to pull the lower extremity along the ground, at the rate of 2 feet per second; at what rate does the other extremity *begin* to descend along the face of the wall?...
>
> 13. A man whose height is 6 feet, walks from under a lamp post, at the rate of 3 miles per hour, at what rate is the extremity of his shadow travelling, supposing the height of the light to be 10 feet above the ground? [5, p. 20–24]

Connell died suddenly on March 26, 1847, leaving a wife and six children. The obituary in the *Glasgow Courier* observed that "he had the rare merit of communicating to his pupils a portion of that enthusiasm which distinguished himself. The science of numbers ... in Dr. Connell's hands ... became an attractive and proper study, and ... his great success as a teacher of children depended on his great attainments as a student of pure and mixed mathematics" [26]. It would be interesting to learn of any contact between Ritchie and Connell, but so far we have found none.

The rates reform movement in America

Related rates problems first appeared in America in an 1851 calculus text by Elias Loomis (1811–1889), professor of mathematics at Yale University. Loomis was also concerned to simplify calculus, writing that he hoped to present the material "in a more elementary manner than I have before seen it presented, except in a small volume by the late Professor Ritchie" [17, p. iv]. Indeed, the initial portion of Loomis's text is essentially the same as Ritchie's. Loomis presents ten related rates problems, nine of which are Ritchie's; the one new problem asks for the rate of change of the volume of a cone whose base increases steadily while its height is held constant [17, p. 113]. Loomis's text remained in print from 1851 to 1872; a revision remained in print until 1902.

The next text to base the presentation of calculus on related rates was written by J. Minot Rice (1833–1901), professor of mathematics at the Naval Academy, and W. Woolsey Johnson (1841–1923), professor of mathematics at St. John's College in Annapolis. Where Loomis quietly approved the simplifications introduced by Ritchie, Rice and Johnson were much more enthusiastic reformers, drawing more from Connell than from Ritchie:

> Our plan is to return to the method of fluxions, and making use of the precise and easily comprehended definitions of Newton, to deduce the formulas of the Differential Calculus by a method which is not open to the objections which were largely instrumental in causing this view of the subject to be abandoned [19, p. 9].

In their 1877 text they derive basic differentiation techniques using rates. Letting dt be a finite quantity of time, dx/dt is the rate of x and "dx and dy are so defined that their ratio is equal to the ratio of the relative rates of x and y" [20, p. iv]. This approach has several advantages. First, it allows the authors to delay the definition of dy/dx as the limit of $\Delta y/\Delta x$ until Chapter XI, by which time the definition is more meaningful. Second, "the early introduction of elementary examples of a kinematical character ... which this mode of presenting the subject permits, will be found to serve an important purpose in illustrating the nature and use of the symbols employed" [20, p. iv].

These kinematical examples are related rates problems. Rice and Johnson use 26 related rates problems, scattered throughout the opening 57 pages of the text, to illustrate and explain differentiation. Rice and Johnson credit Connell in their preface and some of their problems resemble Connell's. Several other problems are similar to those of Loomis. However, Rice and Johnson also add to the collection of problems:

> A man standing on the edge of a wharf is hauling in a rope attached to a boat at the rate of 4 ft. per second. The man's hands being 9 ft. above the point of

attachment of the rope, how fast is the boat approaching the wharf when she is at a distance of 12 ft. from it? [**20**, p. 28]

Wine is poured into a conical glass 3 inches in height at a uniform rate, filling the glass in 8 seconds. At what rate is the surface rising at the end of 1 second? At what rate when the surface reaches the brim? [**20**, p. 37–38]

After Rice died in 1901, Johnson continued to publish the text until 1909. He was "an important member of the American mathematical scene ... [who] served as one of only five elected members of the Council of the American Mathematical Society for the 1892–1893 term" [**22**, p. 92–93]. The work of Rice and Johnson is likely to have inspired the several late 19th century calculus texts which were based on rates, focusing less on calculus as an analysis of tangent lines and areas and more on "how one quantity changes in response to changes in another." [**22**, p. 92]

James Morford Taylor (1843–1930) at Colgate, Catherinus Putnam Buckingham (1808–1888) at Kenyon, and Edward West Nichols (1858–1927) at the Virginia Military Institute all wrote texts that remained in print from 1884 to 1902, 1875 to 1889, and 1900 to 1918, respectively. Buckingham, a graduate of West Point and president of Chicago Steel when his text was published was, perhaps, the most zealous of these reformist "rates" authors, believing that limits were problematic and could be avoided by taking rate itself as the primitive concept, much as he believed Newton did [**2**, p. 39].

Newton and precursors of the rates movement

Buckingham was correct that Newton conceived of magnitudes as being generated by motion, thereby linking calculus to kinematics. Newton wrote:

> I consider mathematical quantities in this place not as consisting of very small parts; but as described by a continued motion. Lines are described, and thereby generated not by the apposition of parts, but by the continued motion of points; superficies by the motion of lines .... These geneses really take place in the nature of things, and are daily seen in the motion of bodies .... Therefore considering that quantities which increase in equal times ... become greater or less according to the greater or less velocity with which they increase and are generated; I sought a method of determining quantities from the velocities of the motions ... and calling these velocities ... *fluxions*. [**3**, p. 413]

Since the 19th century reformers drew on Newton in revising the pedagogy of calculus, one wonders whether rates problems were part of an earlier tradition in England. The first calculus book to be published in English, *A Treatise of Fluxions or an Introduction to Mathematical Philosophy* [**13**] by Charles Hayes (1678–1760), published in 1704, treats fluxions as increments or decrements. Motion is absent and there are no related rates problems. But, in 1706, in the second book published in English, *An Institution of Fluxions* [**10**] by Humphrey Ditton (1675–1715), there are several problems which could be seen as precursors of related rates questions. While Ditton is interested in illustrating ideas of calculus using rates, he sticks to geometrical applications, not mechanical ones. He writes:

A vast number of other Problems relating to the Motion of Lines and Points which are directly and most naturally solved by Fluxions might have been propos'd to the Reader. But this Field is so large, that 'twill be besides my purpose to do any more upon this Head than only just give some little Hints. [**10**, p. 172]

He gives one worked example. In FIGURE 2, b and c represent the new positions of points B and C respectively:

If the Line AB, in any moment of Time be supposed to be divided into extream and mean Proportion, as ex. gr in the point C; then the Point A continues fixt, and the Points B and C moving in the direction AB, 'tis requir'd to find the Proportion of the Velocities of the points B and C; so that the flowing line Ab, may still be divided in extream and mean Proportion, e.g. in the Point c." [**10**, p. 171–172]

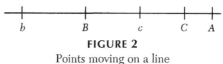

FIGURE 2
Points moving on a line

In his solution he lets $AB = y$, $AC = x$ and $BC = y - x$ and obtains $\frac{y}{y-x} = \frac{y-x}{x}$ giving $3yx = y^2 + x^2$. He differentiates, obtaining $3y\dot{x} + 3x\dot{y} = 2y\dot{y} + 2x\dot{x}$ which gives $\frac{\dot{x}}{\dot{y}} = \frac{2y - 3x}{3y - 2x}$. Ditton summarizes by saying, "the velocity of the Increment of the less Segment AC, must be the velocity of the Increment of the whole line AB as $\frac{2AB - 3AC}{3AB - 2AC}$" [**10**, p. 172]. His concern was to express ratios of rates of change of lengths in terms of ratios of lengths within the context of some geometric invariance. He was not seeking to use an equation involving rates of change as a centerpiece of pedagogy, but rather as a straightforward application of the calculus. His work did not lead to the development of such problems. Of the thirteen 18th century English authors surveyed, only William Emerson (1701–1782), writing in 1743, included a related rates problem, but not in a significant way [**11**, p. 108].

In the early 19th century we find scattered related rates problems. There is a sliding ladder problem in a Cambridge collection: "The hypotenuse of a right-angled triangle being constant, find the corresponding variations of the sides." [**25**, p. 678] John Hind's text included one problem: "Corresponding to the extremities of the *latus rectum* of a common parabola, it is required to find the ratio of the rates of increase of the abscissa and ordinate" [**14**, p. 148]. Neither of these problems plays the important pedagogical role that we find in the works of Ritchie or Connell.

The twilight of related rates

Why did so few books illustrate calculus concepts using related rates problems? One reason is that from the beginning of the 18th century to the middle of the 19th century, the foundations of calculus were hotly debated, and Newton's fluxions did not compete very successfully against infinitesimals, limits, and infinite series. Among those who chose to base calculus on fluxions, many still felt uneasy about including kinematical considerations in mathematics. In *A Comparative View of the Principles of the Fluxional and Differential Methods*, Prof. D.M. Peacock wrote that one of the leading objections to the fluxional approach was that "it introduces Mechanical considerations of *Motion, Velocity*, and *Time*, foreign to the genius of pure Analytics" [**18**, p. 6]. Such concepts were considered by some to be "inconsistent with the rigour of mathematical reasoning, and wholly foreign to science." [**23**, p. 7]

In England, moreover, resistance to Newton's approach to calculus as well as to the French approach as expressed in Lacroix's textbook [**16**] rested in part on the belief that the purpose of mathematics was to train the mind. That meant doing calculus within a Euclidean framework with a clear focus on the properties of geometrical figures [**1**, p. v–xx].

By the end of the 19th century, most authors were developing calculus on the basis of limits. In the works of Simon Newcomb (1835–1909) and Edward Bowser (1845–1910), for example, related rates problems illustrate the derivative as a rate of change, but the problems are not central. In 1904, William Granville (1863–1943) published his *Elements of the Differential and Integral Calculus*, which remained in print until 1957. This text, which introduced concepts intuitively before establishing analytical arguments, became the standard by which other texts were measured. In the 1941 edition, Granville laid out a method for solving related rates problems, but these problems had now become an end in themselves rather than an exciting and pedagogically important method by which to introduce calculus.

Conclusion

An informal survey of ours suggests that for many teachers these problems have lost their significance. One often hears teachers say that related rates problems are contrived and too difficult for contemporary students. It is thus ironic that such problems entered calculus through reformers who believed, much as modern reformers do, that in order for calculus to be accessible, concrete, apt illustrations of the derivative are necessary. Twilight for our 19th century reformers would have suggested a lengthening, accelerating shadow, not the end of an era. They might well have written:

> Related rates, a pump, not a filter; a sail, not an anchor.

NOTE. See `http://www.maa.org/pubs/mm-supplements/index.html` for a more extensive bibliography.

Acknowledgment. The research for this paper was conducted while the authors were at the M.A.A. Institute in the History of Mathematics and its Use in Teaching (funded by the National Science Foundation) at The American University, during the summers of 1996 and 1997. We thank the Institute's directors, V. Frederick Rickey, Victor J. Katz, and Steven H. Schot for their support. We also thank the referees for pertinent suggestions that improved the article.

REFERENCES

1. Rev. Arthur Browne, *A Short View of the First Principles of the Differential Calculus*, J. Deighton & Sons, Cambridge, UK, 1824.
2. C. P. Buckingham, *Elements of the Differential and Integral Calculus by a New Method, Founded on the True System of Sir Isaac Newton, Without the Use of Infinitesimals or Limits*, S.C. Griggs, Chicago, IL, 1875.

3. R. Calinger, *Classics of Mathematics*, Prentice Hall, Englewood Cliffs, NJ, 1995. (This passage is from Newton's Introduction to the *Tractatus de quadratura curvarum* published in 1704.)
4. J. Callahan, K. Hoffman, *Calculus in Context, The Five College Calculus Project*, W. H. Freeman, New York, NY, 1995.
5. James Connell, *The Elements of the Differential and Integral Calculus*, Longman, Brown, Green, and Longman, London, UK, 1844.
6. Augustus De Morgan, *On the Study and Difficulties of Mathematics*, The Open Court Publishing Company, Chicago, IL, 1902. (Originally published in 1831.)
7. Augustus De Morgan, *Elementary Illustrations of the Differential and Integral Calculus*, The Open Court Publishing Company, Chicago, IL, 1909. (Originally published in 1832.)
8. Augustus De Morgan, *The Differential and Integral Calculus*, Baldwin and Cradock, London, UK, 1842.
9. Augustus de Morgan, *A Budget of Paradoxes*, edited by David Eugene Smith, Dover Publications, New York, NY, 1954. (Originally published in 1872.)
10. Humphrey Ditton, *An Institution of Fluxions*, W. Botham, London, UK, 1706.
11. W. Emerson, *The Doctrine of Fluxions*, J. Bettenham, London, UK, 1743.
12. W. A. Granville, *Elements of the Differential and Integral Calculus*, Ginn and Co., Boston, MA, 1941.
13. Charles Hayes, *A Treatise of Fluxions or an Introduction to Mathematical Philosophy*, Edward Midwinter, London, UK, 1704.
14. John Hind, *A Digested Series of Examples in the Applications of the Principles of the Differential Calculus*, J. Smith, Cambridge, UK, 1832.
15. Deborah Hughes-Hallett, et. al., *Calculus*, John Wiley and Sons, New York, NY, 1994.
16. S.F. LaCroix, *An Elementary Treatise on the Differential and Integral Calculus*, transl. by M. Babbage, G. Peacock, Sir J. F. W. Herschel, Deighton & Sons, Cambridge, UK, 1816.
17. Elias Loomis, *Elements of Analytical Geometry and of the Differential and Integral Calculus*, Harper, New York, NY, 1852.
18. Rev. D. M. Peacock, *Comparative View of the Principles of the Fluxional and Differential Methods*, J. Smith, Cambridge, UK, 1819.
19. John Minot Rice and William Woolsey Johnson, *On a New Method of Obtaining Differentials of Functions with esp. reference to the Newtonian Conception of Rates or Velocities*, Van Nostrand, New York, NY, 1875.
20. John Minot Rice and William Woolsey Johnson, *An Elementary Treatise on the Differential Calculus Founded on the Method of Rates or Fluxions*, John Wiley, New York, NY, 1879.
21. William Ritchie, *Principles of the Differential and Integral Calculus*, John Taylor, London, UK, 1836.
22. G. M. Rosenstein, Jr., The Best Method. American calculus textbooks of the nineteenth century, *A Century of Mathematics in America*, ed. Peter Duren, American Mathematical Society, Providence, RI, 1989, 77–109.
23. James Ryan, *The Differential and Integral Calculus*, White, Gallagher, and White, New York, NY, 1828.
24. George Stronach, Ritchie, *The Dictionary of National Biography*, Vol. 16, Oxford U. Press, London, UK, 1960.
25. I. M. F. Wright, *Solutions of the Cambridge Problems from 1800–1820*, Vol. 2, Black, Young and Young, London, UK, 1825.
26. Obituary, *Glasgow Courier*, March 28, 1846, Glasgow, Scotland.

The Falling Ladder Paradox

Paul Sholten and Andrew Simonson

Anyone who has studied calculus has probably solved the classic *falling ladder* problem of related rates fame:

> A ladder L feet long leans against a vertical wall. If the base of the ladder is moved outwards at the constant rate of k feet per second, how fast is the tip of the ladder moving downwards?

The standard solution model for this problem is to assume that the tip of the ladder slips downward, maintaining contact with the wall until impact at ground level, so that if the base and tip of the ladder at any time t have coordinates

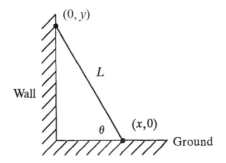

Figure 1. The standard falling ladder model.

$(x(t), 0)$ and $(0, y(t))$, respectively, the Pythagorean theorem gives $x^2 + y^2 = L^2$; see Figure 1. Differentiating with respect to time t yields the formula

$$\dot{y} = \frac{-kx}{y}. \tag{1}$$

The paradox in this solution is that as the ladder nears the ground, $\dot{y}$ attains astronomical proportions. In fact, in [5] the student is lightheartedly asked to find (for a particular k and L) at what height y the ladder's tip is moving at light speed.

149

Of course, the resolution of this paradox is that the ladder's tip leaves the wall at some point in its descent. A few classroom experiments using a yardstick lend observational support for this explanation, for as the base of a stick or ladder is moved away from the wall at constant speed, at the moment of impact it appears as if the tip lands some small distance from the wall, although the action transpires so quickly and catastrophically that it is hard to be certain about what happens. A paper [2] in the physics literature points out the flaw of using (1) and demonstrates the correct model for the falling ladder. Our approach is somewhat simpler, making no use of the force exerted by the wall on the ladder's tip; we furthermore show how to numerically plot the path of the ladder's tip, from the time it leaves the wall until its crash landing.

Let's determine y_c, the critical height at which the ladder leaves the wall and (1) ceases to be valid. We will do this by examining the differential equations governing these two different physical situations: the moving ladder supported by the wall and the unsupported ladder behaving as a stick pendulum.[1]

For the pendulum, recall that the rotational version of Newton's second law of motion states that if a rigid body rotates in a plane about an axis that moves with uniform velocity, then the total torque exerted by all the external forces on the body equals the product of the moment of inertia and the angular acceleration, where the torque and the moment of inertia are computed with respect to this moving axis. We apply this principle for the axis which passes through the point of contact (the *pivot*) of the ladder with the ground and which is perpendicular to the plane in which the ladder falls, since this pivot moves with constant velocity.

The only forces on the freely falling ladder are the upward force from the ground at the pivot point, which produces no torque, and the gravitational force, which produces the same torque τ as a force of magnitude mg acting downward at the center of mass of the ladder, as indicated in Figure 2. That is,

$$\tau = mg\frac{L\cos\theta}{2},$$

a positive value since this torque is counterclockwise. Finding the moment of inertia I of a uniform rod with mass m and length L about its endpoint is a standard exercise in calculus or physics, namely,

$$I = \int_0^L x^2 \frac{m}{L}\, dx = \frac{1}{3}mL^2.$$

The angular acceleration is simply $-\ddot{\theta}$, being the second derivative with respect to time of the angle $\pi - \theta$ between the ground and the ladder, measured counterclockwise. Thus Newton's law for the falling ladder is $\frac{1}{3}mL^2(-\ddot{\theta}) = \frac{1}{2}mgL\cos\theta$, or

$$\ddot{\theta} = -\frac{3g}{2L}\cos\theta, \qquad (2)$$

---

[1] Some texts present an alternative falling ladder problem in which an unfortunate fellow clings to the top of a ladder. Such a problem can be modeled by the motion of a standard pendulum, by neglecting the mass of the (lightweight) ladder and taking the mass of the pendulum to be the mass of the man. This analysis would be a suitable project for students.

which is valid after the ladder loses contact with the wall.

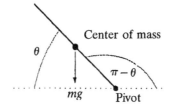

Figure 2. A straight stick pendulum of length L.

On the other hand, when the ladder is in contact with the wall, $y = L \sin \theta$ and differentiation yields $\dot{y} = L \cos \theta \dot\theta = x\dot\theta$. By equation (1)

$$\dot\theta = -\frac{k}{L \sin \theta}, \qquad (3)$$

and another differentiation yields

$$\ddot\theta = \frac{k \cos \theta}{L \sin^2 \theta} \dot\theta = -\frac{k^2 \cos \theta}{L^2 \sin^3 \theta}, \qquad (4)$$

which is valid while the ladder maintains contact with the wall.

Given specific values of L and k, we can determine the critical angle θ_c at which the ladder loses contact with the wall by finding the point of intersection of the graphs of (2) and (4), plotting $\ddot\theta$ versus θ. Figure 3 (page 52) illustrates this idea using the values $L = 41$ ft, $k = 10$ ft/s, $g = 32$ ft/s^2, from [1]. From the graph we see that as θ decreases the ladder falls according to equation (4) until the two curves meet at $\theta_c \approx 0.38$, the critical angle, and thereafter the ladder falls according to equation (2). That is, up until the critical angle the ladder is held up by the wall, but after θ_c it is free to behave as a stick pendulum.

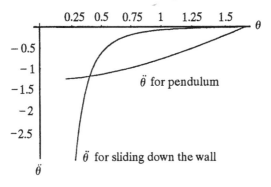

Figure 3. The transition between sliding and swinging.

Leaving L and k as parameters, we equate the right sides of (2) and (4), then simplify, yielding

$$\sin^3 \theta_c = \frac{2k^2}{3gL}. \tag{5}$$

If $2k^2/(3gL) \geq 1$, that is, if $k \geq \sqrt{\frac{3}{2}gL}$, equation (5) is impossible, and we conclude that the tip of the ladder pulls away from the wall immediately when the bottom begins to move away with speed k. Otherwise, since $y_c = L \sin \theta_c$,

$$y_c = \sqrt[3]{\frac{2k^2L^2}{3g}}. \tag{6}$$

It is interesting to find $\ddot{y}_c$, the acceleration of the tip of the ladder at the critical height. Differentiating (1) and simplifying yields $\ddot{y} = (k^2L^2)/y^3$, which is valid while the ladder stays in contact with the wall. By (6), then, the acceleration at the moment of separation is

$$\ddot{y}_c = -\tfrac{3}{2}g. \tag{7}$$

To find the path of the ladder's tip after it leaves the wall, first observe that at the moment of separation the base is at $x_c = L \cos \theta_c$. Since the base moves away at constant speed k, its distance from the wall t seconds later will be $x_c + kt$, so the distance from the wall to the upper end of the ladder will be $d = x_c + kt - L \cos \theta$ at this time. Thus if we solve the differential equation (2) to find $\theta(t)$, the path of the ladder's tip will be given by the parametric equations

$$\begin{aligned} d(t) &= x_c + kt - L \cos \theta(t), \\ y(t) &= L \sin \theta(t). \end{aligned} \tag{8}$$

Figure 4 shows the trajectory generated by *Mathematica*, which numerically solves (2) and plots the parametric curve, with $L = 41$ ft, $k = 10$ ft/s, and $g = 32$ ft/s². The initial values are

$$\theta(0) = \theta_c = \arcsin \sqrt[3]{\frac{2k^2}{3gL}} \approx 0.379428, \quad \text{from (5), and}$$

$$\dot{\theta}(0) = \dot{\theta}_c = -\frac{k}{L \sin \theta_c} \approx -0.658503, \quad \text{from (3).}$$

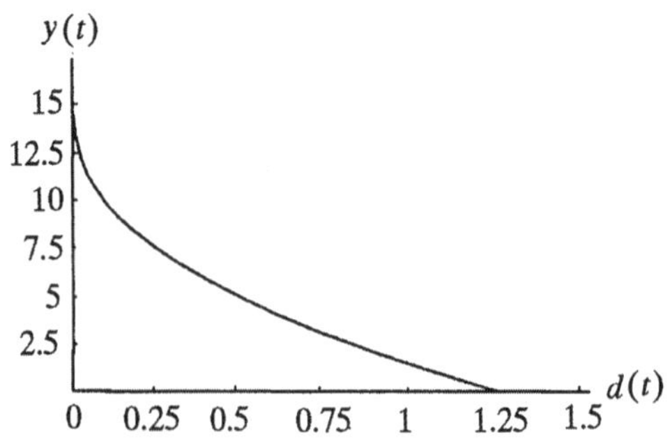

Figure 4. The path of the ladder's tip.

Note that $y_c = L \sin \theta_c \approx 15.19$ ft in this example. The solution is computed as long as $\theta(t) \geq 0$, which turns out to be about 0.42 second, and at this moment of impact the distance of the tip of the ladder from the wall is $d \approx 1.32$ ft.

To contrast these results with a typical textbook solution, consider the problem in [1], where the student is asked to find $\dot{y}$ at the instant when $y = 9$ ft, with L and k as before. Since the ladder separates from the wall when $y = y_c \approx 15.19$ ft, we can use *Mathematica*'s numerically generated solution of the differential equation (2) to find the correct value $\dot{y} \approx -35.49$ ft/s when $y = 9$, rather than the value of -44.44 as given by (1).

So what should be the status of the falling ladder problem in introductory calculus texts? Here are a few possibilities:

- Remove such problems from the textbooks [4].
- Instead of asking for $\dot{y}$, ask for $\dot{x}$ for a ladder falling under the force of gravity, with no friction at either end. But this is a classic mechanics problem, probably best left for a physics course.
- Leave the problems in the text, but ensure that the exercises have $k \geq \sqrt{\frac{3}{2}gL}$ and ask for $\dot{y}$ when y is larger than the y_c of (6), so that the standard approach rings true physically. Mention as a marginal note that the standard model breaks down once $y < y_c$ or if $k \geq \sqrt{\frac{3}{2}gL}$.

An interesting variant that avoids the separation pathology has a 15-foot ladder sliding down a wall while its base slides at 4 ft/s across a 9-foot-wide alley, bounded on the other side by another wall [3]. Equation (1) faithfully models this situation, and would do so up to an alley width of 14.4 feet.

Acknowledgments. The second author gratefully acknowledges a grant from the Michael and Margaretha Sattler Foundation for a copy of *Mathematica*, the use of which provides one with an extra measure of hopeful expectations in the exploration of problems like the one discussed in this paper.

References

1. C. H. Edwards and D. Penney, *Calculus with Analytic Geometry*, Prentice Hall, Englewood Cliffs, NJ, 1994, exercise 45, p. 171.
2. M. Freeman and P. Palffy-Muhoray, On mathematical and physical ladders, *American Journal of Physics* 53:3 (1985) 276–277.
3. P. Gillett, *Calculus and Analytic Geometry*, Heath, Lexington, MA, 1984, pp. 194–195.
4. D. Hughes-Hallett et al., *Calculus*, Wiley, New York, 1994.
5. G. Strang, *Calculus*, Wellesley Cambridge Press, Wellesley, MA, 1991, p. 164.

Solving the Ladder Problem on the Back of an Envelope

Dan Kalman

How long a ladder can you carry horizontally around a corner? Or, in the idealized geometry of FIGURE 1, how long a line segment can be maneuvered around the corner in the *L*-shaped region shown? This familiar problem, which dates to at least 1917, can be found in the max/min sections of many calculus texts and is the subject of numerous web sites. The standard solution begins with a twist, transforming the problem from maximization to minimization. This bit of misdirection no doubt contributes to the appeal of the problem. But it fairly compels the question, *Is there a direct approach?*

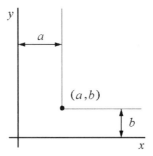

Figure 1 Geometry of ladder problem

In fact there is a beautifully simple direct approach that immediately gives new insights about the problem. It also gives us an excuse to revisit a lovely topic—envelopes of families of curves. This topic was once a standard part of the calculus curriculum, but seems to be largely forgotten in the current generation of texts. A generalized ladder problem considered in [17] can also be analyzed using the direct approach.

Ladder problem history

It is not easy to discover when the ladder problem first appeared in calculus texts. Singmaster [24] has compiled an extensive chronology of problems in recreational mathematics. There, the earliest appearance of the ladder problem is a 1917 book by Licks [15]. As Singmaster notes, this version of the problem concerns a stick to be put up a vertical shaft in a ceiling, rather than a ladder and two hallways, but the two situations are mathematically equivalent. Licks gives what is today the standard solution, finding the maximum length stick that gets stuck in terms of the angle the stick makes with the floor. He concludes *This is a simple way to solve a problem which has proved a stumbling block to many*. Whether this implies an earlier provenance in recreational problem solving, or a more mundane history of people actually putting long sticks up vertical shafts, who can say?

American University is fortunate to possess an extensive collection of mathematical textbooks dating to the 18th century. Haphazardly selecting a sample of eight calculus textbooks published between 1816 and 1902, I searched without success for mention of the ladder problem, or the equivalent, in discussions of maxima and minima. Many of these texts did have quite a number of max/min exercises, including several that our students would recognize. Of particular note is the text by Echols [7], published in 1902. Among the 56 max/min exercises in this work, nearly all of today's standard exercises appear, but not the ladder problem. More than half of the books also have a section on envelopes. Coincidentally, in the 1862 work of Haddon [13], the envelope we will discuss below appears in an example about a ladder sliding down a wall, but not in connection with any max/min problem.

In modern times, a variation on the ladder problem has been the subject of ongoing research. This is the sofa problem, which seeks the region of greatest area that can go around a corner between halls of given widths. The sofa problem appears in a volume on unsolved problems in geometry [6], and gave birth to the *Moving a Sofa Constant* [9]. For more on this open problem, see [27].

The main focus here is the use of envelopes to solve the ladder problem. The earliest record I have found for this approach is an anecdote of Cooper [3], who reports meeting a variant of the ladder problem in 1959 on a physics quiz at Princeton and solving it via envelopes. There is also one reference from Singmaster's compilation in which envelopes are used. Fletcher [10] provides five solutions of the ladder problem, with no explicit use of calculus. One of these methods uses envelopes. However, in order to avoid calculus, Fletcher depends on geometric properties of envelopes that are obscure by present day standards. Moreover, the exposition is rather terse, and says almost nothing about what the envelope is, how it arises, or why it provides a solution to the ladder problem.

The direct approach

As mentioned earlier, the standard solution to the ladder problem begins with a restatement: the goal is shifted from finding the longest ladder that will go around the corner to finding the *shortest* ladder that will get stuck. In seeking a direct approach, we consider actually moving a segment around a corner, trying to use as little of the space as possible. Begin with the segment along one of the outer walls, say with the left end at the origin and the right end at the point $(L, 0)$. Slide the left end up the y axis, all the while keeping the right end on the x axis. Intuitively, this maneuver keeps the line segment as far as possible from the corner point (a, b). Surely, if a segment of length L cannot get around the corner using this conservative approach, then it won't go around no matter what we do.

Now as you slide the segment along the walls it sweeps out a region Ω, as illustrated in FIGURE 2. The outer boundary of Ω is part of curve called an *astroid*, with equation

$$x^{2/3} + y^{2/3} = L^{2/3}. \tag{1}$$

The full astroid is in the shape of a four pointed star (hence the name), but for the ladder problem, we are concerned only with the part of the curve that lies in the first quadrant. Our line segment will successfully turn the corner just when Ω stays within the hallways. And that is true as long as the corner point (a, b) is outside Ω. The extreme case occurs when (a, b) lies on the boundary curve, whereupon

$$a^{2/3} + b^{2/3} = L^{2/3}.$$

This shows that the longest segment that can go around the corner has length $L = (a^{2/3} + b^{2/3})^{3/2}$.

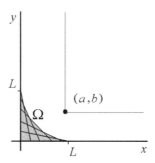

Figure 2 Swept out region Ω

Of course, this solution depends on knowing the boundary curve for Ω. Once you know that, the problem becomes transparent. We can easily visualize the region for a short segment that will go around the corner, and just as easily see what happens if we increase the length of the segment. In contrast with the usual approach to this problem, we are led to a direct understanding of the maximization process. And in the context of the equation for the astroid, we understand *why* the formula for the extreme value of L takes the form that it does.

In fact, in this direct approach, the optimization part of the problem becomes trivial. It is akin to asking "What is the longest segment that can be contained within the unit interval?" This is nominally a max/min problem, but no analysis is needed to solve it. In the same way, the direct approach to the ladder problem renders the solution immediately transparent, once you have found the boundary curve for Ω. But it is not quite fair to claim that this approach eliminates the need for calculus. Rather, the point of application of the calculus is shifted from the optimization question to that of finding the boundary curve.

Envelopes of families of curves

So how is the boundary curve found? The key is to observe that it is the *envelope* of a family of curves. For the current case, notice that each successive position of the line segment can be identified with a linear equation. Let the angle between the segment and the positive x axis be α, as shown in FIGURE 3. Then the x and y intercepts of the line segment are $L \cos \alpha$ and $L \sin \alpha$, respectively, so the line is defined by the equation

Solving the Ladder Problem on the Back of an Envelope

$$\frac{x}{\cos\alpha} + \frac{y}{\sin\alpha} = L. \qquad (2)$$

This equation defines a family of lines in terms of the parameter α.

The region Ω is the union of all the lines in the family. To be more precise, we restrict α to the interval $[0, \pi/2]$, and intersect all the lines with the first quadrant. Visually, it seems apparent that the boundary curve is tangent at each point to one of the lines. This observation, which will be proved presently, shows that the boundary curve is an envelope for the family of lines.

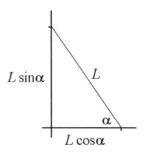

Figure 3 Parameter α

In general, an equation of the form

$$F(x, y, \alpha) = 0 \qquad (3)$$

defines a family of plane curves with parameter α if for each value of α the equation defines a plane curve C_α in x and y. An *envelope* for such a family is a curve every point of which is a point of tangency with one of the curves in the family.

There is a standard method for determining the envelope curve: Differentiate (3) with respect to α, and then use the original equation to eliminate the parameter. Technically, (x, y) is a point of the envelope curve only if it satisfies both (3) and

$$\frac{\partial}{\partial \alpha} F(x, y, \alpha) = 0 \qquad (4)$$

for some α. Combining equations (3) and (4) to eliminate α produces an equation in x and y. This will be referred to as the *envelope algorithm*.

Obviously, the envelope algorithm depends on certain assumptions about F, requiring at the very least differentiability with respect to α. Also, in the general case, the condition is necessary but not sufficient, so there may exist curves which satisfy (3) and (4), but which are not part of the envelope. For the moment, let us gloss over these issues, and move straight on to applying the algorithm for the ladder problem. A more careful discussion of the technicalities will follow.

The first step is to differentiate (2) with respect to α. That gives

$$\frac{x \sin\alpha}{\cos^2\alpha} - \frac{y \cos\alpha}{\sin^2\alpha} = 0$$

and after rearrangement we obtain

$$x \sin^3 \alpha = y \cos^3 \alpha. \tag{5}$$

By combining this equation with (2), we wish to eliminate the parameter α. With that in mind, rewrite (5) in the form

$$\tan \alpha = \frac{y^{1/3}}{x^{1/3}}.$$

This leads to

$$\cos \alpha = \frac{x^{1/3}}{\sqrt{x^{2/3} + y^{2/3}}} \quad \text{and} \quad \sin \alpha = \frac{y^{1/3}}{\sqrt{x^{2/3} + y^{2/3}}}.$$

Now we can substitute these expressions in (2), and so derive the following equation in x and y alone.

$$x^{2/3}\sqrt{x^{2/3} + y^{2/3}} + y^{2/3}\sqrt{x^{2/3} + y^{2/3}} = L$$

Simplifying, we have

$$(x^{2/3} + y^{2/3})\sqrt{x^{2/3} + y^{2/3}} = (x^{2/3} + y^{2/3})^{3/2} = L$$

and so we arrive at (1).

We can also derive a parameterization of the envelope. In the equations above for $\cos \alpha$ and $\sin \alpha$, replace $\sqrt{x^{2/3} + y^{2/3}}$ with $L^{1/3}$. Solving for x and y produces

$$\begin{aligned} x(\alpha) &= L \cos^3 \alpha \\ y(\alpha) &= L \sin^3 \alpha. \end{aligned} \tag{6}$$

In this parameterization, $(x(\alpha), y(\alpha))$ is the point of the envelope that lies on the line corresponding to parameter value α.

Pedagogy

The foregoing computation is an intriguing way to deduce the boundary curve for the region Ω. From that curve we can immediately find the solution to the ladder problem as discussed earlier. As elegant as this solution is, it may be inaccessible to today's calculus students. Interestingly, there is some evidence to suggest that the computation of envelopes via the method above was once a standard topic in calculus. This is certainly the impression left by [5, 8, 12], all of which date to the 1940's and 1950's. On the other hand, anecdotal reports by colleagues who were students and teachers of calculus during that time are inconsistent on this point.

In today's calculus texts (or more precisely, in their indices), one finds no mention of envelopes. The topic is covered in older treatments of calculus [4, 14] and advanced calculus [25] and the expositions in these sources tend to be very similar. My informal survey (as described above) suggests that the topic of envelopes was common in calculus texts throughout the 19th century. Was the topic common enough in the calculus curriculum in the first half of the twentieth century to be considered standard? If so, when and why did this topic fall out of favor? These are interesting historical questions.

If the topic of envelopes has been forgotten in calculus texts, it has not disappeared from the mathematical literature. Indeed, in expository publications like this MAGAZINE, one readily finds recent mention of envelopes and the envelope algorithm. See, for example, [1, 11, 16, 20, 22, 23]. There is also an application of envelopes in the field of economics, referred to as the *Envelope Theorem* [18, 26]. Nevertheless, I have a feeling that this topic is not as widely known among college mathematics faculty as it should be. Accordingly, a rather detailed discussion of envelopes is presented in the next section.

Outside of calculus courses, where might envelopes by found? The topic appears in works on properties of plane curves (see [19, 28]), another subject that seems to have been much more common in an earlier era. To a previous generation of mathematicians who were well acquainted with such terms as *involute*, *evolute*, and *caustic*, the boundary curve (1) would be familiar indeed. It is known not only as an astroid, but more generally as an instance of hypocycloid, the locus of a point on a circle rolling within a larger circle. We obtained it as the envelope of the family of lines (2), identified in [28] as the *Trammel of Archimedes*. The same curve can also be obtained as the envelope of a family of ellipses, the sum of whose axes is equal to L [28, p. 2]. See FIGURE 4.

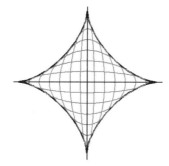

Figure 4 The astroid as envelope of a family of ellipses

The treatment of envelopes in [28] implies that this topic is properly a part of the study of differential equations. Perhaps it is in this context that envelopes once were considered a standard calculus topic, although that is certainly not the case in [4, 14, 25].

However the historical questions are answered, it is something of a shame that envelopes are not included in modern curricula, even for enrichment. The topic has obvious visual appeal, and the method is an attractive application of differentiation. In addition, the consideration of why and how the method works leads to interesting insights. And of course, if our students knew about envelopes, the solution of the ladder problem would be much simplified. Still, taking everything into consideration, this topic is probably too great a digression for most calculus classes. No doubt, embarking on such a digression just to reach an elegant solution to the ladder problem would be (dare I say it) *pushing the envelope*.

As a compromise, it might be reasonable to guide students through a construction of the boundary curve of Ω, without using the general method of envelopes. Here is one approach. Consider sweeping out the region Ω using a segment of length L. For each value of α, there will be one position of the line segment, given by (2). Now for a fixed value of x_0, consider the points $(x_0, y_0(\alpha))$ that lie on the various line segments. Evidently, the maximum value $y_0(\alpha)$ defines the point of the boundary curve corresponding to x_0. From (2), we have

$$y_0(\alpha) = L \sin \alpha - x_0 \tan \alpha$$

and the maximum value for $0 \le \alpha \le \pi/2$ is easily found to be

$$y_0 = (L^{2/3} - x_0^{2/3})^{3/2}.$$

In this way, the boundary curve is obtained. But to solve the ladder problem, we do not really need the entire boundary curve. All we need to know is where the point (a, b) lies relative to the boundary curve. So, in the preceding argument, simply take $x_0 = a$. This provides another approach to the ladder problem.

Technicalities

In deriving the envelope algorithm, one generally assumes that locally the envelope is a curve smoothly parameterized by α. By definition, each point P of the envelope is a point of tangency to some member of the family of curves, and each member of the family is tangent to the envelope at some point P. This suggests that P can be defined as a function of α. However, some caution is necessary. If a curve in the family touches the envelope in multiple points, there will be ambiguity in defining $P(\alpha)$. This is the situation when we generate the astroid as the envelope of a family of ellipses, as in FIGURE 4. In this case there are many functions $P(\alpha)$ that map the parameter domain to the envelope, not all of which are continuous. Of course in this example it is possible to choose $P(\alpha)$ consistently to obtain a smooth parameterization of the envelope. But it is not clear how this can be done in general. Accordingly, we assume the envelope has a smooth parameterization $P(\alpha) = (x(\alpha), y(\alpha))$ such that $P(\alpha)$ is the point of tangency between the envelope and the curve C_α.

With that assumption, observe that the equation $F(x(\alpha), y(\alpha), \alpha) = 0$ holds identically, so the derivative with respect to α is zero. Viewing F as a function of three variables, the chain rule gives

$$\frac{\partial F}{\partial x}\frac{dx}{d\alpha} + \frac{\partial F}{\partial y}\frac{dy}{d\alpha} + \frac{\partial F}{\partial \alpha} = 0. \tag{7}$$

But we can also view F as a function of two variables, thinking of α as a fixed parameter. In this view, the xy gradient of F is normal to the curve $F(x, y, \alpha) = 0$ at each point. Meanwhile, the parameterization of the envelope provides a tangent vector $(\frac{dx}{d\alpha}, \frac{dy}{d\alpha})$ at each point of that curve. At the point $(x(\alpha), y(\alpha))$, the two curves are tangent, so the normal vector $\nabla_{xy} F = (\frac{\partial F}{\partial x}, \frac{\partial F}{\partial y})$ is orthogonal to the velocity vector $(\frac{dx}{d\alpha}, \frac{dy}{d\alpha})$. This shows that the first two terms on the left side of (7) add to 0, and hence $\frac{\partial F}{\partial \alpha} = 0$.

The preceding argument is the basis for the envelope algorithm. It shows that at each point (x, y) of the envelope there is a value of α for which both (3) and (4) hold. This is a necessary condition, and it can be satisfied by points which are not on the envelope. Indeed, we can construct an example of this phenomenon by reparameterizing the family of curves. The general idea behind this construction will be clear from the specific case of the family of lines for the ladder problem.

Let $s(\alpha)$ be any differentiable function from $(0, \pi/2)$ onto itself. Then the equation

$$\frac{x}{\cos s(\alpha)} + \frac{y}{\sin s(\alpha)} = L$$

parameterizes the same family of lines as (2), and so has the same envelope. Applying the envelope algorithm, we compute the partial derivative with respect to α, obtaining

$$\frac{x \sin s(\alpha)\, s'(\alpha)}{\cos^2 s(\alpha)} - \frac{y \cos s(\alpha)\, s'(\alpha)}{\sin^2 s(\alpha)} = 0.$$

Clearly, this equation will be satisfied for any value of α where $s'(\alpha) = 0$. If $\alpha*$ is such a point, then every point of the corresponding line

$$\frac{x}{\cos s(\alpha*)} + \frac{y}{\sin s(\alpha*)} = L$$

satisfies the two conditions of the envelope algorithm. That is, the entire line segment corresponding to $\alpha*$ will be produced by the envelope algorithm. No such line is actually included in the boundary of Ω, nor can any such line be tangent to all the lines in the family. This illustrates how the envelope algorithm can produce extraneous results.

A more complete discussion of these technical points can be found in [4, 19]. In particular, conditions that can give rise to extraneous results from the envelope algorithm are characterized. As Courant remarks, once the envelope algorithm produces a curve, "it is still necessary to make a futher investigation in each case, in order to discover whether it is really an envelope, or to what extent it fails to be one." In practice, graphing software can often give a clear picture of the envelope of a family of curves, and so guide our understanding of the results of the envelope algorithm.

A technical point of a slightly different nature concerns the relationship between the envelope of a family of curves, and the boundary of the region that family encompasses. By definition, the envelope is a curve which is tangent at each of its points to some member of the family. This is the definition used to justify the envelope algorithm. But the curve we are interested in for the ladder problem is defined as a boundary curve. How are these two concepts related? Visually, it appears obvious that at each point of the boundary of Ω, the tangent line is a member of the family of lines defining Ω. We substantiate this appearance as follows.

Since each line segment is contained within the region Ω, none of the lines can cross the boundary curve. On the other hand, each point of the boundary must lie on one of the lines. To see this, consider a boundary point P, and a sequence of points P_j in Ω converging to P. Each point P_j is on a line for some parameter value α_j, and these values all lie in the interval $[0, \pi/2]$. So there is a convergent subsequence α_{j_k} with limit $\alpha*$. Now by the continuity of (2), $P = \lim P_{j_k}$ is a point on the line with parameter $\alpha*$. Since this line cannot cross the boundary curve at P, it must be tangent there.

This suggests as a general principle that the boundary of a region swept out by a family of curves lies on the envelope for that family. As in the earlier discussion, some caution is necessary. Here, it is sufficient to assume that the boundary curve is smoothly parameterized by α, the parameter defining the family of curves. On any arc where this is true, the boundary curve will indeed fall along the envelope. On the other hand, consider the following:

$$F(x, y, \alpha) = x^2 + y^2 - \sin^2 \alpha.$$

This describes a family of circles centered at the origin, with radius varying smoothly between 0 and 1. The region swept out by the family of circles is the closed unit disk $x^2 + y^2 \leq 1$, and the unit circle is the boundary curve. But the unit circle is not an envelope for the family of circles. What went wrong? Arguing as above, we can again assign a value of α to each point of the boundary curve. But to do this continuously, we have to take α to be constant, say $\alpha = \pi/2$. Then the entire boundary is one of the curves in the family, but it is not parameterized by α. Notice as well that in this example, the curve for $\alpha = \pi/2$ also satisfies the equation $\frac{\partial F}{\partial \alpha} = 0$. Thus, while the envelope algorithm fails to produce the envelope in this example, it does locate the boundary of the region.

Relating the envelope to the boundary also gives a different insight about why the envelope technique works. View (3) as defining a level surface S of the function F in $xy\alpha$ space. The family of curves is then defined as the set of level curves for this surface. At the same time, the region Ω swept out by the family of curves is the projection of S on the xy plane. Now suppose A is a point on S that projects to a point P on the boundary of Ω. The tangent plane to S at A must be vertical and project to a line in the xy plane. Otherwise, there is an open neighborhood of A on the tangent plane that projects to an open neighborhood of P in the xy plane, and that puts P in the interior of Ω. So the tangent plane is vertical. That implies a horizontal normal vector to S at A. But that means that the gradient of F, which is normal to the surface at each point, must be horizontal at A. This shows that the partial derivative of F with respect to α vanishes at A, which is the derivative condition of the envelope algorithm.

As a final topic in this section, I cannot resist mentioning one more way to think about the envelope of a family of curves. The idea is to consider a point on the envelope as the intersection point of two *neighboring* members of the family. Each α gives us one member of the family of curves, and the intersection of curves for two *successive* values of α gives a point on the envelope. Of course, that is not literally possible, but we can implement this idea using limits. Just express the intersection of the curves C_α and $C_{\alpha+h}$ as a function of h and α, and take the limit as h goes to 0. It is instructive to carry this procedure out for the example of the ladder problem. It once again yields the envelope as the astroid (1). In the process, one can observe differentiation with respect to α implicitly occuring in the calculation of the limit as h goes to 0. Indeed Courant [4] uses this idea to provide a heuristic derivation of the envelope algorithm, before developing a more rigorous justification. In contrast, Rutter [19] terms this the *limiting position* definition of *envelope*, one of three closely related but distinct definitions that he considers. He also provides the following interesting example of a family of circles with an envelope, but for which neighboring circles in the family are disjoint.

Begin with the ellipse

$$\frac{x^2}{a^2} + \frac{y^2}{b^2} = 1$$

and at each point, compute the osculating circle. This is the circle whose curvature matches that of the ellipse at the specified point, and whose center and radius are the center and radius of curvature of the ellipse. The family of osculating circles for all the points of the ellipse is the focus for this example. This situation is illustrated for an ellipse with $a = 8$ and $b = 4$ in FIGURE 5, which shows the ellipse and several members of the family of circles.

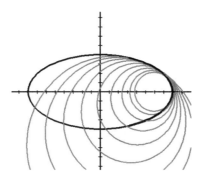

Figure 5 Circles of curvature for an elliptical arc

It is apparent from the construction of this example that the original ellipse is tangent to each of the circles in the family, and so is an envelope for the family. But the ellipse can not be obtained as the limiting points of intersections of neighboring circles. In fact, the neighboring circles are disjoint! This surprising state of affairs is shown in FIGURE 6, with an enlarged view in FIGURE 7.

This same example also exhibits some of the other exceptional behaviors that have been discussed above. One can show that the original ellipse does satisfy the two conditions specified in the envelope algorithm, and so the algorithm would properly identify the envelope for this family of circles. But the entire circles of curvature for each

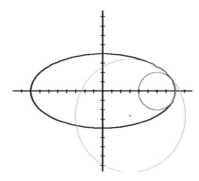

Figure 6 Neighboring circles are disjoint

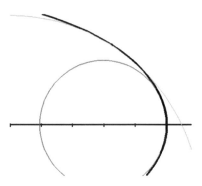

Figure 7 Closeup view of neighboring circles

of the vertices $((\pm a, 0), (0, \pm b))$ also satisfy the conditions of the envelope algorithm, although these circles are *not* part of the envelope. And these circles also contain the boundary of the region swept out by the family of circles. These properties can be observed in the next two figures, defined by two different ellipses. Note also how visually striking these figures are.

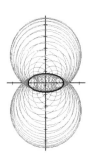

Figure 8 Family of circles for an ellipse

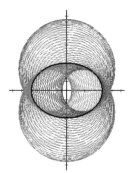

Figure 9 Family of circles for another ellipse

Experimenting with figures of this sort can convey a good deal of understanding of the properties of envelopes discussed above, and I highly recommend it. Modern graphical software is ideally suited to this purpose. For the family of circles in the preceding example, graphical exploration is abetted by the following formulae ([**19**, p. 192]). Identify one point of the ellipse as $(a \cos \alpha, b \sin \alpha)$. Then the radius of curvature at that point is given by

$$\frac{(a^2 \sin^2 \alpha + b^2 \cos^2 \alpha)^{3/2}}{ab}$$

and the center of the circle of curvature is

$$\left(\frac{a^2 - b^2}{a} \cos^3 \alpha, \frac{b^2 - a^2}{b} \sin^3 \alpha \right).$$

Extending the ladder problem

A slight variation on the ladder problem is illustrated in FIGURE 10, with a rectangular alcove in the corner where the two hallways meet. The same configuration might occur if there is some sort of obstruction, say a table or a counter, in one hallway near the corner. As before, the problem is to find how long a line segment will go around this corner.

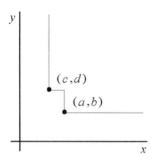

Figure 10 Family of circles for another ellipse

Here, the envelope method again provides an immediate solution. Consider again the region Ω swept out by a family of lines of fixed length L. If this region avoids both points (a, b) and (c, d), then the segment can be moved around the corner. As L increases, the envelope (1) expands out from the origin. The maximal feasible L occurs when the envelope first touches one of the corner points (a, b) and (c, d). This shows that the maximal value of L is given by

$$L_{\max} = \min\left\{(a^{2/3} + b^{2/3})^{3/2}, (c^{2/3} + d^{2/3})^{3/2}\right\}.$$

Going a bit further in this direction, we might replace the inside corner with any sort of curve C (see FIGURE 11). The ladder problem can then be solved by seeking the point (x, y) of C for which $f(x, y) = x^{2/3} + y^{2/3}$ is minimized. Unfortunately, this plan is not so easy to execute. For example, an elliptical arc is a natural choice for the curve C. But even for that simple case the analytic determination of the minimal value of f appears quite formidable, if not impossible. On the other hand, if C is a polygonal path, we need only find the minimum value of f at the vertices.

The couch problem. Extending the problem in a different direction, we can make the situation a bit more faithful to the real world by recognizing that a ladder actually has some positive width. Thus, in the idealized geometry of the problem statement, perhaps we should try to maneuver a rectangle rather than a line segment around the corner. If the width of the rectangle is fixed at w, what is the greatest length L that permits the rectangle to go around the corner?

This version of the problem also provides a reasonable model for moving bulkier objects than ladders. For example, trying to push a desk or a couch around a corner in a corridor is naturally idealized to the problem of moving a rectangle around the corner in FIGURE 1. This is the motivation given by Moretti [17] in his analysis of the rectangle version of the ladder problem. In honor of his work, we refer to the

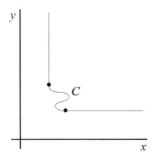

Figure 11 Ladder problem with a curve in the corner

rectangular version hereafter as the couch problem. It should not be confused with the *sofa* problem, which concerns the *area* of a figure to be moved around a corner.

Moretti's analysis mimics the standard solution to the ladder problem. Thus, rather than looking for the longest couch that will go around the corner, he seeks the shortest couch that will get stuck. This occurs when the outer corners of the rectangle touch the outer walls of the corridor, and the inner edge touches the inside corner point, as illustrated in FIGURE 12. Using the slope as a parameter, Moretti reduces the problem to finding a particular root of a sixth degree polynomial.

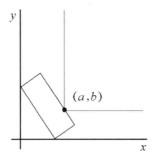

Figure 12 This couch is stuck

For the couch problem, as for the ladder problem, the direct approach using envelopes is illuminating. Indeed, we again make use of the astroid, and one of its *parallel* curves. Here, a *parallel* curve means one whose points are all at a fixed distance from a given curve. From each point P on the given curve C, move a fixed distance w along the normal vector to locate a point Q, being careful to choose the direction of the normal vector consistently. The locus of all such Q is a curve parallel to C at distance w. Parallel curves are discussed in [**19**].

For the couch problem, the envelope we need is parallel to the envelope we found for the ladder problem. This leads to the following appealing geometric interpretation. Let L be the longest rectangle of width w that can be moved around a corner as in the original ladder problem. Then the boundary of Ω (as in FIGURE 2) must be tangent to the circle of radius w centered at the point (a, b). That is, w must be the distance from the corner point (a, b) to the astroid (1). Algebraically, this approach has an appealing simplicity, up to the point of actually finding a solution. Unfortunately, that requires solving a sixth degree equation, which is essentially equivalent to the one considered by Moretti.

On the other hand, the geometric setting of the envelope approach provides a simple method for parameterizing a family of rational solutions to the couch problem. That is, we can specify an infinite set of triples (a, b, w) such that the couch problem has an exact rational solution L. This partially answers one of Moretti's questions. In fairness, though, a similar parameterization can be developed using Moretti's method.

To apply the envelope method to the couch problem, we adopt the same strategy as for the ladder problem. Consider the following process for moving a rectangle around a corner. Initially, the rectangle is aligned with the walls of the corridor, so that the bottom of the rectangle is on the x axis and the left side is on the y axis. Slide the rectangle in such a way that the lower left-hand corner follows the y axis, while keeping the lower right-hand corner on the x axis. Thus, the bottom edge of the rectangle follows the exact trajectory of the segment in the ladder problem, sweeping out the region Ω, as before. But now we want to look at the region swept out by the entire rectangle. The upper boundary of this region is the envelope of the family of lines corresponding to the motion of the *top* edge of the rectangle. For a couch with length L and width w, these lines are characterized as follows. Begin with a line in the family for the original ladder problem, whose intersection with the first quadrant has length L. Construct the parallel line at distance w (and in the direction away from the origin). We seek the envelope of the family of all of these parallel lines.

As before, we parameterize the lines in this new family in terms of the angle α between such a line and the (negatively directed) x axis. The parallel unit vector is given by $\mathbf{m} = (-\cos\alpha, \sin\alpha)$, and the normal unit vector (pointing into the first quadrant) is $\mathbf{n} = (\sin\alpha, \cos\alpha)$. These vectors provide a simple way to define a line at a specified distance d from the origin: begin with the line through the origin parallel to $\mathbf{m}$, and translate by $d\mathbf{n}$. That defines a point on the line as

$$(x, y) = t\mathbf{m} + d\mathbf{n}.$$

Taking the dot product of both sides of this equation with $\mathbf{n}$ thus gives

$$\sin\alpha\, x + \cos\alpha\, y = d.$$

Lines in the original family are described by (2), which we rewrite as

$$\sin\alpha\, x + \cos\alpha\, y = L\sin\alpha\cos\alpha.$$

This line is at a distance $L\sin\alpha\cos\alpha$ from the origin. Now we want the parallel line that is w units further away. The equation for that line is evidently

$$\sin\alpha\, x + \cos\alpha\, y = L\sin\alpha\cos\alpha + w.$$

To make use of the envelope algorithm, let us define the function

$$G(x, y, \alpha) = \sin\alpha\, x + \cos\alpha\, y - L\sin\alpha\cos\alpha - w.$$

Then, thinking of α as a fixed value, the equation $G(x, y, \alpha) = 0$ defines one line in the family. Similarly, with

$$F(x, y, \alpha) = \sin\alpha\, x + \cos\alpha\, y - L\sin\alpha\cos\alpha$$

we obtain the lines in the original family by setting $F(x, y, \alpha) = 0$. It will be convenient in what follows to express these functions in the form

$$F(x, y, \alpha) = \mathbf{n} \cdot (x, y) - L \sin\alpha \cos\alpha$$
$$G(x, y, \alpha) = \mathbf{n} \cdot (x, y) - L \sin\alpha \cos\alpha - w$$

Our goal is to find the envelope for the lines defined by G, (hereafter, the envelope for G). According to the envelope algorithm, we should eliminate α from the equations

$$G(x, y, \alpha) = 0$$
$$\frac{\partial}{\partial \alpha} G(x, y, \alpha) = 0$$

But rather than apply this directly, we can use the fact that we know the envelope for F. In fact, since each line in G's family is parallel to a corresponding line in F's family, and at a uniform distance w, it is not surprising that the envelope of G is parallel to the envelope of F, and at the same distance. That is, if (x, y) is on the envelope of F, then the corresponding point of the envelope of G is w units away in the normal direction.

To make this more precise, let us consider a point (x, y) on the envelope of F. There is a corresponding α such that (x, y, α) is a zero of both F and $\frac{\partial F}{\partial \alpha}$. Then (x, y) is on the line with parameter α, which is tangent to the envelope of F at (x, y). Thus, at this point, the line and the envelope share the same normal direction. As observed earlier, the unit normal is given by $\mathbf{n} = (\sin\alpha, \cos\alpha)$. We will now consider a new point $(x', y') = (x, y) + w\mathbf{n}$. We wish to show that (x', y') is on the envelope of G.

To that end, observe that $F(x, y, \alpha) = G(x', y', \alpha)$ and $\frac{\partial F}{\partial \alpha}(x, y, \alpha) = \frac{\partial G}{\partial \alpha}(x', y', \alpha)$. To justify the first of these equations,

$$G(x', y', \alpha) = \mathbf{n} \cdot (x', y') - L \sin\alpha \cos\alpha - w$$
$$= \mathbf{n} \cdot [(x, y) + w\mathbf{n}] - L \sin\alpha \cos\alpha - w$$
$$= \mathbf{n} \cdot (x, y) + w - L \sin\alpha \cos\alpha - w$$
$$= F(x, y, \alpha)$$

To justify the second equation, we observe first that since F and G differ by a constant, they have the same derivatives. Also, note that $\frac{\partial \mathbf{n}}{\partial \alpha} = -\mathbf{m}$. This gives $\frac{\partial G}{\partial \alpha}(x, y, \alpha) = \frac{\partial F}{\partial \alpha}(x, y, \alpha) = -\mathbf{m} \cdot (x, y) - L(\cos^2\alpha - \sin^2\alpha)$. Now we can write

$$\frac{\partial G}{\partial \alpha}(x', y', \alpha) = -\mathbf{m} \cdot [(x, y) + w\mathbf{n}] - L(\cos^2\alpha - \sin^2\alpha)$$
$$= -\mathbf{m} \cdot (x, y) - L(\cos^2\alpha - \sin^2\alpha)$$
$$= \frac{\partial F}{\partial \alpha}(x, y, \alpha).$$

Together, these results show that (x, y) is on the envelope of F if and only if (x', y') is on the envelope of G, and that in each case the points (x, y) and (x', y') correspond to the same value of α. In fact, this result reflects a more general situation: If the family G consists of parallels of the curves in F, all at a fixed distance w, then the envelope for G is the parallel of the envelope of F, at the same distance w. In the context of the couch problem, we can find the needed envelope of G as a parallel to the known envelope of F.

Based on earlier work, we know that the envelope of F is parameterized by the equations
$$x = L \cos^3 \alpha$$
$$y = L \sin^3 \alpha.$$

That leads immediately to the following parametric description of the envelope of G:
$$x = L \cos^3 \alpha + w \sin \alpha$$
$$y = L \sin^3 \alpha + w \cos \alpha.$$

For the solution L of the couch problem, the point (a, b) must lie on the envelope of G. Therefore, we can find L (and also find the critical value of α) by solving the system
$$a = L \cos^3 \alpha + w \sin \alpha$$
$$b = L \sin^3 \alpha + w \cos \alpha.$$

This leads readily enough to an equation in α alone:
$$a \sin^3 \alpha - b \cos^3 \alpha = w(\sin^2 \alpha - \cos^2 \alpha). \tag{8}$$

At this point, finding α appears to depend on solving a sixth degree polynomial equation. To derive such an equation, substitute x for $\sin \alpha$ and $\sqrt{1-x^2}$ for $\cos \alpha$ in (8) to obtain
$$ax^3 - b(1-x^2)^{3/2} = w(2x^2 - 1).$$

Isolating the term with the fractional exponent and squaring both sides then leads to the equation
$$(a^2 + b^2)x^6 - 4awx^5 + (4w^2 - 3b^2)x^4 + 2awx^3 + (3b^2 - 4w^2)x^2 + w^2 - b^2 = 0.$$

It is is easy to solve this equation numerically (given values for a, b, and w), and very likely impossible to solve it symbolically.

As mentioned, the foregoing analysis leads to a nice geometric interpretation for the solution. If (a, b) is on the envelope of G, then there is a corresponding point (x, y) on the envelope of F. We know that (x, y) is w units away from (a, b), and that the vector between these two points is normal to the envelope of F. This shows that the circle centered at (a, b) of radius w is tangent to the envelope of F at (x, y).

Visually, we can see how to find the maximum value of L. Start with a small enough L so that the astroid (1) stays well clear of the circle about (a, b) of radius w. Now increase L, expanding the astroid out from the origin, until the curve just touches the circle. When that happens, the corresponding value of L is the solution to the couch problem. See FIGURE 13.

The visual image of solving the couch problem in this way is reminiscent of Lagrange Multipliers. Indeed, what we have is the dual of a fairly typical constrained optimization problem: find the point on the curve (1) that is closest to (a, b). The visual image for that problem is to expand circles centered at (a, b) until one just touches the astroid. Our dual problem is to hold the circle fixed and look at level curves for increasing values of the function $f(x, y) = x^{2/3} + y^{2/3}$. We increase the value of f until the corresponding level curve just touches the fixed circle. This geometric conceptualization is associated with the following optimization problem: Find the minimum value of $f(x, y)$ where (x, y) is constrained to lie on the circle of radius w centered at (a, b). We will return to the idea of dual problems in the last section of the paper.

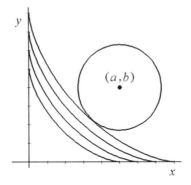

Figure 13 Maximizing L geometrically

Let us examine more closely the Lagrangian-esque version of the couch problem. We wish to find a point of tangency between the following two curves

$$x^{2/3} + y^{2/3} = L^{2/3}$$
$$(x - a)^2 + (y - b)^2 = w^2.$$

For each curve, we can compute a normal vector as the gradient of the function on the left side of the curve's equation. Insisting that these gradients be parallel leads to the following additional condition

$$x^{1/3}(x - a) = y^{1/3}(y - b).$$

In principle, solving these three equations for x, y, and L would produce the desired solution L to the couch problem. Or, solving the second and third for x and y, and then substituting those values in the first equation, also would lead to the value of L. Unfortunately, every approach appears to lead inevitably to a sixth degree equation.

While the envelope method does not seem to provide a symbolic solution to the couch problem, it does provide a nice procedure for generating solvable examples. Here, we will begin with a value of L and produce a triple (a, b, w) so that L is a solution to the (a, b, w) couch problem. To begin, we generate some *nice* points on the astroid $x^{2/3} + y^{2/3} = L^{2/3}$ using Pythagorean triples. Specifically, if $r^2 + s^2 = t^2$, we can take $x = r^3$, $y = s^3$ and $L = t^3$ to define a point on an astroid. In particular, we can generate an abundance of rational points on astroids. Notice that the original Pythagorean triple need not be rational. For example, if $(r, s, t) = (3, 4, 5)/\sqrt[3]{5}$, we find $(27/5, 64/5)$ as a rational point on the astroid curve for $L = 25$.

Now the equations $x = r^3$, $y = s^3$ are closely related to the parameterization

$$x = L \cos^3 \alpha \quad y = L \sin^3 \alpha$$

of the astroid. As a result, we can recover α from the equations

$$\cos\alpha = \frac{r}{t} \quad \sin\alpha = \frac{s}{t}.$$

This in turn gives us the normal vector $\mathbf{n} = (\frac{s}{t}, \frac{r}{t})$, and hence, for any value of w, leads to the point (x', y'). Define that point to be (a, b). It necessarily lies on the envelope for G. This shows that the L for the astroid constructed at the outset solves the (a, b, w) couch problem. We formalize these arguments in the following theorem.

THEOREM. *For any positive pythagorean triple (r, s, t) and any positive w define*

$$a = r^3 + w\frac{s}{t}$$

$$b = s^3 + w\frac{r}{t}$$

$$L = t^3$$

Then L is the solution to the (a, b, w) couch problem.

For example, with $(r, s, t) = (3, 4, 5)/\sqrt[3]{5}$ and $w = 2$, the equations above give $(a, b) = (7, 14)$ and $L = 25$. So for a rectangle of width 2, 25 is the maximum length that will fit around the corner defined by the point $(7, 14)$. In general, if (r', s', t') is a rational Pythagorean triple, and if u^3 is rational, then taking $(r, s, t) = u(r', s', t')$ and rational w produces rational values of a, b, and L, as well as a rational point (x, y) where the astroid meets the circle centered at (a, b) of radius w.

The preceding example, where $L = 25$ is the solution of the $(7, 14, 2)$ couch problem, was given by Moretti. He mentioned that such examples are relatively rare, and asked for conditions on a, b, and w that make the (a, b, w) couch problem exactly solvable. The theorem above provides a partial answer to Moretti's question, by providing an infinite family of such triples. It would be nice to know whether every rational (a, b, w) with rational solution L to the couch problem arises in this way. If the critical value of α corresponds to a rational point (x, y) on the astroid (1), then a, b, w, and L are related as in the theorem. But there might be rational (a, b, w) for which the solution to the couch problem is also rational, but which does not correspond to a rational point (x, y).

The envelope approach leads in a natural way to the theorem, and provides a nice geometric interpretation of the couch problem solution. But it should be observed that Moretti's approach can also lead to an equivalent method for parameterizing triples (a, b, w) with rational solution L. He formulates the problem in terms of a variable m (corresponding to $\cot\alpha$ in this paper) and derives a sixth degree equation in m with coefficients that depend on a, b, and w. If that equation is solved for w, one can again parameterize solutions in terms of Pythagorean triples. From this standpoint, the envelope method does not seem to hold any advantage over Moretti's earlier analysis.

Duality in the ladder problem

In discussing the tangency condition for an astroid and a circle, the idea of dual optimization problems was briefly mentioned. As a concluding topic, we will look at this idea again.

Segalla and Watson [21] discuss what they call the flip side of a constrained optimization problem in the context of Lagrange multipliers. For example, in seeking to maximize the area of a rectangle with a specified perimeter, we have an objective function (the area) and a constraint (the perimeter). At the solution point, the level curve for the extreme value of the objective function is tangent to the given level curve of the constraint function. Here, the roles of the objective and constraint are symmetric, and can be interchanged. Given the maximal area, we can ask what is the minimal perimeter that can enclose a rectangle having this area. The solution corresponds to the same point of tangency between level curves of the objective and constraint functions. Thus, we see that the problem of maximizing area with a fixed perimeter, and minimizing the perimeter with a fixed area are linked.

Maximizing area with fixed perimeter is the famous isoperimetric problem (see Blåsjö [2] for a beautiful discussion), and in that context minimizing the perimeter with fixed area is referred to as the *dual* problem. Duality in this sense corresponds to Segalla and Watson's idea of flip side symmetry. It is also reminiscent of the idea of duality in linear and non-linear programming. There, although the primal and dual optimization problems occur in different spaces, one again finds the idea of linked problems whose solutions somehow coincide.

Duality permits information about one problem to be inferred from information about its dual problem. This property is important in both the isoperimetric problem and in linear programming. As Segalla and Watson point out, a solution to an initial optimization problem immediately leads to a corresponding statement and solution of a dual problem. Thus, discovering that a rectangle with perimeter 40 has maximal area 100, also tells us at once that a rectangle with area 100 has minimal perimeter 40. But there is another way to use duality. If you are unable to solve an optimization problem, try to solve the dual.

Here is how this works for the perimeter-area problem. Imagine that we do not know how to maximize the area of a rectangle with perimeter 40. The dual problem is to minimize the perimeter subject to a given area, but of course we do not know what that fixed area should be. So we solve the general problem for a fixed area of A. That is, we prove that area A occurs with a minimal perimeter of $4\sqrt{A}$. Now relate this to the original problem by insisting on a perimeter of 40. That forces $A = 100$, and tells us this: for area 100, the minimal perimeter is 40. The dual statement now solves the original problem.

Segalla and Watson give several examples of pairs of dual problems. In addition to the area-perimeter example mentioned above, they discuss the *milkmaid problem*: find the minimum distance the milkmaid must walk from her home to fetch water from a river and take it to the barn. In this case the dual problem fixes the length of the milkmaid's hike, and minimally shifts the river to accommodate her. They also give the example of the ladder problem, but do not describe the dual. Indeed, they ask for an interpretation of the dual ladder problem. Let us answer this question using envelopes, and see how the dual version leads to another solution of the ladder problem.

Segalla and Watson use the standard approach to the ladder problem—finding the minimal length segment that will get stuck in the corner. We formulate this as a constrained optimization problem in terms of variables u and v, interpreted as intercepts on the x and y axes of a line segment in the first quadrant. The objective function, $f(u, v) = \sqrt{u^2 + v^2}$, is the distance between the intercepts. The goal is to minimize this distance subject to the constraint that the line must pass through (a, b).

For the dual problem, if we hold f fixed, and look at varying values of the constraint function, what does that mean? An answer will depend, naturally, on how the constraint g is formulated. Here is one approach. A line with intercepts u and v satisfies the equation
$$\frac{x}{u} + \frac{y}{v} = 1.$$
From this equation, the condition that (a, b) lie on the line is
$$\frac{a}{u} + \frac{b}{v} = 1.$$
Accordingly, define $g(u, v) = a/u + b/v$. Now observe that $g(u, v) = t$ means that $(a/t, b/t)$ lies on a line with intercepts u and v. That is, g measures the (reciprocal of the) distance from the origin to the line for u and v along the ray through (a, b). This gives the following meaning to the dual problem: Look at all the lines with intercepts u and v, where fixing the value of f means that the distance between these intercepts is constant. Among all these lines, find the one whose distance from the origin, measured in the direction of (a, b) is a maximum.

The by-now-familiar astroid appears once again as the envelope of a family of lines. Holding f constant with value L, we are again considering the family of line segments of length L with ends on the positive x and y axes, filling up the region Ω. The point that maximizes g will now be the furthest point you can reach in Ω traveling on the ray from the origin to (a, b). That is, the solution occurs at the intersection of the ray with the envelope (1).

The solution point will be of the form $(a/t, b/t)$ where t is the optimal value of g. Substituting in (1), we find $t = ((a/L)^{2/3} + (b/L)^{2/3})^{3/2}$. This gives us the optimal value of g for the dual problem, in terms of L. To return to the primal problem, we have to choose the value of L that gives us the original constraint value for g, namely $g = 1$. So with $t = ((a/L)^{2/3} + (b/L)^{2/3})^{3/2} = 1$ we again find $L = (a^{2/3} + b^{2/3})^{3/2}$.

It is interesting that the ladder problem has so many formulations. The usual approach is to reverse the original problem, so that we seek a minimal line that cannot go around the corner rather than a maximal line that will go around the corner. The envelope approach presented here deals directly with the problem as stated, finding the maximum line that will fit around the corner. A third approach is to take the dual of the reversed version, viewed as an example of constrained optimization. Although all of these approaches are closely related, each contributes a slightly different understanding of the problem.

REFERENCES

1. Leah Wrenn Berman, Folding Beauties, *College Math. J.* **37** (2006) 176–186.
2. Viktor Blåsjö, The Isoperimetric Problem, *Amer. Math. Monthly* **112** (2005) 526–566.
3. John Cooper, *Ladder Problem Query*, private correspondence, 2006.
4. Richard Courant, *Differential and Integral Calculus, Volume 2*, tranlated by E. J. McShane, Interscience, New York, 1949.
5. L. M. Court, Envelopes of Plane Curves, *Amer. Math. Monthly* **57** (1950) 168–169.
6. H. T. Croft, K. J. Falconer, and R. K. Guy, *Unsolved Problems in Geometry*, Springer-Verlag, New York, 1994.
7. William Holding Echols, *An Elementary Text-Book on the Differential and Integral Calculus*, Holt, New York, 1902.
8. Howard Eves, A Note on Envelopes, *Amer. Math. Monthly* **51** (1944) 344.

9. S. R. Finch, *Moving Sofa Constant,* section 8.12 (pp. 519–523) in Mathematical Constants, Cambridge University Press, Cambridge, England, 2003.
10. T. J. Fletcher, Easy Ways of Going Round the Bend, *Mathematical Gazette* **57** (1973) 16–22.
11. Peter J. Giblin, Zigzags, this MAGAZINE, **74** (2001) 259–271.
12. J. W. Green, On the Envelope of Curves Given in Parametric Form, *Amer. Math. Monthly* **59** (1952) 626–628.
13. James Haddon, *Examples and Solutions of the Differential Calculus,* Virtue, London, 1862.
14. Morris Kline, *Advanced Calculus,* 2nd ed., Wiley, New York, 1972.
15. H. E. Licks, *Recreations in Mathematics,* D. Van Nostrand, New York, 1917, p. 89.
16. Brian J. Loe and Nathanial Beagley, The Coffee Cup Caustic for Calculus Students, *College Math. J.* **28** (1997) 277–284.
17. Christopher Moretti, Moving a Couch Around a Corner, *College Math. J.* **33** (2002) 196–201.
18. The Economics Professor, *Envelope Theorem,* Arts & Sciences Network, http://www.economyprofessor.com/economictheories/envelope-theorem.php.
19. John W. Rutter, *Geometry of Curves,* Chapman & Hall/CRC, Boca Raton, 2000
20. Mark Schwartz, The Chair, the Area Rug, and the Astroid, *College Math. J.* **26** (1995) 229–231.
21. Angelo Segalla and Saleem Watson, The Flip-Side of a Lagrange Multiplier Problem, *College Math. J.* **36** (2005) 232–235.
22. Andrew Simoson, An Envelope for a Spirograph, *College Math. J.* **28** (1997) 134–139.
23. Andrew Simoson, The Trochoid as a Tack in a Bungee Cord, this MAGAZINE **73** (2000) 171–184.
24. David Singmaster, *Sources in Recreational Mathematics, an Annotated Bibliography,* article 6.AG. http://us.share.geocities.com/mathrecsources/.
25. Angus E. Taylor and W. Robert Mann, *Calculus: an Intuitive and Physical Approach,* 2nd ed., Wiley, New York, 1977.
26. D Thayer Watkins, *The Envelope Theorem and Its Proof,* San Jose State University, http://www2.sjsu.edu/faculty/watkins/envelopetheo.htm.
27. Eric W. Weisstein, *Moving Sofa Problem,* From MathWorld—A Wolfram Web Resource, http://mathworld.wolfram.com/MovingSofaProblem.html.
28. Robert C. Yates, *Curves and their Properties,* National Council of Teachers of Mathematics, Reston, VA, 1974

How Not to Land at Lake Tahoe!

Richard Barshinger

The following problem gives a simplified model of landing an airplane. It is adapted and extended from Trim [1] and is regularly presented in first semester calculus at my campus, where it is unanimously enjoyed and wins some converts to the methods of calculus.

Problem. An aircraft landing approach pattern is shaped generally as in Figure 1 below. The following conditions are imposed:

a) The cruising altitude is h when descent begins at a horizontal distance L from the airstrip.
b) A constant horizontal airspeed U must be maintained throughout descent (somewhat unrealistic).
c) At no time must the vertical component of acceleration exceed (in absolute value) some fixed constant k, $0 \le k \ll g$, where g is the acceleration constant for gravity; i.e., $g = 32$ ft/sec^2 (English units).

Model the plane's approach path by means of a cubic polynomial, using a coordinate system with origin at the beginning of the runway, so that descent starts at the point $(x, y) = (-L, h)$, in units of your choice. Impose suitable conditions at the beginning of descent and at touchdown. Discuss the implications of condition c) above, in the cases: 1) transcontinental flight; and, 2) peculiar airport situations (such as at South Lake Tahoe, CA).

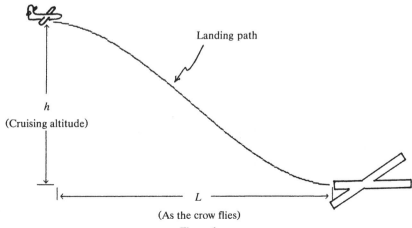

Figure 1

Solution. We let the landing path have the form:
$$y(x) = ax^3 + bx^2 + cx + d.$$

The following reasonable conditions are imposed:

$$\left.\begin{array}{ll} y(0) = 0 & \text{(touchdown)} \\ \dfrac{dy}{dx}\bigg|_{x=0} = 0 & \text{(no crash)} \end{array}\right\} \text{ imply } c = d = 0;$$

$$\left.\begin{array}{ll} y(-L) = h & \text{(descent)} \\ \dfrac{dy}{dx}\bigg|_{x=-L} = 0 & \text{(no dive)} \end{array}\right\} \text{ imply } \begin{array}{l} a = 2h/L^3 \\ b = 3h/L^2. \end{array}$$

Thus these conditions give:
$$y(x) = h\{2(x/L)^3 + 3(x/L)^2\},$$

where x/L is a dimensionless coordinate.

By using the chain rule (with the simplification of constant horizontal airspeed component $dx/dt = U$), we obtain:

$$v_y = \frac{dy}{dt} = \frac{6Uh}{L}\{(x/L)^2 + (x/L)\}$$

and

$$a_y = \frac{d^2y}{dt^2} = \frac{6U^2h}{L^2}\{2(x/L) + 1\}.$$

Now,

$$(a_y)_{\text{max(min)}} = (\overset{+}{-})\frac{6U^2h}{L^2},$$

which occur at $(0,0)$ and $(-L, h)$, respectively. [Hence the airport approach resembles a ride in an elevator, where we "feel" the motion only at the top and bottom of descent.] Since we want $|a_y|_{\text{max}} \leq k \ll g$, we have:

$$\frac{6U^2h}{L^2} \leq k.$$

Implications. 1) Los Angeles to New York (LAX to JFK) transcontinental flight aboard a jumbo ("heavy") jet.

$$L \geq \sqrt{\frac{6U^2h}{k}}.$$

If U and h are large, while k is small, L (the distance from the airport where descent begins) must be relatively big. On such a flight, with an airspeed of $U = 600$ mph and a cruising altitude of $h = 37{,}000$ ft, the author discovered, from his own experience, that descent began at his home near Scranton, PA, about 130 miles from New York! This will make the value of k, which is given by:

$$k = \frac{6U^2h}{L^2 \cdot (3600)^2},$$

come out to $k = 0.36$ ft/sec^2. [The value $(3600)^2$ converts k from ft/hr^2 to ft/sec^2, since a mix of units such as mph and ft is actually in use by airlines (as opposed to mathematicians?)!]

2) San Francisco to South Lake Tahoe. Here we solve for U and obtain:

$$U \leq \sqrt{\frac{kL^2}{6h}}.$$

If L and k are small but h is relatively large, and if we don't want our coffee or the flight attendant to go floating about the cabin, then the airspeed must be kept low.

A few years ago the author had occasion to visit his two sisters-in-law (who are both in applied mathematics, dealing blackjack in the casinos) at Lake Tahoe. As our "gamblers's special" aircraft crossed the last peak of the Sierra Nevada mountains ($h = 11,000$ ft), there was the airport, seemingly directly below us ($L = 20$ mi), and we almost dove into a landing (see Figure 2)!

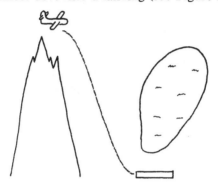

Figure 2. Landing at Lake Tahoe!
(Vertical scale exaggerated)

Our plane was, in fact, a two engine prop plane with an airspeed of about $U = 175$ mph. With the above values for U, L, and h, $k = 0.39$ ft/sec^2, not much different from the value of k for the transcontinental flight discussed above! Parenthetically, because of noise restrictions aircraft are not allowed to land from over the lake to the north of the airport, and, consequently, jets cannot land at the airport at Lake Tahoe.

[Actually, I fudged a bit on the values for L and h in the example above, for the descent was somewhat more harrowing than I made it out to be. So therein lies a research project for the calculus class: to write letters and contact flight engineers at TWA and Golden West Airways for more accurate values of L, h, and U for the flights discussed.]

In practice, aircraft decrease their airspeed when landing and often engage in a banked loop around the airport in order to slow down further before touchdown. Nevertheless, the above simplistic model for the approach pattern qualitatively agrees with actual flying experience.

REFERENCE

1. D. W. Trim, *Calculus and Analytic Geometry*, Addison-Wesley Publishing Company, Reading, MA, 1983, p. 124.

The Best Shape for a Tin Can

P. L. Roe

Some time ago, I came across a book intended to popularize mathematics, whose last chapter dealt with the calculus of one variable. Its final section, evidently intended to climax the whole work, solved the problem of designing the proportions of a tin can so as to obtain the greatest volume out of a given amount of material. The well-known solution is of course that the material used is proportional to

$$M = 2\pi r^2 + 2\pi rh, \tag{1}$$

whereas the volume is

$$V = \pi r^2 h, \tag{2}$$

so that

$$h = \frac{V}{\pi r^2}. \tag{3}$$

Inserting (3) into (1) and setting $dM/dr = 0$ gives

$$4\pi r - \frac{2V}{r^2} = 0, \tag{4}$$

or, in view of (2)

$$h = 2r. \tag{5}$$

---

*This article first appeared in *The Mathematical Gazette*, Vol. 75, 472 (1991) 147–150 and is reprinted with the kind permission of The Mathematical Association.

The Best Shape for a Tin Can

In other words, the most economical shape has its height equal to its diameter. The authors then drew attention to the fact that most cans are not 'square' and sought to explain the discrepancy. They concluded that tradition and design fashion must count for more than rational thought in the commercial world. Their parting message to the reader was to the effect that intellectual beauty was its own reward.

Such a patrician view of mathematics is, these days, a luxury that few can afford. The irony is that the chosen example, and the discrepancy between the mathematical model and the real world, actually illustrate rather nicely the true character and value of applied mathematics.

Consider first that when the lid and base of the can are cut from sheet there must be wastage, which is presumably returned for recycling, but has little value to the can makers. If we suppose that the sheet is divided first into squares of sides $2r$, and that one circle is cut from each square, equation (1) should be replaced by

$$M = 8r^2 + 2\pi rh, \tag{6}$$

leading to

$$\frac{h}{r} = \frac{8}{\pi} \simeq 2.55. \tag{7}$$

A better strategy (from a mathematical viewpoint, anyway) would be to divide the sheet into a honeycomb of hexagons. Neglecting the waste at the edge of the sheet, we find

$$\frac{h}{r} = \frac{4\sqrt{3}}{\pi} \simeq 2.21. \tag{8}$$

Although this may be interesting, neither (7) nor (8) describe very well the usual proportions of a can. We are still forgetting things. For example, examination of a real tin can shows that the top and bottom are formed from discs somewhat larger than r in radius, which are then shaped over the ends. Allowing for this would increase (h/r), as would any extra costs associated with forming a lid, or making the lid of thicker material. Also importantly, the cost of a can needs to include its fabrication as well as its materials. If the most costly operation consists of joining the side and two rims of the can, the total cost is proportional, with most economical cutting, to

$$c = 4\sqrt{3}\, r^2 + 2\pi rh + K(4\pi r + h), \tag{9}$$

where K is the length that can be joined for the price of buying unit area of material (with this definition (9) is dimensionally consistent). Then repeating the earlier maneuvers leads to

$$4\sqrt{3}\, r - \frac{V}{r^2} + 2\pi K - \frac{KV}{\pi r^3} = 0 \tag{10}$$

as the condition to be satisfied by an optimum design, together of course with (3).

Extracting the information from (10) requires some ingenuity. Dimensional analysis suggests a relationship between (h/r) and (V/K^3) and indeed (10) will yield it. After dividing by r, the terms are regrouped as

$$4\sqrt{3} - \frac{V}{r}\left(\frac{1}{r}\right)^2 + \frac{2\pi K}{r^{1/3}}\left(\frac{1}{r}\right)^{2/3} - \frac{KV}{\pi r^{4/3}}\left(\frac{1}{r}\right)^{8/3} = 0,$$

and the terms in $(1/r)$ can be substituted by $(\pi h/V)^{1/2}$, leading to

$$4\sqrt{3} - \frac{V}{r}\left(\frac{\pi h}{V}\right) + \frac{2\pi K}{r^{1/3}}\left(\frac{\pi h}{V}\right)^{1/3} - \frac{KV}{\pi r^{4/3}}\left(\frac{\pi h}{V}\right)^{4/3} = 0.$$

From this, finally,

$$\left(4\sqrt{3} - \frac{\pi h}{r}\right) + \left(\frac{\pi K^3}{V}\right)^{1/3}\left(\frac{h}{r}\right)^{1/3}\left(2\pi - \frac{h}{r}\right) = 0. \tag{11}$$

This can be thought of as a quartic for $(h/r)^{1/3}$, with K^3/V as a parameter, but several things are immediately clear. First, if joining is cheap (K small) or the can is large (V large) then we have the original design ($h/r = 4\sqrt{3}/\pi$). In the opposite cases where joining is expensive, or the can is small, we find $h/r = 2\pi$. This corresponds to manufacturing costs greatly exceeding material costs. Also, for any other situation there is by Rolle's theorem a value of (h/r) between these limits that satisfies (11).

A graph of the complete relationship (11) is easily made by inverting it to read

$$\frac{V}{K^3} = \pi\left(\frac{h}{r}\right)\left[\frac{2\pi - h/r}{\pi h/r - 4\sqrt{3}}\right]^3, \tag{12}$$

which allows Figure 1 to be plotted.

The predicted trend, that big cans should be nearly square, whereas small cans should be tall and thin, can be verified in a supermarket. Compare, for example, cans of marmalade oranges with cans of cocktail olives. However there are cans that do not fit the trend. Sometimes there is an explanation deriving from the nature of the product (pineapple rings, for example). Very small cans are often squarer than one would expect, perhaps for convenience of handling the tin opener. Convenience would also be a consideration for any can designed to be drunk from.

To summarize, the failure of the original model to predict the real shape of a tin can arises from its being a very naive model. More complete models, still within the range of school mathematics, began to reveal the real issues. Without complicating the analysis, many other questions could be explored. Would the argument be affected if the sheet material is only available in standard sizes? Or if we knew

The Best Shape for a Tin Can

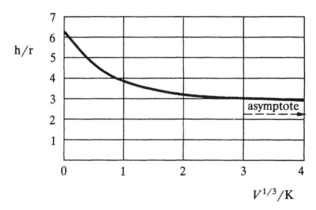

Figure 1

how the can-making machine actually worked? Are we designing to a given volume? Or to a given weight of contents? How are the cans stacked for transport? How much does it cost to be slightly off the optimal proportions? Is this offset by any other saving?

There is plenty of scope here for developing a true appreciation of how mathematics contributes to technology.

To Build a Better Box

Kay Dundas

Most calculus students have encountered the problem of finding the maximum volume of a box that is constructed from a rectangular piece of cardboard by cutting equal squares from each corner and folding up the sides. Have you ever asked your students to actually construct such a box? I have. The students soon discover that the most practical part of this "application of calculus" is the fact that it opens the door to more practical methods of construction. To begin with, removing the corners is ridiculous. If you just cut along one side of each square and use the squares to reinforce the sides, the result is a much stronger box. Another thing they notice, with a little gentle persuasion, is that a box without a top is not very useful. This observation gives me a chance to suggest the construction method shown in Figure 1.

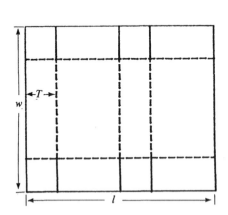 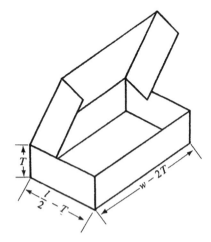

Figure 1.

If you cut along the solid lines and fold along the dotted lines, four well-placed staples will secure a fairly useable box. My students have dubbed this one the *Pizza Box*.

With the no-top construction, cutting out T by T squares from the corners of a rectangle with length l and width w ($w \leqslant l$), the volume is given by

$$V(T) = T(l - 2T)(w - 2T)$$

for $T < w/2$. With the Pizza Box construction, the volume is given by

$$V(T) = T(l/2 - T)(w - 2T) = T(l - 2T)(w - 2T)/2 \quad \text{for} \quad T < w/2.$$

Therefore, for any value of T, the volume is half as large using the Pizza Box method, and the maximum occurs at the same value of T in each case.

With a little more prodding, some students will come to the conclusion that restricting the shape of the rectangular piece of cardboard limits the maximum volume of the box. They also can see that this is not a reasonable "real world" restriction. To allow variable dimensions for the rectangle and variable corner sizes would usually require the calculus of several variables. Since this is not available to students when I want to cover this topic, I suggest the following approach.

Suppose A square inches of cardboard is used to construct a box using the Pizza method. Fixing the height at T inches, find the dimensions of the rectangle that will maximize the box's volume.

Taking $w = A/l$ in Figure 1, we have

$$V(l) = T\left(\tfrac{l}{2} - T\right)\left(\tfrac{A}{l} - 2T\right).$$

Then $V'(l) = 0$ when $l = \sqrt{A}$. The cardboard's required dimensions are therefore $l = w = \sqrt{A}$. Using the $\sqrt{A}$ by $\sqrt{A}$ cardboard, we want to find the height T that will maximize this volume. Thus, we begin with

$$V(T) = T\left(\tfrac{1}{2}\sqrt{A} - T\right)\left(\sqrt{A} - 2T\right).$$

Then $V'(T) = 0$ for $T = \sqrt{A}/6$ and the maximum volume of the Pizza Box is $V_{\text{pizza}} = A^{3/2}/27$.

For classroom development, use $A = 144$ square inches because it gives a nice maximum volume of 64 cubic inches when $l = w = 12$ and $T = 2$.

By the time we have solved the Pizza Box problem, some of the students will usually have discovered another commonly used construction method. This method, dubbed the *Popcorn Box*, is shown in Figure 2.

When students first look at this method, they usually choose a box with a square horizontal cross section and the 12 by 12 piece of cardboard that worked for the Pizza Box. Without calculus, they discover that this produces 81 cubic inches of volume—quite an improvement over the previous maximum of 64 cubic inches.

Next, they usually try one of two methods: either they keep the 12 by 12 piece of cardboard and allow the width to vary, or they keep the square base on the box and allow the dimensions of the 144 square inch cardboard to vary. Surprisingly, both methods produce the same maximum volume, $48\sqrt{3} \approx 83.14$. Is this true in general?

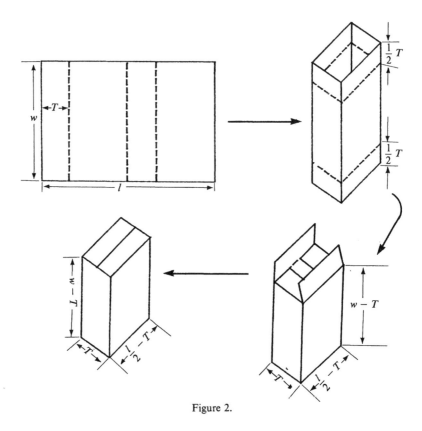

Figure 2.

To check this out for the general case, assume that the cardboard has area A square inches. For $l = w = \sqrt{A}$, we have

$$V(T) = T(\sqrt{A} - T)(\tfrac{1}{2}\sqrt{A} - T)$$

and the maximum volume occurs for $T = (3 - \sqrt{3})\sqrt{A}/6$. For the second alternative, let $T = l/4$ (the box has a square base) giving

$$V(l) = (l/4)^2(\tfrac{A}{l} - \tfrac{l}{4})$$

and maximum volume when $l = 2\sqrt{A/3} = 4w/3$. In both cases, the maximum volume is $V = \sqrt{3}\,A^{3/2}/36$.

The volume of 83.14 isn't much better than the volume of 81 that was obtained without using any calculus. However, when both the shape of the original piece of cardboard and the width T of the box are allowed to vary, the improvement is more dramatic.

To see this, assume again that the area of the cardboard is A square inches. As with the Pizza Box, first fix the box's width at T and let the length of the cardboard vary. This gives

$$V(l) = T(\tfrac{l}{2} - T)(\tfrac{A}{l} - T),$$

and $V'(l) = 0$ when $l = \sqrt{2A}$. Using this $\sqrt{2A}$ by $\sqrt{2A}/2$ cardboard (recall that the area was fixed at A square inches), allow the box width T to vary. Under these conditions,

$$V(T) = T\left(\frac{\sqrt{2A}}{2} - T\right)\left(\frac{A}{\sqrt{2A}} - T\right) = T\left(\tfrac{1}{2}\sqrt{2A} - T\right)^2$$

and $V'(T) = 0$ when $T = \sqrt{2A}/6$. Thus, the Popcorn Box has maximum volume $V_{popcorn} = \sqrt{2}\, A^{3/2}/27$. It is now clear that $V_{popcorn}$ is approximately 41% larger than V_{pizza}.

After spending a class period and a daily assignment on box problems, I like to include a box problem on the next unit test. Usually I give them a specific l by w rectangle, tell them which method of construction to use, and ask them to find the maximum volume. As a test question, I prefer integers for l and w, and rational values for the optimal box dimensions. The following developments show how to choose l and w to accomplish this for the Pizza Box and then for the Popcorn Box.

From Figure 1, we have

$$V(T) = T(l/2 - T)(w - 2T) = (lwT/2) - (l + w)T^2 + 2T^3.$$

Therefore, $V'(T) = 0$ when

$$T = \left(l + w \pm \sqrt{l^2 - lw + w^2}\right)/6.$$

The correct T value will be rational when $l^2 - lw + w^2$ is a perfect square. Choosing correct l and w values to accomplish this result is an interesting problem whose solution has been published by the author in an earlier paper "Quasi-Pythagorean Triples for an Oblique Triangle," the TYCMJ 8 (1977), 152–155. The problem is related to the "ambiguous case" triangle pictured in Figure 3.

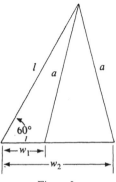

Figure 3.

The cosine law gives $a^2 = l^2 - lw + w^2$ for $w = w_1$ or $w = w_2$. Direct substitution

shows that $a = m^2 + mn + n^2$ when

$$l = 2mn + m^2, \quad w_1 = m^2 - n^2 \quad \text{and} \quad w_2 = 2mn + n^2,$$

where $m > n$. It is more difficult to show that all solutions are generated by multiples of these when m and n are relatively prime and do not differ by a multiple of three. The net result is that for $m > n$, the pairs

$$(l, w) = (2mn + m^2, 2mn + n^2) \quad \text{and} \quad (l, w) = (2mn + n^2, m^2 - n^2)$$

generate all the Pizza Box problems one needs.

For the Popcorn Box, referring to Figure 2, we have

$$V(T) = T(l/2 - T)(w - T) = (lwT/2) - (l/2 + w)T^2 + T^3.$$

Thus, $V'(T) = 0$ when

$$T = \left(l + 2w \pm \sqrt{l^2 - 2lw + 4w^2}\right)/6.$$

In this case, $l^2 - l(2w) + (2w)^2$ needs to be a perfect square. This is the same problem as above with w replaced by $2w$. It follows that $(l, 2w) = (2mn + m^2, 2mn + n^2)$ and $(l, 2w) = (2mn + m^2, m^2 - n^2)$ generate the desired dimensions.

This article would have ended here if I had not recently purchased a three way light bulb. It was packaged in an interesting box whose construction is indicated in Figure 4.

The horizontal cross-section of this box is hexagonal, and the ends are folded over just enough to reach the center. This provided a new direction to go in search of a better box.

I assigned an extra credit problem to my class to find the maximum volume using this construction method and 144 square inches of cardboard. Nobody solved the problem, but I'll try again next semester. In my solution, the volume is computed by multiplying the area of six equilateral triangles by the height. This gives the formula

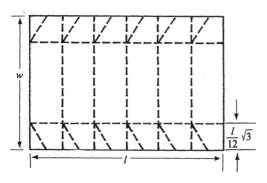
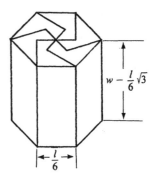

Figure 4.

$$V(l) = 6(1/2)(l/6)(l\sqrt{3}/12)(144/l - l\sqrt{3}/6).$$

Then $V'(l) = 0$ when $l = 4\sqrt[4]{108}$, and the maximum volume is $48\sqrt[4]{12} \approx 89.34$.

Many questions could be asked at this point. If we retain the A square inch rectangular piece of cardboard, what is the maximum volume possible using the hexagonal cross section? If more sides are used, will the maximum volume increase? Is some number of sides optimum, or does some smooth curve eventually produce the "best" box?

These questions can be answered under the following restrictions: The polygonal cross-section must be equiangular and have $2n$ sides for some natural number $n > 1$; each of $2n - 2$ sides have length y and the remaining 2 sides have length $(l/2) - (n-1)y$, where l is the length of the original rectangle. Thus, in the cross-section (see Figure 5), each of the $2n - 2$ isosceles triangles has its vertex angle equal to π/n and its altitude of length $(y/2)\cot(\pi/2n)$.

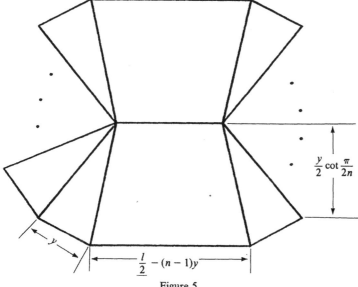

Figure 5.

For the case where $A = 144$, the box has volume

$$V = (y/2)(\cot(\pi/2n))(l - ny)(144/l - y\cot(\pi/2n)).$$

To see this, observe first that the last factor is the height of the box. The cross sectional area is seen when you place the two trapezoids and the $2n - 2$ triangles side by side, with half of the triangles and one trapezoid in the inverted position. This gives a parallelogram with length $l - ny$ and altitude $(y/2)\cot(\pi/2n)$. Taking partial derivatives with respect to l and y, we find that maximum volume

$$V = 128\sqrt{\cot(\pi/2n)}/\sqrt{n}$$

occurs when $l = 12\sqrt{n}\sqrt{\tan(\pi/2n)}$ and $y = 4\sqrt{\cot(\pi/2n)}/\sqrt{n}$.

The hexagonal cross section ($n = 3$) produces a maximum volume of $128\sqrt[4]{27} \approx 97.26$ cubic inches, while an octagonal cross section ($n = 4$) produces a maximum volume of $64\sqrt{\sqrt{2}+1} \approx 99.44$ cubic inches.

The preceding remarks show the maximum volume V is a function of n. Since $V'(n) > 0$, we see that V is an increasing function. Rewriting $V = 128\sqrt{\cot(\pi/2n)}/\sqrt{n}$ as

$$V = 128\sqrt{\frac{\pi/2n}{\sin(\pi/2n)} \cdot \frac{2\cos(\pi/2n)}{\pi}},$$

we see that $\lim_{n\to\infty} V = 128\sqrt{2/\pi} \approx 102.13$. If we could construct such a box with infinitely many sides, it would have a cross-section in the form of a rectangle with a semicircle on each end. The radius of the semicircles would be $2\sqrt{2/\pi}$, the dimensions of the rectangle would be $4\sqrt{2/\pi}$ and $2\sqrt{\pi/2}$, and the height would be $8\sqrt{2/\pi}$.

The Curious 1/3

James E. Duemmel

The United States Postal Service will accept for delivery only packages that conform to this rule: the length plus the girth must not exceed 108 inches. This rule generates a number of maximization problems in calculus texts [1, p. 260], [2, p. 240], [3, p. 220]. The answers to these problems contain an interesting surprise.

To generalize slightly we may replace 108 by any positive constant c. The typical textbook problem asks for the maximum volume of a right cylindrical package when the cross section of the package is a square or when it is a circle. The length of the box of maximum volume is, in both cases, $c/3$.

In a recent class I asked the students to try using equilateral triangles for the cross section. The length of the box of maximum volume was still $c/3$. The reader might want to try other shapes, perhaps an isosceles right triangle.

Is this just a strange coincidence? Or is the length of the box of maximum volume actually independent of the shape of the cross section?

Suppose we decide on a (reasonable) shape for the cross section of the box. Consider one example of that shape with perimeter P and area A. We may use a "magnification factor" x to describe all similar shapes, which will have perimeter Px and area Ax^2. (It should be easy to convince students of the existence of such a magnification factor, at least in the case of figures that can be decomposed into a finite number of triangles. For more complicated figures and more advanced students an argument using line integrals for the perimeter and area should be convincing.)

Let z be the length of the box. We seek the maximum volume $V = Ax^2 z$ subject to the restriction $Px + z = c$. This constraint can be solved for x to yield $V = (A/P^2)z(c-z)^2$ for $0 \le z \le c$.

The maximizing z is clearly independent of the shape and always $c/3$. However, the maximum volume $(4A/P^2)(c/3)^3$ does depend on the shape through the ratio $4A/P^2$. Indeed, the appearance of the isoperimetric ratio $4A/P^2$ in the solution clarifies the situation and opens up a whole new avenue for classroom discussion.

Note that two-thirds of c is left for the perimeter of the cross section so that maximum dimension of the cross section cannot exceed $c/3$, one-half of the perimeter. The largest dimension of the box is indeed what we have called the length.

One last question. What is the "correct" generalization to other dimensions?

References

1. Ross L. Finney and George B. Thomas, *Calculus*, Addison-Wesley, Reading, MA, 1990.
2. Phillip Gillett, *Calculus and Analytic Geometry*, 3rd Edition, Heath, Lexington, MA, 1988.
3. Al Shenk, *Calculus and Analytic Geometry*, 3rd Edition, Scott, Foresman, Glenview, IL, 1984.
4. W. H. Bussey, Maximum parcels under the new parcel post law, *American Mathematical Monthly* 20 (1913) 58–59, reprinted in *Selected Papers on Calculus*, MAA, 1969, pp. 232–233.

Hanging a Bird Feeder: Food for Thought

John W. Dawson, Jr.

Calculus instructors who have grown weary of the usual maximum/minimum problems may find the following example of interest. Its context is one familiar to students, yet unlike most geometric optimization problems they will have encountered, the optimal configuration depends in an unexpected way on the numerical values chosen for the parameters. It is a thinly veiled variant of Steiner's problem, a classic problem in geometry which has been unduly neglected by authors of calculus texts.

The Problem. In the autumn, many people put up feeders for wild birds—and thereby initiate the annual round of "squirrel wars." Seasoned veterans of the combat have learned to thwart the acrobatic rodents by suspending the feeders from wires, which raises the question: What configuration will minimize the length of wire needed?

The wire is strung between two trees a distance D apart and is attached to each of them at a common height above the ground—high enough so a person can walk under the wire near the trees. The feeder is suspended midway between the trees but must be a distance d below the height at which the wires are attached to the trees so a person can reach the feeder easily. There are three configurations to consider, whose shapes resemble the letters T, V, and Y. The first two are special cases of the third. Indeed, if we take the length h of the "tail" of the Y as the independent variable, then $0 \leq h \leq d$ and the V and T configurations correspond to the endpoints of this interval. Letting L_C denote the length of wire required for configuration C, we have $L_T = D + d$, $L_V = (D^2 + 4d^2)^{1/2}$, and a straightforward calculation shows that $L_Y = h + (D^2 + 4(d-h)^2)^{1/2}$ has but one critical value, namely $h = d - D/2\sqrt{3}$. In order for this quantity to be positive, we must have $D < 2\sqrt{3}\, d$; if so, then we find that $(L_Y)_{\min} = (\sqrt{3}/2)D + d$. Hence $(L_Y)_{\min} < L_T$, and a bit more calculation shows that $(L_Y)_{\min} < L_V$. So, if $D < 2\sqrt{3}\, d$, the Y

configuration is best. On the other hand, $L_V < L_T$ whenever $D > \frac{3}{2}d$, so if $D > 2\sqrt{3}\,d$, the V configuration is best.

It is instructive to have students draw the optimal configurations to scale for several different values of the parameters. Better yet, to obtain a physical solution to the minimization problem, wedge three thin pegs between two transparent plates and then dip the apparatus into soapy water. The film makes a configuration that minimizes the total distance connecting the three pegs [Richard Courant and Herbert Robbins, *What is Mathematics?*, Oxford University Press, New York, 1941, p. 392]. By projecting the image of the soap film onto a screen with the aid of an overhead projector, students may then notice, as Steiner did, that the angles between the pegs in the Y configuration are equal. Having made that observation, the students can verify it by computing

$$\frac{D/2}{d-h} = \sqrt{3} = \tan 60°.$$

For an overview of Steiner's problem, see the article by H. W. Kuhn in G. B. Dantzig and B. C. Eaves, *Studies in Optimization*, MAA, Washington, D.C., 1974.

"I don't understand it. Rodney feels he can't use the feeder since he flunked calculus!"

Honey, Where Should We Sit?

John A. Frohliger and Brian Hahn

There are times when, in their haste to solve a particular problem, students (and their instructors) miss an opportunity to notice some interesting mathematics. For example, when calculus students are introduced to the derivatives of inverse trigonometric functions, they frequently run across a classic problem that goes something like this:

There is a 6-foot tall picture on a wall, 2 feet above your eye level. How far away should you sit (on the level floor) in order to maximize the vertical viewing angle θ? (See FIGURE 1.)

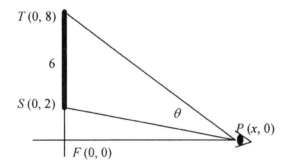

Figure 1 Find where θ is a maximum

This problem can be solved using the standard calculus technique for maximization. First, on the coordinate plane, we could set the top and bottom of the picture at $T(0, 8)$ and $S(0, 2)$, respectively. Then it is easy to show that if your eye is at a point $P(x, 0)$ on the positive x-axis, the viewing angle would be $\theta = \tan^{-1}(8/x) - \tan^{-1}(2/x)$. From the derivative,

$$\frac{d\theta}{dx} = \frac{6(16 - x^2)}{(x^2 + 8^2)(x^2 + 2^2)},$$

you can easily show that the only critical number for $x > 0$ occurs at $x = 4$. Finally, (the part that many students like to skip) the first or second derivative test can provide arguments that θ must be an absolute maximum at $P(4, 0)$.

At this point, many calculus students declare that the greatest viewing angle occurs 4 feet from the wall, express some relief and gratitude for having solved the problem, and move on to the next assignment. In doing so, unfortunately, they miss some fascinating geometry. Notice that, if we let F represent the origin, then at the point P of maximum θ, $PF/FS = 2 = TF/PF$ (FIGURE 2). This makes $\triangle PFS$ and

△*TFP similar* right triangles. Thus, the viewing angle is largest at the point P where $\angle FPS \cong \angle FTP$!

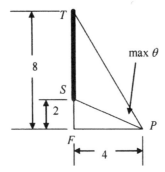

Figure 2 Similar triangles $\triangle PFS$ and $\triangle TFP$

So now a mathematician starts to wonder: is this result just a coincidence (if there is such a thing as a mathematical coincidence)? What if we change the y-coordinates of S and T? How about if, instead of being level, the floor were slanted and P were on a line $y = mx$? (Stewart gives a numerical approach to a variation of this problem [1, p. 478].)

Curiously enough, even in these cases the answer is that the viewing angle is a maximum where $\angle FPS \cong \angle FTP$. (This could be a good assignment for a bright student.) In fact, we can generalize even further and consider the case where the floor is curved rather than straight. The result is the following:

THEOREM. *Let $S(0, a)$ and $T(0, b)$ be points on the y-axis with $a < b$, and let $y = f(x)$ be a continuous function on $[0, \infty)$ and, without loss of generality, $f(0) < a$. Then there is point $P(x, f(x))$, $x > 0$, on the graph of f such that the measure of $\angle TPS$ is a maximum. Furthermore, if f is differentiable at P, then $\angle FPS \cong \angle FTP$, where F is the point where the tangent to $f(x)$ at P intersects the y-axis* (FIGURE 3).

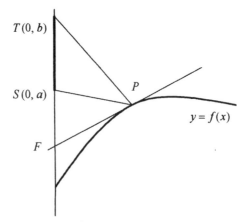

Figure 3 The generalized case

Note: In the original problem P is on the x-axis $y = 0$, and in the variation P is on the line $y = mx$. Both times, the point F is given as the origin. This notation is consistent with our generalized property since, in those cases, the tangent line to the graph of $y = f(x)$, which is simply the graph itself, intersects the y-axis at $(0, 0)$. Also, when we refer to a maximum θ, or θ being maximized, we shall implicitly restrict ourselves to the domain $(0, \infty)$.

Proof. The property that $\angle FPS \cong \angle FTP$ at maximum θ can be proved using standard calculus. Suppose f is differentiable at the maximum angle. We will assume for the time being that a greatest θ exists. It is straightforward to show that, if point P has coordinates $(x, f(x))$, then $\angle TPS$ has measure

$$\theta = \tan^{-1}\left(\frac{b - f(x)}{x}\right) + \tan^{-1}\left(\frac{f(x) - a}{x}\right).$$

Differentiating and simplifying, we can see that

$$\frac{d\theta}{dx} = (a - b)\frac{\left[x^2 + (xf'(x))^2\right] - [a - (f(x) - xf'(x))][b - (f(x) - xf'(x))]}{[x^2 + (b - f(x))^2][x^2 + (a - f(x))^2]}.$$

Since the denominator involves products of sums of perfect squares, and since $f(0)$ is neither a nor b, we can see that the denominator is never zero; hence, $d\theta/dx$ is never undefined. It follows that at the maximum, the derivative must be zero. At this point then,

$$x^2 + (xf'(x))^2 = [a - (f(x) - xf'(x))][b - (f(x) - xf'(x))]. \tag{1}$$

All we need to do is interpret this in terms of lengths. The slope of the tangent to $f(x)$ at P is $f'(x)$. If we follow the tangent line back to the y-axis, we see that F has coordinates $(0, f(x) - xf'(x))$, as in FIGURE 4.

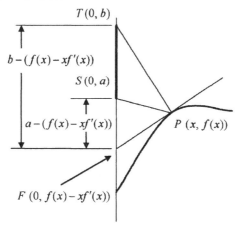

Figure 4 Where the tangent hits the y-axis

From (1), we see that $PF^2 = SF \cdot TF$, that is,

$$\frac{PF}{SF} = \frac{TF}{PF}.$$

Since they share a common angle and have two pairs of proportional sides, it follows that $\triangle SFP$ and $\triangle PFT$ are similar triangles. Therefore, we can conclude that $\angle FPS \cong \angle FTP$ when P is chosen to make $\angle TPS$ largest. ∎

Geometric approach Now we turn to some more general questions: Assuming f is continuous, not necessarily differentiable, on $[0, \infty)$, are we guaranteed that there is a point P where the viewing angle is greatest? If there is such a point P, is it necessarily unique or might the maximum angle occur at more than one point on the graph? We can answer these questions by taking a different approach to the problem. Let's leave calculus and its potentially messy computations and turn instead to geometry (with just a pinch of topology).

Recall that, in a circle, the measure of an inscribed angle is one-half that of the intercepted arc [3]. A corollary of this property is that every inscribed angle that intercepts the same arc has the same measure. Conversely, given fixed points T and S and an angle θ, the set of all points Q on one side of $\overline{ST}$ satisfying $m(\angle SQT) = \theta$ is a portion of a circle passing through S and T.

Now let's return to our problem. Again, we let S and T represent the top and bottom of our picture. For a fixed positive measure c, consider the set of points Q on the right half-plane such that $m(\angle SQT) = c$. From our discussion above, we can easily see that this level curve is the right-hand portion of a circle passing through S and T (FIGURE 5).

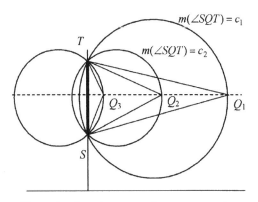

Figure 5 Level curves of constant angles

Moreover, the smaller the value of c, the farther the center of the circle is to the right. For instance, if Q_1, Q_2, and Q_3 are placed on the perpendicular bisector of $\overline{ST}$ as shown in FIGURE 5, it is easy to see that $m(\angle SQ_1T) < m(\angle SQ_2T) < m(\angle SQ_3T)$. Also notice that the regions bounded by $\overline{ST}$ and these circular curves are nested: If $0 < c_1 < c_2$, then the region bounded by $\overline{ST}$ and the curve $m(\angle SQT) = c_2$ is contained in the region bounded by $\overline{ST}$ and $m(\angle SQT) = c_1$.

Now we can answer the questions we posed earlier. Must there be a point P along the graph of $y = f(x)$ at which $m(\angle SPT)$ is a maximum? If so, where is P? The answer to the second question is that P occurs where $y = f(x)$ intersects the circular arc $m(\angle SQT) = c$ for the largest value of c, that is, the leftmost curve $m(\angle SQT) = c$

(FIGURE 6). It is probably obvious that there must be such a point; however, to be safe, we could turn to a little topology. (If this result is obvious, feel free to skip the next paragraph.)

Let G represent the graph of $y = f(x)$. For each positive c, let D_c be the closed bounded region in the right closed half-plane bounded by $\overline{ST}$ and the arc $m(\angle SQT) = c$. Then define G_c to be the intersection of G with D_c. Now consider the nonempty collection $A = \{G_c : G_c \neq \emptyset\}$ of nonempty intersections of G with the sets D_c. The continuity of f implies that G is closed; hence, each G_c is compact. Furthermore, since the D_cs are nested, it follows that the G_cs satisfy the finite intersection

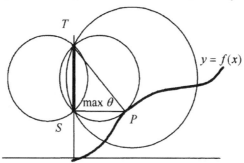

Figure 6 Where θ is maximized

property [2]. Therefore, $\bigcap_{G_c \in A} G_c \neq \emptyset$ and $m(\angle SPT)$ is a maximum at any point P in $\bigcap_{G_c \in A} G_c$.

We can see that this result is consistent with our earlier findings about similar triangles. If the tangent to the circle at P intersects the y-axis at F (FIGURE 7) then, since $\angle SPF$ and $\angle PTF$ intercept the same arc, they are congruent. Consequently, $\triangle SFP$ and $\triangle PFT$ are similar triangles and $PF/SF = TF/PF$, as before.

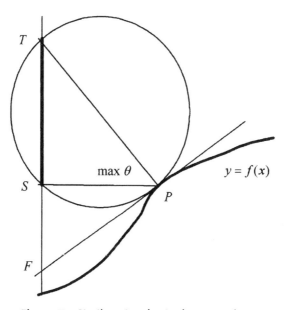

Figure 7 Similar triangles in the general case

This geometric approach allowed us to see, without ugly computations, that there must be a point P on G such that the viewing angle, $m(\angle SPT)$, is maximized. Furthermore, an easy construction allows us to show that, depending upon G, this point of greatest angle may occur at more than one point (FIGURE 8a). In fact, if G moves along a section of one such circular arc, there would be an infinite number of such points (FIGURE 8b).

We now address one final question: How do we construct such a point P? As we showed earlier, sometimes you can find P using possibly cumbersome calculus computations. In the special cases where the graph G is a line, however, we can use the geometry of the situation to physically construct the point of maximum angle using a compass and straightedge. In these situations the smallest circle through S and T that

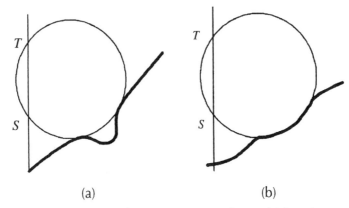

Figure 8 Cases where θ is maximized at multiple points

intersects G must be tangent to G at that point. Thus, all we need to do is find this tangent circle and determine the point P of tangency.

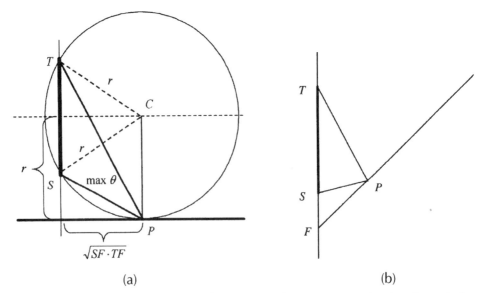

Figure 9 (a) Constructing the point of maximum θ (b) Constructing the slant line solution

This task is especially easy if G is a horizontal line (FIGURE 9a). In this situation, the one we started with, the smallest circle through S and T that intersects G must be tangent to G at that point. Thus, all we need to do is find this tangent circle and determine the point P of tangency. First we find the distance r from the perpendicular bisector of $\overline{ST}$ to G. Next we locate the point C on the right side of this perpendicular bisector that is r units from both S and T. The maximum angle then occurs at the foot P of the perpendicular from C to G. Notice that, from our previous discussion, $PF/SF = TF/PF$; hence, $PF = \sqrt{SF \cdot TF}$, so PF is the geometric mean of SF and TF.

Now that we've constructed the solution for a horizontal line, the solution for the slant line situation becomes easy. At the point of greatest angle measure, we still have the similar triangles, so the distance from P to F is still $PF = \sqrt{SF \cdot TF}$. We constructed this distance in the horizontal line case. All we need to is to construct a circle with center F and radius $\sqrt{SF \cdot TF}$. The desired point P is the intersection of this circle and the slant line (FIGURE 9b).

REFERENCES

1. James Stewart, *Calculus*, 4th ed., Brooks/Cole, Pacific Grove, CA, 1999.
2. James R. Munkres, *Topology, A First Course*, Prentice-Hall, Englewood Cliffs, NJ, 1975.
3. James R. Smart, *Modern Geometries*, 5th ed., Brooks/Cole, Pacific Grove, CA, 1998.

A Dozen Minima for a Parabola

Leon M. Hall

On the parabola defined by $y = x^2$, let $P = (a, a^2)$ be any point except the vertex. Because of symmetry, we can assume that $a > 0$. The normal line to the parabola at P will intersect the parabola again at, say, Q. The region bounded by the parabola and its normal line will be called the *parabolic segment*. See Figure 1.

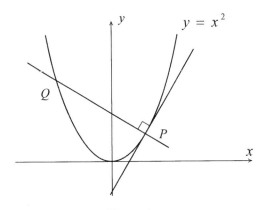

Figure 1.

This setting is a rich source of minimization problems involving several calculus concepts. Among these problems there are at least twelve different values of a that minimize some interesting quantity such as the distance between P and Q. All the problems can be set up by hand, and some can be finished by hand, but for others a computer algebra system is eventually a welcome aid. In several of the problems, the function to be minimized is defined by a definite integral, with the independent variable a in one or both limits of integration, and usually in the integrand as well. In such cases, the calculations are sometimes simpler using the Fundamental Theorem of Calculus or its extension, Leibniz' Rule, as opposed to evaluating the integrals first.

The Missouri Section of the MAA uses one of these problems each year as the lead question on the Missouri MAA Collegiate Mathematics Competition, a team problem solving event held annually in conjunction with the section meeting since 1996. Information about the competition and lists of problems and solutions can be found on the Missouri Section webpage, momaa.math.umr.edu/. The author first heard version 2b of the problem from the late Bob Krueger when we were colleagues at the University of Nebraska.

Here are the problems. Those sharing the same number have the same solution.

For the parabola $y = x^2$ and $a > 0$, consider the normal line to the parabola at $P = (a, a^2)$, which intersects the parabola again at Q. Find the value of a that minimizes:

1a. *The y-coordinate of Q.*

1b. *The length of line segment PQ.*

2a. *The horizontal distance between P and Q.*

2b. *The area of the parabolic segment. (The area turns out to be one-sixth of the cube of the horizontal distance in 2a.)*

2c. *The volume of the solid formed by revolving the parabolic segment around the vertical line k units to the right of P or k units to the left of Q, where $k \geq 0$. (Refer to problem 3 and try using Pappus' Theorem instead of the usual integral.)*

3. *The y-coordinate of the centroid of the parabolic segment. (The x-coordinate of the centroid is always $x = -\frac{1}{4a}$.)*

4. *The length of the arc of the parabola between P and Q.*

5. *The y-coordinate of the midpoint of the line segment PQ.*

6. *The area of the trapezoid bounded by the normal line, the x-axis, and the vertical lines through P and Q.*

7. *The area bounded by the parabola, the x-axis, and the vertical lines through P and Q.*

8. *The area of the surface formed by revolving the arc of the parabola between P and Q around the vertical line through P.*

9. *The height of the parabolic segment (i.e., the distance between the normal line and the tangent line to the parabola that is parallel to the normal line).*

10. *The volume of the solid formed by revolving the parabolic segment around the x-axis.*

11. *The area of the triangle bounded by the normal line, the vertical line through Q, and the x-axis.*

12. *The area of the quadrilateral bounded by the normal line, the tangent line, the vertical line through Q, and the x-axis.*

Exact answers can be found, sometimes with the aid of the computer, for all except problem 8, but some of the exact answers come from solving cubic or quartic equations and are not simple. For instance, in one of the problems, the minimum occurs when

$$a = \frac{1}{12}\sqrt{\frac{1}{2}(-24 + (276480 - 69120\sqrt{11})^{1/3} + 24(5(4 + \sqrt{11}))^{1/3})},$$

which has the much more useful decimal form $a = 0.564641\ldots$.

Here are the twelve distinct solutions (values of a) in increasing order. Readers and their students are invited to match the solutions with the problems.

$$\left\{ \frac{1}{2\sqrt{2}}, \frac{1}{2}, \sqrt{\frac{3}{10}}, 0.558480, 0.564641, 0.569723, 0.574646, \frac{1}{\sqrt{3}}, \frac{1}{\sqrt[4]{8}}, \frac{1}{\sqrt[4]{6}}, 0.644004, \frac{1}{\sqrt{2}} \right\}$$

Maximizing the Area of a Quadrilateral

Thomas Peter

In this note, we use calculus to characterize the quadrilateral Q with side lengths a, b, c and d that has the largest possible area. Think of Q as having hinges at its vertices, A, B, C and D. We can assume that Q is convex, because every non-convex quadrilateral is contained in a convex quadrilateral having the same edge lengths and larger area.

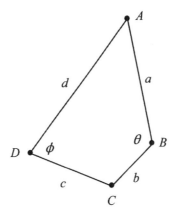

Figure 1.

Assume, by relabelling the sides if necessary, that $a + b \leq c + d$. Then by pushing on the hinge at B, we can deform the quadrilateral into a triangle having A, B, C collinear. In this triangle, $\theta + \phi > \pi$. Assume without loss of generality that $b + c \leq a + d$. Then by pulling on the hinge at B, we can deform the quadrilateral into a triangle having B, C, D collinear. In this triangle $\theta + \phi < \pi$. Therefore, we can find a position for B where $\theta + \phi = \pi$. A quadrilateral Q is said to be *cyclic* if its vertices lie on a circle. Using the fact that Q is cyclic if and only if opposite angles are supplementary, we have proven the following theorem.

Theorem 1. *For any quadrilateral with given edge lengths, there is a cyclic quadrilateral with the same edge lengths.*

Theorem 2. *The cyclic quadrilateral Q has the largest area of all quadrilaterals with sides of the same length as those of Q.*

Proof. Let $x = \cos\theta$ and $y = \cos\phi$, where θ and ϕ are in $(0, \pi)$. By applying the law of cosines to triangles ABC and ACD, we have $a^2 + b^2 - 2abx = c^2 + d^2 - 2cdy$, from which it follows that

$$\frac{dy}{dx} = \frac{ab}{cd}.$$

Adding the areas of triangles ABC and ACD, the area of the quadrilateral is

$$K = \frac{1}{2}ab\sin\theta + \frac{1}{2}cd\sin\phi = \frac{1}{2}\left(ab\sqrt{1-x^2} + cd\sqrt{1-y^2}\right).$$

Using

$$\frac{dy}{dx} = \frac{ab}{cd},$$

it follows that

$$\frac{dK}{dx} = \frac{-ab}{2}\left[\frac{x}{\sqrt{1-x^2}} + \frac{y}{\sqrt{1-y^2}}\right]. \tag{1}$$

Therefore, $\frac{dK}{dx} = 0$ if and only if $x = \pm y$. If $x = y$, then (1) and $\frac{dK}{dx} = 0$ imply that $x = 0 = y$ and $\theta = \frac{\pi}{2} = \phi$. If $x = -y$, then $\cos\theta = -\cos\phi = \cos(\pi - \phi)$ and, since both θ and ϕ are in $[0, \pi]$, we again have $\theta + \phi = \pi$.

Since

$$\frac{d^2K}{dx^2} = \frac{-ab}{2}\left[\frac{1}{(\sqrt{1-x^2})^3} + \frac{1}{(\sqrt{1-y^2})^3}\frac{dy}{dx}\right],$$

and $\frac{dy}{dx} = \frac{ab}{cd} > 0$, we have that $\frac{d^2K}{dx^2} < 0$ for all values of $x \in (0, 1)$. Therefore, the area K of quadrilateral Q is maximized when $\theta + \phi = \pi$. Hence, the area is maximized when Q is cyclic.

A Generalization of the Minimum Area Problem

Russell A. Gordon

Let (a, b) be a point in the plane with positive coordinates. An elementary and interesting optimization problem is to find the line that passes through the point (a, b) and cuts off the least area in the first quadrant. As the reader may verify, the optimal line is the tangent line to the hyperbola $xy = ab$ at the point (a, b). Consider the following generalization of this problem: find a downward opening parabola that passes through both (a, b) and the origin so that the area under the parabola and above the x-axis is minimized. An initial guess might be that the point (a, b) is the vertex of the parabola, but this turns out to be incorrect. The surprising answer is that the desired parabola has the same tangent line at (a, b) as does the hyperbola $xy = ab$. In fact, we will show that this feature of curves through (a, b) which generate a minimum area is valid for a wide variety of curves.

To restate the problem, we are searching for curves of a given type that pass through a fixed point (a, b) in the first quadrant and minimize the area under the curve. For the linear case, it is not difficult to write down an expression for the area in terms of the slope of the line, then use differential calculus to determine the minimum area. Another way to state the linear problem is as follows: find positive numbers s and t so that the line $y = s - tx$ passes through (a, b) and minimizes $\int_0^{s/t} (s - tx)\, dx$. To get a sense for the area problem when other curves are considered, we will briefly look at two specific curves. As with the linear problem just considered, the numbers s and t are positive parameters. For the curve $y = sx - tx^2$ (these are parabolas that pass through the origin), the problem is

$$\text{minimize} \quad A = \int_0^{s/t} (sx - tx^2)\, dx = \frac{s^3}{6t^2} \quad \text{subject to} \quad b = sa - ta^2.$$

For the curve $y = se^{-tx}$ (exponential curves that decay rapidly), the problem is

$$\text{minimize} \quad A = \int_0^{\infty} se^{-tx}\, dx = \frac{s}{t} \quad \text{subject to} \quad b = se^{-ta}.$$

We leave the reader with the problem of solving these minimization problems, showing that the slopes of the optimal curves at (a, b) are $-b/a$, and finding the values of k for which the minimum areas are kab. Thus, all the optimal curves have the same slope, namely $-b/a$, at (a, b) (see Figure 1).

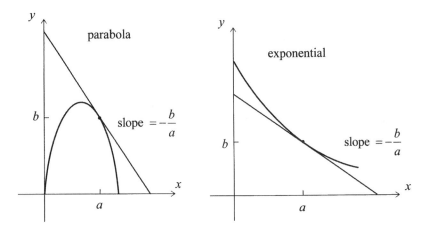

Figure 1.

As indicated by these examples, there are two types of problems here; one in which the integral is improper and the other in which it is not. To consider the latter problem in general terms, we need a function that yields an area under its graph and a method to represent families of curves based on the original curve. After experimenting with various functions, the following method became evident. Let β be a positive number and let f be a function that is continuous on $[0, \beta]$, differentiable on $(0, \beta)$, positive on $(0, \beta)$, and satisfies $f(\beta) = 0$. We want to find a curve of the form $y = sf(tx)$, where s and t are positive numbers, that passes through (a, b) and minimizes the quantity $A = \int_0^{\beta/t} sf(tx)\,dx$. (For the curves considered thus far, we have $f(x) = 1 - x$ for the linear case and $f(x) = x - x^2$ for the parabolic case.) Making the substitution $u = tx$ and using the fact that $b = sf(at)$, we find that

$$A = \frac{s}{t}\int_0^{\beta} f(u)\,du = \frac{sA_f}{t} = \frac{bA_f}{tf(at)},$$

where $A_f = \int_0^{\beta} f$ is a constant that depends only on f. (The graph of $y = sf(tx)$ represents a scaling of the x and y axes for the graph of $y = f(x)$; the area under the graph is scaled by a factor of s/t.) Since A is a continuous function of t defined on the interval $(0, \beta/a)$ and $\lim_{t \to 0^+} A(t) = +\infty = \lim_{t \to \beta/a^-} A(t)$, the function A must assume its minimum value at some point in $(0, \beta/a)$. Since A is differentiable, this value will occur at a point t_m for which $A'(t_m) = 0$. Differentiating A and substituting t_m for t yields $at_m f'(at_m) + f(at_m) = 0$. The optimal curve is thus $y = bf(t_m x)/f(at_m)$, and we find that

$$\left.\frac{dy}{dx}\right|_{x=a} = \frac{b}{f(at_m)} \cdot t_m f'(at_m) = -\frac{b}{a}.$$

Hence, the slope of the optimal curve at (a, b) is independent of the function f.

Another way to view this problem is to note that

$$A = \frac{bA_f}{tf(at)} = \frac{abA_f}{atf(at)} = \frac{A_f}{xf(x)} \cdot ab,$$

where the variable $x = at$ can assume values in $(0, \beta)$. Consider the continuous function P defined on $[0, \beta]$ by $P(x) = xf(x)$. The minimum value of A occurs when P has its maximum value. If $c \in (0, \beta)$ is the value of x that maximizes P on $[0, \beta]$, then $t_m = c/a$ is the value of t that minimizes A on $(0, \beta/a)$. Hence, the optimal curve and minimum area are

$$y = \frac{b}{f(c)} f(cx/a) \quad \text{and} \quad A = \frac{A_f}{P(c)} ab,$$

respectively. (The point $(c, f(c))$, which maximizes the area of the $x \times f(x)$ rectangle, is scaled to the point (a, b). In other words, the $x \times y$ rectangle for the optimal curve is maximized at (a, b). This same point minimizes the area cut off by the tangent line to the curve $y = f(x)$; see [1].)

While on this particular topic, we leave the reader with the problem of finding curves of the form $y = sf(tx)$ that pass through the point (a, b) and minimize the volume of the solid that is generated when the region under the curve is revolved around either the x-axis or the y-axis. For the x-axis volume, all of the optimal curves have slope $-b/2a$ at (a, b); for the y-axis volume, all of the optimal curves have slope $-2b/a$ at (a, b).

Returning to the area problem, we will now consider the case in which the integral is improper. Let f be a function that is continuous on $[0, \infty)$, differentiable on $(0, \infty)$, positive on $(0, \infty)$, and satisfies $\lim_{x \to \infty} xf(x) = 0$, and for which the integral $\int_0^\infty f(x)\,dx$ converges. As previously, we want to find a curve of the form $y = sf(tx)$, where s and t are positive numbers, that passes through (a, b) and minimizes the quantity $A = \int_0^\infty sf(tx)\,dx$. Making the substitution $u = tx$ and using the fact that $b = sf(at)$, we find that

$$A = \frac{s}{t} \int_0^\infty f(u)\,du = \frac{sA_f}{t} = \frac{bA_f}{tf(at)},$$

where $A_f = \int_0^\infty f$ is a constant that depends only on f. Note that A is a continuous function of t defined on the interval $(0, \infty)$ and that $\lim_{t \to 0^+} A(t) = +\infty = \lim_{t \to \infty} A(t)$. It can be shown as above that the slope of the optimal curve at (a, b) is $-b/a$.

This two-dimensional area problem can be extended to a three-dimensional volume problem. Given a point (a, b, c) in $\mathbb{R}^3$ with positive coordinates, find the plane that passes through (a, b, c) and cuts off the least volume in the first octant. As the reader may verify, the optimal plane is $\frac{x}{a} + \frac{y}{b} + \frac{z}{c} = 3$, which happens to be the tangent plane to the surface $xyz = abc$ at the point (a, b, c). The normal to this plane is the vector $\langle 1/a, 1/b, 1/c \rangle$. One can then search for other surfaces (such as a hemisphere or a paraboloid) that pass through the given point and minimize the volume that lies beneath the surface and above the first quadrant of the xy-plane. Using ideas similar to those in the two-dimensional case, it can be shown that all of the optimal surfaces have the same tangent plane at (a, b, c).

Reference

1. H. Bailey, A surprising max-min result, *The College Mathematics Journal* **18** (1987) 225–229.

Constrained Optimization and Implicit Differentiation

Gary W. DeYoung

One of my favorite topics of first semester calculus is optimization using derivatives. Here students can really begin to see derivatives as powerful tools to solve many practical problems that occur in a wide variety of areas. This capsule presents an approach to constrained optimization problems that avoids much of the algebraic difficulties associated with the standard method of solving constrained optimization problems. This approach often leads to relations, not immediately available from the standard method, that give deeper insight into solutions. Hopefully the students will remember the insights even after they have forgotten how to take derivatives.

The standard approach to constrained optimization problems may be summarized as follows: (1) Find a function $z = f(x, y)$ that describes the quantity z to be optimized and find a relation $0 = g(x, y)$ that forms the constraint on x and y; (2) Use $0 = g(x, y)$ to write z as a single variable function $z(x) = f(x, y(x))$; (3) Find values of x where $z'(x) = 0$; and (4) finally find the optimal values of z.

The approach presented here differs from the standard approach by using implicit differentiation in place of substitution in step 2 and by finding a relation that holds between x and y instead of finding values for x in step 3. The relation found can then be used to find optimal variable or function values. In problems with one constraint, the solution can often be visualized by graphing the constraint and the newly found relation as illustrated in problem 1 below.

Although there are multivariable techniques for constrained optimization (Lagrange multipliers or differential forms [1]) that yield results similar to those presented here, students in first semester calculus generally are not ready for them. In many texts, implicit differentiation is covered before optimization, so using this approach should be within the reach of first semester students. It also gives students a very good application of implicit differentiation.

The following problems, found in many calculus texts, will illustrate the use of implicit differentiation in constrained optimization problems.

Problem 1. Using a fixed length of fence, find the maximum area enclosed in a rectangular field that is to be subdivided into rectangular plots. The sides of the field may be river bank, cliff walls, *etc*. See Figure 1.

Constrained Optimization and Implicit Differentiation

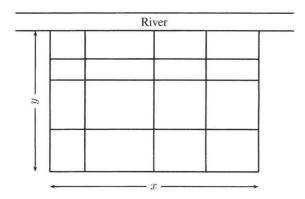

Figure 1.

Solution. The mathematical problem is to maximize $A = xy$ subject to $L = mx + ny$, where n is the number of vertical sections of fence, m is the number of horizontal sections of fence, and L is the fixed length of fence available.

We proceed by using implicit differentiation to compute dA/dx and then set it to zero,

$$0 = \frac{dA}{dx} = y + x\frac{dy}{dx}.$$

Using $L = mx + ny$, we implicitly compute

$$\frac{dy}{dx} = -\frac{m}{n}$$

and substitute dy/dx into the first equation to obtain

$$ny = mx. \tag{1}$$

From (1) and $L = mx + ny$, it follows that $y = L/(2n)$ and $x = L/(2m)$ for a maximum area of $A = L^2/(4mn)$. Computation of d^2A/dx^2, using implicit differentiation, shows that the area is maximized since $d^2A/dx^2 = -2m/n$.

The result can be visualized easily by graphing (1) and $L = mx + ny$. The coordinates of the intersection point provide the dimensions of the (shaded) field. See Figure 2.

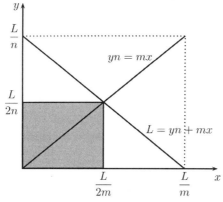

Figure 2.

An important point to note is that (1) gives us far more insight into the solution than merely the dimensions of the optimal field. Together with the constraint equation, (1) leads to a general principle that for a given length of fence, the partitioned rectangular field having maximum enclosed area is formed by using half of the fence for horizontal divisions and half for vertical divisions. The farmer in your class will remember this answer.

Problem 2. What is the minimal cost can (right circular cylinder) for a fixed volume, where the cost of the ends vary from the costs of the side?

Solution. The problem boils down to minimizing $C = A(2\pi r^2) + B(2\pi rh)$ subject to $V = \pi r^2 h$, where A and B are the respective costs per unit of surface area for the ends and the sides, r and h are the radius and height of the can, and V is an unknown fixed volume.

Proceeding as in the previous problem, we use implicit differentiation to differentiate C with respect to one of the variables, say h, and then set dC/dh equal to zero,

$$0 = \frac{dC}{dh} = 4\pi Ar \frac{dr}{dh} + 2\pi B \left(r + h \frac{dr}{dh} \right).$$

Implicitly differentiating $V = \pi r^2 h$ to find dr/dh, we substitute

$$\frac{dr}{dh} = -\frac{r}{2h}$$

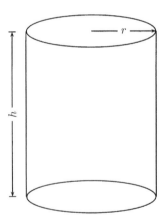

Figure 3.

into the optimization equation and simplify to obtain

$$2Ar^2 = Brh. \tag{2}$$

Constrained Optimization and Implicit Differentiation

Looking back at the formula for C, we see (2) implies that in the optimal can the ends cost half as much as the side. If the $A = B$, then the shape that minimizes the can's surface area for a fixed volume has diameter equal to its height. The entrepreneur will remember this.

Problem 3. Two towns are to build a water pumping station on a river with pipelines running straight from the pump station to each of the towns. Where should the pump station be located to minimize the total length of the pipelines?

Solution. For fixed lengths A, B and D, the problem's essence is minimize $L = L_1 + L_2$ subject to the constraints $L_1^2 = A^2 + x^2$ and $L_2^2 = B^2 + (D - x)^2$. See Figure 3.

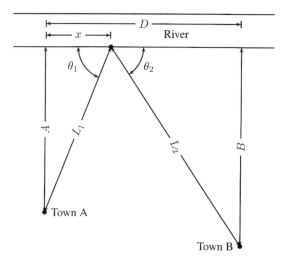

Figure 4.

Beginning with the optimization equation

$$0 = \frac{dL}{dx} = \frac{dL_1}{dx} + \frac{dL_2}{dx}, \tag{3}$$

we implicitly differentiate both constraint equations and substitute

$$\frac{dL_1}{dx} = \frac{x}{L_1} \quad \text{and} \quad \frac{dL_2}{dx} = \frac{-(D - x)}{L_2}.$$

into (3) to obtain

$$\frac{x}{L_1} = \frac{(D - x)}{L_2}.$$

Therefore, $\cos\theta_1 = \cos\theta_2$ and $\theta_1 = \theta_2$ (since θ_1 and θ_2 are less than π). In other words, the pumping station should be positioned so that the angles between the river and the towns are equal. The civil engineer will remember this result.

As you can see from the problems above, using implicit differentiation in place of substitution in constrained optimization problems leads to a delightfully straightforward method with insightful solutions. It also begins (or continues) the process of leading students to understanding equations as relations between variables rather than objects to be solved and that relations do actually have meaning!

Hopefully the farmer, entrepreneur, and civil engineer will remember these results long after they have completed their calculus course.

Reference

1. Frank Zizza, Differential forms for constrained max-min problems, *The College Mathematics Journal* **29** (1998) #5, 387–396.

For Every Answer There Are Two Questions

A.M. Fink and Juan A. Gatica

Duality is a concept that dominates modern optimization theory. The most familiar one is that associated with linear programming. Here we look at a different one. Most students are introduced to optimization in a first course in calculus. We propose to show that the idea and use of duality can be introduced in such a course.

In a general way, duality asserts the existence of two solutions of extrema problems that are intimately related. Ideally, the two problems should use the same data; one problem should have a solution if and only if the other one does; and the solution of one should give information about the solution of the other.

Consider by way of example two problems that are in every calculus text.

I. Find the rectangle of maximum area when the perimeter is fixed.

II. Find the rectangle of minimum perimeter when the area is fixed.

An alert calculus student will observe that the extremal rectangle in both problems is a square. Is this an accident?

One way to begin is to observe that "Every time you solve an optimization problem you have proved an inequality." This is hardly a deep statement, but it is a useful principle. For example, if one has solved Problem I and discovered that the answer is a square then one can formulate a useful inequality. For let A be the area of any rectangle of perimeter P. Then $A \leq A_1$ where A_1 is the area of the square of perimeter P. But $A_1 = (P/4)^2$ so that

$$A < P^2/16 \text{ unless the rectangle is a square, when equality holds.} \quad (1)$$

When (1) is rewritten as

$$4\sqrt{A} < P \text{ unless the rectangle is a square, when equality holds,} \quad (2)$$

then we also have a solution of problem II. That is, problems I and II have the same answer because the inequality (1) solves both of them simultaneously. They may be called dual problems. The procedure shows the value of solving problem 1 with P as a parameter.

This idea is introduced in Niven [1] with the statement that it (duality) is true for geometric problems and will be used (without a proof being given). See also Fink [2]. One might ask what the principle is and how widely it is applicable.

The simple problem above serves as a special case of the general principle. Let X be a set and let f and g be real-valued functions defined on X. Suppose that there is a subset $Y \subset X$ such that

i) $f|Y$ and $g|Y$ are onto $\mathbb{R}^+ = (0, \infty)$
ii) for $y_1, y_2 \in Y$, $f(y_1) < f(y_2)$ if and only if $g(y_1) < g(y_2)$.

---

[1]This research was partially supported by AFSOR87-1040.

We then consider the two statements (which may be true or false):

A) For every $c > 0$, $\max_{x \in X} f(x)$ under the constraint $g(x) = c$ is attained at a $y_0 \in Y$ (with $g(y_0) = c$).
B) For every $d > 0$, $\min_{x \in X} g(x)$ under the constraint $f(x) = d$ is attained at a $y_0 \in Y$ (with $f(y_0) = d$).

If $X = \{\text{rectangles}\}$, $Y = \{\text{squares}\}$, f is the area function and g is the perimeter function then we have the above situation and both A and B are true.

THEOREM. (*Duality Principle.*) *With X, Y, f, g as above, satisfying i) and ii), then A holds if and only if B holds.*

As a preliminary to the proof of the theorem we observe that $g \circ (f|Y)^{-1} : \mathbb{R}^+ \to \mathbb{R}^+$ (and also $f \circ (g|Y)^{-1} : \mathbb{R}^+ \to \mathbb{R}^+$) is a strictly increasing function. To see this, consider $g \circ (f|Y)^{-1}$ and suppose that $c \in \mathbb{R}^+$ and $y_1, y_2 \in Y$, with $f(y_1) = f(y_2) = c$. Then we must have that $g(y_1) = g(y_2)$ since otherwise we would contradict ii); thus $g \circ (f|Y)^{-1}$ is well defined. For ease of notation we will write $g \circ f^{-1}$ for $g \circ (f|Y)^{-1}$.

If $c_1, c_2 \in \mathbb{R}^+$, $c_1 < c_2$ and $y_1 \in f^{-1}(c_1), y_2 \in f^{-1}(c_2)$ then, since $f(y_1) < f(y_2)$ we must have $g(y_1) < g(y_2)$ by ii) so that $g \circ f^{-1}(c_1) < g \circ f^{-1}(c_2)$, and $g \circ f^{-1}$ is increasing.

On the other hand if we assume that $g \circ f^{-1} : \mathbb{R}^+ \to \mathbb{R}^+$ is well defined and strictly increasing, then ii) must hold. For if $y_1, y_2 \in Y$ with $f(y_1) < f(y_2)$, then it must be the case that $g \circ f^{-1}(f(y_1)) < g \circ f^{-1}(f(y_2))$ and hence, since the function is well defined, we must have $g(y_1) < g(y_2)$.

Conversely, if $g(y_1) < g(y_2)$ then it cannot be the case that $f(y_1) \geq f(y_2)$, since if it were then $g \circ f^{-1}(f(y_2)) \geq g \circ f^{-1}(f(y_2))$ and $g(y_1) \geq g(y_2)$, a contradiction.

The conclusion we have reached is that ii) is equivalent to

ii)' $g \circ f^{-1}$ is well defined and increasing on $\mathbb{R}^+$.

By a similar argument one can show that ii) is also equivalent to

ii)'' $f \circ (g|Y)^{-1}$ is well defined and increasing on $\mathbb{R}^+$.

Proof of the Theorem. Suppose that A holds. We use the principle that we should write the inequality that is proved. It is: For all $x \in X$ such that $g(x) = c$, $f(x) \leq f(y_0)$ for some $y_0 \in Y$ such that $g(y_0) = c$. Thus

$$f(x) \leq f \circ g^{-1}(c) \quad \text{when } g(x) = c. \tag{3}$$

Now let d be given and consider the problem of minimizing $g(x)$ when $f(x) = d$. If $x \in X$ with $f(x) = d$, set $c = g(x)$ and we have (3) and $d \leq f(g^{-1}(c))$. Since $g \circ f^{-1}$ is increasing we have

$$g \circ f^{-1}(d) \leq c.$$

But there is a $y_1 \in Y$ such that $f(y_1) = d$, and with this substitution we have

$$g(y_1) \leq c = g(x) \text{ for all } x \text{ such that } f(x) = d, \tag{4}$$

which proves that B holds. The reverse implication is proved in the same way.

The key to the argument is the inequality (3) that corresponds to (1) for the special problem.

It should be noted that this theorem gives a strong version of duality. At the risk of being redundant we can say this in several ways; (a) the "answers" are the same in both cases, namely a $y_0 \in Y$; (b) the solution to both problems can be read off a single inequality (3); and (c) the existence of an extremum in Y for one problem implies the existence of an extremum in Y for the other problem.

The last statement (c) is a strong form of duality. It gives the existence of an extremum without invoking any topology. For example, a student in calculus would solve problem I by writing

$$A(x) = xy = x(P/2 - x), \qquad 0 \leq x \leq P/2$$

and find the solution by using the existence of the maximum on a closed interval. But for problem II, eliminating a variable leads to

$$P(x) = 2x + \frac{2A}{x}, \qquad 0 < x < \infty,$$

and one needs a more sophisticated argument (to a first-year student) to prove the existence of a minimum. The duality principle gives the existence directly.

It should be noted that if $f|Y$ and $g|Y$ are one-to-one then we have uniqueness in both statements.

By way of giving more examples we choose some that might be of interest.

EXAMPLE 1. Minimize $x^2 + y^2$ subject to $x^2 y = 1$ and $x > 0$, $y > 0$. Since the constraint set is not compact we choose instead to maximize $x^2 y$ subject to $x^2 + y^2 = d$ and $x \geq 0$, $y \geq 0$. Calculus shows that the maximizing point of $y(d - y^2)$, $0 \leq y \leq \sqrt{d}$ is at $y^2 = d/3$. Here we have $X = R_+^2$, $Y = \{(x, y): x = \sqrt{2} y\}$, $f(x, y) = x^2 y$ and $g(x, y) = x^2 + y^2$. Then $f|_Y = 2y^3$, $g|_Y = 3y^2$ and $(f \circ g|_Y)^{-1}(c) = 2(c/3)^{3/2}$ and we can apply the duality principle. The original minimization problem has a solution with the minimizing point satisfying $x = \sqrt{2} y$. It follows that $y = 2^{-1/3}$ and $x = 2^{1/6}$ and the minimum value is $3(2)^{-2/3}$.

As an alternate way of thinking about the problem, we may let $D = x^2 y$ and write the solution of the maximizing problem in Example 1 as

$$D < 2\left(\frac{d}{3}\right)^{3/2} \quad \text{unless } x = \sqrt{2} y. \tag{5}$$

Rewriting this as

$$3\left(\frac{D}{2}\right)^{2/3} < d \quad \text{unless } x = \sqrt{2} y, \tag{6}$$

we solve the original problem by taking $d = 1$ in (6). This way of thinking about the problem highlights the inequality (3) instead of the theorem.

EXAMPLE 2. Let $a > 0$ be given, $X = \{\text{rectangles}\}$ and f the area function and g be the function $2al + 2w$ with l being the "length" and w the "width". We can interpret g as a cost function when two kinds of fence are used for fencing a rectangle.

Consider the minimum cost problem with the area given. Again the constraint $xy = A$ is noncompact so we choose to solve instead the problem of maximizing the area subject to the cost being a constant C. One finds that the maximizing point lies in

$Y = \{$rectangles whose width is a times the length$\}$

so that $f|_Y = al^2$ and $g|_Y = 4al$. The inequality (3) is

$$A \leq C^2/16a$$

from which one can solve the original problem. We have obtained a rule of thumb for the prospective fence buyer: In order to minimize cost of fencing a rectangle with given area, spend half of your money on each kind of fence.

EXAMPLE 3. Let $X = \{$right circular cylinders$\}$ and let V and S be the volume and surface functions. The easier of the two problems to solve is to maximize volume when the surface area is given. One finds the answer lies in $Y = \{$square cylinders$\}$ = $\{$right circular cylinders with diameter equal to the height$\}$. Then $V|_Y = 2\pi r^3$ and $S|_Y = 6\pi r^2$ for r the radius and $V \circ S^{-1}(c) = c^{3/2}/3\sqrt{6\pi}$. Thus the problem of minimizing the surface area with volume fixed is also solved by a square cylinder. The inequality (3) in this case is

$$V \leq \frac{1}{3\sqrt{6\pi}} S^{3/2}.$$

EXAMPLE 4. Let $X = R_+^n$, $f(x) = [\prod_{i=1}^n x_i]^{1/n}$, and $g(x) = (1/n)\sum_{i=1}^n x_i$. Again the problem of maximizing f subject to $g = c$, $x_i \geq 0$ has a solution in $Y = \{x | x_1 = x_2 = \cdots = x_n\}$ = the "diagonal in X". In this case $f|Y = g|Y = f \circ g^{-1}$ = identity and (3) is the arithmetic-geometric mean inequality.

REFERENCES

1. I. Niven, *Maxima and Minima without Calculus*, MAA, Washington, DC, 1981.
2. A. M. Fink, Max-Min without calculus, *The Math Log*, XV 3 (1971).

Old Calculus Chestnuts: Roast, or Light a Fire?

Margaret Cibes

George Polya warned us a long time ago about the necessity to recognize the many opportunities presented by and among problems:

> Even fairly good students, when they have obtained the solution of the problem and written down neatly the argument, shut their books and look for something else. Doing so, they miss an important and instructive phase of the work. ...A good teacher should understand and impress on his students the view that no problem whatever is completely exhausted.
>
> One of the first and foremost duties of the teacher is not to give his students the impression that mathematical problems have little connection with each other, and no connection at all with anything else. We have a natural opportunity to investigate the connections of a problem when looking back at its solution. [1]

Year after year calculus classes work on optimization problems involving geometric figures. Are we teachers, like most textbook authors, limiting our expectations of students to procedurally correct write-ups of case-specific results with answers that match those in the back of the book? Or are we encouraging our students to reflect upon their answers, discover patterns in them, and connect these patterns to those in other problems? I offer here three familiar problems that will reward extra reflection.

First is the classic problem of *maximizing the area of a rectangular field* that can be enclosed on 3 or 4 sides and/or subdivided with a given length of fence. Consider the ordinary case in which fence lengths for opposite sides, if they both exist, are equal and any fence length that subdivides the interior is parallel to a side and equal in length to it. (The existence of a maximum area, of course, depends upon at least one "vertical" side and one "horizontal" side being fenced. Otherwise, there is no upper limit to the area that could be fenced in.)

One pattern that becomes evident from looking at one case and then removing or adding fence lengths for other cases is that the maximum area is achieved when the sum of the lengths of the "vertical" pieces of fence is equal to the sum of the lengths of the "horizontal" pieces of fence. Each sum equals one-half of the total fence length, and so in a sense every maximizing rectangle has one of the regularities of a square, a striking result. But there is more. The fence problem can be connected to our second problem: *maximizing the volume of a square-base box*, with or without a "lid," given a total surface area. One might conjecture an analogous regularity in the relationship between the total surface area of the "horizontal" square end(s) and that of the other four "vertical" sides. One might further investigate this relationship when the box is without one or more "vertical" sides.

The first table below gives the results after removal of 0, 1, or 2 faces, as well as the general case of covering m ends and n sides.

In order to relate our optimal configurations to the cube, we have expressed values in terms of the length of the edge of an ordinary cube with given surface area A, that is, $x = \sqrt{A/6}$. The term "end" refers to one of the two "horizontal" square faces of the box; "side" refers to one of the four "vertical" rectangular faces; and "Total Area" refers to the total area of the faces that have not been removed but are to be covered with the given material.

Optimization of Volume of Box

Face(s) Removed	Critical Values		Total Area of End(s)	Total Area of Side(s)	Maximum Volume
	End Length	Side Height			
None	x	x	$2x^2$	$4x^2$	x^3
One End	$\sqrt{2}\,x$	$\dfrac{1}{\sqrt{2}}x$	$2x^2$	$4x^2$	$\sqrt{2}\,x^3$
One Side	x	$\dfrac{4}{3}x$	$2x^2$	$4x^2$	$\dfrac{4}{3}x^3$
Two Ends	none	none			none
One End & One Side	$\sqrt{2}\,x$	$\dfrac{4}{3}\dfrac{1}{\sqrt{2}}x$	$2x^2$	$4x^2$	$\dfrac{4}{3}\sqrt{2}\,x^3$
Two Sides	x	$2x$	$2x^2$	$4x^2$	$2x^3$
GENERAL CASE $m, n \neq 0$	$\sqrt{\dfrac{2}{m}}\,x$	$\dfrac{4}{n}\sqrt{\dfrac{m}{2}}\,x$	$2x^2$	$4x^2$	$\dfrac{4}{n}\sqrt{\dfrac{2}{m}}\,x^3$

The conjecture about surface areas is substantiated. For an optimal configuration, we find that the ratio of total surface area of the covered ends to the total surface area of the covered sides is 1 to 2, just as for the cube!

Where could the calculus student go from here? The next connection might be to a third classic problem: *maximizing the volume of a cylinder* with a given surface area. The basic results are shown in the table below.

In order to connect this cylinder problem to the preceding box problem, we again use $x = \sqrt{A/6}$ as our unit of length.

	Optimization of Volume of Cylinder				
Faces Removed	Critical Values		Total Area of Ends	Total Area of Side	Maximum Volume
	Radius	Height			
None	$\dfrac{1}{\sqrt{\pi}} x$	$\dfrac{2}{\sqrt{\pi}} x$	$2x^2$	$4x^2$	$\dfrac{2}{\sqrt{\pi}} x^3$
One End	$\sqrt{\dfrac{2}{\pi}}\, x$	$\sqrt{\dfrac{2}{\pi}}\, x$	$2x^2$	$4x^2$	$2\sqrt{\dfrac{2}{\pi}}\, x^3$

Here again in each case the surface area ratio of ends to side is 1 to 2.

One might inquire further about the relative sizes of the boxes and cylinders for a given surface area. Scale drawings or models might prove very fruitful here. One might also vary the box or cylinder by subdividing the interior with pieces of the given material to provide "shelves." See the general case in the table for the box.

Inquiries such as those described above keep mathematics alive as an ongoing, creative process in our students and not dead as a list of theorems, problems, and solutions in textbooks. Leon Henkin described it aptly:

> One of the big misapprehensions about mathematics that we perpetrate in our classrooms is that the teacher always seems to know the answer to any problem that is discussed. This gives students the idea that there is a book somewhere with all the right answers to all of the interesting questions, and that teachers know those answers. And if one could get hold of the book, one would have everything settled. That's so unlike the true nature of mathematics. [2]

References

1. G. Polya, *How To Solve It*, Princeton University Press, Princeton, NJ, 1946, p. 14.
2. L. A. Steen and D. J. Albers, eds., *Teaching Teachers, Teaching Students*, Birkhauser, Boston, 1981, p. 89.

Cable-laying and Intuition

Yael Roitberg and Joseph Roitberg

In almost every calculus book there appears a problem isomorphic to the following:

Points A and B are opposite each other on the shore of a straight river that is w feet wide. Point C is on the same side of the river as B, l feet down the river. A telephone company wishes to lay a cable from A to C. It costs \$$a$ per foot to run the cable underwater and \$$b$ per foot to run the cable on land. Which path would be least expensive for the company?

Other versions of the problem have a person walking along the shore and swimming, or rowing a boat, across the river and wanting to minimize the time to get from A to C. All versions lead to the picture in Figure 1.

If $a \leq b$, it is intuitively clear that the least expensive path is the straight line joining A to C. If a is much greater than b, it is plausible to guess that the least expensive path would be to go from A to B and then from B to C; that is, one would think that

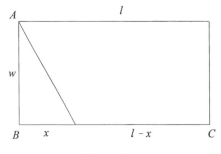

Figure 1.

minimizing the amount of cable to be laid underwater would be best. Surprisingly, this intuition is unreliable: the piecewise linear path joining A to B and B to C is *never* the least expensive path, at least if $b > 0$.

Cable-laying and Intuition

To see this, the problem is to minimize

$$f(x) = a\sqrt{w^2 + x^2} + b(l - x), \quad 0 \le x \le l, \quad a, b, w, l > 0.$$

Computing $f'(x)$ shows that there is exactly one critical value,

$$x_0 = \frac{bw}{\sqrt{a^2 - b^2}},$$

which can occur only when $a > b$. Substituting,

$$f(0) = aw + bl, \quad f(l) = a\sqrt{w^2 + l^2}, \quad f(x_0) = w\sqrt{a^2 - b^2} + bl.$$

If $a \le b$ then there is no critical value and we need only compare $f(0)$ with $f(l)$. Since

$$a^2(w^2 + l^2) < a^2 w^2 + 2abwl + a^2 l^2 \le (aw + bl)^2,$$

$f(l) < f(0)$, so f is minimized at l. The least expensive path in this case is the linear path joining A and C.

If $a > b$, we must compare $f(0)$, $f(x_0)$, and $f(l)$. It is easy to see that

$$f(x_0) = w\sqrt{a^2 - b^2} + bl < w\sqrt{a^2} + bl = f(0)$$

which seems to justify the assertion that f cannot be minimized at 0.

But are we really finished? It is easy to overlook the possibility (and we are willing to bet that many calculus students would be guilty of the oversight) that x_0 might not lie in the domain $0 \le x \le l$ of f. Thus, for f to be minimized at 0, the two conditions $x_0 > l$ and $f(0) \le f(l)$ must be met. The first is

$$\frac{bw}{\sqrt{a^2 - b^2}} > l, \text{ so } w > \frac{l\sqrt{a^2 - b^2}}{b}.$$

The second is

$$aw + bl \le a\sqrt{w^2 + l^2} \text{ or } a^2 w^2 + 2abwl + b^2 l^2 \le a^2 w^2 + a^2 l^2, \text{ so } w \le \frac{l(a^2 - b^2)}{2ab}.$$

These imply

$$\sqrt{a^2 - b^2} < \frac{a^2 - b^2}{2a}.$$

But this says

$$2a\sqrt{a^2 - b^2} < a^2 - b^2, \text{ or } 2a < \sqrt{a^2 - b^2} < \sqrt{a^2} = a,$$

which is impossible.

To summarize, if $0 \le x_0 \le l$, then $f(x_0) < f(0)$; if $x_0 > l$, then $f(l) < f(0)$. Hence, the piecewise linear path joining A to B and B to C is never the least expensive path.

If a/b is very large, then $x_0 = w/\sqrt{(a/b)^2 - 1}$ is close to zero, so the incorrect intuition that f is minimized at 0 becomes correct "in the limit."

It is interesting to note that when $a > b$ the *maximum* of f occurs either at 0 or l: at 0 if $w \geq ((a^2 - b^2)/2ab)l$ and at l if $w \leq ((a^2 - b^2)/2ab)l$.

Descartes Tangent Lines

William Barnier and James Jantosciak

Given a particular cubic polynomial function, can you find a tangent line that intersects the curve exactly once; that is, only at the point of tangency? This was the main question driving a project given to students in a Calculus I class. Using a computer algebra system, *Mathematica*, the students experimented with different tangent lines to the given cubic, plotted the curve along with the tangent line, and finally found a tangent line that intersects the curve exactly once. They also noted that the point of tangency for such a tangent line is the inflection point. Finally, they tried to justify whether (or not) there is only one such tangent line using arguments based on slope and concavity including relative growth rates at either end of the cubic and the tangent line.

As an extra credit problem the students were asked to find a curve that had exactly two tangent lines that intersected the curve exactly once. Several students found a quintic polynomial with such tangent lines; again, each at an inflection point.

Question. For n a positive integer, can you find a curve that has exactly n tangent lines that intersect the curve exactly once?

Needing terminology and recalling that Descartes' method of equal roots [2] for finding slopes of tangent lines depended on these lines intersecting the given curve exactly once, we will refer to tangent lines that intersect the curve exactly once as *Descartes tangent lines* or more briefly *Descartes lines*. If a given curve has exactly n Descartes tangent lines, we will say the curve has a *Descartes tangent number* or *Descartes number* of n.

Descartes Tangent Number Problem. For n a positive integer, find a curve with a Descartes number of n.

We will search for a solution to the Descartes tangent number problem among polynomial curves of degree greater than 1. After some thought, it is clear that a polynomial curve of even degree (for example, a parabola) is not a candidate because at either end of the curve beyond all inflection points and far enough away every point admits a Descartes line.

For an odd-degree polynomial, any point that admits a Descartes line must be an inflection point. To see why, suppose the point of tangency is not an inflection point and consider the case where the leading coefficient of the polynomial is positive. Then the polynomial is either concave up or down in an open interval containing the point of tangency. In the former case the curve will be above the tangent line in that interval, so above it to the left of the tangency point. Since the leading coefficient is positive, the curve will be below the tangent line for all points far enough to the left. Hence, by the Intermediate Value Theorem, the curve and the tangent line must intersect at some point to the left of the point of tangency. Similar arguments can be given for the other cases.

Our class project established that there is a cubic with a Descartes number of 1 and a quintic with a Descartes number of 2. Before giving a solution for the Descartes tangent number problem, let us consider an example of a quintic.

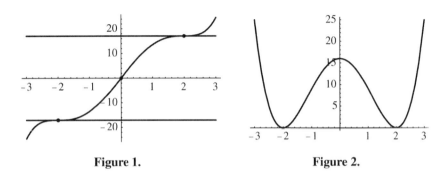

Figure 1. **Figure 2.**

Figure 1 shows the graph of $F(x) = \frac{1}{5}x^5 - \frac{8}{3}x^3 + 16x$, its three inflection points, and its two Descartes lines. Graphically or using an argument like the one given above, it is easy to verify that the tangent line through the inflection point $(0, 0)$ is not a Descartes line. The Descartes number for $F(x)$ is indeed 2.

Figure 2 shows the graph of the derivative $f(x) = F'(x) = (x^2 - 4)^2$. The fact that the x-values of the local extrema for $f(x)$ and the inflection points for $F(x)$ are equal is the key to a solution for the Descartes tangent number problem. To construct a solution, we found that it was helpful to restrict our search to odd-degree polynomials that, as in the example, were symmetric with respect to the origin, increasing, and had Descartes lines with zero slopes.

The strategy for a given value of n then is to first construct the derivative polynomial $f(x)$ of even degree $2n$ with $f(x) \geq 0$ and local minima at n points symmetrically placed on the x-axis. Hence a general solution is

$$f(x) = \begin{cases} \prod_{k=1}^{n/2} (x^2 - a_k^2)^2, & \text{for } n \text{ even;} \\ x^2 \prod_{k=1}^{(n-1)/2} (x^2 - a_k^2)^2, & \text{for } n \text{ odd,} \end{cases}$$

where the a_k are distinct positive real numbers and

$$F(x) = \int_0^x f(t)\, dt.$$

It is not hard to prove the following: Each of the n local minima for $f(x)$ corresponds to an inflection point for $F(x)$ admitting a Descartes line; and each of the $n-1$ local maxima for $f(x)$ corresponds to an inflection point for $F(x)$ that does not admit a Descartes line. Hence, the polynomial $F(x)$ of degree $2n+1$ has a Descartes number of n. Note that, given the restrictions on $F(x)$ of symmetry, increasing values, and Descartes lines with zero slopes, the solution above is unique up to multiplication by a positive constant and the choice of values a_k.

Every cubic has exactly one inflection point, which admits a Descartes line, and so every cubic has a Descartes number of 1. Moreover, the polynomial x^{2n+1} has a Descartes number of 1 for all n. Is there for each n, a polynomial of degree greater (less) than $2n+1$ with a Descartes number of n?

The reader may think of other questions regarding Descartes lines but we will pose just two more in this paper. How does a vertical shift, either up or down, of $f(x)$ affect $F(x)$? In particular, how does such a shift affect the slope of the Descartes lines? Is there another family of continuous curves such that for each n there is a curve that admits a Descartes number of n?

The students' solution to the project relied heavily on *Mathematica* for generating examples and providing graphical insight. Judging from their written explanations of the role that concavity played in their solutions, the students thereby gained a better understanding of concavity, especially inflection points. They understood how their visual impressions of a curve bending up or down was related to the tangent line lying below or above the curve. Their appreciation for the relationship between the first and second derivatives of a given function and the shape of the corresponding curve was also enhanced. All who solved the extra credit problem conjectured that some quintic should have two Descartes lines and then used graphical reasoning; that is, examining the concavity and slope and placing tangent lines in the plot of a candidate quintic until they found one that worked. One student first constructed a quartic and took the antiderivative.

See [1] for a website from which you can download two *Mathematica* notebooks, one illustrating solutions to the Descartes tangent number problem and the other containing the calculus project. Also, see [3] for more on Descartes' method.

References

1. William Barnier, http://www.sonoma.edu/users/b/barnierb/DTL.html
2. Allan Cruse and Millianne Granberg, *Lectures on Freshman Calculus*, Addison-Wesley, 1971.
3. Jeff Suzuki, The Lost Calculus (1637–1670): Tangency and Optimization without Limits, *Mathematics Magazine* **78** (2005) 339–353.

Can We Use the First Derivative to Determine Inflection Points?

Duane Kouba

Points on the graph of a function f where the concavity changes are called *points of inflection*, and because concavity is determined by the sign of the second derivative, finding the points of inflection is a typical application of the second derivative in introductory calculus courses. But more than one attentive student has suggested a plausible shortcut that uses only the first derivative to find certain inflection points.

Example. Let $f(x) = 3x^4 - 8x^3$. The first derivative is $f'(x) = 12x^3 - 24x^2 = 12x^2(x-2)$ and its sign is

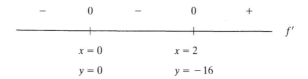

We conclude that f achieves its absolute minimum value of $y = -16$ at $x = 2$. Proceed to compute f'' in order to determine the inflection points for f.

But wait! At this point an enthusiastic student interjects that the *first* derivative f' tells us that f must have an inflection point at $x = 0$, since the graph of f is flat at $x = 0$ and decreases on either side of $x = 0$. The student argues that the sketch in Figure 1 must look somewhat like the graph of f near $x = 0$. The student

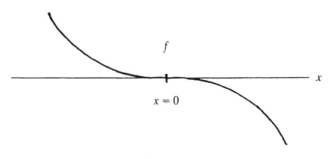

Figure 1

Can We Use the First Derivative to Determine Inflection Points?

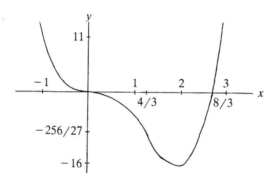

Figure 2

certainly *appears* to be correct. It looks as if $x = 0$ does in fact determine an inflection point for f.

Let's check by computing the second derivative, which is $f''(x) = 36x^2 - 48x = 12x(3x - 4)$, and its sign:

```
     +      0      −      0      +
  ─────────┼─────────────┼──────────── f''
         x = 0         x = 4/3
         y = 0         y = −256/27
```

Indeed, there is an inflection point at $x = 0$! There is an additional one at $x = 4/3$. We can now sketch a detailed graph of f; see Figure 2.

Question. Assume that f is twice differentiable on (a, b) and suppose the first derivative $f'(x)$ is strictly positive at all points x in (a, b) (or strictly negative at all such points), except that $f'(c) = 0$ at one point c, $a < c < b$. Does it then follow that f has an inflection point at $x = c$?

A counterexample. Unfortunately, unless we make additional assumptions about the function f, the answer is no! Consider the function

$$f(x) = \begin{cases} x^3 + x^4 \sin(1/x) & \text{for } x \neq 0 \\ 0 & \text{for } x = 0. \end{cases}$$

For $x \neq 0$ the derivative is $f'(x) = 3x^2 - x^2 \cos(1/x) + 4x^3 \sin(1/x)$, and using the limit definition of the derivative at $x = 0$, we get

$$f'(0) = \lim_{h \to 0} \frac{f(0+h) - f(0)}{h} = \lim_{h \to 0} \frac{h^3 + h^4 \sin(1/h)}{h}$$

$$= \lim_{h \to 0} \left[h^2 + h^3 \sin(1/h) \right] = 0.$$

The second derivative is $f''(x) = 6x - \sin(1/x) - 6x \cos(1/x) + 12x^2 \sin(1/x)$ for

$x \neq 0$; at $x = 0$

$$f''(0) = \lim_{h \to 0} \frac{f'(0+h) - f'(0)}{h}$$
$$= \lim_{h \to 0} \frac{3h^2 - h^2 \cos(1/h) + 4h^3 \sin(1/h)}{h}$$
$$= \lim_{h \to 0} [3h - h \cos(1/h) + 4h^2 \sin(1/h)] = 0.$$

Thus this function is twice differentiable. (Note that f'' is not continuous at $x = 0$ since the limit of $f''(x)$ does not exist as x approaches zero.)

We next show that the sign of f' is positive on some two-sided neighborhood of zero. Since the first derivative is $f'(x) = 3x^2 - x^2 \cos(1/x) + 4x^3 \sin(1/x)$, we can insure that $f'(x) > 0$ for $x > 0$ if an appropriate inequality is solved for x. Since $-1 \leq \cos(1/x), \sin(1/x) \leq 1$, then for positive x,

$$f'(x) = 3x^2 - x^2 \cos(1/x) + 4x^3 \sin(1/x)$$
$$> 3x^2 - x^2 - 4x^3 = 2x^2(1 - 2x).$$

Thus, $f'(x) > 0$ if $0 < x < 1/2$. Similarly, for negative x,

$$f'(x) = 3x^2 - x^2 \cos(1/x) + 4x^3 \sin(1/x)$$
$$> 3x^2 - x^2 + 4x^3 = 2x^2(1 + 2x),$$

so that $f'(x) > 0$ for $-1/2 < x < 0$. Hence, f' is positive throughout $(-1/2, 1/2)$ except that $f'(0) = 0$.

The last point to be verified is that this function does *not* have an inflection point at $x = 0$. We are motivated by the fact that the graph of the derivative f' appears to have infinitely many horizontal tangent lines in every neighborhood of zero. The graph of f' is shown in Figure 3.

Recall that the second derivative is $f''(x) = 6x - \sin(1/x) - 6x \cos(1/x) + 12x^2 \sin(1/x)$ for $x \neq 0$. Consider the sequence of x-values given by $x_n = 1/2n\pi$ and satisfying $-1/2 < x_n < 1/2$ for $n = 1, 2, 3, \ldots$. Then

$$f''(x_n) = 6(x_n) - \sin\left(\frac{1}{x_n}\right) - 6(x_n)\cos\left(\frac{1}{x_n}\right) + 12(x_n)^2 \sin\left(\frac{1}{x_n}\right)$$
$$= \frac{6}{2n\pi} - \sin(2n\pi) - \frac{6}{2n\pi}\cos(2n\pi) + 12\left(\frac{1}{2n\pi}\right)^2 \sin(2n\pi)$$
$$= \frac{6}{2n\pi} - 0 - \frac{6}{2n\pi}(1) + 12\left(\frac{1}{2n\pi}\right)^2 (0)$$
$$= 0.$$

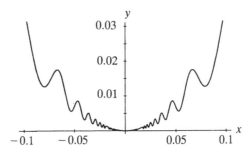

Figure 3

Since the sequence x_n converges to zero, it follows that there is no number t such that f is strictly concave up or concave down on $(0, t)$. Since f'' is an odd function, the same argument applied to the sequence $-x_n$ shows that f is not strictly concave up or down on any interval $(-t, 0)$. We conclude that f does not have an inflection point at $x = 0$.

Remark. Even if we require that the second derivative be continuous, this shortcut is still not valid. Consider the function

$$f(x) = \begin{cases} x^3 + x^6 \sin(1/x) & \text{for } x \neq 0 \\ 0 & \text{for } x = 0. \end{cases}$$

It can be shown (with some difficulty) that f is twice continuously differentiable, $f'(0) = 0$, and $f'(x) > 0$ for x-values near zero. However, $x = 0$ does not determine an inflection point for f.

The key property of both our counterexamples is that the second derivative has zeros arbitrarily close to the origin. It is reassuring to note that such functions are rarely encountered in a standard calculus textbook, and the following theorem shows that, except for such anomalies, the students were right all along about their shortcut for identifying inflection points where the tangent line to the graph is horizontal.

Theorem. *Let function f be twice continuously differentiable on the interval $[a, b]$, and suppose the second derivative f'' has only a finite number of zeros on this interval. Suppose $f'(x)$ is strictly positive at all points x in (a, b) (or strictly negative at all such points), except that $f'(c) = 0, a < c < b$. Then f has an inflection point at $x = c$.*

Proof. It suffices to show that there exist numbers $s > 0$ and $t > 0$ so that

$$f''(x) > 0 \quad \text{for all } x \text{ in } (c, c + t), \tag{1}$$
$$f''(x) < 0 \quad \text{for all } x \text{ in } (c - s, c). \tag{2}$$

We will prove (1) by contradiction, assuming $f'(x)$ is positive at all points $x \neq c$ in (a, b). The other cases are entirely similar.

Since f'' has only a finite number of zeros in $[a, b]$, there exists a number $t > 0$ such that $c + t < b$ and f'' is never zero in $(c, c + t)$. The continuity of f'' together with the intermediate value theorem guarantees that f'' has constant sign on the interval $(c, c + t)$, and we wish to show that it is in fact positive. (If f'' changed sign, the intermediate value theorem would imply that it had a zero somewhere in this interval, contradicting our choice of t.)

That f'' must be positive on $(c, c + t)$ now follows by applying the mean value theorem to the first derivative: $[f'(c + t) - f'(c)]/t = f''(z)$ for some z in $(c, c + t)$. Since by our assumptions $f'(c + t)$ and t are positive, and $f'(c) = 0$, it follows that $f''(z) > 0$, which completes the proof. ∎

Differentiate Early, Differentiate Often!

Robert Dawson

In first-year calculus, constrained-optimization and related-rates word problems are two of the biggest stumbling blocks. In this note, I contrast the methods suggested in calculus textbooks for the solution of these two types of problems, and conclude that a different approach to constrained-optimization problems, similar to that widely used for related-rates problems, would be advantageous.

Let us first consider related-rates problems. Traditional textbooks (see, for instance, Adams [**1**, p. 235]; Edwards and Penney [**3**, p. 193]; Finney, Weir, and Giordano [**5**, p. 209]; Johnston and Mathews [**6**, p. 316]; Stewart [**8**, p. 258], and Strauss *et al.* [**9**, p. 158]) introduce these shortly after implicit differentiation. These texts all suggest that implicit differentiation of the equation relating the rates should be an early step in the solution of such a problem. Nonetheless, many students, faced with a related-rates problem, persistently avoid implicit differentiation by eliminating a variable. For instance:

Problem 1. *A ladder of length 5 m is sliding with one end on the ground and the other on a vertical wall. The end on the ground is sliding away from the wall at a constant rate of 1 m/sec. How fast is the end on the wall moving when it is 4 m off the ground?*

Solution A (standard). By the Pythagorean theorem, the distance x from the foot of the wall to the ladder and the height y of the top of the ladder are linked by the relation

$$x^2 + y^2 = 25; \qquad (1)$$

differentiating implicitly with respect to t yields

$$x\, dx/dt + y\, dy/dt = 0. \qquad (2)$$

We can now substitute the instantaneous value $y = 4$ into (1) to obtain $x = 3$; substituting these values and $dx/dt = 1$ into (2) we obtain $3 + 4\,dy/dt = 0$, so that $dy/dt = -3/4$ m/sec.

Solution B (avoiding implicit differentiation). Solving (1) for y, we obtain

$$y = \sqrt{25 - x^2}. \qquad (3)$$

Differentiating with respect to x gives

$$\frac{dy}{dx} = \frac{-2x}{2\sqrt{25-x^2}}$$

and by the Chain Rule we have

$$\frac{dy}{dt} = \frac{dy}{dx}\frac{dx}{dt} = \frac{-x}{\sqrt{25-x^2}}\frac{dx}{dt}.$$

Substituting the instantaneous value $x = 3$ and $dx/dt = 1$, we obtain the answer.

The second approach is more difficult. Moreover, it admits two sources of error that are avoided by the first. The most common mistake on such a problem involves "freezing" one or more instances of a variable by substituting an instantaneous value before differentiation; the earlier the student differentiates, the less likely this is to happen. The other standard mistake, of course, involves incorrect differentiation of the comparatively complicated expression in (3).

All of this suggests that the traditional approach to related-rates problems is valid, and that students should be strongly encouraged to follow it. Like many other instructors, I usually take the view that students who prefer to use a certain technique, and get the right answer, should be permitted to do so. However, in this case, I feel that students who insist on avoiding implicit differentiation are not making an informed decision, even though they will probably be able to grind out the solutions to many problems.

A few weeks after related rates (depending on the textbook and course plan), students will usually encounter constrained-optimization problems. These resemble related-rates problems, not only in being presented as "word problems", but also in involving two variables on an equal footing. The usual approach in most textbooks (see, for instance, [1, p. 264], [4, p. 292], [5, p. 288], [8, pp. 331–2], and [9, p. 238])—and that favored by many instructors—is to use the constraint equation to eliminate one variable from the objective function, differentiate the resulting one-variable function, and find the extremum.

Problem 2. *Find the dimensions of the largest rectangle that can be inscribed in a semicircle of radius R.*

Solution A (traditional). From the constraint $x^2 + y^2 = R^2$ we get

$$y = \sqrt{R^2 - x^2}.$$

Substituting this into the area $A = 2xy$ of the rectangle, we obtain

$$A = 2x\sqrt{R^2 - x^2},$$

and differentiating this yields

$$\frac{dA}{dx} = 2\sqrt{R^2 - x^2} - \frac{2x^2}{\sqrt{R^2 - x^2}},$$

which simplifies to

$$\frac{dA}{dx} = \frac{2(R^2 - x^2) - 2x^2}{\sqrt{R^2 - x^2}}.$$

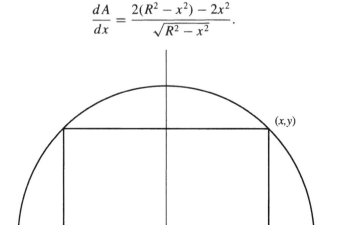

Figure 1. Maximize the area of the rectangle

Setting the numerator equal to 0 and solving, we get $2x^2 = R^2$ or $x = R/\sqrt{2}$; the edges of the rectangle are thus $R/\sqrt{2}$ and $\sqrt{2}R$.

This problem, like some others, can be simplified somewhat by maximizing the square of the area rather than the area itself. This trick, mentioned in some books, has limited application, though it is certainly worth knowing. A much more generally applicable technique is found in a few books. Implicit differentiation is used at the outset, on both the constraint and objective functions; the derivative is then eliminated to obtain the solution.

Solution B (early differentiation). Implicitly differentiating both the constraint function

$$x^2 + y^2 = R^2 \tag{4}$$

and the objective function $A = 2xy$, and setting the latter equal to 0, we get:

$$2x + 2yy' = 0$$
$$A' = 2y + 2xy' = 0.$$

Combining these, we get $2(x^2 - y^2) = 0$, whence the solution

$$x = \pm y \tag{5}$$

readily follows. Substituting this into (4) gives us $x = R/\sqrt{2}$ as before.

This is faster for two reasons. It is usually easier to differentiate a relation than to differentiate the function obtained by solving it for one variable; and the resulting equation is always linear in the derivative, so the step of eliminating the derivative is straightforward. Moreover, differentiation lowers the degree of a polynomial function, often simplifying the algebra. It is still possible that the resulting system of equations in x and y cannot be solved, but the odds are improved.

The functions in textbook constrained-optimization problems rarely go much beyond quadratics. As the complexity of the functions rises, so do the difficulties of eliminating a variable, or finding the zeros of the derivative finally obtained. Fortunately, quadratic relations are common in real-world applications, so the techniques learned do have some value.

With the early-differentiation method, it is possible to go somewhat further. The reader may like to try and compare the two methods on the problem of determining the point at which the curve $2x^3 + 3xy^2 + y^3 = 6$ comes closest to the origin. The above-mentioned trick of minimizing $x^2 + y^2$ rather than the distance itself will help here.

Early differentiation is given as an alternative method for one problem in Adams [1, pp. 266–7], and one in Stewart [8, p. 334]; but neither author suggests it as a method of first choice. Most textbooks examined do not mention it at all. Interestingly, *Schaum's Outline of Calculus*, while characteristically sparing of explanation, gives three examples of this technique [2, pp. 50–53] among ten worked constrained optimization problems (cf. [7, pp. 237–242].)

It is also worth noting that (5) gives not only a solution to the entire family of equations with different values of R, but also the general solution to the dual family of problems in which a rectangle of specified area must be inscribed in a semicircle with the smallest possible radius (this is mentioned in Adams, [1]). These are general features of this approach whenever the objective and constraint are both specified as values of functions; this duality will be familiar to the student who has studied linear programming, but is not commonly mentioned in first-year calculus. With early differentiation, little extra effort is needed to do so.

Implicit differentiation is, of course, an important technique in its own right, and is used heavily in subjects such as thermodynamics, mechanics, and economics. It is usually only "in the spotlight" for a comparatively short period during the first year calculus course, and students may consider it as an unimportant diversion from the main thrust of the course. Stressing it as a technique for both related-rates and constrained-optimization problems should emphasize its true importance.

Finally, the student who continues into multivariate calculus will learn to solve more advanced optimization problems using the method of Lagrange multipliers. Here, too, an important part of the technique is to do the differentiation first, rather than eliminating variables; the student who is already confident with operating in this order should find Lagrange multipliers less intimidating.

Acknowledgments. I thank the editors and the anonymous referee for various helpful suggestions. I also thank my first year calculus students over the last few years, who tried this method out. Work on this article was supported by a NSERC research grant.

References

1. R. A. Adams, *Calculus: A Complete Course*, 5th ed., Addison-Wesley Longman, 2003.
2. F. Ayres, *Theory and Problems of Differential and Integral Calculus, Schaum's Outline Series*, McGraw-Hill, 1964.
3. C. H. Edwards and D. E. Penney, *Calculus: Early Transcendentals, Matrix Version*, 6th ed., Prentice-Hall, 2002.
4. R. L. Finney, G. B. Thomas, F. D. Demana, and B. K. Waits, *Calculus: Graphical, Numerical, Algebraic*, Addison-Wesley, 1994.
5. R. L. Finney, M. D. Weir, and F. R. Giordano, *Thomas' Calculus*, 10th ed., Addison-Wesley, 2001.
6. E. J. Johnston and J. C. Mathews, *Calculus*, Pearson, 2002.
7. G.-C. Rota, *Indiscrete Thoughts*, Birkhäuser, 1997.
8. J. Stewart, *Single Variable Calculus (Early Transcendentals)*, 5th ed., Thomson Brooks/Cole, 2003.
9. M. J. Strauss, G. L. Bradley, and K. J. Smith, *Calculus*, 3rd ed., Prentice-Hall, 2002.

A Calculus Exercise For the Sums of Integer Powers

Joseph Wiener

Let $S_{k,n} = 1^k + 2^k + \cdots + n^k$, where k, n are positive integers. The usual method of finding $S_{k,n}$ for $k = 1, 2, \ldots$ is by means of the identity

$$\sum_{i=0}^{k-1} \binom{k}{i} S_{i,n} = (n+1)^k - 1, \qquad (1)$$

where $\binom{k}{i}$ denotes the binomial coefficient $k!/i!(k-i)!$. Recently, D. Acu has obtained (1) and similar formulas involving only even or only odd values of i from certain binomial identities [1].

We shall derive a generalization of (1) by differentiating the function

$$F(x) = e^{ax} + e^{(a+d)x} + e^{(a+2d)x} + \cdots + e^{(a+(n-1)d)x}.$$

Indeed,

$$F^{(k)}(0) = \sum_{m=1}^{n} (a + (m-1)d)^k, \qquad (2)$$

where $F^{(k)}(0)$ is the kth derivative of $F(x)$ at $x = 0$. Furthermore, the formula for the sum of terms of this finite geometric sequence yields

$$F(x) = \frac{(e^{(a+nd)x} - e^{ax})}{(e^{dx} - 1)},$$

whence

$$(e^{dx} - 1) F(x) = e^{(a+nd)x} - e^{ax}.$$

Now, differentiate this identity k times and set $x = 0$ to obtain

$$\sum_{i=0}^{k-1} \binom{k}{i} d^{k-i} F^{(i)}(0) = (a+nd)^k - a^k. \qquad (3)$$

Denote

$$S_{k,n}(a,d) = \sum_{m=1}^{n} (a+(m-1)d)^k; \qquad (4)$$

then, by virtue of (2), (3), (4), we get the formula

$$\sum_{i=0}^{k-1} \binom{k}{i} d^{k-i} S_{i,n}(a,d) = (a+nd)^k - a^k, \qquad (5)$$

which generalizes (1) since $S_{k,n} = S_{k,n}(1,1)$. This method of deriving identity (5) has been first discovered in [2] and used later [3] to evaluate some other sums involving integer powers.

Furthermore, to obtain formulas of type (5), where the index i varies only over even integers, we introduce the even function $G(x) = F(x) + F(-x)$ and differentiate the identity

$$(e^{dx}-1)G(x) = e^{(a+nd)x} - e^{-(a+(n-1)d)x}$$
$$- e^{ax} + e^{(d-a)x},$$

k times at $x = 0$ to obtain

$$\sum_{i=0}^{k-1} \binom{k}{i} d^{k-i} G^{(i)}(0) = (a+nd)^k$$
$$+ (-1)^{k-1}(a+(n-1)d)^k - a^k + (d-a)^k.$$

Since $G^{(i)}(0) = F^{(i)}(0) + (-1)^i F^{(i)}(0)$ and $F^{(i)}(0) = S_{i,n}(a,d)$, then

$$\sum_{i=0}^{k-1} \binom{k}{i} d^{k-i} (1+(-1)^i) S_{i,n}(a,d)$$
$$= (a+nd)^k + (-1)^{k-1}(a+(n-1)d)^k - a^k + (d-a)^k.$$

The substitution $i = 2p$ changes this formula to

$$2 \sum_{p=0}^{[(k-1)/2]} \binom{k}{2p} d^{k-2p} S_{2p,n}(a,d)$$
$$= (a+nd)^k + (-1)^{k-1}(a+(n-1)d)^k - a^k + (d-a)^k, \qquad (6)$$

where $[\cdot]$ denotes the greatest integer function. For $a = d = 1$, formula (6) yields

$$2 \sum_{p=0}^{[(k-1)/2]} \binom{k}{2p} S_{2p,n} = (n+1)^k + (-1)^{k-1} n^k - 1.$$

To obtain formulas of type (5), where the index i varies only over odd integers, we introduce the odd function $H(x) = F(x) - F(-x)$ and differentiate the identity

$$(e^{dx} - 1)H(x) = e^{(a+nd)x} + e^{-(a+(n-1)d)x}$$
$$- e^{ax} - e^{(d-a)x}$$

k times at $x = 0$. Since $H^{(i)}(0) = F^{(i)}(0) - (-1)^i F^{(i)}(0)$, then

$$\sum_{i=0}^{k-1} \binom{k}{i} d^{k-i}(1 - (-1)^i) S_{i,n}(a, d)$$
$$= (a + nd)^k + (-1)^k (a + (n-1)d)^k - a^k - (d-a)^k.$$

The substitution $i = 2p - 1$ changes this formula to

$$2 \sum_{p=1}^{[k/2]} \binom{k}{2p-1} d^{k-2p+1} S_{2p-1,n}(a, d)$$
$$= (a + nd)^k + (-1)^k (a + (n-1)d)^k - a^k - (d-a)^k. \qquad (7)$$

For $a = d = 1$, formula (7) yields

$$2 \sum_{p=1}^{[k/2]} \binom{k}{2p-1} S_{2p-1,n} = (n+1)^k + (-1)^k n^k - 1.$$

Acknowledgements. Research partially supported by U.S. Army Grant DAAL03-89-G-0107. I am thankful to the referees for a number of useful suggestions.

REFERENCES

1. D. Acu, Some algorithms for sums of integer powers, this MAGAZINE 61 (1988), 189–191.
2. J. Wiener and A. Wojczechowsky, Evaluating certain finite sums by differentiation, *U.S.S.R. Journal Mathematics in School* 2 (1968), 84–87.
3. J. Wiener, Solution of Problem E. 2770, *Amer. Math. Monthly* 87 (1980), 491.

L'Hôpital's Rule Via Integration

Donald Hartig

In elementary calculus texts L'Hôpital's rule is usually proven only for the case $0/0$, $x \to x_0$ (finite), by applying the Cauchy mean value theorem. Extension to $x \to \infty$ is then accomplished by replacing x with $1/x$. Verification of the rule for the ∞/∞ indeterminate form is regarded as too difficult and may be discussed in an exercise, an appendix, or not at all. In this note we give a proof for the ∞/∞ case that does not make use of the Cauchy mean value theorem. Instead, we require that the functions have continuous derivatives and take advantage of the order properties of the definite integral. The argument adapts nicely to the case $0/0$ as well.

L'Hôpital's Rule. ∞ / ∞. *Let f and g have continuous derivatives with $g'(x) \neq 0$. If $\lim_{x \to \infty} f(x) = \infty$, $\lim_{x \to \infty} g(x) = \infty$, and $\lim_{x \to \infty} f'(x)/g'(x) = L$, then $\lim_{x \to \infty} f(x)/g(x) = L$ also.*

Proof. We assume that L is finite; the other case can be handled in a similar fashion. The limit hypothesis on g allows us to assume that it is a positive function. Moreover, since g' is continuous and nonvanishing it too must be positive.

Let ε be some positive number. Choose M so that

$$\left| \frac{f'(x)}{g'(x)} - L \right| < \varepsilon$$

whenever $x > M$. Since $g'(x)$ is positive we have

$$|f'(x) - Lg'(x)| < \varepsilon g'(x),$$

so that

$$\left| \int_a^b [f'(x) - Lg'(x)] \, dx \right| \leq \int_a^b |f'(x) - Lg'(x)| \, dx < \int_a^b \varepsilon g'(x) \, dx$$

whenever $M < a < b$. Therefore, for such a and b,

$$|f(b) - f(a) - L[g(b) - g(a)]| < \varepsilon [g(b) - g(a)]. \tag{*}$$

Dividing through by the positive number $g(b)$ we obtain

$$\left| \frac{f(b)}{g(b)} - \frac{f(a)}{g(b)} - L\left[1 - \frac{g(a)}{g(b)}\right] \right| < \varepsilon \left[1 - \frac{g(a)}{g(b)}\right] < \varepsilon.$$

It follows easily that

$$\left|\frac{f(b)}{g(b)} - L\right| < \varepsilon + \frac{|f(a)|}{g(b)} + |L|\frac{g(a)}{g(b)}.$$

As b increases, $g(b)$ grows larger and larger without bound. Consequently, both $|f(a)|/g(b)$ and $|L|g(a)/g(b)$ will eventually become (and remain) smaller than ε, implying that

$$\left|\frac{f(b)}{g(b)} - L\right| < 3\varepsilon$$

for all b sufficiently large. This shows that $\lim_{x \to \infty} f(x)/g(x) = L$. □

Since our proof of this version of L'Hôpital's rule makes no use of the assumption $\lim_{x \to \infty} f(x) = \infty$, that condition can be dropped from the hypotheses. A straightforward variation of the preceding proof works when $x \to x_0$; alternatively, that case can be derived from the $x \to \infty$ case by considering $F(x) = f(x_0 + 1/x)$ and $G(x) = g(x_0 + 1/x)$.

This type of proof also applies to the indeterminate form $0/0$. For example, if we assume that L is finite and g' is a positive function (as was the case above), then allowing b to increase without bound in inequality $(*)$ will force $f(b)$ and $g(b)$ towards 0, implying that

$$|-f(a) - L[-g(a)]| \leq \varepsilon[-g(a)].$$

Dividing by the (positive) number $-g(a)$ reveals that

$$\left|\frac{f(a)}{g(a)} - L\right| \leq \varepsilon$$

whenever $a > M$.

As you can see, the algebra for $0/0$ is a bit simpler making this proof even more suitable for popular consumption.

Acknowledgment. The author wishes to thank the referee for helpful comments on the arrangement of the proofs.

Indeterminate Forms Revisited

R. P. Boas

1. Introduction You must all have seen at least one calculus textbook. It may surprise some of you that three centuries ago no such book existed: the very first book that was in any sense a calculus text was published, anonymously, in 1696, under the rather forbidding title *Analysis of the Infinitely Small* [5]. It was well known in European mathematical circles that the author was a French marquis, Guillaume de L'Hôpital. (I give him the modern French spelling, which at least keeps students from pronouncing the silent s in L'Hospital.) The book was hardly easy reading. It began with propositions like this: "One can substitute, one for the other, two quantities which differ only by an infinitely small quantity; or (what amounts to the same thing) a quantity that is increased or decreased only by another quantity infinitely less than it, can be considered as remaining the same." This sort of presentation gave calculus a reputation, which has survived to modern times, of being unintelligible.

Sylvester, writing in about 1880 [10, vol. 2, pp. 716-17], said that when he was young (around 1830) "a boy of sixteen or seventeen who knew his infinitesimal calculus would have been almost pointed out in the streets as a prodigy like Dante, who had seen hell." (Here and now, we would very likely find students of the same age feeling much the same; but Sylvester, in 1870, was teaching students who dealt casually with topics that we would now describe as advanced calculus.) When I was in high school (somewhat later), calculus was thought of, by otherwise well-educated people, as being as deep and mysterious as (say) general relativity is thought of today. My parents knew somebody who was reputed to know calculus, but they had no idea what that was (and they were college teachers—of English). Nowadays there are perhaps too many calculus books, but some of the answers that students give to examination questions make me wonder whether the subject has even now become sufficiently intelligible.

In his own time, and for long afterwards, L'Hôpital had an impressive reputation. Today he is remembered only for "L'Hôpital's rule," which evaluates limits like

$$\lim_{x \to 1} \frac{(2x - x^4)^{1/2} - x^{1/3}}{1 - x^{3/4}}$$

(L'Hôpital's own example) by replacing the numerator and denominator by their derivatives and hoping for the best.

L'Hôpital's rule seems to have fallen somewhat out of favor; I have heard it claimed that all it is useful for is as an exercise in differentiation.

It has been known for some time that many of L'Hôpital's results, including the rule, were purchased (quite literally) from John (= Jean = Johann) Bernoulli. Immediately after L'Hôpital's death in 1704, Bernoulli published an article claiming that he had communicated the rule for 0/0 to L'Hôpital, along with other material, before L'Hôpital had published it. This claim was disbelieved for some two hundred years; sceptics wondered why Bernoulli had not advanced his claim earlier. The reason for the delay eventually became clear when Bernoulli's correspondence with L'Hôpital came to light in the early 1900s. Bernoulli gave the rule to L'Hôpital only after L'Hôpital had promised to pay for it, had repeatedly asked for it, and had finally come across with the first installment. We now also know that there are records of Bernoulli's having lectured on the rule before L'Hôpital's book was published.

In the preface to his book, L'Hôpital says, "I acknowledge my debt to the insights of MM Bernoulli, above all to those of the younger [John], now Professor at Groningen. I have unceremoniously made use of their discoveries and of those of M Leibnis [sic]. Consequently I invite them to claim whatever they wish, and will be satisfied with whatever they may leave me." Considering what we now know, this seems somewhat disingenuous, especially since L'Hôpital was clearly unable to discover for himself how to prove the rule of which, as Plancherel once said of his own theorem, he had "the honor of bearing the name."

You can find the whole story in the 1955 volume of Bernoulli's correspondence [7], or in Truesdell's review of the volume [11].

I used to wonder, from time to time, what kind of proof L'Hôpital had used, but never when I was where I could find a copy of his book. Recently I happened to mention this question to Professor Alexanderson—who promptly produced his own copy. Professor Underwood Dudley, who is more resourceful than I am, also found a copy, and has translated it into modern terminology [4], but retaining its geometric character (L'Hôpital thought of functions as curves). L'Hôpital actually considered only $\lim_{x \to a} f(x)/g(x)$, where a is finite, $f(a) = g(a) = 0$, and both $f'(a)$ and $g'(a)$ exist, are finite, and not zero. In analytical language, what L'Hôpital did amounts to writing

$$\lim_{x \to a} \frac{f(x)}{g(x)} = \lim_{x \to a} \frac{f(x) - f(a)}{g(x) - g(a)} = \lim_{x \to a} \frac{f'(a) + \varepsilon(x)}{g'(a) + \delta(x)} (\varepsilon \to 0, \delta \to 0) = \frac{f'(a)}{g'(a)}.$$

It is not trivial to extend such a proof to the cases when $f'(a)$ and $g'(a)$ do not exist (but have limits as $x \to a$), or are both zero, or $f(a) = g(a) = \infty$, or $a = \infty$. I do not know when or by whom these refinements were added, but the complete theory was in place by 1880 [8, 9].

2. A common modern proof If you saw a proof of L'Hôpital's rule in a modern calculus class, the probability is about 90% that it is Cauchy's proof. This proof appeals to mathematicians because it is elegant, but often fails to appeal to students because it is subtle. It depends on knowing Cauchy's refinement of the mean-value theorem, namely that (with appropriate hypotheses)

$$\frac{f(x) - f(a)}{g(x) - g(a)} = \frac{f'(c)}{g'(c)}, \quad c \text{ between } a \text{ and } b. \tag{1}$$

Given this, L'Hôpital's rule becomes obvious.

In spite of its elegance, Cauchy's proof seems to me to be inappropriate for an elementary class. Any proof that begins with a lemma like Cauchy's mean value theorem, that says "Let us consider...," repels most students. Students are also uncomfortable with the nebulous point c. They want to know where it is, and feel that the instructor is deliberately keeping them in the dark. Of course, the exact location of c is completely irrelevant (although numerous papers have been written about it).

3. A caution Cauchy's proof tacitly assumes that there is a (one-sided) neighborhood of the point a in which $g'(x) \neq 0$. Strictly speaking, if there is no such neighborhood, the limit in (1) is not defined, and we would have no business talking about it. However, if f' and g' are given by explicit formulas, they may happen to share a common factor that is zero at a, and the temptation to cancel this factor is irresistible. One can obtain a spurious result in this way [8, 9; 3, p. 124, ex. 24; 1].

Let me give you a specific example, just to emphasize that there is a reason for the requirement that $g'(x) \neq 0$. Let $f(x) = 2x + \sin 2x$, $g(x) = x \sin x + \cos x$; $a = +\infty$. Then $f'(x) = 4\cos^2 x$, $g'(x) = x \cos x$, and $f'(x)/g'(x) \to 0$, whereas $f(x)/g(x)$ does not approach a limit. The trouble comes from cancelling a factor that changes sign in every neighborhood of the point a; it would have been legitimate to cancel a quadratic factor.

Some writers think that the difficulty arises only in artificial cases that would never occur in practice. But then, what happens to our claim to be giving correct proofs?

You might not guess from Cauchy's proof that there is a discrete analog of L'Hôpital's rule; see, for example, [6]. This was known to Stolz in the 1890's, and has often been rediscovered.

4. A more satisfactory proof I want now to show you a proof of L'Hôpital's rule that avoids the difficulties of Cauchy's and establishes a good deal more. It may seem more complicated, but not if you include a proof of Cauchy's mean value theorem as part of Cauchy's proof. This proof is also quite old; Stolz knew it, but preferred Cauchy's proof, perhaps because of Cauchy's reputation. It has been published several times by people (including me) who failed to search the literature.

Let us suppose that $f(x)$ and $g(x)$ approach 0 as $x \to a$ from the left, where a might be $+\infty$; it does no harm to define (if necessary) $f(a) = g(a) = 0$. We may suppose that $g'(x) > 0$ (otherwise consider $-g(x)$). Now let $f'(x)/g'(x) \to L$, where $0 < L < \infty$. Then, given $\varepsilon > 0$, we have, if x is sufficiently near a, and a is finite,

$$L - \varepsilon \leqslant \frac{f'(x)}{g'(x)} \leqslant L + \varepsilon,$$

$$(L - \varepsilon) g'(x) \leqslant f'(x) \leqslant (L + \varepsilon) g'(x) \quad \text{(since } g'(x) > 0\text{)}.$$

Integrate on (x, a) to get

$$-(L - \varepsilon) g(x) \leqslant -f(x) \leqslant -(L + \varepsilon) g(x)$$

(notice that since g increases to 0, we have $g(x) < 0$). Since $-g$ and $-f$ are positive near a,

$$L - \varepsilon \leqslant \frac{f(x)}{g(x)} \leqslant L + \varepsilon,$$

$$\lim_{x \to a} \frac{f(x)}{g(x)} = L.$$

Only formal changes are needed if $a = +\infty$ or if $L = 0$ or ∞.
For the ∞/∞ case, we get, in the same way, with $\delta > 0$,

$$L - \varepsilon < \frac{f(a-\delta) - f(x)}{g(a-\delta) - g(x)} < L + \varepsilon,$$

$$L - \varepsilon < \frac{\dfrac{f(a-\delta)}{f(x)} - 1}{\dfrac{g(a-\delta)}{g(x)} - 1} \cdot \frac{f(x)}{g(x)} < L + \varepsilon.$$

Letting $x \to a$, we obtain

$$L - \varepsilon \leqslant \liminf_{x \to a} \frac{f(x)}{g(x)} \leqslant \limsup_{x \to a} \frac{f(x)}{g(x)} \leqslant L + \varepsilon.$$

Letting $\varepsilon \to 0$, we obtain $f(x)/g(x) \to L$.

If it happens that $f'(a) = g'(a) = 0$, one repeats the procedure with f'/g', and so on. If $f^{(n)}(a) = g^{(n)}(a) = 0$ for every n (which can happen with f and g not identically zero), the procedure fails. Otherwise, the limit can be handled more simply in a single step, as we shall see below.

5. Generalizations If f and g are defined only on the positive integers, we can reason in a similar way with differences instead of derivatives to conclude that if the differences of g are positive, and $f(n)$ and $g(n)$ approach zero as $n \to \infty$, then if

$$\frac{f(n) - f(n-1)}{g(n) - g(n-1)} \to L \quad \text{as } n \to \infty,$$

it follows that $f(n)/g(n) \to L$. This is sometimes called Cesàro's rule. For more detail, and illustrations, see [6]. A possibly more familiar version is as follows:
If $a_n \to 0$ and $b_n \to 0$, and $a_n/b_n \to L$, then

$$\frac{\sum_{k=1}^{n} a_k}{\sum_{k=1}^{n} b_k} \to L.$$

The key point in the proof of L'Hôpital's rule is the principle that the integral of a nonnegative function ($\neq 0$) is positive. More precisely, if $f(x) \geqslant 0$ on $p \leqslant x \leqslant q$ then

$$\int_p^q f(t)\,dt > 0 \quad \text{if } p < x < q \text{ and } f(t) \not\equiv 0.$$

Repeated integration with the same lower limit has the same property, as we see by rewriting the n-fold iterated integral as a single integral:

$$\frac{1}{(n-1)!} \int_p^x (t-p)^{n-1} f(t)\,dt.$$

This suggests the appropriate treatment of the case of L'Hôpital's rule when $f'(a) = g'(a) = 0$, or more generally when $f^{(k)}(a) = g^{(k)}(a) = 0, k = 1, 2, \ldots, n-1$, but not both of $f^{(n)}(a)$ and $g^{(n)}(a)$ are 0. The positivity of iterated integration then yields the conclusion of L'Hôpital's rule in a single step.

An operator that carries positive functions to positive functions is conventionally called a positive operator. If F is an invertible operator whose inverse is positive, we can conclude that if $F[f(x)]/F[g(x)] \to L$ and $F[g] > 0$, then $f(x)/g(x) \to L$.

As an example of the use of operators, consider $D + P(x)I$, where $D = d/dx$ and I is the identity operator. This is the operator that occurs in the theory of the linear first-order differential equation $y' + P(x)y = Q(x)$. The solution of this differential equation, with $y(a) = 0$, is

$$y = \exp\left\{\int_a^x P(t)\,dt\right\} \int_a^x Q(t) \exp\left\{\int_a^t P(u)\,du\right\} dt. \qquad (2)$$

In other words, (2) provides the inverse Λ of $D + P(x)I$.

The explicit formula shows that if $Q(x) \geq 0$ we have $\Lambda[Q] > 0$, so that Λ is a positive operator. Hence we may conclude that if

$$\frac{[D + PI]f}{[D + PI]g} \to L$$

and $[D + PI]g > 0$ then

$$(L - \varepsilon)[D + PI]g < [D + PI]f < (L + \varepsilon)[D + PI]g.$$

A positive linear operator evidently preserves inequalities. Consequently, if we apply Λ to both sides, we obtain

$$(L - \varepsilon)g(x) < f(x) < (L + \varepsilon)g(x)$$

and hence

$$f(x)/g(x) \to L.$$

Thus $D + P(x)I$ can play the same role as D in L'Hôpital's rule. It is at least possible that $D + P(x)I$ might be simpler than D.

Since some forms of fractional integrals and derivatives are defined by positive operators, one could also formulate a fractional L'Hôpital's rule.

This article is the text of an invited address to a joint session of the American Mathematical Society and the Mathematical Association of America, January 14, 1989.

REFERENCES

1. R. P. Boas, Counterexamples to L'Hôpital's rule, *Amer. Math. Monthly* 94 (1986), 644–645.
2. _____, *Indeterminate Forms Revisited*, videotape, Amer. Math. Soc., Providence, RI, 1989.
3. R. C. Buck and E. F. Buck, *Advanced calculus*, 3rd ed., McGraw-Hill, New York, etc., 1978.
4. U. Dudley, Review of Calculus with Analytic Geometry, *Amer. Math. Monthly* 95 (1988), 888–892.
5. [G. de L'Hôpital], *Analyse des infiniment petits, pour l'intelligence des lignes courbes*, Paris, 1696; later editions under the name of le Marquis de L'Hôpital.
6. Xun-Cheng Hwang, A discrete L'Hôpital's rule, *College Math. J.* 19 (1988), 321–329.
7. [O. Spiess, ed.], *Die Briefwechsel von Johann Bernoulli*, vol. 1, Birkhäuser, Basel, 1955.
8. O. Stolz, Über die Grenzwerthe der Quotienten, *Math. Ann.* 15 (1889), 556–559.
9. _____, *Grundzüge der Differential- und Integral-rechnung*, vol. 1, Teubner, Leipzig, 1893.
10. J. J. Sylvester, *The Collected Mathematical Papers*, Cambridge University Press, 1908.
11. C. Truesdell, The new Bernoulli edition, *Isis* 49 (1958), 54–62.

The Indeterminate Form 0^0

Louis M. Rotando and Henry Korn

Consider the following limit problems often encountered in elementary calculus textbooks:

$$\lim_{x \to 0^+} (\sin x)^{\tan x} \qquad \lim_{x \to 0^+} (e^{x+1} - e)^x$$

$$\lim_{x \to 0^+} (\arctan x)^x \qquad \lim_{x \to 1^-} (1-x)^{\sinh(x-1)}.$$

In each of the above problems the limit is 1. Curiously enough the limit is also 1 for most similar problems typically included in the exercise sets devoted to indeterminate forms. Yet the limit process does not yield 1 for every example of the type 0^0: G. C. Watson [1] discusses a generalization of the counterexample

$$\lim_{x \to 0^+} x^{a/\log x},$$

and conditions on f wherein $\lim_{x \to 0^+} x^{f(x)} = 1$ are investigated by L. J. Paige [2].

The purpose of this note is to further study of the indeterminate form 0^0 by looking at examples of the more general form

$$\lim_{x \to 0^+} f(x)^{g(x)}$$

in which f and g are real functions **analytic** at $x = 0$, that is, representable there by a Taylor series. In such cases we can show that the limit is 1.

THEOREM. *Suppose that f and g are nonzero real analytic functions at $x = 0$ for which $f(x) \geq 0$ for all positive x sufficiently close to 0. If $\lim_{x \to 0^+} f(x) = \lim_{x \to 0^+} g(x) = 0$, then $\lim_{x \to 0^+} f(x)^{g(x)} = 1$.*

Proof. Applying L'Hospital's Rule, we have

(1) $$\lim_{x \to 0^+} f(x)^{g(x)} = \exp\left[\lim_{x \to 0^+} \frac{\log f(x)}{1/g(x)}\right] = \exp\left[\lim_{x \to 0^+} \frac{f'(x) g^2(x)}{-f(x) g'(x)}\right].$$

Since f and g are analytic and approach zero as x approaches zero, it follows by continuity that $f(0) = g(0) = 0$, and in some neighborhood of $x = 0$, f and g have the form

$$f(x) = x^m F(x), \quad F(0) \neq 0$$
$$g(x) = x^n G(x), \quad G(0) \neq 0$$

where m and n are positive integers, while F and G are analytic at $x = 0$. Substituting these into (1) we obtain, after simplification,

(2) $$\lim_{x \to 0^+} f(x)^{g(x)} = \exp\left[\lim_{x \to 0^+} \frac{x^n [xF'(x) + mF(x)] G^2(x)}{-F(x)[xG'(x) + nG(x)]}\right].$$

The numerator $N(x) \equiv x^n [xF'(x) + mF(x)] G^2(x)$ as well as the denominator $D(x) \equiv -F(x)[xG'(x) + nG(x)]$ in this last limit are analytic at $x = 0$. Since $D(0) \neq 0$, it follows from continuity at $x = 0$ that $\lim_{x \to 0^+} D(x) \neq 0$, and therefore $\lim_{x \to 0^+} [N(x)/D(x)] = 0$. Hence $\lim_{x \to 0^+} f(x)^{g(x)} = 1$.

We note in conclusion that a simple linear transformation of variables will permit coverage of the case $\lim_{x \to a^+} f(x)^{g(x)}$. The theorem may also be easily modified to include the case $\lim_{x \to 0^-} f(x)^{g(x)}$ by requiring $f(x) \geq 0$ for negative x sufficiently close to 0. Finally, if the restriction of "analytic" is weakened to "infinitely differentiable" at $x = 0$ then the theorem would be false. A nice counterexample to illustrate this is given by letting $g(x) = x$ and

$$f(x) = \begin{cases} e^{-1/x^2} & \text{if } x \neq 0 \\ 0 & \text{if } x = 0. \end{cases}$$

In this case the reader can easily verify that $\lim_{x \to 0^+} f(x)^{g(x)} = 0$ and $\lim_{x \to 0^-} f(x)^{g(x)} = \infty$.

The author is indebted to the referee for several suggestions that led to substantial clarification of the proof.

References

[1] G. C. Watson, A note on indeterminate forms, Amer. Math. Monthly, 68 (1961) 490–492.
[2] L. J. Paige, A note on indeterminate forms, Amer. Math. Monthly, 61 (1954) 189–190.

On the Indeterminate Form 0^0

Leonard J. Lipkin

All calculus books have a section on L'Hôpital's Rule and its application to study various indeterminate forms. One of the forms is "0^0." In most books, examples such as $\lim_{x \to 0^+} x^x$ or $\lim_{x \to 0^+} (\tan x)^{\sin x}$ invoke logarithms to show that the limit is 1. Typically the same thing happens in the exercises given for this indeterminate form. So the question arises: Why is this form indeterminate if the limit is always 1?

In this capsule, we point out why the typical examples have limit 1, and we show how to construct examples with limits other than 1. As we will see below, the key is that most examples in calculus courses are either (real) analytic functions or are products of analytic functions with x^α for some $\alpha > 0$. The two theorems below show why most of the limits that appear in the calculus books are 1.

Theorem 1. Let $f(x) = x^\alpha \phi(x)$, where $\alpha > 0$ and $0 < m \leq \phi(x) \leq M < \infty$ for constants α, m, and M. Let $g(x)$ be any function that satisfies $\lim_{x \to 0^+} g(x) = 0$ and $\frac{g(x)}{x}$ bounded on $(0, a)$ for some $a > 0$. Then $\lim_{x \to 0^+} f(x)^{g(x)} = 1$.

Proof. Using standard techniques, we obtain

$$\log f(x)^{g(x)} = g(x) \log f(x)$$
$$= g(x) (\log x^\alpha + \log \phi(x))$$
$$= \frac{g(x)}{x} \alpha x \log x + g(x) \log \phi(x).$$

Since $\frac{g(x)}{x}$ is bounded and $x \log x \to 0$ as $x \to 0$, the first term has limit 0. Since $\lim_{x \to 0^+} g(x) = 0$ and $\log \phi(x)$ is bounded (by hypothesis on ϕ), the second term has limit 0. Exponentiation yields the conclusion $\lim_{x \to 0^+} f(x)^{g(x)} = 1$. ∎

In Theorem 1, we assumed that $g(x)/x$ is bounded near 0. This assumption rules out, for example, $g(x) = \sqrt{x}$. Theorem 1 is actually a special case of the following more general theorem. The proof is similar to that given before.

Theorem 2. Let $f(x)$ be as in Theorem 1 and let $g(x) = x^\beta \gamma(x)$, where $\beta > 0$ and $\gamma(x)$ is bounded on $(0, a)$. Then $\lim_{x \to 0^+} f(x)^{g(x)} = 1$.

Theorems 1 and 2 take care of the analytic functions and much more. Notice that differentiability is not assumed. For example, we can take $g(x) = x^\beta Q(x)$, where $\beta > 0$ and $Q(x)$ is the characteristic function of the rationals. We now know where *not* to look in order to find cases in which $\lim_{x \to 0^+} f(x)^{g(x)} \neq 1$. The following examples show that any limit may occur.

Example 1. Let $f(x) = x$ and $g(x) = 1/\ln(x)$ for $x > 0$. Since the function

$$f(x)^{g(x)} = x^{1/\ln(x)}$$

has the property that $\ln(x^{1/\ln(x)}) = 1$ for all $x > 0$, we see that $\lim_{x \to 0^+} (x^{1/\ln(x)}) = e$. Therefore, $\lim_{x \to 0^+} x^{1/\ln(x^\tau)} = e^{1/\tau}$ for every $\tau > 0$. More generally, let $f(x) = x^\alpha \phi(x)$ where $\alpha > 0$ and $0 < m \le \phi(x) \le M < \infty$ for some positive constants α, m, and M. Then

$$\lim_{x \to 0^+} f(x)^{1/\ln(x^\tau)} = e^{\alpha/\tau}.$$

Example 2. Let $f(x) = e^{-1/x}$ and $g(x) = -1/\ln(x)$. Then

$$\lim_{x \to 0^+} f(x)^{g(x)} = \lim_{x \to 0^+} e^{1/(x \ln(x))} = 0.$$

Variations on a Theme of Newton

Robert M. Corless

We use a particularly simple example function[1], and the computer algebra system Maple, to try to learn something about Newton's method. The discussion here presumes only a minimal amount of calculus—including the standard introduction to Newton's method, such as is found in [2, Sec. 2.10]—and some algebraic fluency. This discussion, though aimed at undergraduate students, contains surprises (perhaps even for instructors), items not found in the usual calculus course, and pointers to many more such items. The intention is to provoke or reinforce an interest in pure and applied mathematics. If this works, *everyone* will take something new away.

Newton's method

Newton's method is for approximately solving nonlinear equations $f(x) = 0$. Applied mathematics problems usually lead to nonlinear equations—we cannot rely on everything being linear. Some examples of applied problems requiring Newton's method or an equivalent are:

- so-called "implicit" numerical methods for the solution of ordinary differential equations.
- practically any engineering design problem, where instead of being asked to calculate the behavior of a machine or system as given, you are asked to calculate the design parameters that will make the system behave in a certain desired way. For example, many problems in robotic control fall into this category.
- computer-aided design uses piecewise polynomials to model physical objects. Calculating their intersection points requires the solution of systems of polynomial equations. Even if initial approximations to the solutions are arrived at by other means, Newton's method can be used to "polish" the roots.

The basic idea behind Newton's method is that if you can't solve $f(x) = 0$ for x, replace f with a simpler function F, namely, the best linear approximation to $f(x)$ near some initial guess point x_0. This approximation is $F(x) = f(x_0) + f'(x_0)(x - x_0)$, and we *can* solve $F(x) = 0$ to get $x_1 = x_0 - f(x_0)/f'(x_0)$, provided $f'(x_0) \neq 0$. Repeating this with the new approximation x_1 to get x_2 and so on gives us the iterative formula

$$x_{n+1} = x_n - \frac{f(x_n)}{f'(x_n)}.$$

---

[1] Our example function $f(x) = x^2 - a$ is indeed particularly simple, and this is important: if it were not so simple, we wouldn't be able to go anywhere near as far as we do. Hold on to your seat!

We will explore this formula with an extremely simple nonlinear function, namely $f(x) = x^2 - a$, in order to learn something about Newton's method (and some computer tools). It is clear that the zeros of $f(x) = x^2 - a$ are just $x = \sqrt{a}$ and $x = -\sqrt{a}$, so Newton's method is not really required for this problem. Even worse, we are later going to specify $a = 1$, so we will be using Newton's method to find the square root of 1! Our iteration is, for general a,

$$x_{n+1} = x_n - \frac{x_n^2 - a}{2 x_n} \tag{1}$$

or, mathematically equivalent but slightly less numerically stable,

$$x_{n+1} = \frac{1}{2}\left(x_n + \frac{a}{x_n}\right).$$

A Maple program The following program, written in the computer algebra language Maple (see [1], for example, for an accelerated introduction to Maple), will be used to compute iterates of Newton's method for the rest of this discussion. The routine normal just simplifies expressions.

```
Newton := proc(a, x0, n)  local xn;
  xn := x0;
  to n do
    xn := normal(xn- (xn^2 - a) / (2*xn))
  od
end:
```

Numerical tests If we choose $a = 2$, then our function is $f(x) = x^2 - 2$ and we are looking for $\sqrt{2}$. Choosing an initial guess of $x_0 = 1$, the program Newton produces Table 1.

TABLE 1 Newton iterates of $f(x) = x^2 - 2$.

n	x_n	error
0	1	-1.0
1	3/2	$2.5 \cdot 10^{-1}$
2	17/12	$6.0 \cdot 10^{-3}$
3	577/408	$6.0 \cdot 10^{-6}$
4	665857/470832	$4.5 \cdot 10^{-12}$
5	886731088897/627013566048	$2.5 \cdot 10^{-24}$

REMARKS

1. The error reported in the above table is the so-called "residual" error $r_n = f(x_n)$. If r_n is zero, then of course x_n is a root; if r_n is "small," then, in some sense, x_n is "close" to a root. This type of measure of accuracy is always available, even when the exact answer is not known. For "well-conditioned" problems it gives the same information as the difference between the approximate answer and the true answer; this problem is well-conditioned because $x_n - \sqrt{a} = (x_n^2 - a)/(x_n + \sqrt{a}) \approx r_n/(2\sqrt{a})$ and so the *relative* error here ($a = 2$) is about $(x_n - \sqrt{a})/\sqrt{a} \approx r_n/4$.

2. Exact arithmetic *costs a lot*. We notice that the *length* of the answer approximately doubles each time; a quick calculation shows that the answer after 30 iterations would take a few gigabytes of memory to store. This is why people instead use arithmetic with a fixed number of decimals (*i.e.*, floating-point).
3. We can simplify our problem by the *nondimensionalization*[2] $u_n = x_n/\sqrt{a}$, at least for the purpose of understanding what is happening. Of course, for actual calculations we can't nondimensionalize by $\sqrt{a}$ which we don't know. If we use this conceptual scaling, then the Newton iteration becomes

$$u_{n+1} = \frac{1}{2}\left(u_n + \frac{1}{u_n}\right).$$

This is exactly the same iteration but with $a = 1$. Thus the scaled iteration uses Newton's method to compute the square root of 1. But the relative error in x_n is $(x_n - \sqrt{a})/\sqrt{a} = u_n - 1$ and so this iteration really does tell us something about Newton's method, and we will keep it in mind. It is easy to see that if $u_n = 1 + e_n$ where e_n represents the error after n iterations, then

$$e_{n+1} = \frac{e_n^2}{2(1+e_n)} \approx \frac{1}{2}e_n^2.$$

This is called *quadratic convergence*. Using this formula shows that after about 30 iterations we will have about 1 billion digits of $\sqrt{a}$ correct, if we start with roughly one correct digit.

4. If we convert these rational numbers to "continued fraction form" (using the Maple routine `convert(17/12, confrac)`) where a *continued fraction* is something of the form

$$n_0 + \cfrac{1}{n_1 + \cfrac{1}{n_2 + \cfrac{1}{\ddots}}}$$

we see the quite remarkable patterns

$$1 = 1$$

$$3/2 = 1 + \frac{1}{2}$$

$$17/12 = 1 + \cfrac{1}{2 + \cfrac{1}{2 + \cfrac{1}{2}}}$$

$$577/408 = 1 + \cfrac{1}{2 + \cfrac{1}{\ddots + \cfrac{1}{2}}}$$

---

[2] If a has units, say square meters, this scaling removes them.

where the length of the continued fraction is 2^n, and every entry is 2. This is the beginning of an interesting foray into number theory.

Symbolic initial guess If the program Newton given earlier had been written in C or FORTRAN, then calling it with a symbol (say g) for the initial guess would generate an error message. But here,

```
> Newton(1, g, 1);
```

in Maple, returns $(g^2 + 1)/(2g)$. We can ask Maple to continue, giving the results in Table 2.

TABLE 2 Newton iterates for $x^2 - 1$ with a symbolic initial guess, g.

n	x_n
0	g
1	$\dfrac{g^2 + 1}{2g}$
2	$\dfrac{1}{4} \dfrac{g^4 + 6g^2 + 1}{g(g^2 + 1)}$
3	$\dfrac{1}{8} \dfrac{g^8 + 28g^6 + 70g^4 + 28g^2 + 1}{g(g^4 + 6g^2 + 1)(g^2 + 1)}$

In FIGURE 1 we plot the first few results from Newton. We see that these rational functions are trying to approximate a step function; as n increases, we see clear evidence that these functions converge. The moral of this section is that the error message that FORTRAN would have given us would have concealed an insight, namely that the result of n iterations of Newton's method is a rational function of the initial guess g. Further, we have learned that this rational function looks (for large n) rather like a step function with heights $\pm \sqrt{a}$. Note that the graph in FIGURE 1 works for all a because the axes are scaled—the horizontal axis is the $g/\sqrt{a}$ axis and the vertical axis is the $x_n/\sqrt{a}$ axis.

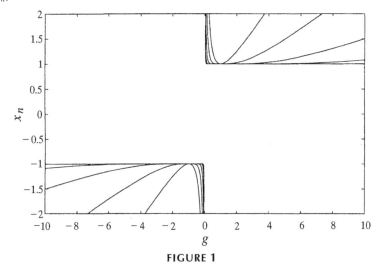

FIGURE 1

Newton iterates with a symbolic initial guess, plotted together. As n increases we must have $x_n/\sqrt{a} \to \pm 1$, and we can see that the convergence is rapid near $g/\sqrt{a} = \pm 1$, as we expect.

Variations on a Theme of Newton

Symbolic *a* Now let us choose instead $x_0 = 1$ (we will discuss this choice of initial guess in a moment) and look at the results from Maple if we input a symbolic *a* to the program. The first few of these are presented in Table 3.

TABLE 3 Rational approximations obtained by using a symbolic a.

n	x_n
0	1
1	$\frac{1}{2} + \frac{1}{2}a$
2	$\frac{1}{4} \frac{1 + 6a + a^2}{1+a}$
3	$\frac{1}{8} \frac{1 + 28a + 70a^2 + 28a^3 + a^4}{(1 + 6a + a^2)(1+a)}$

When we plot these[3], we get a sequence of rational (in *a*) approximations to $\sqrt{a}$, as is quite evident in FIGURE 2.

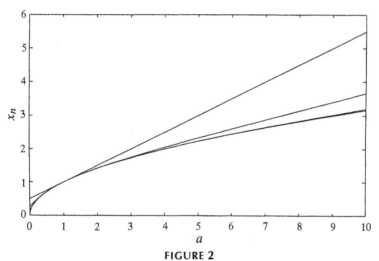

FIGURE 2
The first few iterates of Newton's method on $f(x) = x^2 - a$ with symbolic *a* give quite good rational approximations to $\sqrt{a}$.

REMARKS

1. Nondimensionalization shows that choosing $x_0 = 1$ is perfectly general. Put $x_n = x_0 v_n$ in equation 1, and simplify to get

$$v_{n+1} = \frac{1}{2}\left(v_n + \frac{a/x_0^2}{v_n}\right).$$

---

[3] Both FIGURE 1 and FIGURE 2 were actually prepared using Matlab, not Maple, because Matlab plots look slightly nicer; moreover, the graphs were generated by giving a *vector* of g values and a *vector* of a values to a Matlab implementation of Newton's method, much like the Maple symbolic version.

This is just the iteration for finding the square root of a/x_0^2. Therefore, the graph in FIGURE 2 differs from the graph of the approximations we would get with some other initial guess (say $x_0 = 2$) *only in the scale of the axes*—in particular the shape of the curves remains the same. If we label the y-axis with x_0 where 1 is now, and likewise $2 x_0$ for 2 and so on, and label the x-axis with x_0^2 where 1 is now, etc., then FIGURE 2 represents the first few iterates of the general case. That is, all the curves with general initial guess *collapse onto the same curve*. This shows the true power of nondimensionalization.

2. We can replace `normal` in the routine `Newton` with a call to Maple's `series` command, and execute Newton's method in the domain of power series. Quadratic convergence in this domain means that the number of correct terms in the power series doubles each time.

3. We can show with Maple that the error in our rational functions of a above are proportional to $(a-1)^{2^k}$; for example, after three iterations the error is

$$f_3(a)^2 - a = \frac{1}{64} \frac{(a-1)^8}{\left(1 + 6a + a^2\right)^2 (1+a)^2}.$$

As before the difference between $f_3(a)$ and $\sqrt{a}$ will be about $1/(2\sqrt{a})$ times this.

4. We can convert the rational approximations in Table 3 to continued fraction form; indeed these approximations are one step towards *approximation theory* which underlies much of scientific computing.

5. Again FORTRAN would give us an error message if we tried this in that language. We begin to suspect that whenever a language gives us an error message, there is something to learn.

Chaotic dynamics

Now we choose $a = -1$ and see what happens. We are trying to find an x such that $x^2 + 1 = 0$, and if we start with a real x_0 we are doomed to failure. However, the failure is very interesting.

A few experiments show us that some initial guesses ($x_0 = 0$, $x_0 = 1$, $x_0 = 1 - \sqrt{2}$, etc.) lead to division by zero. We ignore these minor annoyances. A few more experiments show that most initial guesses don't lead (immediately) to division by zero, but rather wander all over the x-axis, without showing any kind of pattern.

Since the x_n appear random in this case, we consider looking at a *frequency distribution* of them. We divide the axis up into bins—the bins are chosen according to a rule given by an advanced theory, namely a rule depending on the theoretical probability density function—and count the number of x_n that appear in each bin. The results appear in Table 4.

To explain the theoretical probability density function would take us to the boundaries of *ergodic theory*, which is a "main artery," if you will, of statistical mechanics, dynamical systems, and indeed probability theory.

Symbolic n If we call the Maple program not with symbolic a or x_0 but rather with symbolic n, the number of iterations, we get the error message

```
Error, (in Newton) unable to execute for loop.
```

Variations on a Theme of Newton

TABLE 4 Frequency distribution for x_n where $x_{n+1} = (x_n - 1/x_n)/2$ and $x_0 = 0.4$ (10,000 iterates). The bin boundaries b_k, $0 \le k \le 10$ are chosen so that $b_0 = -\infty$ and $\int_{b_{k-1}}^{b_k} 1/(\pi(x^2+1))\,dx = 1/10$. According to theory, there should be roughly the same number of x_n in each bin.

k	b_k	number of x_n in (b_{k-1}, b_k)
1	−3.0777	1001
2	−1.3764	999
3	−0.7265	1006
4	−0.3249	1000
5	0.0000	986
6	0.3249	1000
7	0.7265	1007
8	1.3764	980
9	3.0777	986
10	∞	1035

As we have discovered, an error message indicates that we have something to learn. Maple might not be able to do this problem for a symbolic n, *but we can* (in this case). Assume first that $u_0 > 1$ (this corresponds to $x_0 > \sqrt{a}$). Put $u_n = \coth \theta_n$. (The hyperbolic functions $\sinh \theta = (\exp(\theta) - \exp(-\theta))/2$, $\cosh \theta = (\exp(\theta) + \exp(-\theta))/2$, $\tanh \theta = \sinh \theta / \cosh \theta$ and so on, are strongly related to the ordinary trig functions.) We have

$$\coth \theta_{n+1} = u_{n+1} = \frac{1}{2}\left(\frac{\cosh \theta_n}{\sinh \theta_n} + \frac{\sinh \theta_n}{\cosh \theta_n} \right)$$

$$= \frac{\cosh 2\theta_n}{\sinh 2\theta_n}$$

$$= \coth 2\theta_n$$

where we have used $\cosh^2 \theta + \sinh^2 \theta = \cosh 2\theta$ and $2\sinh \theta \cosh \theta = \sinh 2\theta$ to simplify. Taking $\coth^{-1}$ of both sides, we see $\theta_{n+1} = 2\theta_n$, which is easily solved to get

$$\theta_n = 2^n \theta_0.$$

Therefore $u_n = \coth(2^n \theta_0)$, if $u_0 > 1$.

For the case when $0 < u_0 < 1$, we note that we will immediately have $u_1 = (u_0 + 1/u_0)/2 > 1$ (for example, by elementary calculus we see the minimum of u_1 occurs when $u_0 = 1$). Thereafter the previous analysis applies. The case of $u_0 < 0$ is symmetric to the positive case. So we can say $u_n = \coth 2^{n-1}\theta_1$, regardless of what u_0 is.

Similarly, it is an elementary exercise to show in the complex case, with $a = -1$, that $u_n = \cot(\theta_n)$ gives $\theta_{n+1} = 2\theta_n$ or

$$u_n = \cot(2^n \theta_0).$$

This lays bare all of the chaotic dynamics of this iteration in the complex case. See [3] for more discussion of this case.

REMARKS

1. Now we have the solution for symbolic n, we can answer the question "What do you get if you do half a Newton iteration?" For this problem, we get $u_{3/2} = \coth(\sqrt{2}\,\theta_1)$ (by definition). This doesn't have any apparent application, but in more complicated dynamical systems finding such an interpolation is very useful indeed.

2. The *Lyapunov exponent* in the chaotic case is ln 2. The formula (3) also tells us how to find the theoretical probability density function alluded to earlier.
3. No "fractals" appear in this problem, unless it is on the imaginary axis (where the chaos is). However, looking at Newton's method for solving $f(x) = x^3 - 1 = 0$, we get fractals in $\mathbb{C}$ immediately. See [3], and the other papers in that same issue of the *College Mathematics Journal*.
4. The "asymptotics" of $\coth 2^n \theta_0$ tell us how quickly the iterates approach 1. By Maple,

$$u_n = 1 + 2e^{-1 \cdot 2^n \theta_1} + 2e^{-2 \cdot 2^n \theta_1} + O(e^{-3 \cdot 2^n \theta_1})$$

which tells us everything about how fast u_n approaches 1 (and by extension how fast x_n approaches $\sqrt{a}$).

Concluding remarks

In this discussion we have stepped outside the normal route to mathematics. By asking just slightly different questions about Newton's method than is usual in a calculus class —using a very simple example, just trying to understand it better—we have used or discovered links to nondimensionalization, numerical analysis, complexity theory, continued fractions, approximation theory, series algebra, asymptotics, ergodic theory, and dynamical systems (chaos and fractals). One hopes the student will be stimulated to search out other references on these subjects (one might begin with the references in [3], and the other papers in that same issue of the *College Mathematics Journal*).

The discussion in this paper also suggests that it might have been premature to drop Newton's method (for computing the square root) from the high-school curriculum, as it has been dropped in some districts, merely because calculators can compute square roots with the press of a button. The important thing may not ever have been to compute a square root, but rather to provide a nice introduction to Newton's method, from which "central trunk" we may move on to other significant areas of modern mathematics.

Probably the most significant concept used in this discussion is nondimensionalization. From a practical viewpoint, it is an invaluable tool in the management of large numbers of variables; from the pure mathematical viewpoint it is an overture to the theory of symmetry, itself a vigorous and powerful branch of modern mathematics.

But even just on its own, Newton's method is an extremely important and well-studied tool in applied mathematics, used every day for the solution of systems of nonlinear equations. It is surprising how easy it is to find new questions to ask about it.

Acknowledgment. Many of these ideas are due to Charles M. Patton, and I first heard them in his workshop at the 4th International Conference on Technology in Education, in Portland, Oregon, 1991. This paper also benefited from discussions with Peter Poole, David Jeffrey, and Bob Bryan.

REFERENCES

1. Robert M. Corless, *Essential Maple*, Springer-Verlag, New York, NY, 1994.
2. J. Stewart, *Calculus*, Brooks/Cole, Pacific Grove, CA, 1991.
3. Gilbert Strang, A chaotic search for i, *College Math. Journal* 22 (1991), pp. 3–12.

A Useful Notation for Rules of Differentiation

Robert B. Gardner

The freshman calculus instructor is often faced with the task of deciphering the freshman calculus student's versions of the product rule, quotient rule and chain rule. Even when correctly computed, it can be difficult to follow the student's steps, especially when the differentiated function involves several trigonometric functions. I propose a notational convention to deal with this problem.

Several popular calculus books (e.g. Swokowski, Larson et al., etc.) represent the differential operator as $D[\cdot]$ or $d/dx[\cdot]$. With this as motivation, I have used the following notation quite successfully in the classroom. Whenever a function is differentiated using the product rule, quotient rule, or chain rule, put the differentiated parts in square brackets. This means that the product rule would be written as

$$D[(f)(g)] = [f'](g) + (f)[g'].$$

In fact, if the instructor introduces the convention of "always differentiate f (the first function in the product) first," then the product rule can be presented pictorially as

$$D[(\ )(\ )] = [\ ](\ ) + (\ )[\ ].$$

This reduces exercises involving the product rule to nothing more than fill-in-the-blank problems.

Similarly, the quotient rule can be represented as

$$D\left[\frac{(\ )}{(\ )}\right] = \frac{[\ ](\ ) - (\ )[\ ]}{(\ )^2}.$$

The habit of "differentiating f first" developed in the product rule, must be carried over to the quotient rule.

The chain rule is a bit harder to "draw"; however the special case known as the power rule can be illustrated in this manner:

$$D[(\ )^n] = n(\ )^{n-1}[\ ].$$

My experience has shown that, not only is the students' work easier to follow, but the material is easier to present clearly. Students have reacted quite positively to this notation. In fact, when encouraged to adopt the notation, but not required to, I have found that practically all students choose to use the notation. This has been the case even when the notation was introduced some time after the rules of differentiation, i.e., Calculus 2 and Calculus 3 students quickly "pick up the habit."

An example illustrates the utility of this method. Suppose

$$f(x) = \frac{\sec x \tan x}{(x^2 + 1)^6}.$$

With the notation,

$$f'(x) = \frac{[[\sec x \tan x](\tan x) + (\sec x)[\sec^2 x]](x^2 + 1)^6 - (\sec x \tan x)\left[6(x^2 + 1)^5[2x]\right]}{(x^2 + 1)^{12}}.$$

Although a formidable derivative, it can be viewed as nothing more than a large fill-in-the-blank problem.

With the square brackets, it is much easier to follow the students' steps and, if applicable, to give partial credit. In fact, give the above problem on your next calculus test. I think, in the absence of the square brackets, you will find it quite a task simply to determine if the students' responses *are* partially correct!

Wavefronts, Box Diagrams, and the Product Rule: A Discovery Approach

John W. Dawson, Jr.

Of the many differentiation rules encountered in calculus, the product rule is perhaps the first to shake students' naive faith in the "simplicity" of mathematics. It is easy to refute their expectation that $(fg)' = f'g'$, but counterexamples often seem more to bewilder than enlighten. Moreover, the correct formula is not easy to guess, except for the special case when one factor is constant; and although many students accept the formula $(cg)' = cg'$ as agreeing with their intuition, few realize that this special case itself contradicts the equation $(fg)' = f'g'$. Later, having learned the correct general formula, some students laboriously apply it to all products, ignoring the erstwhile special case, or failing to recognize it as such!

Such observations suggest that few students develop any real understanding of the product rule. Rather, having no idea *why* the rule takes the form it does, most students simply memorize it, adding it to their growing stockpile of unrelated facts. Of course, at some point they have probably seen the product rule derived through algebraic manipulation of the defining difference quotient. The usual proof is concise and rigorous, and students will probably assent to the correctness of each step. Nevertheless it is a hollow demonstration, one that *verifies* rather than *explains*, for it conveys no idea of how the formula, or the proof itself, might ever have been discovered. In particular, the rationale for transforming $f(x + \Delta x)g(x + \Delta x) - f(x)g(x)$ to $[f(x + \Delta x) - f(x)]g(x + \Delta x) + f(x)[g(x + \Delta x) - g(x)]$ via the simultaneous addition and subtraction of the term $f(x)g(x + \Delta x)$ is mystifying to students; too often the step is justified merely as a clever trick that works.

In short, the product rule provides a classic example of how a *logically* rigorous "proof" may fail to be *psychologically* convincing or enlightening. There is, however, a more heuristic approach to the product rule that leads to discovery of both the formula and its proof. This approach invokes the geometric representation of products as areas, an idea later to prove so important in integral calculus, while stressing certain physical considerations important in their own right. At the same time the mystery surrounding the algebraic manipulations is dispelled. Indeed, the only mystery left unexplained is the conspicuous absence of such an approach from most calculus texts.

The very word "calculus" provides a fitting point of departure: Latin for "pebble," it alludes to the counting stones used in early abacus-like devices, but it also calls to mind the ripple created by a pebble cast into water, and so suggests the familiar problem of computing the rate of increase of the area enclosed by a radially expanding circular wavefront. Although this problem appears in nearly every calculus text, it is usually considered only after introduction of the chain rule, and all too often it is presented as a routine exercise devoid of any particular significance. However, merely by factoring a difference of two squares, it is easy to show directly that

$$\frac{d}{dt}(\pi r^2) = 2\pi r \frac{dr}{dt} \; ;$$

and far from being surprising, the result is readily conjectured, since the area inside the ripple increases through the engulfing of space by the expanding wavefront, which at any given instant has length $2\pi r$ and is moving outward at speed dr/dt (Figure 0). Similarly, a radially expanding spherical membrane engulfs 3-dimensional space at a rate equal to the product of its instantaneous surface area and its instantaneous velocity. The physical intuitions involved are quite natural and general*, and students might well pass on to consider the rate at which area is engulfed by an "arbitrary" radially-expanding simple closed curve. It is again

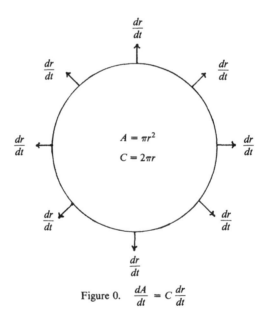

Figure 0. $\dfrac{dA}{dt} = C \dfrac{dr}{dt}$

---

* Similar considerations underlie the basic integral theorems of vector analysis, a classical subject notoriously obscured by modern rigorous treatments. The outstanding text by Schey cited below presents a lucid informal account of those topics from a physicist's point of view, while vividly describing the bewilderment and hostility that such material often evokes from students previously exposed to abstract presentations.

reasonable to conjecture that the rate should be proportional to the length of the curve, but exploration of the assumptions involved could quickly lead to such deeper concepts as rectifiability, orientability, and homology. (As in any discovery approach, the instructor must be well-grounded in his discipline to anticipate and lead discussion of such matters!) In any case, we are here concerned with the less ambitious aim of conjecturing the product rule.

Toward that end, let $h(t)$ and $w(t)$ denote the height and width, varying as functions of time t, of a rectangle having one pair of adjacent sides lying along the axes of a Cartesian coordinate system. Any product may be represented as the (signed) area of such a rectangle, and we may think of the sides not lying along the axes as forming an L-shaped wavefront. By analogy with the ripple, at any time t the side of length $w(t)$ will be moving with speed $h'(t)$, and the side of length $h(t)$ will be moving with speed $w'(t)$ (Figure 1).

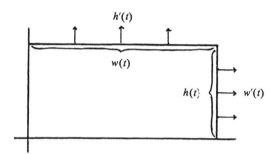

Figure 1. $[h(t)w(t)]' = h'(t)w(t) + h(t)w'(t)$.

Hence the total rate of engulfing (or disgorging) of area should be $w(t)h'(t) + h(t)w'(t)$. To prove the correctness of this result, we draw the rectangles corresponding to times t and $t + \Delta t$. In case all of $h(t)$, $h'(t)$, $w(t)$, and $w'(t)$ are positive, we obtain Figure 2, in which the L-shaped region representing the difference of the areas (numerator of the difference quotient) is left unshaded.

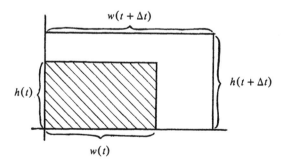

Figure 2. $A(t + \Delta t) - A(t) = h(t + \Delta t)w(t + \Delta t) - h(t)w(t)$.

There are two quite natural ways (Figures 3 and 4) to decompose that region into a pair of rectangles, each yielding an alternative form of the identity invoked in the algebraic proof. Of course, there are other cases to consider, and it would be well for students to draw the figures corresponding to other sign combinations; nonetheless the algebraic proof is perfectly general. The illustration (Figure 5) for the special case is also enlightening, since it is obvious that if one dimension is held fixed, the area of the rectangle can still change through the independent motion of the other dimension.

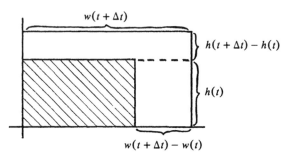

Figure 3. $A(t + \Delta t) - A(t) = [h(t + \Delta t) - h(t)]w(t + \Delta t) + h(t)[w(t + \Delta t) - w(t)]$.

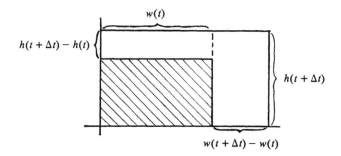

Figure 4. $A(t + \Delta t) - A(t) = [h(t + \Delta t) - h(t)]w(t) + h(t + \Delta t)[w(t + \Delta t) - w(t)]$.

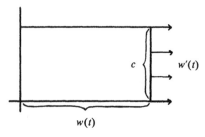

Figure 5. $[cw(t)]' = cw'(t)$.

Two further observations are in order. First, in seeking to discover the form of the product rule, one may wish to consider examples of functions $f \neq F$ and $g \neq G$ for which $f' = F'$ and $g' = G'$ but $(fg)' \neq (FG)'$; such examples show that *no* formula for $(fg)'$ involving only f' and g' can possibly be correct, thereby providing a simple demonstration of a non-existence theorem. Second, box diagrams are useful in many other contexts—in the author's opinion, their range of applicability is at least as wide as the much-touted Venn diagrams. As an example, consider Bayes' theorem. The formula

$$P(A|B) = \frac{P(A)P(B|A)}{P(A)P(B|A) + P(S-A)P(B|S-A)}$$

corresponding to events A, B in a sample space S is difficult for most students to remember, and its analytic derivation means little, if anything, to them. Yet after struggling mightily in vain attempts to train students to solve Bayesian problems, the author was amazed to discover how readily the ideas were assimilated when they were presented in the context of a box diagram rather than a formula. For example, consider the twins problem:

> Roughly 3/4 of all twins are fraternal, while only 1/4 are identical. Among identical twins, half are girls and half boys, while among fraternal twins, half are mixed sets and 1/4 each consist of just boys or just girls. Given that a set of twins consists of two boys, what is the chance those twins are fraternal?

Many students never learned to fit such a problem into Bayes' formula, but after a single lecture most could quickly draw the diagram below and use it to obtain the solution. Many even remarked how easy such problems were to solve!

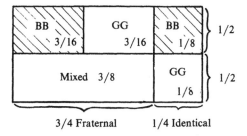

Figure 6. $P(\text{Fraternal} | \text{both boys}) = \dfrac{3/16}{3/16 + 1/8} = \dfrac{3}{5}$.

REFERENCE

H. M. Schey, Div, Grad, Curl, and All That, Norton, 1973.

$(x^n)' = nx^{n-1}$: Six Proofs

Russell Jay Hendel

A perusal of calculus textbooks published during the past twenty years reveals six distinct approaches to the proof that the derivative of x^n is $(x^n)' = nx^{n-1}$ for a positive integer n. These proofs use important techniques that should be in the repertoire of every calculus student.

I. Proof with induction and the product rule. First, $(x)' = 1$ by definition. Assuming that $(x^{n-1})' = (n-1)x^{n-2}$ and using the product rule we have

$$(x^n)' = (x^{n-1}x)' = (x^{n-1})'x + x^{n-1}(x)' = nx^{n-1}.$$

[L. Bers with F. Karal, *Calculus*, Holt, Rinehart, and Winston, 1976, p. 87; N. Friedman, *Basic Calculus*, Scott, Foresman, 1968, p. 112; S. Salas and E. Hille, *Calculus: One and Several Variables with Analytic Geometry*, Wiley, 1974, p. 84]

II. Proof using properties of logarithms. If $y = x^n$, $\ln y = n \ln x$ and logarithmic differentiation yields $y'/y = n/x$ which immediately implies $y' = nx^{n-1}$. [L. Hoffmann, *Calculus for Business, Economics, and the Social and Life Sciences*, McGraw-Hill, 1986, p. 255]

III. Proof by estimation. We use the big O notation common in analytic proofs: We say $f(h) = O(g(h))$ if there is some constant C such that, if h is sufficiently close to 0, then $f(h) < C|g(h)|$. This is slightly nonstandard (h goes to zero instead of infinity) but it clarifies the proof.

Lemma. $(x + h)^n = x^n + nx^{n-1}h + O(h^2)$.

The lemma has a direct inductive proof independent of the binomial expansion. The lemma is used to simplify the numerator of the difference quotient $[(x+h)^n - x^n]/h$. A similar simplification occurs in proofs IV and V. [S. Stein, *Calculus and Analytic Geometry*, McGraw-Hill, 1987, p. 85]—Stein, however, does not use the big O notation.

IV. Proof using factoring. We use the factor formula for the difference of nth powers, $(x+h)^n - x^n = (x+h-x)\Sigma(x+h)^{n-1-i}x^i$, to simplify the difference quotient. [S. Grossman, *Calculus (International Edition)*, Academic Press, 1981, p. 135; L. Loomis, *Calculus*, Addison-Wesley, 1974, p. 100; A. Spitzbart, *Calculus with Analytic Geometry*, Scott, Foresman, 1975, pp. 85–86]

V. Proof using the binomial theorem. We use the binomial expansion, $(x+h)^n = \sum \binom{n}{i} x^{n-i} h^i$, to simplify the difference quotient. This seems to be the most popular method. [H. Anton, *Calculus with Analytic Geometry*, Wiley, 1988, pp. 159–160; J. Fraleigh, *Calculus with Analytic Geometry*, Addison-Wesley, 1980, p. 45; L. Leithold, *The Calculus with Analytic Geometry*, Harper and Row, 1986, p. 189; G. Simmons, *Calculus with Analytic Geometry*, McGraw-Hill, 1985, pp. 63–64; E. Swokowski, *Calculus with Analytic Geometry*, Prindle, Weber and Schmidt, 1983, pp. 99–100; A. Willcox, R. Buck, H. Jacob, and D. Bailey, *Introduction to Calculus 1 and 2*, Houghton Mifflin, 1971, pp. 63–64]

VI. "Proof" by example. Finally, not every formula is totally proven in math courses. In such cases, computation of several examples or proof of several subcases of the main theorem is a welcome procedure [L. Goldstein, D. Lay, and D. Schneider, *Calculus and Its Applications*, Prentice-Hall, 1980, pp. 57–61]. Verification by examples or subcases has intrinsic value even when a general proof is given. For example, some calculus students can prove $(x^n)' = nx^{n-1}$ by the binomial theorem for general n, but still fumble when asked to prove it for the case $n = 2$ or $n = 3$.

Occasionally, a book does present several of the above proofs [H. Flanders and J. Price, *Calculus with Analytic Geometry*, Academic Press, 1978, p. 65; L. Bers, *Calculus*, Holt, Rinehart and Winston, 1969, p. 169]. Some proofs (II, III) are mentioned rarely. Even proof I, despite the importance of induction, occurs in only about 20% of the books. I suggest that freshman calculus courses be enriched by presenting all the above proofs.

Sines and Cosines of the Times

Victor J. Katz

Why does the derivative of the sine equal the cosine? Or the derivative of the tangent equal the square of the secant? One answer, that you learned early in your calculus course, is that these rules can be proved. In fact, your instructor probably proved the first from the definition of derivative, having first convinced you that
$$\lim_{x \to 0} (\sin x)/x = 1,$$
and proved the second by using the quotient rule. But, after all, the trigonometric functions are defined geometrically; one ought to be able to understand their derivatives geometrically as well. If we look back at the history of these functions and their relationship to the history of calculus, we can do exactly that.

Today, we generally consider the sine and the other trigonometric functions as numerical functions of real numbers, where the numbers in the domain can be thought of as measures of angles. But until the time of Euler in the mid-eighteenth century, sines were certain lines in circles of a given radius, the lines being generally associated not to angles but to arcs. (Hence the term arcsine = arc of a given sine for the inverse function of a sine.) Thus, in the figure, we have $y = \sin t$, where t is the measure of the arc and, for convenience, we take the radius equal to 1; similarly, $x = \cos t$.

So suppose we want to take the derivative of the sine function with respect to the arc. In the eighteenth century, this meant that we would find the ratio of the infinitesimal change of $y = \sin t$ to the infinitesimal change in t. Now an infinitesimal change in the arc t can best be represented by drawing an infinitesimal tangent line to the circle at the end of the arc labeled t. If we consider this tangent as the hypotenuse of a right triangle, then the vertical leg represents dy, the infinitesimal change in the sine. Since the infinitesimal triangle is similar to the original large triangle, the laws of similarity show that $dy/dt = x/1$, or $d(\sin t)/dt = \cos t$, as desired. (Exercise: Develop an analogous argument to find the derivative of the tangent.)

The geometrical arguments giving the derivatives of the sine and tangent first appeared in print in aposthumously published paper of Roger Cotes (1682—1716), the editor of the second edition of Newton's *Principia*. The first appearance of the sine argument in a calculus text, however, was in *A New Treatise of Fluxions* (1737) by Thomas Simpson (1710—1761), famous today mostly for Simpson's rule for numerical integration by parabolic approximation. Simpson was one of a group of private teachers in England who

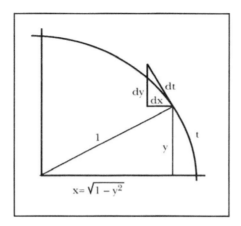

met the growing demand of the English middle class for mathematical knowledge. The textbook grew out of Simpson's membership in the Mathematical Society at Spitalfields, whose rules made it the duty of every member "if he be asked any mathematical or philosophical question by another member, to instruct him in the plainest and easiest manner he is able."

Interestingly, although neither Newton nor Leibniz considered explicitly the derivatives of the trigonometric functions, they did deal with their power series and their differential equations. For example, Leibniz used the same figure with its differential triangle to conclude that $dy^2 + dx^2 = dt^2$, or, since $x = \sqrt{1-y^2}$ and $dx = y\,dy/\sqrt{1-y^2}$, that $dy^2 + y^2 dt^2 = dt^2$. Assuming, then, that dt is a constant increment and therefore that its differential is 0, Leibniz applied his differential operator d to both sides of the equation to get $d(dy^2 + y^2 dt^2) = 0$. Using the product rule for differentials on the left side, he simplified this into $2dy(d\,dy) + 2y\,dy\,dt^2 = 0$ or $d^2y + y\,dt^2 = 0$ or, finally, into the familiar differential equation of $y = \sin t$: $d^2y/dt^2 = -y$. Note that Leibniz's method of manipulating with second order differentials explains our seemingly strange placement of the 2's in the modern notation for second derivatives.

Although arguments using infinitesimals were replaced by arguments using limits in the early nineteenth century, their heuristic value in the learning of calculus remains. And in recent decades, the work of Abraham Robinson on non-standard analysis has shown that these arguments can even be modified to meet modern standards of rigor. Rigorous arguments notwithstanding, the history of many of the concepts of calculus helps us to develop an intuitive understanding of the basic ideas of the subject, an understanding necessary for us to apply these techniques to the solving of problems.

The Spider's Spacewalk Derivation of sin´ and cos´

Tim Hesterberg

The usual proofs of the derivatives of sine and cosine in introductory calculus involve limits. I shall outline a simple geometric derivation that avoids evaluating limits, based on the interpretation of the derivative as the instantaneous rate of change. The principle behind this proof is found in a late nineteenth-century calculus textbook by J. M. Rice and W. W. Johnson, *The Elements of the Differential Calculus, Founded on the Method of Rates or Fluxions* (Wiley, New York, 1874).

A spider walks with speed 1 in a circular path around the outside of a round satellite of radius 1, as shown in Figure 1. At time t the spider will have travelled a distance t, which corresponds to a central angle of t radians. The altitude of the spider, in the standard coordinate system, is $y = \sin(t)$ and the spider is $x = \cos(t)$ units to the right (or left) of the origin.

Now look how fast the spider is moving *upward*. Since the altitude of the spider at time t is $y = \sin(t)$, its upward velocity is $y' = \sin'(t)$. Oops!—The spider loses its footing at time t, and since the gravity in outer space is negligible, it continues with the same direction and speed. It moves a distance 1 in additional time $\Delta t = 1$,

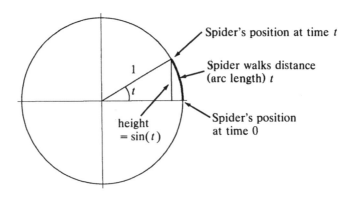

Figure 1

The Spider's Spacewalk Derivation of sin' and cos'

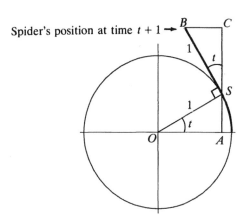

Figure 2

from point S to point B; see Figure 2. The spider's altitude changes by $\Delta y = SC$, so its upward velocity is $y' = \Delta y/\Delta t = SC$. But because triangles OAS and SCB are congruent ($\angle OSA = 90° - t$, so $\angle CSB = t = \angle AOS$), $SC = OA = x = \cos(t)$. Therefore $\sin'(t) = y' = \cos(t)$.

Similarly, the spider is $x = \cos(t)$ units to the right of the center, and its horizontal velocity is $x' = \cos'(t) = \Delta x/\Delta t = -BC = -SA = -\sin(t)$. The minus sign arises because the spider is moving to the left when it is above the x-axis, i.e., $\Delta x/\Delta t$ is negative whenever $y = \sin(t)$ is positive.

Figure 2 describes the case $0 < t < \pi/2$, but it is easily modified to yield the same results when the spider is located anywhere on the unit circle.

Acknowledgment. I thank Bob Gether and George Rosenstein for helpful comments and for the reference to Rice and Johnson's book.

[*Editor's note.* This proof also appeared in C. S. Ogilvy, Derivatives of $\sin \theta$ and $\cos \theta$, *American Mathematical Monthly* 67 (1960) 673.]

Differentiability of Exponential Functions

Philip M. Anselone and John W. Lee

We present a new proof of the differentiability of exponential functions. It is based entirely on methods of differential calculus. No current or recent calculus text gives or cites a proof of the differentiability that depends only on such elementary tools. Our proof makes it possible to give a comprehensive treatment of the derivative properties of exponential and logarithmic functions in that order in differential calculus, building on the standard introduction to these topics in precalculus courses. This is the logical order and has considerable pedagogical merit.

Most calculus books defer the treatment of exponential and logarithmic functions to integral calculus in order to prove differentiability. A few texts introduce these topics in differential calculus under the heading of "early transcendentals" but defer the proof of differentiability to integral calculus. Both approaches have serious pedagogical faults, which are discussed later in this paper.

Our proof that exponential functions are differentiable provides the missing link that legitimizes the "early transcendentals" presentation.

Preliminaries

We assume that a^r has been defined for $a > 0$ and r rational in a precalculus course and that the familiar rules of exponents are known to hold for rational exponents. It is natural to define a^x for $a > 0$ and x irrational as the limit of a^r as $r \to x$ through the rationals. In this way, a^x is defined for all real x.

Basic properties of a^x for real x are inherited by limit passages from corresponding properties of a^r for r rational. These properties include the rules of exponents with real exponents and

$$a^x \text{ is positive and continuous,}$$
$$a^x \text{ is increasing if } a > 1,$$
$$a^x \text{ is decreasing if } a < 1.$$

It is not especially difficult to justify the definition of a^x for x irrational and to derive the foregoing properties of a^x for x real, but there are a lot of small steps. A program along these lines is carried out by Courant in [2, pp. 69–70]. The general idea of each step is well within the grasp of students in typical calculus classes. However, just as properties of a^r with r rational are routinely stated without proof, it is better to give just an overview of the basic properties of a^x with x real, illustrated with graphs, and move on to the question of differentiability, which is more central to differential calculus.

Differentiability of Exponential Functions

A more complete development, beginning with the derivation of properties of a^r with r rational, might be given in an honors class. The properties can be extended to a^x with x real with the aid of the density of the rationals in the reals and the squeeze laws for limits. The conclusion that a^x with $a > 1$ is increasing also relies on the following proposition which should seem evident from graphical considerations:

> If f is a continuous function on a real interval I
> and f is increasing on the rational numbers in I,
> then f is increasing on I.

The same proposition will provide a key step in the proof that a^x is differentiable.

Henceforth, we restrict our attention to properties of a^x with $a > 1$. Corresponding properties of a^x with $0 < a < 1$ follow from $a^x = (1/a)^{-x}$.

The differentiability of a^x

Consider an exponential function a^x with any $a > 1$. In order to prove that a^x is differentiable for all x, the main task is to prove that it is differentiable at $x = 0$. Our proof of this depends only on methods of differential calculus. It is motivated by the fact that the graph of a^x (see Figure 1) is concave up, even though this fact is not assumed a priori.

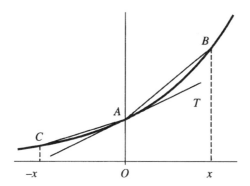

Figure 1. Graph of a^x with $B = (x, a^x)$ and $C = (-x, a^{-x})$ for $x > 0$.

In Figure 1, imagine that $x \to 0$ with $x > 0$ and x decreasing. Then B and C slide along the curve toward A. The upward bending of the curve seems to imply that

$$\text{slope } \overline{AB} \text{ decreases,} \quad \text{slope } \overline{AC} \text{ increases,}$$
$$\text{and} \quad \text{slope } \overline{AB} - \text{slope } \overline{AC} \to 0.$$

It follows that the slopes of $\overline{AB}$ and $\overline{AC}$ approach a common limit, which is the slope of the tangent line T in Figure 1 and the derivative of $f(x) = a^x$ at $x = 0$. This geometric argument will be made rigorous.

The curve in Figure 1 is actually the graph of $f(x) = 2^x$. The following table gives values of the slopes of $\overline{AB}$ and $\overline{AC}$ rounded off to two decimal places. It appears that the slopes of $\overline{AB}$ and $\overline{AC}$ approach a common limit, which is $f'(0) = \text{slope } T \approx 0.7$.

x	1	1/2	1/4	1/8	1/16	1/32
slope $\overline{AB}$	1	.83	.76	.72	.71	.70
slope $\overline{AC}$	.50	.59	.64	.66	.67	.69

With this preparation, we are ready to prove that $f(x) = a^x$ is differentiable at $x = 0$. The foregoing geometric description of the proof and the numerical evidence should be informative and persuasive to students, even if they do not follow all the details of the argument.

Theorem 1. *Let $f(x) = a^x$ with any $a > 1$. Then f is differentiable at $x = 0$ and $f'(0) > 0$.*

Proof. To express our geometric observations in analytic terms, let

$$m(x) = \frac{f(x) - f(0)}{x - 0} = \frac{a^x - 1}{x}.$$

In Figure 1, $x > 0$ and

$$\text{slope } \overline{AB} = m(x),$$
$$\text{slope } \overline{AC} = m(-x).$$

We shall prove that, as $x \to 0$ with $x > 0$ and x decreasing, $m(x)$ and $m(-x)$ approach a common limit, which is $f'(0)$.

To begin with, $m(x)$ is continuous because a^x is continuous. The crux of the proof, and the only tricky part, is to show that

$$m(x) \text{ is increasing on } (0, \infty) \text{ and } (-\infty, 0).$$

We give the proof only for $(0, \infty)$ since the proof for $(-\infty, 0)$ is essentially the same. We show first that m is increasing on the rationals in $(0, \infty)$. Fix rational numbers r and s with $0 < r < s$ and let a vary with $a \geq 1$. Define

$$g(a) = m(s) - m(r) = \frac{a^s - 1}{s} - \frac{a^r - 1}{r}.$$

Then $g(a)$ is continuous for $a \geq 1$ and

$$g'(a) = a^{s-1} - a^{r-1} > 0 \quad \text{for } a > 1.$$

Thus, $g(a)$ increases as a increases and $g(a) > g(1) = 0$ for $a > 1$, so

$$m(r) < m(s) \quad \text{for } 0 < r < s.$$

Thus, $m(x)$ is continuous on $(0, \infty)$ and $m(x)$ increases on the rational numbers in $(0, \infty)$. As noted earlier, this implies that $m(x)$ increases on $(0, \infty)$. The argument for the interval $(-\infty, 0)$ is similar.

For $x > 0$,

Differentiability of Exponential Functions

$$m(-x) = m(x)a^{-x},$$
$$0 < m(-x) < m(x),$$
$$0 < m(x) - m(-x) = m(x)\left(1 - a^{-x}\right).$$

Let $x \to 0$ with x decreasing. Then

$m(x)$ decreases, $\quad m(-x)$ increases, $\quad m(x) - m(-x) \to 0.$

It follows that $m(x)$ and $m(-x)$ approach a common limit as $x \to 0$, which is $f'(0)$. Furthermore, $0 < m(-x) < f'(0) < m(x)$ for $x > 0$, which implies that $f'(0) > 0$. ∎

We believe that this proof is new. We have been unable to find any other proof that depends only on methods of differential calculus. However, the interplay between convexity and differentiability has a long history, and we recommend Chapter 1 of Artin [1] to interested readers.

Theorem 1 and familiar reasoning give the principal result on the differentiability of exponential functions.

Theorem 2. *Let $f(x) = a^x$ with any $a > 1$. Then f is differentiable for all x and $f'(x) = f'(0)a^x$.*

It follows that $f''(x) = f'(0)^2 a^x > 0$, and $f(x) = a^x$ is concave up, as anticipated. By routine arguments,

$$a^x \to \infty \quad \text{as} \quad x \to \infty,$$
$$a^x \to 0 \quad \text{as} \quad x \to -\infty.$$

The intermediate value theorem implies that the range of a^x is $(0, \infty)$.

The natural exponential function e^x

The next step in the development of properties of derivatives of exponential functions is to define e, the base of the natural exponential function e^x, within the context of differential calculus. Different authors define e in various ways. Some of the definitions involve more advanced concepts. We prefer a definition of e based on an important property of e^x, namely that e is the unique number for which

$$\left.\frac{d}{dx}e^x\right|_{x=0} = 1.$$

In view of Theorem 2, an equivalent property is

$$\frac{d}{dx}e^x = e^x \quad \text{for all } x.$$

It is not difficult to justify the definition of e and, at the same time, to find an explicit formula for e. To begin with, consider any base $a > 1$. Since 2^x is increasing and has range $(0, \infty)$, there is a unique number $c > 0$ such that $a = 2^c$. By Theorem 1,

$$\frac{d}{dx}a^x\bigg|_{x=0} = \lim_{x \to 0} \frac{a^x - 1}{x} = c \lim_{cx \to 0} \frac{2^{cx} - 1}{cx} = cm,$$

where $a = 2^c$ and $m = \frac{d}{dx}2^x\bigg|_{x=0} = \lim_{x \to 0} \frac{2^x - 1}{x}$.

Observe that

$$\frac{d}{dx}a^x\bigg|_{x=0} = 1 \text{ only for } c = 1/m \text{ and } a = 2^{1/m}.$$

Therefore, $e = 2^{1/m}$. Since $m \approx 0.7$ by previous calculations, $e \approx 2.7$. Of course, there are much better approximations for e.

Since e is one particular value of a, the function e^x has all the properties mentioned earlier for general exponential functions a^x with $a > 1$. Thus, e^x is increasing and concave up,

$$e^x \to \infty \quad \text{as} \quad x \to \infty,$$
$$e^x \to 0 \quad \text{as} \quad x \to -\infty,$$

and the range of e^x is $(0, \infty)$.

With this foundation, all relevant applications of exponential functions become available in differential calculus.

Logarithmic functions

Once the basic properties of exponential functions have been established, it is easy to introduce logarithmic functions as corresponding inverse functions and to develop their relevant properties within differential calculus.

The natural logarithmic function (or natural log) is defined by

$$y = \ln x \iff x = e^y.$$

The derivative rule for inverse functions implies that $y = \ln x$ is differentiable and

$$\frac{d}{dx}\ln x = \frac{dy}{dx} = \frac{1}{dx/dy} = \frac{1}{e^y} = \frac{1}{x}.$$

Thus,

$$\frac{d}{dx}\ln x = \frac{1}{x}.$$

The familiar algebraic properties and asymptotic properties of logarithmic functions follow easily from corresponding algebraic rules of exponents and asymptotic properties of exponential functions. In typical textbooks that defer exponential and logarithmic functions to integral calculus, proofs of algebraic properties of $\ln x$ are based on the uniqueness of solutions to initial value problems and are less informative for most first-year calculus students.

Comparisons

It is worthwhile to contrast our approach with current practices. Most mainstream calculus texts, such as [5] and [8], defer the entire discussion of exponential and logarithmic functions to integral calculus, where exponential functions are expressed as inverses of logarithmic functions in order to establish their differentiability. As we wrote earlier, this is the reverse of the natural order. It has the unfortunate consequence that exponential functions are often defined in two different ways that ultimately have to be reconciled. The upshot is a circuitous argument that blurs the distinction between definitions and conclusions. Moreover, a substantial block of material about derivatives and rates of change is presented in integral calculus, instead of in its natural place in differential calculus. Exponential and logarithmic functions have many important applications, such as motion with resistance, that belong in differential calculus.

A few books, often called "early transcendentals" texts, introduce exponential and logarithmic functions in differential calculus. Although this arrangement is an improvement over the standard approach, the differentiability of exponential functions remains a stumbling block. Some of these books, such as [3] and [4], simply display a little numerical evidence and/or a plausibility argument in support of differentiability and then assume differentiability thereafter. Others, such as [6] and [7], start out with plausibility arguments in differential calculus and then return to the subject in integral calculus where proofs are given. Neither alternative is really satisfactory. It is much better, both logically and pedagogically, to settle the question of differentiability when the issue arises.

Our proof that exponential functions are differentiable makes it possible to give a mathematically complete "early transcendentals" presentation of exponential and logarithmic functions in differential calculus. Later on, when methods of integral calculus are applied to exponential and logarithmic functions, progress will not be impeded by unfinished business in differential calculus.

References

1. E. Artin, *The Gamma Function*, Holt, Rinehart and Winston, 1964.
2. R. Courant, *Differential and Integral Calculus*, vol. 1, Interscience, 1959.
3. D. Hughes-Hallett, A. M. Gleason, W. G. McCallum, et al., *Calculus*, 3rd ed., Wiley, 2002.
4. R. Larson, R. P. Hostetler, and B. H. Edwards, *Calculus, Early Transcendental Functions*, 3rd ed., Houghton Mifflin, 2003.
5. S. Salas, E. Hille, and E. Etgen, *Calculus, One and Several Variables*, 9th ed., Wiley, 2003.
6. J. Stewart, *Calculus, Early Transcendentals*, 5th ed., Brooks/Cole, 1999.
7. M. J. Strauss, G. L. Bradley, and K. J. Smith, *Calculus*, 3rd ed., Prentice Hall, 2002.
8. D. Varberg, E. J. Purcell, and S. E. Rigdon, *Calculus*, 8th ed., Prentice Hall, 2000.

A Discover-e

Helen Skala

Here is a simple but interesting discovery problem for calculus students who have learned to differentiate exponential functions. Distribute a printout of the graphs of four to six exponential functions for various bases (Figure 1). The students will place a straightedge through the origin and rotate it until it is tangent to one of the curves, then draw the tangent line and mark the point of tangency. Have them do this for each curve (Figure 2).

The points of tangency seem to lie on a horizontal line. Ask your students to prove this conjecture.

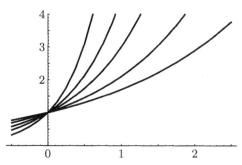

Figure 1. $y = 1.7^x$, 2.1^x, 3^x, 4.5^x, 9.5^x.

To do so, use differentiation to find the point of tangency, which, for the function $y = a^x$, is $(1/\ln(a), a^{1/\ln(a)})$. But from properties of logarithms, it follows that $a^{1/\ln(a)} = e$ for any value of $a \neq 1$. Hence the points of tangency do indeed lie on a horizontal line, namely, the line $y = e$.

A Discover-*e* 277

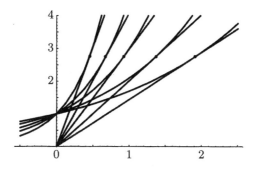

Figure 2. Points of tangency.

An Exponential Rule

G. E. Bilodeau

Students of elementary calculus who are quite comfortable with the power rule

$$\frac{d}{dx}[f(x)]^p = p[f(x)]^{p-1}f'(x)$$

become discouraged to discover that it fails when $(d/dx)p^{g(x)} = p^{g(x)}g'(x)\ln p$, $p > 0$. They become further chagrined to see that a derivative of the form $(d/dx)f(x)^{g(x)}$ requires a different technique. They ask why, when these are structurally similar, are they so different? Insightful students will note that the above constant p may be considered a constant function, so some connection must exist. To reduce the mystery and prompt discussion, one might introduce a general exponential rule

$$\frac{d}{dx}f(x)^{g(x)} = g(x)[f(x)]^{g(x)-1}f'(x) + f(x)^{g(x)}g'(x)\ln f(x), \qquad f(x) > 0$$

Ostrowski, *Differential and Integral Calculus*, Scott, Foresman, 1968, p. 276].

The first term appears as the power rule with $g(x)$ treated as a constant and the second term the rule for functional exponents with $f(x)$ treated as a constant. While not a substitute for the valuable technique of logarithmic differentiation (the rule is a good exercise in it), it allows students to see the harmony in the various rules. Its similarity to the product rule, "take the derivative of f holding g constant plus the derivative of g holding f constant," adds to its appeal for students.

The Derivative of Arctan x

Norman Schaumberger

In a standard first course in calculus, the formulas for the derivatives of the trigonometric functions are derived before the formulas for the derivatives of the inverse trigonometric functions. It is instructive to note that this need not be the case and that these formulas can be developed in the reverse manner.

We offer a simple derivation of the formula for the derivative of arctan x.

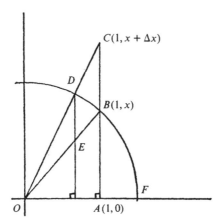

In the Figure, DBF is the arc of a circle with center at the origin and radius equal to $\sqrt{1 + x^2}$. It is clear that

$$\text{area triangle } OED < \text{area sector } OBD < \text{area triangle } OBC. \qquad (*)$$

Area triangle OBC = area triangle OAC − area triangle OAB = $(\Delta x)/2$. Since triangle OED is similar to triangle OBC, it follows that

$$\text{area triangle } OED = \frac{1 + x^2}{1 + (x + \Delta x)^2} \cdot \frac{\Delta x}{2}.$$

Furthermore,
$$\text{area sector } OBD = \tfrac{1}{2}(1 + x^2)(\arctan(x + \Delta x) - \arctan/x).$$

Substituting into (*) and dividing by $1 + x^2$, we readily get

$$\frac{1}{1 + (x + \Delta x)^2} < \frac{(\arctan(x + \Delta x) - \arctan x)}{\Delta x} < \frac{1}{1 + x^2}.$$

Consequently

$$\frac{d(\arctan x)}{dx} = \lim_{\Delta x \to 0} \frac{\arctan(x + \Delta x) - \arctan x}{\Delta x} = \frac{1}{1 + x^2}.$$

Using this result, the chain rule, and the procedure for obtaining the derivative of an inverse function, it is a simple matter to get the formulas for the derivatives of the remaining inverse trigonometric functions and the six trigonometric functions.

Examples:

$$\frac{d(\arcsin x)}{dx} = \frac{d\left(\arctan(x/\sqrt{1 - x^2})\right)}{dx} = 1/\sqrt{1 - x^2}.$$

If $y = \tan x$, then $x = \arctan y$ and $dx/dy = 1/(1 + y^2)$. Thus, $dy/dx = 1 + y^2 = \sec^2 x$.

The Derivative of the Inverse Sine

Craig Johnson

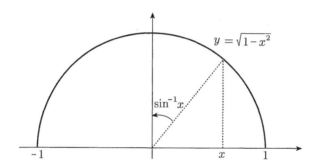

By the integral formula for arc length, $\sin^{-1} x = \int_0^x \frac{1}{\sqrt{1-t^2}}\, dt$. Thus, $\frac{d}{dx}\left(\sin^{-1} x\right) = \frac{1}{\sqrt{1-x^2}}$.

Graphs and Derivatives of the Inverse Trig Functions

Daniel A. Moran

In a calculus course the differentiation formulas for the inverse trig functions are derived by implicit differentiation (at least for two or three of the functions). To avoid tedious repetition, the formulas for the others are merely stated, and their proofs omitted or left as an exercise.

The approach outlined below gives half of the differentiation formulas as immediate consequences of the others. After the inverse functions are defined, it is established that $f^{-1}(x)$ and $\operatorname{cof}^{-1}(x)$ are always complementary when f is sine, tangent or secant. Along the way, there is an opportunity to use graphics (computer-driven or otherwise) and strengthen the students' grasp of the elementary geometry of reflections and translations. And the whole process takes less classroom time than the conventional method!

The archetypical demonstration:

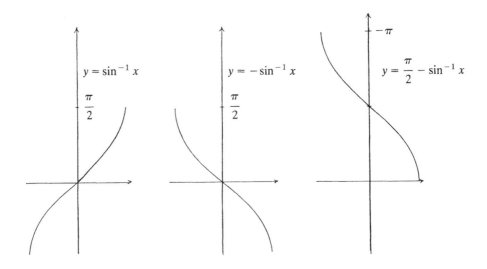

In the figure, the first graph is reflected in the horizontal axis to produce the second; the latter is then translated $\pi/2$ units upward to yield the third (which is evidently congruent to the graph of $y = \cos^{-1} x$). This establishes that $\sin^{-1} x + \cos^{-1} x = \pi/2$. We can now differentiate to discover that $D_x \cos^{-1} x = -D_x \sin^{-1} x$.

The demonstrations for $\tan^{-1}$ and $\sec^{-1}$ require practically no change from the above.

Logarithmic Differentiation: Two Wrongs Make a Right

Noah Samuel Brannen and Ben Ford

You're teaching a Calculus class, and get to the point in the course where it is time to discuss logarithmic differentiation. As usual, you begin by asking the class "If $f(x) = x^x$, what is $f'(x)$?"

One student raises his hand and says "That's just the power rule. It's xx^{x-1}." Another student says "That's just for when the exponent is constant. This has x in the exponent, so it's like e^x, but when it's not e you need to put the log in. So it's $(\ln x)x^x$." Smiling, you point out that just as one is assuming that n is a constant when one uses the formula for the derivative of x^n, one is assuming that a is a constant when one uses the formula for the derivative of a^x. Since neither the base nor the exponent of x^x is constant, the function $f(x) = x^x$ is neither a power function nor an exponential function, and therefore the derivative of x^x cannot be found using either of these rules. Instead, you say, we will use a technique called logarithmic differentiation.

(Pedagogical aside: Of course, here you have the option of using the definition $x^x = e^{x \ln x}$; you have already covered methods for differentiating this. But the technique of using logarithmic differentiation to break the natural log of a function into a sum of easily-differentiable summands is a nice one, and this is a very good context in which to introduce it.)

At this point a student in the back of the class raises her hand and says "Isn't $f(x) = x^x$ a combination of a power function and an exponential function, and therefore shouldn't the derivative be a combination of the derivative of a power function and the derivative of an exponential function?"

Trying not to discourage the student, you attempt to take her question seriously. You ask if she means that the derivative should be the sum of the two answers given by the first two students. She replies "sure."

You decide to indulge the student, saying "Let's see what happens. The sum of the two answers is $xx^{x-1} + (\ln x)x^x$. The real answer can be found as follows: First, we let $y = x^x$. Then we take the natural logarithm of both sides, obtaining $\ln y = \ln(x^x) = x \ln x$. Differentiating both sides of this equation with respect to x gives us

$$\frac{y'}{y} = (1)(\ln x) + x\left(\frac{1}{x}\right) = \ln x + 1,$$

and multiplying both sides by y yields

$$y' = y(\ln x + 1).$$

Substituting x^x for y, we see that the derivative y' of $y = x^x$ is

$$y' = x^x(\ln x + 1)."$$

You are about to say "As you can see, the answers are completely different," when it occurs to you that they are not different at all! Flustered, you try to assure the class that it is just a coincidence, but you are not so sure.

After class, you hurry back to your office where you decide to see if this example of "two wrongs making a right" is indeed just a coincidence. What about functions of the form $f(x)^{g(x)}$? Reasoning as your student had above, you first treat the exponent $g(x)$ as a constant, obtaining $g(x)f(x)^{g(x)-1}f'(x)$, using the chain rule. Next, you treat $f(x)$ as a constant, obtaining $\ln(f(x))f(x)^{g(x)}g'(x)$, again using the chain rule.

Adding the two "wrongs" gives

$$\left(f(x)^{g(x)}\right)' = g(x)f(x)^{g(x)-1}f'(x) + \ln\left(f(x)\right)f(x)^{g(x)}g'(x),$$

which can be rewritten as

$$\left(f(x)^{g(x)}\right)' = g(x)f(x)^{g(x)}\frac{f'(x)}{f(x)} + \ln(f(x))f(x)^{g(x)}g'(x). \tag{1}$$

You find the "true" derivative by writing $y = f(x)^{g(x)}$, taking natural logs, and differentiating both sides with respect to x:

$$y = f(x)^{g(x)}$$

$$\ln y = \ln\left(f(x)^{g(x)}\right) = g(x)\ln\left(f(x)\right)$$

$$\frac{y'}{y} = g'(x)\ln\left(f(x)\right) + g(x) \cdot \frac{1}{f(x)} \cdot f'(x)$$

$$y' = f(x)^{g(x)}\left(g'(x)\ln\left(f(x)\right) + g(x) \cdot \frac{1}{f(x)} \cdot f'(x)\right)$$

which simplifies to

$$y' = f(x)^{g(x)}g'(x)\ln\left(f(x)\right) + f(x)^{g(x)}g(x)\frac{f'(x)}{f(x)}. \tag{2}$$

You are simultaneously devastated and delighted to find that the expressions in (1) and (2) are equal!

It is clear now that it was not a coincidence that the "two wrongs made a right." Instead, you realize that what the student wanted to do was indeed legitimate. But why does it work? Why can you first treat $g(x)$ as a constant, and then treat $f(x)$ as a constant? As soon as this thought enters your head, you realize the answer. If $y = f(x)^{g(x)}$, then y depends on two unknowns, f and g, which in turn depend on the parameter x. Thus to take the derivative of y with respect to x, you must calculate

$$\frac{\partial y}{\partial f}\frac{df}{dx} + \frac{\partial y}{\partial g}\frac{dg}{dx},$$

by the chain rule for a function of two variables (see [2, p. 673]). When calculating $\frac{\partial y}{\partial f}$, one treats $g(x)$ as if it were constant, and when one calculates $\frac{\partial y}{\partial g}$, one treats $f(x)$ as if it were constant. Since this is precisely what the student in your class was doing, you now realize that it is indeed a valid technique.

Usefulness? The reader may believe that an event such as the one described above is unlikely to occur. However, a situation in which a student actually did something equivalent on a test is related in [4], and is addressed in slightly greater generality in [1].

The standard sum, product, and quotient rules from differential calculus also lend themselves to a similar interpretation: To differentiate $f(x)g(x)$ (or $f(x) + g(x)$ or $f(x)/g(x)$), first differentiate as if $g(x)$ was constant, then as if $f(x)$ was constant; add the results. Indeed, a function $f(x)$ which is comprised of an algebraic combination of an arbitrary number of simpler functions $f_i(x)$ can be differentiated in this way, differentiating as if all the f_i are constant except for f_1; repeat for all f_j; add the results.

This point of view on the standard "differentiation rules" might be a useful unifying idea. Students typically see these rules as a disparate collection of formulas to memorize. At the least, pointing out this connection between the sum, product, quotient, and exponentiation rules raises the question of "why?" The answer—the proof of the multivariable chain rule—is a semester or two away, however. When it does arrive, these rules are nice examples to have ready.

A possible approach would be to teach the differentiation of functions of the form $y = f(x)^{g(x)}$ using this point of view. An advantage might be that students would not have to learn yet another technique (logarithmic differentiation), and could instead simply combine two formulas that they have already learned. However, the technique of logarithmic differentiation is probably more important to most instructors than simply the ability to differentiate $f(x)^{g(x)}$.

This observation about differentiation rules does raise an interesting grading question: If a student writes "the derivative of x^x is xx^{x-1}," how much partial credit will you give? Shouldn't you give the answer half credit, since it is half right?

Acknowledgments. The authors thank the referees for alerting them to the references and for suggestions which improved the exposition. For an interesting exploration of the function x^x, see [3].

References

1. G. E. Bilodeau, An exponential rule, *College Math. J.* **24** (1993), 350–351.
2. Deborah Hughes-Hallett, *et al.*, *Calculus: Single and Multivariable* (3rd ed.), Wiley, 2002.
3. Mark D. Meyerson, The x^x spindle, *College Math. J.* **69** (1996), 198–206.
4. Gerry Myerson, A natural way to differentiate an exponential, *College Math. J.* **22** (1991), 404.

A Comparison of Two Elementary Approximation Methods

Harvey Diamond and Louise Raphael

1. Introduction There are two familiar methods one learns in calculus for approximating the value of a differentiable function near one or more data points where the function is known and/or easy to evaluate. The first method is linear interpolation and the second is the differential approximation. A natural question about the the two methods is: "Which is better?" This question was posed by Leon Henkin during a visit to a seminar on curve fitting given by one of the authors in the Summer Mathematics Institute for talented minority students at Berkeley.

In what follows, we study this question from several viewpoints, using only elementary methods. What we find interesting is that, starting with an intuitive but mathematically ill-formulated question, we are lead to a fairly rich investigation touching on several mathematical approaches, including a result on existence/uniqueness that applies to convex functions, an exact/local analysis based on the quadratic case, and a discussion of numerical methods.

2. Formulation and preliminary observations The function f is assumed to be differentiable. We begin by setting out the two alternatives:

Differential approximation: Given $f(a), f'(a)$ we approximate

$$f(x) \cong f(a) + f'(a)(x - a) \equiv f_D(x). \tag{1}$$

Linear interpolation: Given $f(a), f(b)$, we approximate

$$f(x) \cong f(a) + \frac{f(b) - f(a)}{b - a}(x - a) \equiv f_L(x). \tag{2}$$

FIGURE 1 illustrates $f(x)$ and its two approximations.

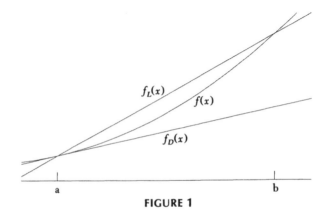

FIGURE 1

The *approximation error* of an approximation method at a particular value of x is simply the difference of $f(x)$ and its approximate value, in this case either $f(x)-f_D(x)$ or $f(x)-f_L(x)$. The *size* of the approximation error is its absolute value. We would be entitled to say that one method is better than the other if for all $a \le x \le b$ its approximation error had smaller size than that of the other method. We will see shortly, however, that this situation never occurs in our problem. The approximation method having the error of smaller size depends on the value of x, and one must try to say something about *where* (i.e. for which values of x) one method is better than another, and for which functions.

We begin with some simple observations. If x is near a, f_D is better, while if x is near b, f_L is better. In fact, the only way this could not be true is if the two approximations are identical. To explain this, note first that both approximations are the same at $x=a$. Near $x=b$, if $f_D(b) \ne f(b)$ then f_L is better near b, since $f_L(b) = f(b)$. On the other hand, if $f_D(b) = f(b)$ then both approximations agree at $x=a$ and $x=b$ and hence are identical since they are linear functions. Next, using the definition of the derivative, one can show that $f_D(x)$ is the best linear approximation of $f(x)$ near $x=a$. Consider all lines through the point $(a, f(a))$ that have the form $y = f(a) + m(x-a)$. The error in approximating $f(x)$ using such a line, at $x \ne a$, is then

$$f(x) - (f(a) + m(x-a)) = (x-a)\left\{\frac{f(x)-f(a)}{(x-a)} - m\right\},$$

so that the term in brackets goes to zero as x goes to a (and hence gives the smallest size error) if, and only if, m is equal to $f'(a)$. Thus either f_D is better than f_L near a or else f_L also has derivative $f'(a)$, in which case it is identically equal to f_D.

Having shown that f_D is better near $x=a$ and f_L is better near $x=b$, the question of which approximation method is better becomes, "For what values of x is f_D better than f_L and for what values is the opposite true?" In turn, the answer to this question hinges on the answer to the question "At what points are the approximation errors of the two methods the same size?" We note that at any point where the approximation errors have the same size, the approximation errors must be of opposite sign or else (again) the approximations are identical. This occurs because if the approximation errors are the same at some point then $f_D = f_L$ at that point and, along with $f_D(a) = f_L(a)$, which is always true, we see that both approximations correspond to the same line. Thus if the functions f_D and f_L are not identical, the set of points at which the approximation errors have the same size is therefore precisely where the approximation errors sum to zero, which is precisely the solution set of the equation

$$f(x) - f_D(x) + f(x) - f_L(x) = 0 \quad \text{or} \tag{3}$$
$$f(x) = [f_D(x) + f_L(x)]/2. \tag{4}$$

A geometric interpretation of (4) is shown in FIGURE 2, namely the solution set of (4) is the set of points where the line $y = [f_D(x) + f_L(x)]/2$ intersects the curve $y = f(x)$. The label c in FIGURE 2 is the location of x at the point of intersection in the interval (a, b).

3. Qualitative results Continuing with the graph in FIGURE 2, we observe that $f(x)$ is concave up and there is exactly one value of x, $x = c$ in the figure, where the sizes of the two approximation errors are the same. The following theorem shows that this is true for any function that is strictly concave up or concave down and has two continuous derivatives.

THEOREM 1. *Let $f''(x)$ be continuous and either strictly positive or strictly negative for all x in some open interval I. Then for any a, b in I, f_D and f_L cannot be identical, and, aside from the point $x = a$, there is exactly one other point in I, lying between a and b, at which the approximation errors have the same size.*

Proof. Motivated by equation (3), we define

$$g(x) = f(x) - f_D(x) + f(x) - f_L(x).$$

First, f_D and f_L cannot be identical for if they are, $g(a) = 0$, $g(b) = 0$, and $g'(a) = 0$. The first two facts (by Rolle's theorem) imply the existence of a point α between a and b for which $g'(\alpha) = 0$. Since $g'(a) = 0$ there exists another point β between a and α at which $g''(\beta) = 0$. But $g''(x) = 2f''(x)$ since f_D and f_L are linear, so $f''(\beta) = 0$. This contradicts the hypothesis that $f''(x)$ is either strictly positive or strictly negative in I.

Since f_D and f_L are not identical, the approximation errors have the same size if, and only if, (3) is satisfied, i.e., $g(x) = 0$. Now $g(a) = 0$ is always true. If g has two other distinct roots in I there are, by Rolle's theorem, at least two distinct points in I where $g'(x)$ is zero, and so at least one point in I where $g''(x) = 0$, again contradicting the hypothesis on $f''(x)$. Thus g has at most one other root in I, but since we showed that f_D and f_L are not identical, by previous arguments there is at least one point between a and b where $g(x) = 0$. Thus, aside from $x = a$, there is a unique root of g and it lies between a and b. This proves the theorem.

We might ask what can happen if the hypotheses of the theorem do not hold and f has one inflection point, say $x = c$. In this case f_D and f_L can be identical, but if not, there is, interestingly, still only one point where the approximation errors have the same size.

THEOREM 2. *Let $f''(x)$ be continuous and suppose that the closed interval $I = [\alpha, \beta]$ contains an interior point c with $f''(c) = 0$ and $f''(x) > 0$ for $x > c$, $f''(x) < 0$, for $x < c$, where $x \in I$ is assumed in each case. Then the following holds:*

i) *There are infinitely many pairs a, b, with $a < c < b$, for which $f_D(x)$ and $f_L(x)$ are identical.*

ii) *If f_D and f_L are not identical for a particular choice of $a, b \in I$ then there is exactly one point between a and b at which the approximation errors have the same size.*

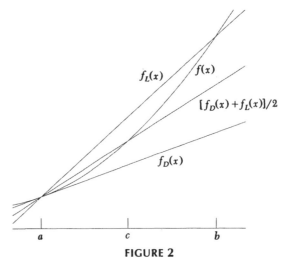

FIGURE 2

A Comparison of Two Elementary Approximation Methods

Proof. We sketch a proof of part i) with the aid of FIGURE 3. The idea is that, using values of a close to but less than c, the graph of $f_D(x)$ (the tangent from $(a, f(a))$), will intersect the graph of $f(x)$ at some $x = b$ with $b > c$ and $b \in I$. Then f_L will be identical to f_D. To effect this idea as a proof, we begin by explicitly exhibiting the dependence of f_D on the point a by using the notation $f_D(x; a)$. Now we can show, using Rolle's theorem, that

$$f_D(x; c) < f(x) \quad \text{for } x > c, \tag{5}$$

i.e., the tangent through the turning point is below the right, concave-up, portion of the curve. Similarly, we can show that

$$\text{if } a < c, \, f_D(x; a) > f(x) \quad \text{for } x \leq c, \tag{6}$$

i.e., a tangent constructed on the left, concave-down, portion of the curve is above the left portion of the curve. Now consider $f_D(\beta; a)$ as a function of a; it is clearly a continuous function of a. When $a = c$, we have $f_D(\beta; c) < f(\beta)$, by virtue of (5), so for $a < c$ and $|a - c|$ sufficiently small, $f_D(\beta; a) < f(\beta)$ holds by continuity. On the other hand, $f_D(c; a) > f(c)$ by virtue of (6). The last two inequalities imply, by the intermediate value theorem, the existence of a point b with $c < b < \beta$, such that $f_D(b; a) = f(b)$, completing part i).

The proof of part ii) is more difficult and is left to the reader, although elementary techniques are still adequate. FIGURE 4 provides a "generic" picture of why the theorem is true. The unique value of x between a and b where the approximation errors have the same size is denoted by c; we observe that at $x = c'$ the approxima-

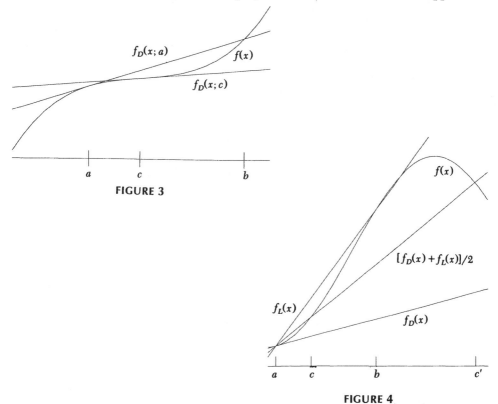

FIGURE 3

FIGURE 4

tion errors also have the same size, but this value of x lies outside the interval $[a, b]$.

4. An important special case In investigating our problem further, we will completely and explicitly solve the problem in the case when f is quadratic. "Solving the problem" means finding the solution of (3), which we know from Theorem 1 is unique, but in any case will be observed to be unique from the calculations below.

If f is quadratic, say $f(x) = px^2 + qx + r$, then $f(x)$ can be written as $f(x) = f(a) + f'(a)(x - a) + p(x - a)^2$.

This representation can most easily be obtained from Taylor's theorem after noting that $p = f''(a)/2$, since the coefficient of x^2 must be p. It then follows from (1) that

$$f(x) - f_D(x) = p(x - a)^2. \tag{7}$$

On the other hand, with f quadratic and f_L linear, $f(x) - f_L(x)$ is a quadratic function that is zero at $x = a$ and $x = b$ so we must have

$$f(x) - f_L(x) = p(x - a)(x - b). \tag{8}$$

The constant factor can be identified as p since subtracting a linear function from f does not change the coefficient of x^2.

At the x we seek, the solution of (3), the sum of (7) and (8) is zero,

$$p\left[(x - a)^2 + (x - a)(x - b)\right] = 0 \text{ or } p(x - a)(2x - a - b) = 0$$

with solutions $x = a$ and $x = (a + b)/2$. Thus for quadratic functions, differentials are better until halfway between a and b and then the linear interpolant is better. This is a pleasing result and provides a good rule of thumb, as we briefly explain next.

The solution of the quadratic case provides a local analysis of the general problem, applicable when a and b are close together and the hypotheses of Theorem 1 hold for some open interval I containing $[a, b]$. Under these conditions, f is well-approximated by a quadratic function and $x = (a + b)/2$ will be a good approximation to the unique solution of (3). A precise formulation and proof of this fact is somewhat difficult, however, and will not be attempted here.

5. Numerical considerations As discussed in section 3, if $f(x)$ is concave up or down, the problem of determining where each approximation method is better than the other reduces to finding the unique solution of (3) for values of x in the interval (a, b). If we define

$$g(x) \equiv f(x) - f_D(x) + f(x) - f_L(x), \tag{9}$$

then we are seeking the solution of $g(x) = 0$ in the interval (a, b). It turns out that Newton's method is guaranteed to converge to the desired solution if we use $x_0 = b$ as the initial guess for the root of g.

THEOREM 3. *If $f(x)$ satisfies the hypotheses of Theorem 1, then Newton's method applied to the function $g(x)$ in (9), namely, the iteration*

$$x_{n+1} = x_n - g(x_n)/g'(x_n),$$

with starting value $x_0 = b$, will converge to the unique solution $x = c$ of $g(x) = 0$ in the interval (a, b).

Proof. We will not provide a proof here. Results of this sort for Newton's method are well known; see for instance [1, p. 62]. The result basically follows from the observations depicted graphically in FIGURE 5:

i) The sequence $\{x_n\}$ is a decreasing sequence with $x_n > c$. This comes from the concavity of f (and therefore g) in the interval (a, b).

ii) As a decreasing sequence bounded from below, $\{x_n\}$ has a limit, say x^*, and this limit must satisfy $x^* = x^* - g(x^*)/g'(x^*)$ so that $g(x^*) = 0$. It follows that $x^* = c$.

Finally, we briefly consider the possibility of solving (4) by iteration. If we solve (4) for x on the (linear) right side, we can obtain an equation of the form $x = h(x)$. It is tempting to try to solve this equation numerically by the iteration $x_{n+1} = h(x_n)$, or what is the same thing, the iteration

$$[f_D(x_{n+1}) + f_L(x_{n+1})]/2 = f(x_n). \tag{10}$$

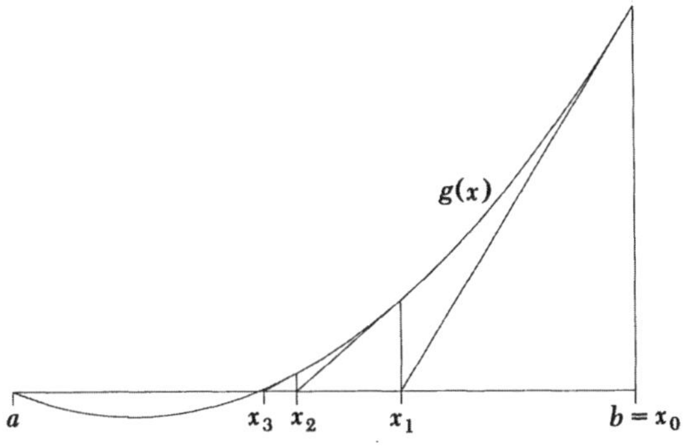

FIGURE 5

This turns out to be rather dangerous, even in the simplest case, where f satisfies the hypotheses of Theorem 1. In the situation depicted in FIGURE 2, for instance, it is easy to see graphically that for the iteration (10), $x = a$ is a stable fixed point, whereas the solution we seek, $x = c$, is an unstable fixed point. Two other possibilities, both with $f'(a) > 0$, $f''(x) < 0$, are depicted in FIGURE 6. If $f'(x) > 0$, $a < x < b$ (see FIGURE 6a), then the solution $x = c$ is a stable fixed point and any x_0 in the interval (a, b) results in a monotone sequence converging to $x = c$. Perhaps most interesting is the case depicted in FIGURE 6b, in which f has a maximum to the left of $x = c$. In this case $x = a$ is an unstable fixed point, but $x = c$ may or may not be stable. In fact this generic situation leads, under certain conditions, to the well-known chaotic maps of an interval onto itself, of which the most famous example is

$$x_{n+1} = 4\lambda x_n(1 - x_n), \tag{11}$$

where chaos occurs for certain values of the parameter λ lying between 0 and 1. (See [2] for a recent discussion at an elementary level.) Indeed, this example arises if we take $f(x) = x - x^2$, $a = 0$ and b a parameter, so that $f_D(x) = x$, $f_L(x) = (1 - b)x$ and (10) becomes (11) with $4\lambda = (1 - b/2)^{-1}$.

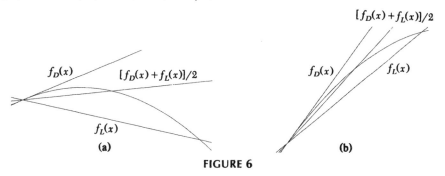

FIGURE 6

6. Concluding remarks We have shown in the preceding discussion, how a simple question concerning the comparison of linear interpolation and the differential approximation leads to some interesting mathematics, using only elementary methods at the calculus level. We discovered classes of functions for which the answer to the question could be well-characterized qualitatively; and obtained a nice exact answer in the case of quadratic functions, which also serves as a local approximation. Finally we showed that Newton's method can be reliably used to find the "cross-over" point in the case of concave up/down functions, while the method of iteration, naively applied, can converge, diverge, or lead to chaotic behavior even in the simplest examples.

The authors were partially supported by an Air Force Office of Scientific Research contract 91MN062.

REFERENCES

1. Kendall E. Atkinson, *An Introduction to Numerical Analysis*, 2nd edition, John Wiley & Sons, New York, 1989.
2. Robert L. Devaney, The orbit diagram and the Mandelbrot set, *The College Math. J.* 22 (1991), 23–38.

Part 3. Integrals

How Should We Introduce Integration?

David M. Bressoud

A perennial problem in first-year calculus is how to introduce integration. What is the key characteristic of integration that we want to lay as the foundation? The reason why it is a problem is that integration has several basic natures: it is the inverse operation to differentiation, enabling us to go from knowledge of the rate at which something is changing to knowledge of where it will be at some future time, and it is a mensuration of area and, by extension, of volume and length. The fundamental theorem of calculus promises us the equivalence of these two natures. This theorem is usually approached by means of a third and far less intuitive understanding of the integral, as the limit of a sum of vanishing quantities. Occasionally (see Bers [2]), yet a fourth tack is introduced. Looking toward the Lebesgue integral, integration is described in terms of its action on step functions. We cannot teach more than one of these viewpoints at a time without confusing our students. How do we approach this plurality? How should we approach it?

One can find textbooks presenting integration in almost any order. Area and volume problems are the most ancient aspect of calculus. Archimedes' method of exhaustion is recognizably equivalent to what we teach today. A good argument can be made that we need to follow the historical order and begin with the integral as the limit of sums of areas of rectangles. This carries some weight if we really are following the historical order and launch into calculus with the problem of areas as in Apostol's *Calculus* [1], but most textbooks already distort the historical order by presenting differentiation before integration. I have no objection to this. Differentiation is conceptually and operationally much simpler than integration.

The common approach is to begin with the indefinite integral as the inverse process of differentiation, often called "antidifferentiation." It is easy to do simple antidifferentiation and to motivate this material through the natural problem of predicting future position from knowledge of acceleration as a function of time together with initial position and velocity.

Where I have difficulties with the usual approach is in how the integral as area is now tackled. Many contemporary authors begin the problem of measuring areas as if it had nothing to do with anything that appeared in the course up to that point. Students struggle with summation notation and the formula for the sum of the first n squares. They are back with Archimedes when, *hey presto!*, the fundamental theorem of calculus makes its appearance. We expect amazement and relief from our students. They do not really need the formula for the sum of the first n squares after all.

In reality, most of my students have seen enough calculus in high school that they know what is coming. They are baffled by the insistence on calculating definite integrals by means of complicated summation formulas, and unimpressed when the rabbit finally appears from the hat.

I would like to offer an alternate tactic that follows the historical development of this concept and meets the overarching concern that should rule our calculus pedagogy:

How will the students use a given concept outside the calculus classroom, and how can we best prepare them to make the necessary transfer?

The computation of areas and volumes is a side issue. The integral as the limit of a Riemann sum is not the primary application. Most of our students will be using integrals to solve differential equations. Integration will largely consist of moving from knowledge of the derivative of a function back to a function from which it is derived. This is where we need to put our emphasis.

I claim that this is also the lesson of history. Archimedes was not doing calculus. We must look to the 17th and 18th centuries for our historical precedent. While the mathematicians of that era understood the integral as area, they worked with it as antiderivative. Jean Bernoulli, Euler, L'Huilier, Lagrange, Lacroix, and Bolzano all defined the integral as the inverse of the derivative (page 278 in [3]). It was Joseph Fourier who changed the emphasis to area and incidentally introduced the notation of the definite integral (page 61 in [5]). This was necessary if he was to make sense of the coefficients in an arbitrary cosine series:

$$\int_0^\pi f(x) \cos nx \, dx.$$

Leibniz understood the integral as a limit of a sum, but only in a very heuristic and intuitive sense. It was not until 1823 that Cauchy began the modern formulation of the integral. The Riemann sum with its arbitrary partition of the interval and the function evaluated at an arbitrary point of each subinterval was introduced in 1854 specifically to deal with problems of Fourier series. Riemann's definition of the integral is the most modern mathematics we teach in the first year of calculus. I believe we should not.

This is not to say that we should not teach the connection between integration and area or deal with integrals as limits of sums. The power of integration rests precisely in its multifaceted nature. But we need to be clear about which approach we are teaching and how we are building the bridges between these concepts. The integral as the limit of a sum is the most difficult of them. That Cauchy wrestled with it and came to a very imperfect understanding should alert us to the fact that the Cauchy or Riemann integral is not appropriate fare for first-year calculus. Judith Grabiner [4] has demonstrated that Cauchy built his understanding of the integral on the extensive work on integral approximations that was developed by Newton, Euler, Legendre, and others. These approximations also provide a natural bridge to link antidifferentiation and the measurement of areas. This is the approach that I have been using in my classes.

After studying several simple differential equations that we could solve with our existing techniques, I moved on to some we could not such as

$$f'(x) = \sqrt{1+x^3}.$$

If we know that $f(0) = 0$, how can we use this knowledge of the derivative to approximate the value of $f(2)$? If we know the average rate of change of $f(x)$ as x moves from 0 to 2, then we can multiply it by 2, the step size or change in x, to obtain the value of $f(2)$. After pointing out that the rate of change of $f(x)$ does not stay the same—it is 1 at $x = 0$, $\sqrt{2}$ at $x = 1$, and 3 at $x = 2$—the students volunteered 2 as a first approximation to the average rate of change, yielding

$$f(2) \simeq 4.$$

They readily saw that if we used $(1 + \sqrt{2})/2$ as the rate of change from 0 to 1 and $(\sqrt{2} + 3)/2$ for the interval from 1 to 2, then we would get a closer approximation, namely

$$f(2) \simeq \frac{1 + \sqrt{2}}{2} + \frac{\sqrt{2} + 3}{2} \simeq 3.414.$$

Each student possessed a graphics calculator which we programmed to compute and plot a succession of these points, yielding a graph of the antiderivative. As we made the step size smaller, it was intuitively evident that we were getting closer to the true graph of the antiderivative. No one had problems with the idea that we could *define* the antiderivative to be the limiting graph as the step size approached 0. This was an opportunity to reinforce the distinction between a function and an algebraic expression. They saw before them the graph of a function for which there is no closed algebraic expression. Once they had a program to graph antiderivatives, there were many interesting examples that could be done such as the Fresnel integral: $f'(x) = \sin x^2$ (see [6] for a worksheet based on this integral).

I next examined our approximation. If we are calculating $f(2)$ using steps of size 0.1, then our technique for finding approximation tells us that

$$f(2) \simeq \frac{f'(0) + f'(.1)}{2} \times 0.1 + \frac{f'(.1) + f'(.2)}{2} \times 0.1 + \cdots + \frac{f'(1.9) + f'(2)}{2} \times 0.1.$$

This sum can also be interpreted as an approximation to the area under the curve $g(x) = \sqrt{1 + x^3}$ from $x = 0$ to $x = 2$ (the term "trapezoidal rule" is naturally introduced here). As the step size gets smaller, the sum approaches the true area. *The antiderivative can be interpreted as area.*

This is the fundamental theorem of calculus. We have demonstrated it, not by a rigorous proof, but by a convincing argument. Now is the time to emphasize the uses of the integral as a measurement of area and the insights this yields in evaluating integrals. Eventually, numerical integration can provide the foundation on which to construct other uses of the integral that rely on understanding it as the limit of a sum. The summation formulas are dispensed with entirely and so can be treated in courses where they are truly useful. Riemann's definition of the integral can wait until the student is comfortable with integral approximations, knows how to work with error terms, and can appreciate the need for this more careful treatment.

References

1. Tom M. Apostol, *Calculus*, two volumes, second edition, Blaisdell, Waltham, MA, 1969.
2. Lipman Bers, *Calculus*, Holt, Rinehart and Winston, New York, 1969.
3. Carl B. Boyer, *The History of the Calculus and Its Conceptual Development*, Dover, New York, 1959.
4. Judith Grabiner, *The Origins of Cauchy's Rigorous Calculus*, MIT Press, Cambridge, MA, 1981.
5. Ivor Grattan-Guinness, *The Development of the Foundations of Mathematical Analysis from Euler to Riemann*, MIT Press, Cambridge, MA, 1970.
6. David Lomen, David Lovelock, and William McCallum, *Calculus Projects*, Department of Mathematics, University of Arizona, Tucson.

Evaluating Integrals Using Self-Similarity

Robert S. Strichartz

1. INTRODUCTION. It is generally believed that there are two elementary methods for evaluating definite integrals:

1. the method of Archimedes, in which you partition the interval, form an approximating sum (often called a Riemann sum), and take a limit;
2. the method of Newton and Leibniz, involving the Fundamental Theorem of the Calculus, in which you find an antiderivative by inspired guesswork, and then compute the increment over the interval.

There are more advanced methods, especially Cauchy's residue calculus using complex variable theory, but a typical calculus course mentions only these two methods. Students quickly learn from experience that Archimedes really was a genius, but for practical purposes the Fundamental Theorem is the way to go.

But there is a third method, which is quite elementary although not well known. It is derived from the theory of integration on fractals, and is based on a self–similarity property of the unit interval. It is not a truly practical method, since it gives exact answers only for integrals of polynomials, but it illustrates the important mathematical article of faith that the study of exotic structures can produce new insights into old and commonplace subjects. After presenting the method in the context of ordinary integrals, I indicate how it can be adapted to the context of integrals on fractals, where it is essentially the only method available.

It is necessary to assume that the integral has a few elementary properties:

(i) linearity,
$$\int_a^b (c_1 f_1(x) + c_2 f_2(x))\, dx = c_1 \int_a^b f_1(x)\, dx + c_2 \int_a^b f_2(x)\, dx$$

(ii) additivity,
$$\int_a^b f(x)\, dx = \int_a^c f(x)\, dx + \int_c^b f(x)\, dx$$

(iii) linear change of variable,
$$c \int_a^b f(cx + d)\, dx = \int_{ca+d}^{cb+d} f(x)\, dx$$

(iv) integration of constants,
$$\int_a^b c\, dx = c(b - a).$$

These properties are easily derived from any reasonable definition of the integral. Of course (iii) is a special case of the more general change of variable formula, which is not as simple.

2. SELF-SIMILARITY OF THE INTERVAL. For simplicity of notation we work with the unit interval [0, 1], but the same considerations apply to any interval. If we break the interval in half, then each half is similar to the whole: the left half $[0, \frac{1}{2}]$

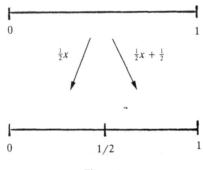

Figure 1

is the image of [0, 1] under the mapping $\frac{1}{2}x$, while the right half $[\frac{1}{2}, 1]$ is the image of [0, 1] under the mapping $\frac{1}{2}x + \frac{1}{2}$. Both mappings are similarities because they scale down all distances by a factor of $\frac{1}{2}$; see Figure 1.

For any function f, we have

$$\int_0^1 f(x)\,dx = \int_0^{1/2} f(x)\,dx + \int_{1/2}^1 f(x)\,dx$$

by additivity, while

$$\int_0^{1/2} f(x)\,dx = \frac{1}{2}\int_0^1 f\left(\frac{1}{2}x\right) dx$$

and

$$\int_{1/2}^1 f(x)\,dx = \frac{1}{2}\int_0^1 f\left(\frac{1}{2}x + \frac{1}{2}\right) dx$$

by the linear change of variable property. Altogether we obtain the *self-similar identity*

$$\int_0^1 f(x)\,dx = \frac{1}{2}\int_0^1 f\left(\frac{1}{2}x\right) dx + \frac{1}{2}\int_0^1 f\left(\frac{1}{2}x + \frac{1}{2}\right) dx. \qquad (1)$$

There is a simple geometric explanation of this identity: the region under the graph of f is cut in half and each half is stretched horizontally by a factor of 2, which explains why we have to compensate by multiplying by the factor of $\frac{1}{2}$; see Figure 2.

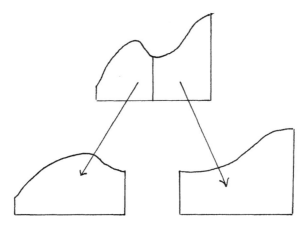

Figure 2

The self-similar identity expresses the consequences for the integral of the self-similarity of the interval. Aside from its intrinsic interest, it can be used as a tool for evaluating integrals. Before doing this, I want to mention a small point that the reader may have already noticed: there are many other ways to express the self-similarity of the interval. You can break up the interval into more than two intervals (they do not even have to be the same length), each similar to the whole interval. For each such decomposition there is a corresponding self-similar identity. For the calculations we are going to do, the single identity (1) suffices, so we leave the general form as an exercise for the reader.

3. INTEGRALS OF POLYNOMIALS. We want to obtain the basic integration formula

$$\int_0^1 x^n \, dx = \frac{1}{n+1}, \quad n = 0, 1, 2, \ldots \tag{2}$$

as a consequence of the self-similar identity (1). We observe that (2) and properties (ii) and (iii) imply the more general integral formula

$$\int_a^b x^n \, dx = \frac{b^{n+1} - a^{n+1}}{n+1},$$

and we can integrate any polynomial using (i). Note that for $n = 0$, (2) is just our assumption (iv), so the first non-trivial case is $n = 1$. We can obtain this by taking $f(x) = x$ in (1). Then we have

$$\int_0^1 x \, dx = \frac{1}{2} \int_0^1 \left(\frac{1}{2} x \right) dx + \frac{1}{2} \int_0^1 \left(\frac{1}{2} x + \frac{1}{2} \right) dx,$$

and using linearity

$$\int_0^1 x \, dx = \frac{1}{2} \int_0^1 x \, dx + \frac{1}{4} \int_0^1 dx.$$

Note that $\int_0^1 x\,dx$ appears on both sides of the equation, but multiplied by different constants, and $\int_0^1 dx = 1$ by the $n = 0$ case. Thus we may bring all the $\int_0^1 x\,dx$ terms to the left side to obtain

$$\frac{1}{2}\int_0^1 x\,dx - \frac{1}{4}\int_0^1 dx = \frac{1}{4}$$

which implies (2) for $n = 1$. While this may not seem like such a big deal (we could actually evaluate this integral by interpreting it as the area of a triangle), you should notice that the answer $\frac{1}{2}$ was obtained as $(1/4)/(1 - 1/2)$, and this is different from the computation that occurs if you use method 1) or 2).

So what happens if you substitute $f(x) = x^2$ in (1)? You get

$$\int_0^1 x^2\,dx = \frac{1}{2}\int_0^1 \left(\frac{1}{2}x\right)^2 dx + \frac{1}{2}\int_0^1 \left(\frac{1}{2}x + \frac{1}{2}\right)^2 dx$$

$$= \frac{1}{8}\int_0^1 x^2\,dx + \frac{1}{8}\int_0^1 x^2\,dx + \frac{1}{4}\int_0^1 x\,dx + \frac{1}{8}\int_0^1 dx.$$

Figure 3 shows the decomposition of the region under the graph of $f(x) = x^2$ corresponding to this identity. Bring all the $\int_0^1 x^2\,dx$ terms to the left side, and use

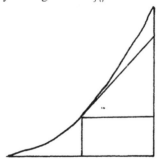

Figure 3

the previous evaluations ($n = 0$ or 1) to obtain

$$\frac{3}{4}\int_0^1 x^2\,dx = \frac{1}{4}\cdot\frac{1}{2} + \frac{1}{8}\cdot 1 = \frac{1}{4},$$

so the correct answer $\frac{1}{3}$ emerges as $(1/4)/(1 - 1/4)$.

Shall we try $n = 3$?

$$\int_0^1 x^3\,dx = \frac{1}{2}\int_0^1 \left(\frac{1}{2}x\right)^3 dx + \frac{1}{2}\int_0^1 \left(\frac{1}{2}x + \frac{1}{2}\right)^3 dx$$

$$= \frac{1}{16}\int_0^1 x^3\,dx + \frac{1}{16}\int_0^1 x^3\,dx + \frac{3}{16}\int_0^1 x^2\,dx + \frac{3}{16}\int_0^1 x\,dx + \frac{1}{16}\int_0^1 dx$$

hence

$$\frac{7}{8}\int_0^1 x^3\,dx = \frac{3}{16}\cdot\frac{1}{3} + \frac{3}{16}\cdot\frac{1}{2} + \frac{1}{16}\cdot 1 = \frac{7}{32}.$$

It is clear that we can continue indefinitely computing $\int_0^1 x^n \, dx$ for higher values of n, at each stage using the results previously obtained, but it is not obvious that this effort can produce the correct answer. This requires a little bit of algebra, and of course the finite binomial formula.

Let's try to prove (2) by induction, assuming that it holds for all integers less than n. Substituting $f(x) = x^n$ in (1) yields

$$\int_0^1 x^n \, dx = \frac{1}{2}\int_0^1 \left(\frac{1}{2}x\right)^n dx + \frac{1}{2}\int_0^1 \left(\frac{1}{2}x + \frac{1}{2}\right)^n dx$$

$$= \frac{1}{2^{n+1}}\left(\int_0^1 x^n \, dx + \int_0^1 (x+1)^n \, dx\right)$$

$$= \frac{1}{2^{n+1}}\left(2\int_0^1 x^n \, dx + \sum_{k=0}^{n-1}\binom{n}{k}\int_0^1 x^k \, dx\right).$$

The induction hypothesis ensures that $\int_0^1 x^k \, dx = 1/(k+1)$, since $k \leq n - 1$. We thus obtain

$$\left(\frac{2^n - 1}{2^n}\right)\int_0^1 x^n \, dx = \frac{1}{2^{n+1}}\sum_{k=0}^{n-1}\binom{n}{k}\left(\frac{1}{k+1}\right).$$

To complete the proof of (2) we need to verify only the following algebraic lemma:

Lemma. $\sum_{k=0}^{n-1}\binom{n}{k}\left(\frac{1}{k+1}\right) = \left(\frac{2}{n+1}\right)(2^n - 1)$.

Proof: Note that

$$\binom{n}{k}\left(\frac{n+1}{k+1}\right) = \left(\frac{n!}{k!(n-k)!}\right)\left(\frac{n+1}{k+1}\right) = \frac{(n+1)!}{(k+1)!(n-k)!} = \binom{n+1}{k+1}$$

so that

$$(n+1)\sum_{k=0}^{n-1}\binom{n}{k}\left(\frac{1}{k+1}\right) = \sum_{k=0}^{n-1}\binom{n+1}{k+1} = \sum_{j=1}^{n}\binom{n+1}{j}.$$

Since $\sum_{j=0}^{n+1}\binom{n+1}{j} = 2^{n+1}$ and we are missing the two terms $j = 0$ and $j = n + 1$ that each contribute 1, we have $\sum_{j=1}^{n}\binom{n+1}{j} = 2^{n+1} - 2$.

4. ITERATION, AND THE CONNECTION WITH RIEMANN SUMS. There do not seem to be any other integrals that can be evaluated exactly by the self-similarity method, but we can use it to gain some insight into the integral of a general function. The self-similar identity (1), like many other aspects of self-similarity, seems to invite iteration. This just means that we substitute the same identity, for the function $f(\frac{1}{2}x)$ and $f(\frac{1}{2}x + \frac{1}{2})$, into the right side of (1). This yields

$$\int_0^1 f(x) \, dx = \frac{1}{4}\int_0^1 f\left(\frac{1}{4}x\right) dx + \frac{1}{4}\int_0^1 f\left(\frac{1}{4}x + \frac{1}{4}\right) dx + \frac{1}{4}\int_0^1 f\left(\frac{1}{4}x + \frac{2}{4}\right) dx$$

$$+ \frac{1}{4}\int_0^1 f\left(\frac{1}{4}x + \frac{3}{4}\right) dx.$$

If we repeat the process again and again, we arrive at the expression

$$\int_0^1 f(x)\,dx = \frac{1}{2^m}\sum_{k=0}^{2^m-1}\int_0^1 f\!\left(\frac{1}{2^m}(x+k)\right)dx \tag{3}$$

for any positive integer m.

This resembles a Riemann sum. If we subdivide the interval into 2^m subintervals of length $1/2^m$ each, then the Riemann sums take the form

$$\frac{1}{2^m}\sum_{k=0}^{2^m-1} f(x_k)$$

where x_k lies in the interval $[k2^{-m},(k+1)2^{-m}]$. So the only difference is that instead of choosing a point x_k in the interval to evaluate the function at, we take the average value $\int_0^1 f((x+k)/2^m)\,dx$ of f over the interval. In contrast to the Riemman sum, we get the exact value of the integral at each stage, so it is not necessary to take the limit as $m \to \infty$. But we certainly are allowed to take this limit, and if we don't know how to evaluate the integrals on the right side of (3), then this is the best thing to do. If the function f is assumed to be continuous on [0, 1], then the values on each of the subintervals do not vary by very much for large values of m, so the average value and any typical value $f(x_k)$ are very close; strictly speaking, this argument requires uniform continuity, and the non-elementary theorem that a continuous function in [0, 1] is automatically uniformly continuous. The formula

$$\int_0^1 f(x)\,dx = \lim_{m\to\infty}\frac{1}{2^m}\sum_{k=0}^{2^m-1}\int_0^1 f\!\left(\frac{1}{2^m}(x+k)\right)dx$$

is almost the same as the usual definition of the integral as the limit of Riemann sums. This is just method 1) for evaluating integrals, nothing really new. However, the Riemann sums came out of the self-similar method. Also, the evaluation of integrals of polynomials in Section 3 did not use iteration and limits, and is genuinely different.

5. EXOTIC INTEGRALS ON THE INTERVAL. It is not necessary to pass to exotic fractal geometry to obtain exotic integrals. We can stay with the plain old interval, but consider exotic averages in place of the usual integral. The technical term is *integration with respect to self-similar measures*, but we can explain the ideas on an intuitive level by referring to probability. It is best to put aside the area interpretation of the integral, and think only about the average interpretation: $\int_0^1 f(x)\,dx$ is the average value of f on the interval. We also say that it is the *expected value* of f with respect to the uniform probability distribution on the interval. If we choose numbers x in [0, 1] at random, then the integral gives the average value of $f(x)$ that we find.

How do we choose a number at random? Let's represent x in binary notation $x = .x_1 x_2 x_3 \ldots$. We may ignore the ambiguity in the binary representation of certain numbers, such as $.1000\ldots = .0111\ldots$ because the set of all such numbers has zero probability. We choose $x_1 = 0$ or 1 with equal probability; then, independently, we choose $x_2 = 0$ or 1 with equal probability; and so on. Indeed, the self-similar identity (1) expresses this fact succinctly. Note that $\frac{1}{2}x = .0x_1 x_2 \ldots$ and $\frac{1}{2}x + \frac{1}{2} = .1x_1 x_2 \ldots$, and the factors $\frac{1}{2}, \frac{1}{2}$ on the right side of (1) are the equal probabilities of choosing 0 or 1 for the first digit. On the left side of (1) we have the average value of $f(x)$ over the interval, while on the right we have the sum of the probability of picking 0 or 1 for the first digit multiplied by the average value of $f(x)$ given that the first digit is 0 or 1.

But there are other ways to pick a number at random. A simple variant of the uniform distribution is to pick 0 and 1 with probabilities p and $1 - p$, for any fixed p in $(0, 1)$. This is called a *Bernoulli distribution*, and we write μ_p for the associated probability, so $\mu_p(A)$ is the probability that x belongs to a subset A of $[0, 1]$ when x is chosen in this manner. We write $\int f \, d\mu_p$ for the average value of f (or the expected value of f) when x is so chosen. When $p = 1/2$ this is the usual integral. In general it is a new kind of exotic integral. It satisfies properties (i), (ii), and (iv) of the integral, but not (iii). In place of (1), we have the self–similar identity

$$\int f(x) \, d\mu_p = p \int f\left(\frac{1}{2}x\right) d\mu_p + (1 - p) \int f\left(\frac{1}{2}x + \frac{1}{2}\right) d\mu_p. \tag{4}$$

The explanation for (4) is the same as the explanation for (1) just given, with the difference that p and $(1 - p)$ are the respective probabilities for the first digit being 0 or 1.

There is no fundamental theorem of calculus for this integral. As far as I know, no one knows how to compute $\int e^x \, d\mu_p$ or $\int \sin x \, d\mu_p$ or $\int \sqrt{x} \, d\mu_p$. But we can compute the integral of polynomials, using the self–similar identity as before. We start with $\int 1 \, d\mu_p = 1$, as this is a property of all averages. Then

$$\int x \, d\mu_p = p \int \left(\frac{1}{2}x\right) d\mu_p + (1 - p) \int \left(\frac{1}{2}x + \frac{1}{2}\right) d\mu_p = \frac{1}{2} \int x \, d\mu_p + \frac{1 - p}{2} \int 1 \, d\mu_p$$

hence

$$\frac{1}{2} \int x \, d\mu_p = \frac{1 - p}{2}$$

or

$$\int x \, d\mu_p = 1 - p.$$

By the way, notice that we also have

$$\int (1 - x) \, d\mu_p = p.$$

A heuristic explanation for these results is that if $p > \frac{1}{2}$ then the probability gives more weight to the interval $[0, \frac{1}{2}]$, where x takes smaller values and $(1 - x)$ takes larger values, so $\int x \, d\mu_p$ should be less than $\frac{1}{2}$. Next

$$\int x^2 \, d\mu_p = p \int \left(\frac{1}{2}x\right)^2 d\mu_p + (1 - p) \int \left(\frac{1}{2}x + \frac{1}{2}\right)^2 d\mu_p$$

$$= \frac{1}{4} \int x^2 \, d\mu_p + (1 - p)\left(\frac{1}{2} \int x \, d\mu_p + \frac{1}{4} \int 1 \, d\mu_p\right)$$

$$= \frac{1}{4} \int x^2 \, d\mu_p + (1 - p)\left(\frac{1}{2}(1 - p) + \frac{1}{4}\right)$$

so

$$\int x^2 \, d\mu_p = \frac{4}{3}(1-p)\left(\frac{1}{2}(1-p) + \frac{1}{4}\right) = \frac{2}{3}(1-p)^2 + \frac{1}{3}(1-p).$$

We can continue the process to higher powers, but the expressions become more and more complicated. It is easy to show that

$$\int x^n \, d\mu_p = \sum_{j=1}^{n} A_{nj}(1-p)^j$$

for positive coefficients A_{nj}, where $A_{n1} = (2^n - 1)^{-1}$, and the recursion relations

$$A_{nj} = \frac{1}{2^n - 1} \sum_{k=j-1}^{n-1} \binom{n}{k} A_{k(j-1)}$$

hold for $2 \leq j \leq n$. However, there does not appear to be any explicit formula for the general coefficient.

6. POLYGASKETS AND OTHER FRACTALS. The familiar Sierpinski gasket SG (Figure 4) is one of the simplest examples of a self-similar set that is truly fractal. Just as in the case of the interval, there are subsets of SG that are similar to the whole. In particular, SG is the union of three similar gaskets of exactly half the size. We can write a self-similar identity for the set SG in the form $K = F_1 K \cup F_2 K \cup F_3 K$ where F_j denotes the contractive similarity $F_j x = \frac{1}{2}(x - q_j) + q_j$ in the plane with contraction ratio $\frac{1}{2}$ and fixed point q_j (so $x = (x_1, x_2)$ denotes a point in the plane), where q_1, q_2, q_3 are the vertices of an equilateral triangle. Here the variable K denotes a closed, bounded, non-empty set in the plane. It can be shown that $K = SG$ is the only solution of this identity. More generally, we can write self-similar identities

$$K = \bigcup_{j=1}^{n} F_j K$$

for any finite family of contractive similarities, called an *iterated function system*.

Figure 4

The unique solution is called a self-similar set. We concentrate on a family of examples called *polygaskets*, obtained as follows. We take the n vertices

$q_1, q_2, \ldots, q_n$ of a regular n-gon, and the contractive similarities $F_j x = r_n(x - q_j) + q_j$ with fixed-points q_j. We choose the contraction ratio r_n so that the n images of the original n-gon just touch. It is an exercise in geometry to show that the correct value is

$$r_n = \frac{\sin \pi/n}{\sin \pi/n + \sin(\pi/n + 2\pi m/n)} \quad \text{for} \quad m = [n/4].$$

Figure 5 shows the pentagasket, hexagasket, and octagasket, corresponding to $n = 5, 6$, and 8. Note that the octagons touch along edges rather than at vertices;

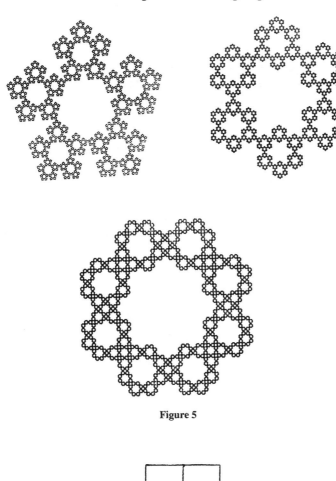

Figure 5

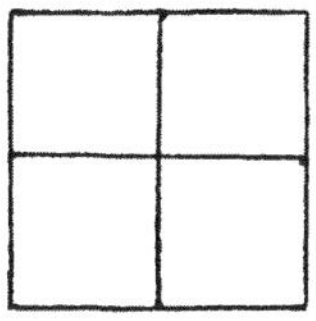

Figure 6

this happens exactly when n is divisible by 4. When $n = 2$ we get the interval, and when $n = 4$ we get the square, as Figure 6 shows.

We want to consider probability measures μ on these fractals, and the associated integrals $\int f d\mu$ for functions f defined on the fractal. Since the fractals are embedded in the plane we can imagine that f is just given by restricting to K a function defined on the plane, but it is not necessary that f be defined at points not in K. The intuition is that $\int f d\mu$ represents the average value of $f(x)$ when x is chosen at random from K according to the probability distribution. Since K is self-similar, we would like the probability distribution to be self-similar as well, so that the probability of choosing a point in a subset A of $F_j K$ is just proportional to the probability of choosing a point in the corresponding subset, which is $F_j^{-1} A$, in K. That is,

$$\mu(A) = p_j \mu(F_j^{-1} A) \quad \text{for} \quad A \subseteq F_j K.$$

Here we need to make an important assumption: the overlaps $F_j K \cap F_k K$ for $j \neq k$ have zero probability. This is true for the polygaskets because these are either points (n not divisible by 4) or subsets of line segments (n divisible by 4), but it is not true in general. With this assumption we see that the constants of proportionality p_j must form a set of finite probabilities, so $0 < p_j < 1$ and $p_1 + \cdots + p_n = 1$; to avoid degenerate cases we require $p_j > 0$. Also, we can combine all n equations into one:

$$\mu(A) = \sum_{j=1}^{n} p_j \mu(F_j^{-1} A) \quad \text{for} \quad A \subseteq F,$$

called a self-similar identity for the probability measure μ. The corresponding identity for the integral is

$$\int f d\mu = \sum_{j=1}^{n} p_j \int f \circ F_j \, d\mu. \tag{5}$$

This is the analog of our original self-similar identity (1). Informally, this self-similar identity says that to find the average value of f on K, take the sum of the product of the probability p_j that a random point x belongs to $F_j K$ times the average value of f on $F_j K$; as x varies over K, $F_j x$ varies over $F_j K$, so $\int f \circ F_j d\mu$ represents this average.

We can iterate (5) just as we did (1), to obtain

$$\int f d\mu = \sum_{j_1=1}^{n} \cdots \sum_{j_m=1}^{n} p_{j_1} \cdots p_{j_m} \int f \circ F_{j_1} \circ \cdots \circ F_{j_m} \, d\mu. \tag{6}$$

We can again interpret this as a kind of Riemann sum. At each level m, we partition K into the n^m subsets $F_{j_1} \circ \cdots \circ F_{j_m} K$, and we assign the weight $p_{j_1} \cdots p_{j_m}$ to each subset. The weights play the role of the lengths of subintervals in a usual Riemann sum. In (6) the weight is multiplied by the average value of f on the subset, but we could replace the average by a typical value of f at some point in the subset, at least for continuous functions f, by taking a limit as $m \to \infty$. In fact this is one way we can define the integral.

For the case of the polygaskets, the most natural choice is to take all the weights to be equal, $p_j = 1/n$ for all j. Then each subset $F_{j_1} \circ \cdots \circ F_{j_m} K$ at level m is a little n-gasket contracted by the ratio r_n^m, and given the probability weight $1/n^m$. Note that the ratio

$$d = \frac{\log(1/n^m)}{\log r_n^m} = \frac{\log(1/n)}{\log r_n} = \frac{\log n}{\log(1/r_n)}$$

is independent of m. We can interpret d as a dimension in the following sense: if we scale a figure by a factor r, we expect its weight to be scaled by the factor r^d, where d is its dimension. In fact this value of d coincides with all the usual dimensions for the polygasket (Hausdorff, box, packing, ...), and the self–similar probability measure with equal weights coincides, up to an unknown constant of proportionality, with the Hausdorff measure in dimension d.

Just as in the case of the interval, the self–similar identity can be used to evaluate the integrals of some functions on polygaskets, including any polynomial function (the restriction to K of a polynomial in 2 variables). The method is exactly the same. It uses the fact that $f \circ F_j$ is also a polynomial of the same degree, explicitly computable. Again it is necessary to proceed inductively on the degree of the polynomial, starting with $\int 1 d\mu = 1$. If f is the monomial $x_1^{k_1} x_2^{k_2}$, then $f \circ F_j$ is the sum of $r_n^{k_1+k_2} x_1^{k_1} x_2^{k_2}$ and terms of lower degree. Thus (5) yields

$$\int x_1^{k_1} x_2^{k_2} d\mu = \sum_{j=1}^{n} p_j r_n^{k_1+k_2} \int x_1^{k_1} x_2^{k_2} d\mu + I$$

where I consists of integrals already computed, and so

$$\int x_1^{k_1} x_2^{k_2} d\mu = \left(1 - r_n^{k_1+k_2}\right)^{-1} I.$$

We do not present the details. For general self–similar fractals we can also use essentially the same method. In cases when the similarities include rotations as well as homothetic contractions, it is necessary to do the computation simultaneously for all monomials of a fixed degree, which leads to a system of linear equations for the integrals.

This method of integration has been used for polynomial functions for a long time [7], and has been used for other, more intrinsic functions, in [4] and [6]. For more information on self–similar fractals, the reader may consult books about fractals, such as [1] or [2]. There is also a theory of differential calculus on fractals, but this is more intricate; see [3] or [5].

REFERENCES

1. M. F. Barnsley, *Fractals everywhere*, 2nd edition, Academic Press, Orlando, FL, 1993.
2. K. J. Falconer, *Fractal geometry: mathematical foundations and applications*, John Wiley and Sons, Chichester, U.K., 1990.
3. J. Kigami, Harmonic calculus on p.c.f. self–similar sets, *Trans. Amer. Math. Soc.* **335** (1993) 721–755.
4. R. Strichartz, Piecewise linear wavelets on Sierpinski gasket type fractals, *J. Fourier Anal. Appl.*, **3** (1997) 387–416.
5. R. Strichartz, Analysis on fractals, *Notices AMS* **46** (1999) 1199–1208.
6. R. Strichartz and M. Usher, *Splines on fractals*, Math. Proc. Cambridge Philos. Soc., to appear.
7. E. R. Vrscay and C. J. Roehrig, Iterated function systems and the inverse problem of fractal construction using moments, in *Computers and Mathematics*, E. Kaltofen and S. M. Watts, Springer–Verlag, 1989, pp. 250–259.

Self-integrating Polynomials

Jeffrey A. Graham

I walked into a colleague's office one day as he was grading calculus papers and he showed me the following mistake that one of his students made:

$$\int_0^1 3x^2 + 2x + 3\,dx = \left(3(1)^2 + 2(1) + 3\right) - \left(3(0)^2 + 2(0) + 3\right) = 5.$$

We see that this student failed to find an anti-derivative before plugging in the limits of integration. The correct computation is given by

$$\int_0^1 3x^2 + 2x + 3\,dx = \left((1)^3 + (1)^2 + 3(1)\right) - \left((0)^3 + (0)^2 + 3(0)\right) = 5.$$

By happy accident this student found a correct answer, but we know this trick won't always work or we wouldn't devote so much time learning to integrate! Let's call polynomials like the one in the example *self-integrating*. A question naturally arises; how many self-integrating polynomials are there over the interval [0, 1]? We will answer that question in this capsule.

A quick check of the definition of a self-integrating polynomial indicates that the only constant polynomial satisfing the conditon is $p_0(x) \equiv 0$. We next check to see if there are any self-integrating linear polynomials. Solving the linear equation

$$\int_0^1 a_1 x + a_0\,dx = ax + b\big|_0^1$$

gives $a_1 = 2a_0$, so we define $p_1(x) = x + 1/2$. A similar computation with a quadratic polynomial

$$\int_0^1 a_2 x^2 + a_1 x + a_0\,dx = a_2 x^2 + a_1 x + a_0\big|_0^1$$

leads to $p_2(x) = x^2 + 2/3$. At this point, we see a pattern and define

$$p_k(x) = x^k + \frac{k}{k+1}.$$

It is easily verified that all polynomials of this form are self-integrating on the interval [0, 1]. We now have a set of self-integrating polynomials

$$S_n = \left\{ p_k(x) = x^k + \frac{k}{k+1} \right\}$$

for $k = 1, 2 \ldots, n$. The linearity property of the integral ensures that any linear combination of polynomials in S_n is self-integrating. In fact, the set of self-integrating polynomials of degree at most n is an n-dimensional subspace of the vector space of all polynomials of degree at most n. An exercise in elementary linear algebra demonstrates that the set S_n is a basis for this subspace, and this description completely characterizes the self-integrating polynomials up to any finite degree n.

For finite dimensions, this is about all we can say. However, if we move into infinite dimensional vector spaces, the situation gets more interesting. Let P be the set of all polynomials and let K be the set of all self-integrating polynomials on the interval [0, 1]. Clearly K is an infinite-dimensional subspace of P. If we equip P with the inner product,

$$\langle f, g \rangle = \int_0^1 f(x)g(x)\, dx,$$

we can measure distance using the norm

$$\|f - g\|_2 = \langle f - g, f - g \rangle^{1/2}. \tag{1}$$

Recall that $L^2[0, 1]$ is the set of all functions f such that $\|f\|_2 < \infty$ and that P and K are subspaces of $L^2[0, 1]$. It is known (see [1]) that P is dense in $L^2[0, 1]$. Is K dense too? If the polynomial $p(x) \equiv 1$ can be approximated to any accuracy by elements in K, so can any positive integer power of x. It follows that K is dense in P and hence also in $L^2[0, 1]$.

Taking our cue from the basis S_N, we define the sequence

$$S = \left\{ x^n + \frac{n}{n+1} \right\}_{n=1}^{\infty}.$$

It is easy to verify that S is contained in K. In Figure 1, we see the graph of some of the terms of this sequence. These pictures suggest that this sequence converges to 1, but we must be careful about the *type* of convergence. For all $0 < x < 1$, we have pointwise convergence of

$$x^n + \frac{n}{n+1}$$

to 1 since $x^n \to 0$ for these values of x. At $x = 1$, we see that

$$x^n + \frac{n}{n+1} = 1 + \frac{n}{n+1} \to 2,$$

so S does not converge in the pointwise sense to 1 in the interval [0, 1]. This sequence does converge in $L^2[0, 1]$ to 1 as $n \to \infty$ since this type of convergence only requires that square of the distance between

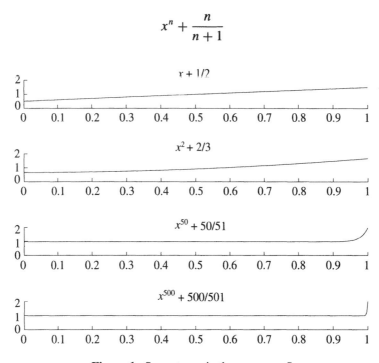

Figure 1. Some terms in the sequence S

and 1 as measured in $L^2[0, 1]$ tends to zero. Using the definition of distance given in (1), the calculation below demonstrates this fact.

$$\left\| 1 - \left(x^n + \frac{n}{n+1} \right) \right\|_2^2 = \int_0^1 \left(1 - \left(x^n + \frac{n}{n+1} \right) \right)^2 dx,$$

which can be shown to equal

$$\frac{n^2}{(2n+1)(n+1)^2},$$

and this expression vanishes as $n \to \infty$.

Acknowledgment. Many thanks to Ken Brakke who asked, "Have you tried graphing those functions?" and to the anonymous referee whose comments made this a better paper.

Reference

1. W. E. Cheney, *Approximation Theory*, Chelsea, 1982.

Symmetry and Integration

Roger Nelsen

Problem A-3 on the 1980 Putnam exam asked contestants to evaluate

$$\int_0^{\pi/2} \frac{dx}{1 + (\tan x)^{\sqrt{2}}}.$$

It appears that most of the contestants found this to be a rather difficult problem, for of the top 207 (of 2043) contestants, 141 did not submit a solution, and only 23 scored 3 or more points (out of a possible 10) [L. F. Klosinski, G. L. Alexanderson, and A. P. Hillman, The William Lowell Putnam Mathematical Competition, *American Mathematical Monthly* 88 (1981), 605–612].

But I think if this problem were given to today's students, armed with graphing calculators, the performance would be quite different. Upon seeing that the graph of the function in the integrand looks something like Figure 1, many students could now exploit the symmetry and proceed to show analytically that the value of the integral must be $\pi/4$.

Of course, using symmetry to simplify integration problems is nothing new. Virtually every calculus text published in the past 20 years has, when considering

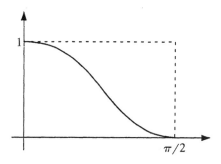

Figure 1
The graph of $y = 1/[1 + (\tan x)^{\sqrt{2}}]$ on $[0, \pi/2]$.

integrals of even and odd functions, used the symmetry they exhibit to simplify the integration—e.g., if f is odd and continuous on an interval (of finite length) centered at 0, then the integral of f over that interval is 0. But functions exhibiting the sort of symmetry in the above problem are rarely considered, even in the new generation of calculus texts which make extensive use of graphing technology. With a graphing calculator, it is easy for the student to discover symmetry such as that exhibited in the Putnam problem cited above. My purpose here is to expand upon the usual notions of symmetry encountered in calculus in order to deal with such problems.

The symmetry of the function in Figure 1 is essentially the symmetry of an odd function, but with the "center" of symmetry at a point other than the origin. To be precise, we say that the graph of a plane curve is *symmetric with respect to the point* (c, d) iff whenever $(c - x, d - y)$ is on the curve, so is $(c + x, d + y)$. When the curve is the graph of a function $y = f(x)$, this condition can be expressed as $f(c - x) + f(c + x) = 2f(c)$ (when $c, c - x$, and $c + x$ are all in the domain of the function), as illustrated in Figure 2.

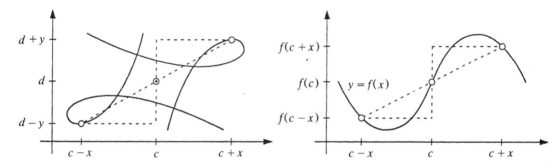

Figure 2
Symmetry with respect to a point.

Now suppose that f is continuous on the interval $[a, b]$ and that the graph of f is symmetric with respect to the point whose x-coordinate is the midpoint $(a + b)/2$ of $[a, b]$; that is, f satisfies $f(x) + f(a + b - x) = 2f((a + b)/2)$ for all x in $[a, b]$. Such functions are easy to integrate:

Theorem. *Suppose f is continuous on $[a, b]$ and that $f(x) + f(a + b - x)$ is constant for all x in $[a, b]$. Then*

$$\int_a^b f(x)\, dx = (b - a)f\left(\frac{a + b}{2}\right) = \tfrac{1}{2}(b - a)[f(a) + f(b)].$$

An analytic proof is straightforward—simply split the interval of integration at its midpoint and use the substitution $u = a + b - x$ in the second integral. But perhaps more instructive is the "proof without words" in Figure 3 (page 42).

Symmetry and Integration

As a simple example, consider $\int_0^{\pi/2} \sin^2 x \, dx$. With a graphing calculator, the student readily notices symmetry with respect to $(\pi/4, 1/2)$, and this is easily verified: $\sin^2 x + \sin^2[(\pi/2) - x] = 1$. Thus $\int_0^{\pi/2} \sin^2 x \, dx = \pi/4$.

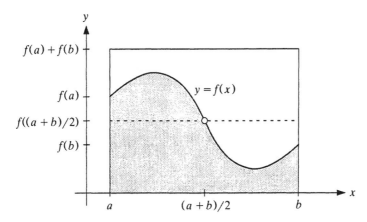

Figure 3
A "proof without words" of the theorem.

The reader is encouraged to try the following exercises, both by the preceding method and by a standard method for comparison. Answers are given below.*

1. $\int_{-1}^{1} \arctan(e^x) \, dx$

2. $\int_{-1}^{1} \arccos(x^3) \, dx$

3. $\int_0^2 \dfrac{dx}{x + \sqrt{x^2 - 2x + 2}}$

4. $\int_0^2 \sqrt{x^2 - x + 1} - \sqrt{x^2 - 3x + 3} \, dx$

5. $\int_0^4 \dfrac{dx}{4 + 2^x}$

6. $\int_0^{2\pi} \dfrac{dx}{1 + e^{\sin x}}$

---

*Answers: **1.** $\pi/2$; **2.** π; **3.** 1; **4.** 0; **5.** 1/2; **6.** π.

Sums and Differences vs Integrals and Derivatives

Gilbert Strang

This article offers one approach to the understanding, and also to the teaching, of the fundamental theorem of calculus. The ideas expressed here are not new—mathematically they cannot be original, and even pedagogically they might already be in use. If so, that is good. The author has become deeply involved in the effective presentation of calculus—in fact he has gone past the fatal point of no return, beyond which a book has to be written [1]. It is impossible to think about explaining this subject without searching for a fresh way to illuminate the relation of derivatives to integrals.

If the hopeful author of a calculus book is viewed with a lean and lively suspicion —as he certainly is—a small request might be made in return. You know this subject too well, and therefore you must forget what you know. Instead of the "suspension of disbelief" that is required of a playgoer, the reader is asked for something more difficult—a suspension of belief, and of a sure familiarity, in the ideas of calculus. All you are permitted is the knowledge that distance equals (constant) velocity multiplied by time. Of course you are encouraged to learn more.

In spite of the happy spirit in which this article is written, its goal is entirely serious. We all want mathematics to be appreciated. I think this only becomes possible when a part of the subject is understood. It might be unwise for us to transform every student into a full-scale mathematician, but there is no reason to fail in our more limited responsibility—to help students see some of the inspiration behind what they learn.

The Goal (in Two Parts) and the Plan

There are two goals, different but complementary. One is abstract, the other is very concrete. The first is to understand the relation of a function f to its derivative v, and the relation of v to its integral f. Starting cold (from the definitions) this is a substantial challenge. Working with graphs is better. Slope and area are tremendous visual supports. But my experience is that $v =$ velocity and $f =$ distance are the best. The intuition that comes from driving a car, and from ordinary use of the speedometer and odometer, is a free gift to calculus teachers. The relation of v to f is understood implicitly, and our contribution is to make it explicit.

The second goal is more special. The applications of calculus are built on a few functions (amazingly few). The student needs to become familiar with those particular functions. *They come in pairs f and v.* Where the first goal was to know the relation between them, the second goal is to know the functions themselves. Certain special pairs arise constantly in mathematical models, and we might as well connect with those v's and f's from the start.

The first pair is the one that everybody knows: *constant v and linear f*. (I don't skip them because they are known. I explain them because they are known!) The step between function and graph is not automatic in general, but that example is clear. Area and slope get a toehold—they are the area of a rectangle and the slope of a straight line. By no means are those ideas held in a firm grip. There is a long way to go, and the choice of the *next pair* is decisive.

A critical moment comes extremely early (often in the first week). Far ahead is the destination, to understand the key ideas and the key functions of calculus. We absolutely need a bridge. The goal cannot be reached from constant v-linear f in a single step. One approach is to start with *definitions* (of real numbers, functions, limits, ...) along with careful examples. I believe that is a seriously wrong start. The approach is taken in good faith, but I am convinced there is a better one. I will concentrate on explaining how a calculus course can go forward from the first pair, to build on the intuition that made it understood before it was expressed.

The second pair is *one step away* from the first pair. It is a piecewise constant v and a piecewise linear f—the best motion seems to be *forward and back*. The velocity is $+V$ up to time T, and then $-V$ from T to $2T$. Every student knows what this motion does, and where it ends up. The distance function f is turned into a graph (Figure 1). The velocity is also expressed by a graph (and negative area for negative v makes sense). The area under the v-graph should be checked at $T/2$ and $3T/2$. It is essential to take time with visual understanding (including the slope) because the next step will grow out of this one.

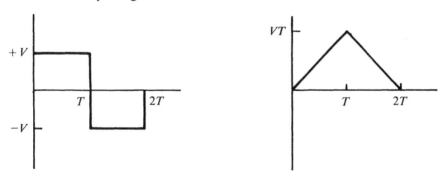

Figure 1
Forward and back: the v-graph and the f-graph.

Now I can describe the plan. *It is to imitate calculus by algebra.* The velocity-distance relation can be made plain when v changes by finite steps. There is a sequence of velocities $v_1, v_2, v_3, \ldots$. For convenience the jumps occur at evenly spaced times $t = 1, 2, 3, \ldots$. The distances reached at those times are $f_1, f_2, f_3, \ldots$. We are dealing with functions, and at the same time with numbers. A typical case is

$$v = 1, 2, 3, 4, \ldots \quad \text{and} \quad f = 1, 3, 6, 10, \ldots.$$

The one thing every student can do is to fill in the next v and f! The pattern is seen; but it has to be written down. Mathematics is about *discovering* patterns and then *describing* them—and the way to teach mathematics is to do it.

Calculus Before Limits

I will propose four specific v's and f's. They are discrete models of familiar functions. The velocity changes, not continuously but in steps. Calculus will deal with continuous change—that is its central idea—and these models simulate that change by a sequence of jumps.

1. A steadily increasing velocity: $v = 1, 3, 5, 7, \ldots$ imitates $v = 2t$.
2. An oscillating velocity: $v = 1, 0, -1, -1, 0, 1, \ldots$ imitates $v = \cos t$.
3. An exponentially increasing velocity: $v = 1, 2, 4, 8, \ldots$ imitates $v = 2^t$.
4. A burst of speed (then stop): $v = 100, 0, 0, \ldots$ makes f imitate a step function.

The oscillation could come ahead of the others, because it is closest to forward-back motion and has the most attractive graph. But the first case of continuous change, when calculus really starts, will be $v = 2t$. Its discrete form yields a neat pattern—comparable to $v = 1, 2, 3, 4$ but with better f's. The presence of squares is concealed by $f = 1, 3, 6, 10$—now it will stand out.

Example 1. **Increasing velocity.** The four velocities $1, 3, 5, 7$ are displayed on the left of Figure 2. The distance function is on the right. The goal is to uncover, numerically and then algebraically, the relation between

$$v = 1, 3, 5, 7, \ldots \quad \text{and} \quad f = 1, 4, 9, 16, \ldots.$$

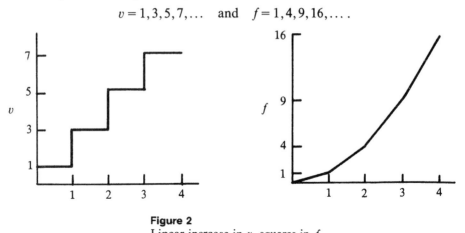

Figure 2
Linear increase in v, squares in f.

At $t = 1$ the distance traveled (starting from zero) is $f_1 = 1$. It equals the velocity (which is 1) multiplied by the time (also 1). At $t = 2$ the distance is $f_2 = 1 + 3$—the velocity in the second time interval is $v_2 = 3$. After three steps f_3 is $v_1 + v_2 + v_3 = 1 + 3 + 5 = 9$. It is a pleasure to prove that the f's are perfect squares, but more important is their relation to the v's.

The first step begins with the f's. We are given $1, 4, 9, 16, 25$ and we are aiming for $1, 3, 5, 7, 9$. Those v's are the *differences* of the f's:

$$v_j \text{ is the difference } f_j - f_{j-1}. \tag{1}$$

The velocity $v_5 = 9$ is the difference $25 - 16$. It is the slope of the f-graph in the fifth interval. Similarly $v_4 = 7$ is $16 - 9$. I admit to a small difficulty in the first interval (at $j = 1$), from the fact there is no f_0. For this example the natural choice is $f_0 = 0$. The need for a starting point will come back to haunt us (or help us) in calculus.

Sums and Differences vs Integrals and Derivatives

Remark 1. The equation $v_j = f_j - f_{j-1}$ asks the student to take a step that we no longer notice. With numbers everything is clear—but now we have switched to letters. There is a risk of muddying up the whole thing by algebra. But we have to do it. The investment in learning algebra is partly justified by our use of it here.

Now comes the crucial step, in the *reverse direction*. How do we recover the f's from the v's? The whole relation between differential calculus and integral calculus rests on the fact that **taking sums is the inverse of taking differences**. We saw the discrete form of differentiation in equation (1), and we will see the discrete form of integration in equation (2).

To repeat: The numbers $1, 3, 5, 7, 9$ are given. The numbers $1, 4, 9, 16, 25$ are wanted. The student sees how to do it, but the professor sees more. This numerical example carries within it—in a limited but absolutely convincing form—the fundamental theorem of calculus. It is still algebra, because v is piecewise constant—but the relation between v and f is the real thing. Why not present it now?

Fundamental Theorem of Calculus (before limits). *If each $v_j = f_j - f_{j-1}$ then*

$$\text{the sum } v_1 + v_2 + \cdots + v_n \text{ equals } f_n - f_0. \tag{2}$$

The area under the graph of v is the change in f.

Proof. We are adding $(f_1 - f_0) + (f_2 - f_1) + (f_3 - f_2) + \cdots + (f_n - f_{n-1})$. The first f_1 is canceled by $-f_1$. Then f_2 is canceled by $-f_2$. At the end only $-f_0$ and $+f_n$ are left. In case $f_0 = 0$, the sum of v's has telescoped into f_n. This is the distance traveled, or the area under v.

Remark 2. It is time for new examples. They provide opportunities to verify the fundamental theorem—but this note will not impose so far on your patience. What the examples offer most of all is something different and more compelling. It is the chance to see simple models of the most crucial functions of mathematics. We are going back to *particular* v's and f's, but not forgetting what they illustrate—a relation that allows us to add v's by discovering f's.

Example 2. **Oscillation.** The six velocities $v = 1, 0, -1, -1, 0, 1$ are on the left of Figure 3. The six distances $f = 1, 1, 0, -1, -1, 0$ are on the right. The f-graph resembles (roughly) a *sine curve*. The v-graph resembles (even more roughly) a *cosine curve*. The f's were formed by the same rule as before, that each f_j equals the sum $v_1 + \cdots + v_j$. It is the total distance, when the velocities are held for one time unit each.

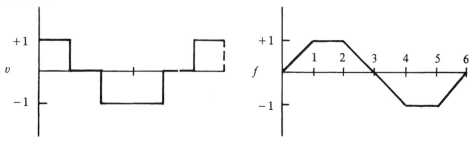

Figure 3
Piecewise constant cosine and piecewise linear sine.

These discrete sines and cosines share three properties (possibly more) with the true waveforms sin t and cos t:

1. The slope of the discrete sine is the discrete cosine, and the area under the discrete cosine is the discrete sine. (Note: This is true for all v-f pairs—it is not particular to this one.)

2. The f_j are one time step behind the v_j. In other words $v_j = f_{j+1}$—the new distance is the old velocity. Since v_j is always $f_j - f_{j-1}$, that cancels f_{j+1} on the left side of a difference equation:

$$(f_{j+1} - f_j) - (f_j - f_{j-1}) = f_{j+1} - 2f_j + f_{j-1} = -f_j. \tag{3}$$

This imitates an important differential equation (second derivative of sin t equals $-\sin t$). The same is true of the cosines v_j. The time lag $f_{j+1} = v_j$ is a close copy of $\sin(t + \pi/2) = \cos t$—but the delay is $1/6$ of the period instead of $1/4$.

3. The motion returns to the start at $t = 6$ because $v_1 + v_2 + \cdots + v_6 = 0$. The numbers v and f could repeat over the next six intervals, and the graphs would be exactly the same. These are *periodic* waveforms. (*Note:* It would be interesting to take equation (3) as fundamental, and show that the period must be 6.)

The true sine and cosine possess other properties that f and v do not imitate. The outstanding example is $(\cos t)^2 + (\sin t)^2 = 1$. If desired, that could be recovered by the faster oscillation $v = 1, -1, -1, 1$ and $f = 1, 0, -1, 0$ (repeated periodically). However we have to choose the *averages* of v at the jumps, and work with $V_1 = 0$, $V_2 = -1$, $V_3 = 0$, $V_4 = 1$. Then $V_j^2 + f_j^2 = 1$. In many ways the averages are more reasonable anyway.

Example 3. **Exponential increase** (powers of 2). In this example v doubles at every step: $v = 1, 2, 4, 8, 16, \ldots$. The velocity in the jth time interval is $v = 2^{j-1}$. The novelty comes for the distance f, *which does not begin at zero*. Its starting value must be $f_0 = 2^0 = 1$, if f is to imitate the exponential. Then the next distances are

$$f_1 = 1 + 1 = 2, \quad f_2 = 1 + 1 + 2 = 4, \quad f_3 = 1 + 1 + 2 + 4 = 8.$$

The formula is $f_j = 2^j$. It jumps out from Figure 4. On a normal day students won't want a proof. So don't call it a proof! It is a beautiful chance to use the fundamental theorem $\Sigma(\Delta f) = f_n - f_0$ as a way to add up the v's by guessing the

        Step 1:   Guess $f_j = 2^j$.

        Step 2:   Verify that $f_j - f_{j-1}$ equals v_j.

        Conclusion: The sum $v_1 + \cdots + v_n$ is $f_n - f_0$.

In language that comes later, f is an antiderivative. The slope of f is v (piecewise constant), so the area under v must be f (piecewise linear). What matters is an idea that need not be so precise: **The best way to add (or to integrate) is to know the answer.** The answer is correct, provided that $f_j - f_{j-1}$ (or df/dx) brings back v. When calculus writes that answer as $f(b) - f(a)$, the definite integral is not a surprise—it is the natural step from $f_n - f_0$.

The exercise of actually summing the powers of 2 is not to be missed.

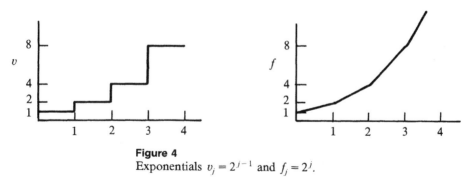

Figure 4
Exponentials $v_j = 2^{j-1}$ and $f_j = 2^j$.

Note. The twin example has exponential *decrease*. The velocities $v = -\frac{1}{2}$, $-\frac{1}{4}, -\frac{1}{8}, \ldots$ are negative. The distances $f = \frac{1}{2}, \frac{1}{4}, \frac{1}{8}, \ldots$ are positive (because $f_0 = 1$). That example gives 2^{-j} a meaning beyond pure manipulation of exponents.

Example 4. **A short burst of speed.** This last example gives a proper importance to the dimension of time. With unit time intervals the slopes were $f_j - f_{j-1}$, and no student is going to notice the division by one. A check on dimensions is the simplest test of any formula, from $f = vt$ to $e = mc^2$, and here are four v's:

$v = 10$	up to	$t = \frac{1}{10}$	(then stop)	$v = 100$	up to	$t = \frac{1}{100}$	(then stop)
$v = $ ____	up to	$t = \frac{1}{10,000}$	(stop)	$v = 10^n$	up to	$t = $ ____	(stop)

Exercise 1 is to fill in the blanks. Exercise 2 is to draw the graphs of f (with $f_0 = 0$). Exercise 3 (the mathematical one) is to say what those f-graphs have in common, and then *what the v-graphs have in common* (equal areas).

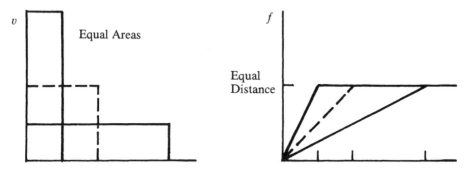

Figure 5
Burst of speed: approach to a step function f and a delta function v.

The areas of these taller and taller rectangles give the same final distance $f = 1$. This seems like the easiest of the four examples, but possibly it will occur to students to *go to the limit*. The f-graph is approaching a step function. Do we ask about the v-graph? I think we should; others may not agree. A step function for f is certainly acceptable (we have been working all along with steps in v). But now the limiting v is *the derivative of a step function*. It is a "*delta function*," totally concentrated at $t = 0$. Its graph is infinitely tall and infinitely thin. The delta function is zero except at one point, but its integral is the jump in f.

In some respects a delta function is painful (but optional). On the other hand, it is a tremendously important idea and why shouldn't we be the first to teach it? In talking about it I use the word *impulse* and the notation $v(x) = \delta(x)$. Here $v = 10\delta$. I freely admit to the class that no function can be zero except at one point, and still give positive area. (They rather like the idea.) The word distribution is never mentioned (not even a dark hint). My main suggestion is just to do the obvious: Ask the class for the slope of a step function.

Conclusion. The author owes the reader a brief accounting. What is achieved, and what is left? Our purpose was to prepare for one of the central ideas of calculus —the relation between v and f. It is not the *fact* of the fundamental theorem, but the *meaning of that fact* and the *application of that fact*, that our teaching has to aim for.

We proposed two goals. One was general (the v-f relation), the other was particular (special v-f pairs). The examples were limited to piecewise constant velocities, but that limitation brought a reward—the functions are accessible to students at the very start of a course. They give something valuable to do (and new functions to create) at that early time when so few tools are available. Psychologically this time is crucial. Like the rest of us, students form quick opinions. If the course can't get itself started, and the ideas are drowned in an ocean of definitions, those opinions will not be what we hope for.

For college freshmen who studied calculus in high school, the first days are decisive. If nothing is new they stop listening. The models suggested here are likely to be new—and the discrete-continuous analogy deserves all the emphasis we can give it.

A note of caution. I am absolutely not proposing that the whole calculus course be built on this discrete approach. In the end, t^2 and $\sin t$ are easier than $1, 4, 9, 16, \ldots$ and $1, 1, 0, -1, -1, 0$. The key rules of differential calculus (product rule and chain rule) are much more satisfying than their discrete counterparts. It is the limiting operation that makes those rules so terrific.

In integral calculus it is the same. Few of us would hesitate, given the choice between integrating x^p and summing j^p. The beauty of calculus is that it kills off the lower-order corrections in a Riemann sum, to leave areas which are really remarkable. We are grown accustomed to them, and take them so much for granted that we insist on pushing ahead to more complicated and exotic examples. To what purpose, when the ideas and functions that are really needed are so few and so beautiful?

I want to emphasize that the new textbook [1] is aimed at careful change, not revolution. The work of calculus is still there to do, but it need not be drudgery. My effort is to make the book lighter, spiritually as well as materially. I hope others are also writing, to make progress on the problems that the 1987 colloquium presented so forcefully [2].

This short article may bring to mind other ideas and other functions. The velocities $1, r, r^2, r^3, \ldots$ connect f to the sum of a geometric series. But "calculus before limits" doesn't last long. The derivative is introduced very soon, and (separately) the integral appears. Functions are minimized and $y' = cy$ is solved. Still the ideas described here do come back for one short and vital moment. Slope and area have to be brought together into the fundamental theorem of calculus. It is there that the early work with sums and differences, which gave the course something to do at the start, makes a deep theorem look true.

It was a moment of pleasure when I learned from [3] that Leibniz had begun with sums and differences. Perhaps they guided his intuition—until Cauchy arrived, everything was calculus without limits. Whether Newton would approve I don't know. He cared less than others about clear expression (at least on the evidence), but then he left one or two priceless compensations. And our own job is not entirely easy. For teachers who have to explain the ideas of Newton in the notation of Leibniz to pupils who may not be quite so apt as Cauchy, every source of illumination is welcome.

Acknowledgment. This research was supported by Army Research Office grant DAAL 03-86-K 0171 and National Science Foundation grant DMS-87-03313.

References

1. Gilbert Strang, *Calculus*, Wellesley-Cambridge Press, Wellesley, MA (to be published in 1990).
2. Lynn Steen, ed., *Calculus for a New Century*, Mathematical Association of America, Washington, 1988.
3. C. H. Edwards, *The Historical Development of the Calculus*, Springer Verlag, New York, 1979.

How Do You Slice the Bread?

James Colin Hill, Gail Nord, Eric Malm and John Nord

When Gail and John make lunches for school, their six-year old twins, Jamie and Michael, frequently ask if they can share a peanut butter sandwich. They always want it cut in half, and always in "triangles." This article is the result of trying to find a method of locating a point on the top curved-crust of a slice of bread that halves the volume of the sandwich, or equivalently the area of the bread-slice face. In addition, we treat the problem of halving the crust of the sandwich with "triangles."

A slice of bread can be modeled in the xy-plane as a rectangle surmounted by a semi-ellipse with semi-major axis parallel to the bottom crust. The origin is located as shown in Figure 1, so that the equation for the semi-ellipse is in simplest form.

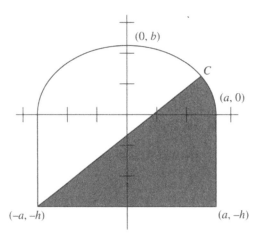

Figure 1. Two-dimensional model of a "triangulated" bread slice.

The line representing the path made by the knife blade passes from $(-a, -h)$ to the point C. C lies on the semi-ellipse with equation $y = \frac{b}{a}\sqrt{a^2 - x^2}$ and has coordinates $(c, \frac{b}{a}\sqrt{a^2 - c^2})$. The equation for the line describing the path of the knife is

$$y = \frac{b/a\sqrt{a^2 - c^2} + h}{c + a}(x + a) - h. \tag{1}$$

We then seek to find values for c such that either the perimeter or the volume of the sandwich is halved. In either case, the x-coordinate of the point C, c, is parameterized in terms of a, b, and h.

The formulation derived above results in ellipse forms most recognizable to students. Dividing by the scalar a results in a non-standard ellipse equation with a cleaner integral form. Free of the dimension a, the problem produces the same relative locations as those determined below. A similar nondimensionalization can be made in the area problem to eliminate both a and b. We leave it up to the instructor to decide which way to present the two problems.

The arc length problem. We first address the problem of where to cut the slice so that the two resulting crust lengths are equal. Using the standard parameterization $x = a\cos\theta$, $y = b\sin\theta$ for the ellipse and the familiar arc length formula $\int \sqrt{dx^2 + dy^2}$ [1], one half of the perimeter of the bread slice is given by

$$\frac{1}{2}\left(2h + 2a + \int_0^\pi \sqrt{a^2\sin^2\theta + b^2\cos^2\theta}\,d\theta\right)$$

$$= h + a + \int_0^{\pi/2} \sqrt{a^2\sin^2\theta + b^2\cos^2\theta}\,d\theta. \qquad (2)$$

The "crust" perimeter of the shaded region in Figure 1 is then given by

$$2a + h + \int_0^{\arccos c/a} \sqrt{a^2\sin^2\theta + b^2\cos^2\theta}\,d\theta. \qquad (3)$$

Equating expressions (2) and (3) and rearranging terms yields

$$a = \int_0^{\pi/2} \sqrt{a^2\sin^2\theta + b^2\cos^2\theta}\,d\theta$$

$$- \int_0^{\arccos c/a} \sqrt{a^2\sin^2\theta + b^2\cos^2\theta}\,d\theta$$

$$= \int_{\arccos c/a}^{\pi/2} \sqrt{a^2\sin^2\theta + b^2\cos^2\theta}\,d\theta. \qquad (4)$$

This trigonometric integral cannot be evaluated in closed form, so we instead consider the following special cases:

Case 1. When $b = a$ in Equation (4), that is, the bread slice is a rectangle surmounted by a semicircle, the following equation results:

$$a = \int_{\arccos c/a}^{\pi/2} a\,d\theta.$$

Carrying out the integration and solving for c yields

$$c = a \sin 1.$$

Case 2. Setting b equal to 0, Equation 4 becomes

$$a = \int_{\arccos c/a}^{\pi/2} \sqrt{a^2 \sin^2 \theta} \, d\theta = \int_{\arccos c/a}^{\pi/2} a \sin \theta \, d\theta.$$

Integrating and solving for c, we have the expected value $c = a$. We expect this result because when $b = 0$, the bread slice is shaped like a rectangle and therefore the cut should extend to the corner.

The area problem. We now address the problem of determining where to cut the slice so that the areas of the two "triangles" are equal. One-half of the area of the entire bread slice is given by

$$\frac{1}{2}\left(\frac{\pi}{2}ab + 2ah\right) = \frac{\pi ab}{4} + ah. \tag{5}$$

The area of the unshaded region in Figure 1 is given by integrating the difference of the y-coordinates of the ellipse and the cut line (1) from $x = -a$ to $x = c$:

$$\int_{-a}^{c} \left[\frac{b}{a}\sqrt{a^2 - x^2} - \left(\frac{b/a\sqrt{a^2 - c^2} + h}{c + a}(x + a) - h\right)\right] dx \tag{6}$$

Equating expressions (5) and (6) results in the integral equation

$$\frac{\pi ab}{4} + ah = \int_{-a}^{c} \left(\frac{b}{a}\sqrt{a^2 - x^2} - \frac{b/a\sqrt{a^2 - c^2} + h}{c + a}(x + a) + h\right) dx$$

Carrying out the integration and simplifying yields the equation

$$-b\sqrt{a^2 - c^2} - h(a - c) + ab \arcsin \frac{c}{a} = 0. \tag{7}$$

In this form, Equation (7) has no closed form solution for c. Assuming that the cut-point c is sufficiently close to a, we can apply the small-angle approximation $\sin \theta \approx \theta$ to obtain $\arcsin(c/a) \approx c/a$. We substitute this into Equation (7) to yield

$$c(b + h) + b\sqrt{a^2 - c^2} - ah = 0$$

with solutions

$$c = \frac{a}{(b + h)^2 + b^2}\left(h(b + h) \pm b\sqrt{2b(b + h)}\right).$$

Rearrangement yields

$$c = \frac{a\sqrt{1+b/h}\left(\sqrt{1+b/h} \pm b/h\sqrt{2b/h}\right)}{(1+b/h)^2 + (b/h)^2}. \tag{8}$$

We now consider two special cases:

Case 1. Taking $\frac{b}{h} \to 0$ in Equation (8) yields $c = a$. As in the second arc length case, this bread slice is rectangular and so we expect a corner-to-corner cut to divide the slice in half.

Case 2. Taking $\frac{b}{h} \to 1$ in Equation (8), we have $c = \frac{4}{5}a$. Using *Mathematica* [2], we solve Equation (7) numerically to obtain the solution $c = 0.767132\,a$ in this case, which is within 5 percent of our approximate solution. Alternately, Newton's method can be used to determine this root of Equation (7).

Conclusion. These two bisection problems present material involving modeling, arc length, and area calculations suited for a first-year calculus course. Instructors looking to avoid parameterizing the ellipse could discuss only the area problem and the semi-circular case of the arc length problem. It is also interesting to note that employing the usual approximation for the perimeter of a semi-ellipse, $\pi\sqrt{(a^2+b^2)/2}$, does not result in a simpler problem. Students and instructors looking to extend this problem could model the bread slices with curves other than ellipses.

References

1. G. B. Thomas, Jr. and R. L. Finney, *Calculus and Analytic Geometry*, 9th ed., Addison-Wesley, 1996.
2. S. Wolfram, *The Mathematica Book*, 3rd ed., Wolfram Media, 1996.

Disks and Shells Revisited

Walter Carlip

It is a common practice in calculus courses to use the definite integral to *define* the area between the graph of a function and the x-axis (see, e.g., [2, p. 252], [3, p. 221], and [4, p. 238]). Soon after, the student is taught two methods to calculate volumes of solids of revolution—the disk method and the shell method—usually with no mention of how *volume* is defined. Most calculus books follow the introduction of disks and shells with several examples in which it is shown that both methods of calculating the volume yield the same answer. The alert student is sure to wonder whether this is always the case.

The equivalence of the disk and shell methods was proven in [1] using integration by parts. We present here a different approach, one that uses only elementary ideas and illustrates an important proof technique.

THEOREM. *Let $f(x)$ be a continuous, invertible function on the interval $[a, b]$, where $a \geq 0$. Suppose the region bounded by $y = f(b)$, $x = a$, and the graph of $f(x)$ is rotated about the y-axis. Then the values of the volume of the resulting solid obtained by the disk and shell methods are equal. That is,*

$$\int_{f(a)}^{f(b)} \left(\pi [f^{-1}(y)]^2 - \pi a^2 \right) dy = \int_a^b 2\pi x [f(b) - f(x)] \, dx.$$

The Ingredients. Early in most calculus curricula, Rolle's Theorem is used to prove the following principle, which is then applied repeatedly.

PRINCIPLE. *If $f(x)$ and $g(x)$ are two functions that satisfy:*
(a) *$f(x)$ and $g(x)$ are differentiable on an interval $[a, b]$,*
(b) *$f'(x) = g'(x)$ for all $x \in [a, b]$, and*
(c) *$f(c) = g(c)$ for one point $c \in [a, b]$,*
then $f(x) = g(x)$ for all $x \in [a, b]$.

Informally, this says that two functions are equal if they are equal at *one* point and have identical derivatives. Although this principle is surprisingly simple, and easily absorbed by students, it has numerous applications and reappears often. Emphasizing this principle helps students see the similarity between proofs that otherwise seem unrelated.

The other ingredients of the proof of our theorem are the Fundamental Theorem of Calculus, the chain rule, and the product rule.

Proof. Let $t \in [a, b]$. We define two functions of t as follows:

$$V(t) = \int_{f(a)}^{f(t)} \left(\pi[f^{-1}(y)]^2 - \pi a^2\right) dy$$

and

$$W(t) = \int_{a}^{t} 2\pi x [f(t) - f(x)] \, dx.$$

We need to prove that $V(b) = W(b)$, as these are the two volumes in the theorem. It is easy to show that $V(t) = W(t)$ for all $t \in [a, b]$ by applying the principle given above.

First simplify $W(t)$:

$$W(t) = 2\pi f(t) \int_{a}^{t} x \, dx - 2\pi \int_{a}^{t} x f(x) \, dx.$$

Now, by the Fundamental Theorem, both $V(t)$ and $W(t)$ are differentiable on the interval $[a, b]$. Furthermore, $V'(t) = (\pi[f^{-1}(f(t))]^2 - \pi a^2) f'(t) = \pi(t^2 - a^2) f'(t)$, and

$$W'(t) = 2\pi f'(t) \int_{a}^{t} x \, dx + [2\pi f(t) t - 2\pi t f(t)]$$

$$= 2\pi f'(t) \left[\frac{t^2 - a^2}{2}\right] = \pi(t^2 - a^2) f'(t).$$

Thus, $W'(t) = V'(t)$ for all $t \in [a, b]$. It remains only to observe that $V(a) = W(a) = 0$, by a fundamental property of integrals. □

REFERENCES

1. Charles A. Cable, The disk and shell method, this MONTHLY, 91 (1984) 139.
2. John B. Fraleigh, Calculus of a Single Variable, Addison-Wesley, Reading, Mass., 1985.
3. Al Shenk, Calculus and Analytic Geometry, Goodyear Publ., Santa Monica, Calif., 1979.
4. Michael Spivak, Calculus, Publish or Perish Inc., Berkeley, Calif., 1980.

Disks, Shells, and Integrals of Inverse Functions

Eric Key

Most calculus students (and many calculus textbooks) take it for granted that the shell and disk methods for computing the volume of a solid of revolution must always give the same result. There are really two problems here: to show that the two methods agree, and that the result is the volume of the solid.

Once one has defined volume to be the result of integrating the constant function 1 over solid regions, it is easy to show that the shell method gives the volume, by using cylindrical coordinates. It remains to be shown that the disk method gives the same result. Several elementary proofs that the two methods agree have been published. In 1955, Parker [4] indicated a proof based on a simple relationship between the integrals of a function and its inverse. Later, Cable [1] gave a short proof using integration by parts. Both of these authors assumed that the function whose graph is revolved to form the boundary of the solid is monotone and has a continuous derivative. Recently Carlip [2] gave a proof that the disk and shell methods agree, assuming only continuity and monotonicity, based on the fundamental theorem of calculus and the principle that two functions with the same derivative that agree at a point are equal. In this note we show that Parker's original proof actually requires only these weaker hypotheses too.

Referring to Figure 1, we would expect that

$$\int_a^b h(x)\,dx + \int_{h(a)}^{h(b)} h^{-1}(y)\,dy = bh(b) - ah(a).$$

This relationship is the basis for a geometric interpretation of integration by parts, that occurs in Courant [3] and many other calculus texts. We sketch the formal proof below. I tell my students this is a "happy theorem" because we are able to formally prove a result that had better be true if our definition of integral successfully captures the notion of area.

Theorem 1. *Suppose that* $h: [a,b] \to [h(a), h(b)]$ *is monotone and continuous. Then* $\int_{h(a)}^{h(b)} h^{-1}(y)\,dy = bh(b) - ah(a) - \int_a^b h(x)\,dx.$

Disks, Shells, and Integrals of inverse Functions

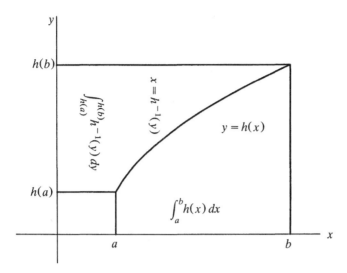

Figure 1

Proof. Observe that the function h gives a one-to-one correspondence between partitions of $[a, b]$ and partitions of $[h(a), h(b)]$. Furthermore, if h is increasing, every upper sum for $I_1 = \int_a^b h(x)\, dx$ corresponds to a lower sum for $I_2 = \int_{h(a)}^{h(b)} h^{-1}(y)\, dy$, and the sum of the two is $bh(b) - ah(a)$ (see Figure 2). Hence the supremum over all partitions of $[h(a), h(b)]$ of the lower sums for I_2 must be $bh(b) - ah(a)$ minus the infimum of the upper sums for I_1 over all partitions of $[a, b]$. Thus it follows from the definition of the integral that $I_2 = bh(b) - ah(a) - I_1$. A similar argument applies if h is decreasing. (In this case, of course, the image of $[a, b]$ under h should properly be written $[h(b), h(a)]$. But the integral formula is identical in either case.)

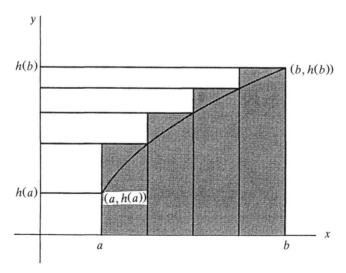

Figure 2
Riemann sums for $\int_a^b h(x)\, dx$ and $\int_{h(a)}^{h(b)} h^{-1}(y)\, dy$.

As an example, we apply this theorem to evaluate $\int_0^1 \arcsin(y)\,dy$. (Note that the usual approach via integration by parts leads to an improper integral.)

$$\int_0^1 \arcsin(y)\,dy = \frac{\pi}{2} - \int_0^{\pi/2} \sin(x)\,dx = \frac{\pi}{2} - 1$$

We now show that the disk and shell methods agree. The following theorem covers a typical case.

Theorem 2. *Suppose that $f: [a,b] \to [f(a), f(b)]$ is continuous and increasing. Let R be the region bounded by the x-axis, the lines $x = a$, $x = b$, and the graph of $y = f(x)$. Let S be the solid obtained by revolving R around the x-axis. Then the shell and disk methods for computing the volume of S give the same result.*

Proof. The shell method gives (make a sketch!)

$$VBS = 2\pi \int_0^{f(a)} y[b-a]\,dy + 2\pi \int_{f(a)}^{f(b)} y\big[b - f^{-1}(y)\big]\,dy$$

$$= \pi(b-a)f(a)^2 + \pi \left[b\big(f(b)^2 - f(a)^2\big) - \int_{f(a)}^{f(b)} 2yf^{-1}(y)\,dy \right]$$

$$= \pi \left[bf(b)^2 - af(a)^2 - \int_{f(a)}^{f(b)} 2yf^{-1}(y)\,dy \right]$$

The disk method says the volume is given by

$$VBD = \pi \int_a^b [f(x)]^2\,dx.$$

We apply Theorem 1 to the function $h: [a,b] \to [f(a)^2, f(b)^2]$ defined by $h(x) = [f(x)]^2$. We see $h^{-1}(y) = f^{-1}(\sqrt{y})$ and then make the substitution $t^2 = y$:

$$\int_a^b [f(x)]^2\,dx = bf(b)^2 - af(a)^2 - \int_{f(a)^2}^{f(b)^2} f^{-1}(\sqrt{y})\,dy$$

$$= bf(b)^2 - af(a)^2 - \int_{f(a)}^{f(b)} f^{-1}(t) 2t\,dt.$$

Thus $VBD = VBS$.

Parker [4] shows how Theorem 1 can also be used to integrate any positive power of a monotone function by computing corresponding moments of its inverse function.

References

1. Charles A. Cable, The disk and shell method, *American Mathematical Monthly* 91 (1984) 139.
2. Walter Carlip, Disks and shells revisited, *American Mathematical Monthly* 98 (1991) 154–156.
3. R. Courant, *Differential and Integral Calculus*, Vol. 1, 2nd ed., Interscience, 1957, p. 219.
4. F. D. Parker, Integrals of inverse functions, *American Mathematical Monthly* 62 (1955) 439–440. Reprinted in *A Century of Calculus*, Part I, Mathematical Association of America, 1969, p. 330.

Characterizing Power Functions by Volumes of Revolution

Bettina Richmond and Tom Richmond

Given a nonnegative increasing function f defined for $x > 0$, let $V(r)$ denote the volume of the solid obtained by revolving about the y-axis the first quadrant region under the curve $y = f(x)$ for $0 \le x \le r$, and let $C(r) = \pi r^2 f(r)$ be the volume of the corresponding right circular cylinder with radius r and height $f(r)$, as shown in Figure 1.

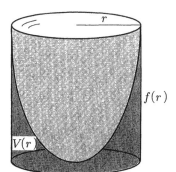

Figure 1. Inner cavity has volume $C(r) - V(r)$.

When $f(x) = kx$ the inner cavity is a cone; thus $V(r)/C(r) = \frac{2}{3}$. If $f(x) = kx^2$ the cavity is a paraboloid with volume equal to $V(r)$, so $V(r)/C(r) = \frac{1}{2}$. It is easy to show that if $f(x) = kx^a$, with $a, k > 0$, this ratio is constant;

$$\frac{V(r)}{C(r)} = \frac{2}{a+2}.$$

We were surprised to discover a sort of converse: The only twice differentiable increasing functions for which the ratio $V(r)/C(r)$ is constant are power functions $f(x) = kx^a$, with $a, k > 0$. Showing that this geometric property characterizes the two-parameter family of power functions provides a nice application of elementary differential equations.

First, if $f(x) = kx^a$, with $a, k > 0$, then using the method of cylindrical shells, we see that

$$V(r) = 2\pi \int_0^r x f(x)\, dx = 2\pi k \int_0^r x^{a+1}\, dx$$
$$= \frac{2\pi k r^{a+2}}{a+2}.$$

Hence,
$$\frac{V(r)}{C(r)} = \frac{2\pi k r^{a+2}}{a+2} \frac{1}{\pi r^2 k r^a} = \frac{2}{a+2}.$$

Proposition. Suppose f is a positive, strictly increasing, twice differentiable function on an interval $I = (0, b)$, and the ratio $V(r)/C(r)$ is constant for $r \in I$. Then $f(x) = kx^a$, for some positive constants k and a.

Proof. By differentiating the constant quotient $V(r)/C(r)$ we see that $C(r)V'(r) - C'(r)V(r) = 0$. Substituting $C(r) = \pi r^2 f(r)$ and applying the fundamental theorem of calculus to find $V'(r)$, we get

$$\left[\pi r^2 f(r)\right] [2\pi r f(r)] - \left[\pi r^2 f'(r) + 2\pi r f(r)\right] \left[2\pi \int_0^r x f(x)\, dx\right] = 0,$$

or

$$\int_0^r x f(x)\, dx = \frac{[rf(r)]^2}{rf'(r) + 2f(r)}. \tag{1}$$

Note that our hypotheses imply that r, $f(r)$, and $f'(r)$ are all positive on I, so the denominator on the right side of (1) is nonzero.

Differentiating (1) gives

$$rf = \frac{(rf' + 2f) 2r f (rf' + f) - (rf)^2 (rf'' + 3f')}{(rf' + 2f)^2},$$

where for simplicity we have suppressed the argument r of the functions. After a bit of algebraic manipulation, this equation becomes

$$rf'^2 - ff' - rff'' = 0 \tag{2}$$

Changing to the customary notation of differential equations, we will let x represent the independent variable (instead of r) and let $y = f(x)$ denote the dependent variable, so that (2) becomes

$$xy'^2 - yy' - xyy'' = 0. \tag{3}$$

Since x, y, and y' are all positive on I, we may divide (3) by $-xyy'$ to get

$$\frac{y''}{y'} - \frac{y'}{y} = -\frac{1}{x}. \tag{4}$$

Integrating yields

$$\ln y' - \ln y = -\ln x + c,$$

or
$$\ln \frac{y'}{y} = c - \ln x.$$

Applying the exponential function to both sides then gives
$$\frac{y'}{y} = \frac{a}{x},$$
where $a = e^c$ is a positive constant.

Integrating again, we conclude that $\ln y = a \ln x + K$, hence $y = kx^a$, where $k = e^K$ is a positive constant.

Gabriel's Wedding Cake

Julian F. Fleron

We obtain the solid which nowadays is commonly, although perhaps inappropriately, known as Gabriel's horn by revolving the hyperbola $y = 1/x$ about the line $y = 0$, as shown in Fig. 1. (See, e.g., [2], [5].) This name comes from the archangel Gabriel who, the Bible tells us, used a horn to announce news that was sometimes heartening (e.g. the birth of Christ in Luke 1) and sometimes fatalistic (e.g. Armageddon in Revelation 8-11).

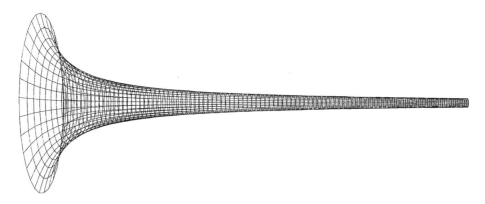

Figure 1. Gabriel's Horn.

This object and some of its remarkable properties were first discovered in 1641 by Evangelista Torricelli. At this time Torricelli was a little known mathematician and physicist who was the successor to Galileo at Florence. He would later go on to invent the barometer and make many other important contributions to mathematics and physics. Torricelli communicated his discovery to Bonaventura Cavalieri and showed how he had computed its volume using Cavalieri's principle for indivisibles. Remarkably, this volume is finite! This result propelled Torricelli into the mathematical spotlight, gave rise to many related paradoxes [3], and sparked an extensive philosophical controversy that included Thomas Hobbes, John Locke, Isaac Barrow and others [4].

This solid is a favorite in many calculus classes because its volume can be readily computed via the method of disks:

$$V = \int_1^\infty \pi (f(x))^2 \, dx = \int_1^\infty \pi \frac{1}{x^2} \, dx = \lim_{n \to \infty} \left(\pi \frac{-1}{x} \Big|_1^n \right) = \pi.$$

The seeming paradox of an infinite solid with a finite volume becomes even more striking when one considers its surface area. The standard method for computing areas of surfaces of revolution gives

$$S = \int_1^\infty 2\pi f(x) \sqrt{1 + (f'(x))^2} \, dx = \int_1^\infty 2\pi \left(\frac{1}{x}\right) \sqrt{1 + \left(\frac{-1}{x^2}\right)^2} \, dx$$

$$= \int_1^\infty 2\pi \frac{\sqrt{x^4 + 1}}{x^3} \, dx.$$

This last integral cannot be evaluated readily, although with the aid of a computer algebra system we find

$$\int 2\pi \frac{\sqrt{x^4 + 1}}{x^3} \, dx = \pi \left[\ln\left(\sqrt{x^4 + 1} + x^2\right) - \frac{\sqrt{x^4 + 1}}{x^2} \right] + C.$$

In lieu of this, typically we estimate $\sqrt{x^4 + 1}/x^3 \geq \sqrt{x^4}/x^3 = 1/x$ so

$$S = \int_1^\infty 2\pi \frac{\sqrt{x^4 + 1}}{x^3} \, dx \geq 2\pi \int_1^\infty \frac{dx}{x} = \lim_{n \to \infty} \left(|\ln(x)|_1^n \right) = \infty.$$

Hence, Gabriel's horn is an infinite solid with finite volume but infinite surface area!

Although Gabriel's horn is an engaging and appropriate example for second semester calculus, analysis of its remarkable features is complicated by two factors. First, many of the new calculus curricula do not include areas of surfaces of revolution. Second, the beauty of the paradox is often obscured by an integral estimate that most students find spurious at best.

In an effort to alleviate these factors, as well as to find an example accessible to less advanced students, we can use a discrete analogue of Gabriel's horn to illustrate the same paradox. To construct it we can replace the function $y = 1/x$ with a step function. Let

$$f(x) = \begin{cases} 1 \text{ for } 1 \leq x < 2 \\ \dfrac{1}{2} \text{ for } 2 \leq x < 3 \\ \cdots \\ \dfrac{1}{n} \text{ for } n \leq x < n+1 \\ \cdots \end{cases}$$

Revolving the graph of f about the line $y = 0$ we obtain the solid of revolution shown in Fig. 2. Notice that, when stood on end, it appears to be a cake with infinitely many layers.

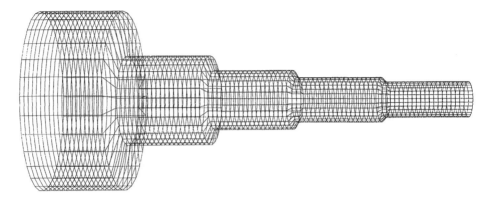

Figure 2. Gabriel's Wedding Cake.

As each layer is simply a cylinder, the volume and surface area of the solid can be readily computed. The nth layer has volume $\pi(1/n)^2$, so the total volume of the cake is

$$V = \sum_{n=1}^{\infty} \pi \left(\frac{1}{n}\right)^2 = \pi \sum_{n=1}^{\infty} \frac{1}{n^2}.$$

This series converges. Calculus students will recognize the series as a p-series with $p = 2$. Less advanced students can see that the series converges by comparison:

$$\sum_{n=1}^{\infty} \frac{1}{n^2} = 1 + \frac{1}{2^2} + \frac{1}{3^2} + \frac{1}{4^2} + \frac{1}{5^2} + \frac{1}{6^2} + \frac{1}{7^2} + \cdots$$

$$\leq 1 + \frac{1}{2}\left(\frac{1}{2} + \frac{1}{2}\right) + \frac{1}{4}\left(\frac{1}{4} + \frac{1}{4} + \frac{1}{4} + \frac{1}{4}\right) + \cdots$$

$$= 1 + \frac{1}{2} + \frac{1}{4} + \cdots = \frac{1}{1 - \frac{1}{2}} = 2.$$

Using Euler's remarkable result that the sum of the series is $\sum_{n=1}^{\infty}(1/n^2) = \pi^2/6$ [1], one can even obtain an exact result: $V = \pi^3/6$.

The surface area is formed by the annular tops and the lateral sides of each layer. The surface area of the nth annular top is $\pi(1/n)^2 - \pi(1/n+1)^2$, so the total area of the annular tops is given by the telescoping series

$$A_T = \sum_{n=1}^{\infty} \left[\pi\left(\frac{1}{n}\right)^2 - \pi\left(\frac{1}{n+1}\right)^2\right] = \pi.$$

Notice this result is obvious if one "collapses" the layers since the resulting top layer will be a complete disk of radius one. The surface area of the nth lateral side is $2\pi(1/n)(1)$, so the total lateral surface area is

$$A_L = \sum_{n=1}^{\infty} 2\pi \left(\frac{1}{n}\right)(1) = 2\pi \sum_{n=1}^{\infty} \frac{1}{n}.$$

This is the harmonic series, among the most important of all the infinite series, which diverges.

Thus, this solid illustrates essentially the same paradox as Gabriel's horn: an infinite solid with finite volume and infinite surface area. In other words: a cake you can eat, but cannot frost.

Regarding a name for this new solid, Gabriel's wedding cake seems appropriate for physical and genealogical reasons. In addition, it seems a bit refreshing since weddings are so unabashedly joyous, and the connotations of the horn have often imposed a heavy burden on Gabriel.

References

1. William Dunham, *Journey Through Genius*, John Wiley & Sons, 1990.
2. P. Gillett, *Calculus and Analytic Geometry*, 2nd ed., D. C. Heath, 1984.
3. Jan A. van Maanen, Alluvial deposits, conic sections, and improper glasses, or history of mathematics applied in the classroom, in F. Swetz, J. Fauvel, O. Bekken, B. Johansson, and V. Katz, eds., *Learn from the Masters*, Mathematical Association of America, 1995.
4. Paolo Mancosu and Ezio Vailati, Torricelli's infinitely long solid and its philosophical reception in the seventeenth century, *Isis*, 82:311 (1991) 50–70.
5. D. W. Varberg and E. J. Purcell, *Calculus*, 7th ed., Prentice Hall, 1997.

Can You Paint a Can of Paint?

Robert M. Gethner

The paradox of "Gabriel's horn" is a favorite topic of many calculus teachers. (See, for example, [1, p. 402].) In this note, I offer two ways to resolve the paradox.

The "horn" is the surface S gotten by revolving the curve $y = 1/x$ for $x \geq 1$ about the x-axis. The surface has infinite area, but the volume of the 3-dimensional region R inside it is finite. Consider how surprising that is: a finite volume of paint is sufficient to fill R, and then every point of S will be in contact with paint—yet no quantity of paint, however large, will be enough to cover S with paint!

Certainly S has infinite area while R has finite volume—the calculations that lead to those results are not in doubt—so any resolution of the paradox requires us to question the interpretation of the calculations. I will argue that the paradox arises when we make erroneous assumptions about the relationships between area and paint, and between area and volume. Once those relationships are clarified, the apparent contradiction dissolves.

(A) *The area of a surface is a measure of the amount of paint needed to paint the surface.*

But paint exists in 3-space. If we are planning to paint a room, we might ask the paint-store clerk for enough paint to cover 1000 square feet, but we wouldn't ask for 1000 square feet of paint. Let's replace (A), then, with the following assumption, which captures more accurately the way real paint behaves.

(B) *There is a minimum thickness t such that, if we cover a surface with a coat of paint having thickness at least t, then we will be unable to see the surface behind the paint.*

"Painting a surface" thus means using a finite volume of paint to cover the surface with a coat of thickness at least t. Under this assumption, it is impossible to paint our surface S since the coat of paint would essentially have to be an infinitely long solid tube, the volume of which would be infinite. (The tube would be hollow, but the hollow space would get extremely narrow.)

More precisely, suppose that the paint covering S occupies the hollow 3-dimensional tube obtained by revolving the region between $y = 1/x$ and $y = 1/x + h(x)$ for $x \geq 1$ about the x-axis. Then by our assumption, there's a constant c such that $h(x) \geq c$ for all $x \geq 1$. (A subtle point: $c \neq t$ except where the graph of h is horizontal.) So the volume of paint is

$$\pi \int_1^\infty \left\{ \left[\frac{1}{x} + h(x) \right]^2 - \frac{1}{x^2} \right\} dx > \pi \int_1^\infty \left(\frac{2c}{x} + c^2 \right) dx = \infty,$$

and so S is unpaintable, as claimed.

But now when we fill the space inside S with paint, there is no contradiction. Even though every point of S is in contact with paint, we do not consider S to have been painted, since the thickness of paint inside S approaches 0. For large x, we don't even see the paint inside S—the space filled by the paint is too narrow to be visible.

But that is not the end of this chapter of the story. Under Assumption (B), *no unbounded surface of revolution can be painted, even if it has finite area!* (Nice examples of functions that revolve to such surfaces are $y = 1/x^2$ and $y = e^{-x}$.) The argument for this unpaintability is simply the one given above, but with $1/x$ replaced by the given function.

That conclusion may seem startling: isn't the volume of paint required simply t times the surface area? The answer is "yes" if the surface is flat, as most walls are, but "no" if the surface is curved, as are all surfaces of revolution. To take a simple example, let's consider a right circular cylinder of radius r and height h. Then the lateral surface area A is $2\pi rh$. But paint to a thickness t forms a shell that has volume $V = \pi h t (2r + t) = At(1 + t/(2r))$. Hence, $V \approx At$ only when t is small when compared to r. This makes sense since then we can estimate V well by cutting the lateral surface along a line parallel to the cylinder's axis, unrolling the surface, and treating the coat of paint as a rectangular prism; not so if t is not small compared to r. Thus the formula At provides a rather poor estimate of the volume of paint even for *bounded* surfaces that are highly curved. (See [2] for a more detailed discussion of the relationship between volume and surface area, along with a geometrically appealing derivation of the standard integral formula for the surface area obtained by revolving the graph of f on $[a, b]$ about the x-axis.)

Second resolution. I formulated Assumption (B) to reflect our ordinary experience that several coats of paint may be needed in order to cover a wall. But you may object that your paint is no ordinary paint—yours is a special, *mathematical* paint that's an opaque, continuous fluid. Any thickness of such paint, however small, is sufficient to hide the wall. Your paint, then, satisfies the following assumption:

(C) *Any thickness of paint will cover a surface.*

"Painting a surface" now means covering it with paint of positive thickness, the thickness possibly varying with the point being covered. Under this assumption, a sufficiently smooth surface of revolution will *always* be paintable (even if the volume inside is infinite). To prove that, we let $g(x)$ be a continuous, positive, decreasing function that approaches 0 as $x \to \infty$, and we consider the hollow tube obtained by revolving the region between the curves $g(x)$ and $g(x)(1 + 1/x^2)$ for $x \geq 1$ about the x-axis. Then we see from elementary calculus that this tube has finite volume, and so to paint the surface of revolution of $g(x)$ we simply fill our tube with paint. There is no paradox. We normally think of the boundary surface of a solid as being "smaller" than the solid itself, so there is nothing surprising about the case where our container has infinite volume yet can be covered with a finite volume of paint.

Mathematical models. Is there any reason to prefer one of our assumptions over the other? Or is there perhaps some other assumption more compelling than either of ours? Why do we need *any* assumption?

My answer to the third question is that the Gabriel's horn paradox is essentially an exercise in mathematical modeling; in making a model we first make assumptions. After all, the paradox offends our sensibilities not because of any internal logical inconsistency but because the conclusion contradicts our everyday experience. If we regard Assumptions (A), (B), and (C) as bases for models of the process of painting a surface, then we can view the paradox as simply pointing out that a certain prediction that follows from Assumption (A) does not agree with experience. We therefore discard (A) in the hope of replacing it by another, more realistic one.

Of the two candidates (B) and (C) offered here, (B) seems to me the more realistic. But both are unrealistic in that they assume that paint is continuous. Since real paint is composed of molecules, and since molecules have small but finite size, we cannot in reality fill Gabriel's horn with paint—and again there is no paradox.

References

1. D. Hughes Hallett, et al., *Calculus: Single Variable*, 3rd ed., Wiley, 2002.
2. K. A. Struss, Exploring the volume-surface area relationship, *College Math. J.* **21** (1990) 40–43.

A Paradoxical Paint Pail

Mark Lynch

We are all familiar with Gabriel's horn from calculus [1], [2]. (See also the preceding Capsule.) It is the object obtained by rotating the graph of $y = 1/x$ for $x \geq 1$ around the x-axis. We tell our students that since it has finite volume and infinite surface area, it can be filled with paint but cannot be painted. However, sometimes clever physics students point out that since Gabriel's horn is unbounded, it would take an infinite amount of time to fill it with paint and so we really can't fill it with paint either. In this note, we answer the critics' charge by constructing a bounded paint pail (hence, it can be filled in finite time) with infinite surface area.

Define a function $f(x)$ on $[0, 1]$ as follows: $f(x) = 1$ if $x = 0$ or $x = 1/n$ for n a positive integer, and in the interval

$$\left(\frac{1}{n+1}, \frac{1}{n}\right),$$

let f be given by a "spike" of length $1/n$. See the figure. Finding expressions for $f(x)$ on the intervals is a good exercise for students.

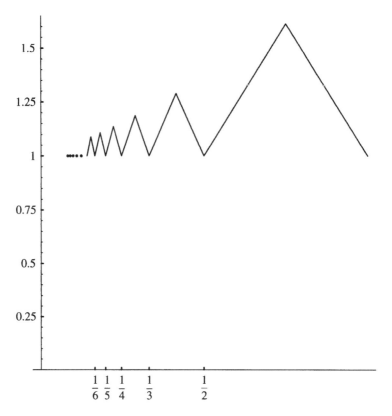

Figure 1. The spike function

1. J. F. Fleron, Gabriel's wedding cake, *College Math. J.* **30** (1999) 35–38.
2. M. Henrikson, *Infinite Acres*, MAA, 1965.

Dipsticks for Cylindrical Storage Tanks—Exact and Approximate

Pam Littleton and David A. Sanchez

A problem often found in pre-calculus and calculus textbooks, of importance in agriculture, is the calibration of a dipstick to measure the volume of liquid in a cylindrical tank lying on its side. The dipstick is inserted into the top of the tank, as in Figure 1.

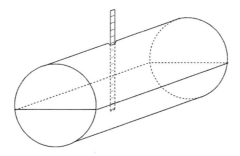

Figure 1.

This gives rise to two different problems:

Problem 1. Determine the volume corresponding to uniformly spaced calibration marks on the dipstick.

Problem 2. Determine the calibration marks on the dipstick corresponding to uniformly spaced volume levels.

In the first case the dipstick might have volume levels marked every six inches whereas in the second case the marks might represent fifty gallon increases in volume.

Let r be the radius of the cylinder and ℓ its length, and suppose that the tank is no more than half full, so the height h of the liquid is at most r. See Figure 2 for a cross section and let A_h denote the area of the shaded sector.

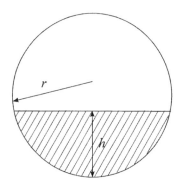

Figure 2.

The volume of liquid in the tank is then $V = k\ell A_h$ where k is the conversion factor from volume in cubic units to liquid volume, e.g., 7.48 gal./ft^3.

If the tank is more than half full we can invert the above picture and the shaded portion then represents the empty part of the tank, so $V = k\ell(\pi r^2 - A_h)$. Thus we need only consider the case $h \leq r$.

Two quantities are of significance in the calculation of the area A_h: the central angle θ in radians subtended by the chord representing the level of the liquid, and the height h (see Figure 3).

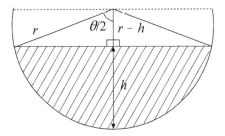

Figure 3.

The area of the pie-shaped circular sector is $\frac{1}{2}r^2\theta$, and the two right triangles have two possible interpretations (Figure 4):

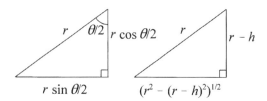

Figure 4.

The area of the isosceles triangle in terms of θ is

$$(r\sin\theta/2)(r\cos\theta/2) = \frac{1}{2}r^2\sin\theta$$

or, in terms of h,

$$(r-h)[r^2-(r-h)^2]^{1/2} = r^2\left(1-\frac{h}{r}\right)\left[1-\left(1-\frac{h}{r}\right)^2\right]^{1/2}$$

$$= r^2\left(1-\frac{h}{r}\right)\left[2\frac{h}{r}-\left(\frac{h}{r}\right)^2\right]^{1/2}.$$

Subtracting this area from that of the circular sector gives two expressions for A_h, one in terms of θ and one in terms of h:

$$A_h = \frac{1}{2}r^2\theta - \frac{1}{2}r^2\sin\theta = r^2\left(\frac{\theta-\sin\theta}{2}\right) = r^2 A(\theta), \quad 0 \leq \theta \leq \pi, \qquad (1)$$

where $\theta = 2\cos^{-1}\left(\frac{r-h}{r}\right)$, or $h = r - r\cos\theta/2$, or

$$A_h = \frac{1}{2}r^2\theta - r^2\left(1-\frac{h}{r}\right)\left[2\frac{h}{r}-\left(\frac{h}{r}\right)^2\right]^{1/2}$$

$$= r^2\left[\cos^{-1}\left(1-\frac{h}{r}\right) - \left(1-\frac{h}{r}\right)\left(2\frac{h}{r}-\left(\frac{h}{r}\right)^2\right)^{1/2}\right] = r^2 B(h), \quad 0 \leq h \leq r.$$

(2)

The last formula is simplified by the rescaling $\mu = \frac{h}{r}$, which will be useful in later calculations, and becomes

$$A_h = r^2[\cos^{-1}(1-\mu) - (1-\mu)(2\mu-\mu^2)^{1/2}] = r^2 B(\mu), \quad 0 \leq \mu \leq 1. \qquad (3)$$

Using (1) and (3) we have the following solutions to the given problems:

Problem 1. a) Given a height h_0 for a dipstick mark, find $\theta_0 = 2\cos^{-1}\left(\frac{r-h_0}{r}\right)$ and compute $A_{h_0} = r^2 A(\theta_0)$. Then $V_0 = k\ell A_{h_0}$ is the volume level associated with h_0, or

b) Given a height h_0 for a dipstick mark, let $\mu_0 = \frac{h_0}{r}$ and compute $A_{h_0} = r^2 B(\mu_0)$. Then $V_0 = k\ell A_{h_0}$ is the volume level associated with h_0.

Problem 2. a) Given a volume V_0 solve $V_0 = k\ell r^2 A(\theta)$ for $\theta = \theta_0$, $0 \leq \theta_0 \leq \pi$. Then $h_0 = r - r\cos\theta_0/2$ is the height of the dipstick mark associated with volume V_0, or

b) Given a volume V_0 solve $V_0 = k\ell r^2 B(\mu)$ for $\mu = \mu_0, 0 \le \mu_0 \le 1$. Then $h_0 = r\mu_0$ is the height of the dipstick mark associated with volume V_0.

Example. Let $r = 2$ ft. and $\ell = 10$ ft., so the total liquid volume of tank is $V = (7.48)(10)\pi(2)^2 \cong 940$ gallons.

Solution (Problem 1). If we mark the dipstick every six inches then marks at height $h_n = n/2$ ft., $n = 1, \ldots, 4$, would correspond to $\mu_n = n/4$, and volumes

$$V_n = (7.48)(10)(2)^2 \left(\frac{\theta_n - \sin \theta_n}{2} \right), \theta_n = 2\cos^{-1}\left(1 - \frac{n}{4}\right),$$

or

$$V_n = (7.48)(10)(2)^2 \left[\cos^{-1}\left(1 - \frac{n}{4}\right) - \left(1 - \frac{n}{4}\right)\left(\frac{n}{2} - \frac{n^2}{16}\right)^{1/2} \right], \quad n = 1, 2, \ldots, 4.$$

The first four marks on the dipstick would be approximately

$$68, \quad 184, \quad 322, \quad \text{and} \quad 470 \text{ gallons.}$$

Since $h_4 = r$ it is a simple arithmetical calculation to compute the remaining marks; for instance the next mark would be $470 + (470 - 322) = 618$ gal.

Solution (Problem 2). If we mark the dipstick at 100 gallon intervals, then $V_n = 100n$, $n = 1, 2, \ldots, 9$ and to obtain the first four marks we must solve

$$\frac{100n}{(7.48)(10)(2)^2} = \frac{\theta_n - \sin \theta_n}{2}, h_n = 2 - 2\cos\theta_n/2$$

or

$$\frac{100n}{(7.48)(10)(2)^2} = \cos^{-1}(1 - \mu_n) - (1 - \mu_n)(2\mu_n - \mu_n^2)^{1/2}, h_n = 2\mu_n, \quad n = 1, \ldots, 4.$$

The first four one hundred gallon marks would be

$$0.6532, \quad 1.0620, \quad 1.4237, \quad \text{and} \quad 1.7655 \text{ feet.}$$

Since the next mark will correspond to a volume of 440 gallons in the empty part of the upper half of the tank, the remaining marks will have heights $4 - h_n$ where h_n corresponds to volumes $440 - 100n$, $n = 0, 1, \ldots, 4$.

While the transcendental equations involving θ in Problems 1 and 2, parts a) are much easier to solve than parts b), the natural parameter is the height h, or the normalized height $\mu = h/r$. The simplest but often impractical solutions to the problems are:

Problem 1. Make the desired calibration marks on the dipstick, insert the stick, and begin filling the tank until the level reaches the first mark. Measure the volume poured and mark the stick; continue filling until the second mark is reached, etc.

Problem 2. Begin filling the tank until the first volume level is reached, insert the stick and mark it. Continue filling until the second volume level is reached, insert the stick, etc.

Readers satisfied with these solutions should probably not have started reading this paper and should stop now. What is needed is an accurate but simpler expression that approximates $B(\mu) = \cos^{-1}(1-\mu) - (1-\mu)(2\mu - \mu^2)^{1/2}$. Its second term is

$$(1-\mu)\sqrt{2\mu}(1-\mu/2)^{1/2} \cong (1-\mu)\sqrt{2\mu}(1-\mu/4) = \sqrt{2\mu}\left(1 - \frac{5}{4}\mu + \frac{1}{4}\mu^2\right). \tag{4}$$

The maximum error in replacing $(1-\mu/2)^{1/2}$ by $1 - \frac{1}{2}\mu/2$ is approximately 0.043 for $0 \leq \mu \leq 1$.

A difficulty with the first part of the expression is that $\cos^{-1}(1-\mu)$ has an infinite derivative at $\mu = 0$ so a Taylor series expansion fails. We use instead an integral representation:

$$\cos^{-1}(1-\mu)$$

$$= -\int_1^{1-\mu} \frac{1}{\sqrt{1-x^2}} dx = \int_0^\mu \frac{1}{\sqrt{1-(1-t)^2}} dt$$

$$= \int_0^\mu \frac{1}{\sqrt{2t - t^2}} dt = \frac{1}{\sqrt{2}} \int_0^\mu \frac{1}{\sqrt{t}\sqrt{1 - t/2}} dt \cong \frac{1}{\sqrt{2}} \int_0^\mu \frac{1}{\sqrt{t}(1 - t/4)} dt \tag{5}$$

$$\cong \frac{1}{\sqrt{2}} \int_0^\mu \frac{1}{\sqrt{t}}(1 + t/4 + t^2/16) dt$$

$$= \frac{1}{\sqrt{2}}\left(2\mu^{1/2} + \frac{1}{6}\mu^{3/2} + \frac{1}{40}\mu^{5/2}\right) = \sqrt{2\mu}\left(1 + \frac{1}{12}\mu + \frac{1}{80}\mu^2\right).$$

Combining (4) and (5) gives the approximation

$$B(\mu) \cong \sqrt{2\mu}\left(\frac{4}{3}\mu - \frac{19}{80}\mu^2\right), \quad 0 \leq \mu \leq 1,$$

which is simpler to compute and is very accurate as the following table indicates:

μ	$B(\mu)$	$B(\mu)$ approx.	Error
0	0	0	0
0.25	0.22666	0.22521	0.00145
0.50	0.61418	0.60729	0.00689
0.75	1.07605	1.06112	0.01492
1.0	$\pi/2$	1.54974	0.02106

The obvious advantage of the approximation is that there are no transcendental expressions to evaluate and the calculation can be done with a simple calculator.

Then the approximate solution to Problem 1 is, given a height $h_0 < r$, the corresponding volume is

$$V_0 \cong k\ell r^2 \left[\sqrt{2\mu_0}\left(\frac{4}{3}\mu_0 - \frac{19}{80}\mu_0^2\right)\right], \quad \mu_0 = h_0/r.$$

For instance, in the example above with $r = 2$ ft., $\ell = 10$ ft. the approximate volume corresponding to a height of $1\frac{1}{2}$ ft. is 318 gallons compared to the actual answer of 322 gallons.

Problem 2 is a much tougher nut to crack, but the above approximation does provide a polynomial to solve instead of a transcendental equation. Given a fixed volume $V_0 < V/2$ then we must solve

$$\frac{V_0}{k\ell r^2} = B(\mu) \cong \sqrt{2\mu}\left(\frac{4}{3}\mu - \frac{19}{80}\mu^2\right) = \frac{\sqrt{2}}{240}(320\mu^{3/2} - 57\mu^{5/2})$$

for μ. Let $z = \mu^{1/2} = \left(\frac{h}{r}\right)^{1/2}$ and $M = V_0/k\ell r^2$, hence $0 < M < \pi/2$, then the previous equation becomes the polynomial equation

$$P(z) = 57z^5 - 320z^3 + \frac{240}{\sqrt{2}}M = 0$$

where $0 \leq z < 1$. A straightforward analysis of $P(z)$ and its derivative shows that there is only one root in the interval $0 < z < 1$.

For example, if a tank is 1/4 full then $V_0 = k\ell\pi r^2/4$ so $M = \frac{1}{4}\pi$ and $P(z) = 57z^5 - 320z^3 + \frac{60\pi}{\sqrt{2}}$ with root $z \cong 0.77556$. Then $h/r = z^2 \cong 0.60149$ so $h \cong 0.60149r$. The exact answer using $A(\theta)$ and solving

$$\pi/4 = \frac{\theta - \sin\theta}{2}, \quad h = r - r\cos\theta/2$$

is $h = 0.59603r$.

The approximation given above for $B\left(\frac{h}{r}\right)$ is useful and has some computational advantage. The reader or students might wish to consider Problems 1 and 2 for a spherical tank, or a cylindrical tank with an elliptical cross section lying on its side. For the spherical tank the exact expression for the volume is a cubic in h, whereas for the elliptical tank the solution is the same expression as (2) with r^2 replaced by ab and $\frac{h}{r}$ by $\frac{h}{a}$, where $2b$ and $2a$ are the major and minor axes, respectively, of the ellipse. Another application would be in solving a problem like

When $h = h_0 \leq r$ the liquid level in the tank is rising at a rate of q ft./min. How fast is the volume increasing at this moment?

Using the two exact expressions (1) and (2) for V we have

Dipsticks for Cylindrical Storage Tanks

$$\frac{dV}{dt} = k\ell r^2 A'(\theta)\frac{d\theta}{dt}, \quad \theta = 2\cos^{-1}(1 - h/r),$$

where

$$\frac{d\theta}{dt} = \frac{2}{r} \frac{1}{(2\frac{h}{r} - (\frac{h}{r})^2)^{1/2}} \frac{dh}{dt},$$

and for (2)

$$\frac{dV}{dt} = k\ell r^2 B'(\mu)\frac{d\mu}{dt} = k\ell r B'(\mu)\frac{dh}{dt}.$$

But replacing $B(\mu)$ with its approximation results in a simpler differentiation. We obtain

$$\frac{dV}{dt} \cong k\ell r \left[\sqrt{2}\left(2\left(\frac{h}{r}\right)^{1/2} - \frac{19}{32}\left(\frac{h}{r}\right)^{3/2}\right)\right]\frac{dh}{dt}.$$

For instance, in the previous example ($r = 2$ ft., $\ell = 10$ ft.) if $h_0 = 3/2$ ft. and $q = 2$ in./min. $= \frac{1}{6}$ ft./min. $= \frac{dh}{dt}$, then $\frac{h_0}{r} = 0.75$ and

$$\frac{dV}{dt} \cong (7.48)(10)(2)\left[\sqrt{2}\left(2(0.75)^{1/2} - \frac{19}{32}(0.75)^{3/2}\right)\right]\left(\frac{1}{6}\right) \cong 47.48 \text{ gal./min.}$$

The exact answer, obtained after considerably more calculation, is 48.28 gal./min.

Acknowledgments. The authors wish to express their appreciation to Dave Engelhard, Michael Kallaher, and Jack Robertson of Washington State University whose wonderful collection of agricultural related mathematics problems, *Tools of the Trade*, (COMAP, 1999) provided a source for the analysis above. They also wish to point out an entertaining discussion of the dipstick problem in the book, *Calculus Mysteries and Thrillers*, (MAA, 1998) by R. Grant Woods.

Finding Curves with Computable Arc Length

John Ferdinands

One of the standard applications of the integral in elementary calculus is to find the arc length of a curve. Having derived the formula

$$L = \int_a^b \sqrt{(f'(x))^2 + 1}\, dx \qquad (1)$$

for the length L of the graph of the differentiable function $f(x)$ for $a \leq x \leq b$, as instructors we often search for examples for which the integral has a nice closed form. Because of that awkward square root, it is apparent that for most elementary functions, the integral in (1) will not be pleasant.

Most calculus books give a few examples of functions for which the integral can be evaluated. The function $f(x) = x^{3/2}$ is a fairly standard one, but more interesting is $f(x) = \frac{x^3}{6} + \frac{1}{2x}$ ([**1**, p. 398]). In this case, $(f'(x))^2 + 1$ turns out to equal $(\frac{x^2}{2} + \frac{1}{2x^2})^2$, from which it follows that the indefinite integral $\int \sqrt{(f'(x))^2 + 1}\, dx$ is just $\frac{x^3}{6} - \frac{1}{2x} + C$. The point, of course, is that $(f'(x))^2 + 1$ is a perfect square. Other such functions in textbooks are $f(x) = \frac{x^4}{8} + \frac{1}{4x^2}$ [**1**, p. 402] and $f(x) = \frac{x^2}{8} - \ln x$ [**2**, p. 465].

While searching for more examples when I was teaching calculus recently, I happened upon an algorithm that produces functions with that "perfect square property." This algorithm is almost surely known to someone, but I have not been able to find it anywhere. Assume that $(f'(x))^2 + 1 = (s(x))^2$. Then it must be the case that $(s(x) + f'(x))(s(x) - f'(x)) = 1$. Now let $g(x) = s(x) + f'(x)$, whence $\frac{1}{g(x)} = s(x) - f'(x)$, and so $f'(x) = \frac{1}{2}\left(g(x) - \frac{1}{g(x)}\right)$. Consequently,

$$f(x) = \frac{1}{2} \int \left(g(x) - \frac{1}{g(x)}\right) dx. \qquad (2)$$

Thus, we can take $g(x)$ to be almost any function we like (as long as we have differentiability of $f(x)$ on our interval $[a, b]$). Observe that, with $f(x)$ as in (2), (1) becomes

$$L = \frac{1}{2} \int_b^a \left(g(x) + \frac{1}{g(x)}\right) dx. \qquad (3)$$

For example, if $g(x) = x^{10}$, then

$$f(x) = \frac{1}{2}\int \left(x^{10} - \frac{1}{x^{10}}\right) dx = \frac{x^{11}}{22} + \frac{1}{18x^9} \ (+C).$$

Or try $g(x) = \tan x$. Then the indefinite integral for $f(x)$ can be computed, using some trig identities, as

$$\frac{1}{2}\int (\tan x - \cot x) \, dx = \frac{1}{2}\left(-\ln\left(\frac{1}{2}\sin 2x\right)\right) + C$$

$$= -\frac{1}{2}\ln(\sin 2x) - \frac{1}{2}\ln\left(\frac{1}{2}\right) + C.$$

By ignoring the constants, we can choose $f(x) = \frac{1}{2}\ln(\sin 2x)$. However, although $f'(x)^2 + 1$ will not then equal $\left(\frac{1}{2}\left(g(x) + \frac{1}{g(x)}\right)\right)^2$, things still come out nicely:

$$\int \sqrt{(f'(x))^2 + 1} \, dx = \int \sqrt{\cot^2 2x + 1} \, dx = \int \csc 2x \, dx$$

$$= -\frac{1}{2}\ln|\csc 2x + \cot 2x| + C \quad (\text{for } 0 \leq x \leq \tfrac{\pi}{2}).$$

We invite the reader to experiment with this algorithm and discover other examples.

References

1. C. H. Edwards and D. E. Penney, *Calculus*, 6th ed., Prentice Hall, 2002.
2. G. B. Thomas, Jr., R. L. Finney, M. D. Weir, and F. R. Giordano, *Thomas' Calculus*, 10th ed., Addison-Wesley, 2001.

Arc Length and Pythagorean Triples

Courtney Moen

In this note we give an example of how a computer algebra system can offer surprises even in the context of a standard calculus topic. When introducing the formula for arc length, some natural examples are the curves C_n which are given parametrically by $x = t^n$, $y = t^{n+1}$, $0 \leq t \leq 1$, (n is a positive integer). Many students have difficulty computing even the length of C_1 by hand, so this is a natural place to use a computer algebra system. The length of C_5, for example, is

$$\frac{3431\sqrt{61}}{20736} + \frac{15625}{124416} \ln 5 - \frac{15625}{124416} \ln(-6 + \sqrt{61}).$$

As n increases, the results become increasingly unpleasant until, surprisingly, we find that the length of C_{20} is rational and equals $\frac{3649566106714513582 9027}{25798674916142804999323}$.

It is easy to see why this is so and to show that infinitely many of the lengths $L(C_n)$ are rational numbers. Using the standard formula $L = \int_a^b \sqrt{(x'(t))^2 + (y'(t))^2}\, dt$ for arc length, we have

$$L(C_n) = \int_0^1 t^{n-1} \sqrt{n^2 + (n+1)^2 t^2}\, dt.$$

After we make the successive substitutions $u = \frac{n+1}{n} t$ and $v = \sqrt{u^2 + 1}$, the integrand becomes $(v^2 - 1)^{(n-2)/2} v^2$.

We now take n to be even, say $n = 2k$. Then our integrand is just a finite sum of integer multiples of powers of v; that is,

$$(v^2 - 1)^{(n-2)/2} v^2 = \sum_{j=1}^k c_{2j} v^{2j}.$$

for some integers $c_2, c_4, \ldots, c_{2k}$. Integrating and substituting back, we find that

$$L(C_n) = \frac{n^{n+1}}{(n+1)^n} \sum_{j=1}^k \frac{c_{2j}}{2j+1} (u^2 + 1)^{(2j+1)/2} \Big|_0^{\frac{n+1}{n}},$$

and hence that

Arc Length and Pythagorean Triples

$$L(C_n) = \frac{n^{n+1}}{(n+1)^n} \sum_{j=1}^{n/2} \frac{c_{2j}}{2j+1} \left(\left(\frac{\sqrt{(n+1)^2 + n^2}}{n} \right)^{2j+1} - 1 \right).$$

This will be rational if $n^2 + (n+1)^2$ is a perfect square; that is, if n and $n+1$ are part of a Pythagorean triple. It is well known that there are infinitely many such pairs (see, for example [1, p. 164, Exer. 17]). In particular, if $(a, a+1, c)$ is a Pythagorean triple, so is $(3a + 2c + 1, 3a + 2c + 2, 4a + 3c + 2)$. Note that the parity of the first term switches, so that by using this result twice we can go from one even case to another. Consequently, not only does C_{20} have a rational length, so does C_{696}, and also infinitely many other curves C_n.

References

1. J. K. Strayer, *Elementary Number Theory*, PWS Publishing, 1994.

Mazimizing the Archlength in the Cannonball Problem

Ze-Li Dou and Susan G. Staples

Every calculus instructor is familiar with some version of the "cannonball problem" in which a cannonball is fired from ground level with an initial speed v and an initial angle of inclination θ, $0 \leq \theta \leq \pi/2$. Typical questions are:

1. For what length of time T is the cannonball in flight?
2. What is the horizontal range R covered by the cannonball?
3. For what angle θ is R maximized?

We suggest a follow-up problem, which may be used after the concept of arclength and techniques of integration are covered.

4. What angle θ maximizes the arclength of the trajectory?

We write, as usual,

$$x = vt \cos \theta \quad \text{and} \quad y = vt \sin \theta - \tfrac{1}{2}gt^2,$$

where g is the gravitational constant. Then the well-known answers to the first three questions are $T = 2v \sin \theta/g$, $R = v^2 \sin 2\theta/g$, and $\theta = \pi/4$, respectively. The arclength of the trajectory is

$$L(\theta) = \int_0^T \sqrt{\left(\frac{dx}{dt}\right)^2 + \left(\frac{dy}{dt}\right)^2}\, dt = \int_0^{2v \sin \theta/g} \sqrt{(v \cos \theta)^2 + (gt - v \sin \theta)^2}\, dt.$$

Making the substitution $u = gt - v \sin \theta$, we obtain

$$L(\theta) = \frac{1}{g} \left(\int_{-v \sin \theta}^{v \sin \theta} \sqrt{u^2 + v^2 \cos^2 \theta}\, du \right).$$

After utilizing a standard trigonometric substitution, we see that, for $0 \leq \theta < \pi/2$,

$$L(\theta) = \frac{1}{g}\left(v^2 \sin \theta + v^2 \frac{\cos^2 \theta}{2} \ln\left|\frac{1 + \sin \theta}{1 - \sin \theta}\right|\right),$$

while the case $\theta = \pi/2$, corresponding to shooting the cannonball straight up, leads to $L(\pi/2) = v^2/g$.

Maximizing the Arclength in the Cannonball Problem

To compute the critical values, we take the derivative of $L(\theta)$ and find

$$L'(\theta) = \frac{1}{g}\left(2v^2\cos\theta - v^2\cos\theta\sin\theta\ln\left|\frac{1+\sin\theta}{1-\sin\theta}\right|\right).$$

The critical value for θ in $(0, \pi/2)$ is the angle θ_0 which satisfies

$$\sin\theta_0 \ln\left|\frac{1+\sin\theta_0}{1-\sin\theta_0}\right| = 2.$$

At the endpoints of the interval $[0, \pi/2]$, we see that $L(0) = 0$ and that $L(\pi/2)$ gives a local minimum. Using a graphing calculator, or by Newton's method, one easily obtains $\theta_0 = .985514738\ldots$ radians or about 56 degrees. It is interesting to notice that this θ_0, which maximizes the arclength, lies strictly between the range-maximizing angle $\pi/4$ and the height-maximizing angle $\pi/2$.

An Example Demonstrating the Fundamental Theorem of Calculus

Bob Palais

When we introduce the Fundamental Theorem of Calculus, it is natural to begin with some simple examples that demonstrate its validity and may be verified without appealing to the theorem itself. The standard examples are usually constant functions and linear functions, for which the geometric area beneath these graphs may be easily determined without computing limits. While in some sense the example of a constant function shows what makes the theorem hold in greater generality, it may seem rather special to beginning students. Here is a function whose graph does not consist of straight lines, for which the theorem may be independently confirmed.

Let $f(x) = \sqrt{1-x^2}$, on the interval $[-1, 1]$, whose graph is the upper half of the unit circle. We will show that the conclusion of the Fundamental Theorem of Calculus holds for f, on the interval $(-1, 1)$. That is, if $A(t)$ is the area of the region $R(t)$ under the graph from $x = -1$ to $x = t$, then a nice computation will verify that $A'(t) = f(t)$.

To do this, we think of $R(t)$ as the union of a sector of the unit circle and a triangle, as shown in Figure 1. We will accept that the area of a triangle is one-half its base times its altitude, and that the area of an angular sector of a disc of radius 1 with central angle θ (in radians) is $\theta/2$. (It is worthwhile to think about how the calculus definition of these areas relates to other derivations.)

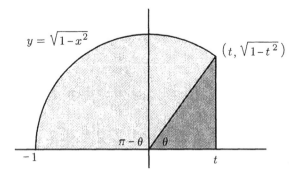

Figure 1. The region $R(t)$.

Since the angle in the triangle is θ, the angle of the shaded sector is $\pi - \theta$, so

$$A(t) = \frac{\pi - \theta}{2} + \frac{1}{2}t\sqrt{1-t^2}$$

An Example Demonstrating the Fundamental Theorem of Calculus

where θ satisfies $\cos\theta = t$ (or $\theta = \arccos(t)$). Observe that this equation for $A(t)$ is valid for both positive and negative values of t, since for $t < 0$, $A(t)$ is the difference of the areas of the sector and the triangle, and this is accounted for by the factor of t in the second term on the right side.

Taking the derivative, we get

$$A'(t) = -\frac{1}{2}\frac{d\theta}{dt} + \frac{1}{2}\left(1 \cdot \sqrt{1-t^2} + t \cdot \frac{1}{2}\frac{1}{\sqrt{1-t^2}} \cdot -2t\right).$$

We apply implicit differentiation or the inverse function rule to the defining equation for θ, which gives

$$-\sin\theta \frac{d\theta}{dt} = 1.$$

Then, using $\sin\theta = \sqrt{1-t^2}$ (from the figure or from $\sin\theta = \sqrt{1-\cos^2\theta}$), we arrive at

$$-\frac{d\theta}{dt} = \frac{1}{\sin\theta(t)} = \frac{1}{\sqrt{1-t^2}}.$$

Simplifying and combining, we find

$$A'(t) = \frac{1}{2}\left(\frac{1}{\sqrt{1-t^2}} + \frac{1-t^2}{\sqrt{1-t^2}} - \frac{t^2}{\sqrt{1-t^2}}\right) = \frac{1-t^2}{\sqrt{1-t^2}} = \sqrt{1-t^2} = f(t),$$

just as promised by the Fundamental Theorem of Calculus.

Acknowledgment. I thank Bill Bynum for several enjoyable conversations regarding integrals known to the ancient Greeks.

Barrow's Fundamental Theorem

Jack Wagner

Somewhere around the middle of the first year of calculus we present our students with the Fundamental Theorem of Calculus, clothed in the majesty of capital letters and demonstrated with the genius of three hundred years of great mathematicians. It is, however, a remarkable fact that its fundamental geometric content is accessible directly from high school plane geometry.

When Sir Isaac Newton said, toward the end of his life, "If I have seen further than others, it is because I have stood on the shoulders of giants," his mentor, Isaac Barrow, who preceded him in the Lucasian chair of mathematics must certainly have been on his mind. In 1670, Newton edited and published Barrow's *Lectiones Geometricae*. They contained, in Lecture X, the following theorem, here presented in modern terminology, which expresses the geometric content of the Fundamental Theorem of Calculus.

Referring to Figure 1, let f be a strictly increasing, positive, continuous function on $[a, b]$ and let F be the function whose value at each x in $[a, b]$ is the area under the graph of f from a to x. For convenience, we draw the graph of $-f$ instead of the graph of f. From any point, A, on the x-axis between $x = a$ and $x = b$, erect a perpendicular. Let K be its intersection with the graph of F and B be its intersection with the graph of $-f$.

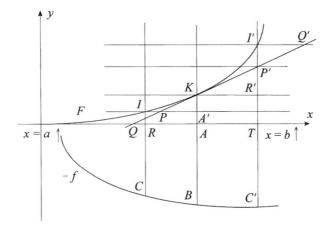

Figure 1.

Theorem. If a line is drawn from the point K to a point on the x-axis, Q, so that

$$\overline{AK}/\overline{AQ} = \overline{AB}$$

(this can always be done with ruler and compass), then this line will be tangent at K to the graph of F.

Proof. Barrow understood a tangent to be a line that intersects a curve at only one point. The proof, therefore, consists in showing that the line $\overline{QK}$ is below, or to the right of, the graph of F.

Again referring to Figure 1, $\overline{AB} = \frac{\overline{AK}}{\overline{AQ}} = \frac{\overline{A'K}}{\overline{A'P}}$ by similar triangles. $\overline{AB} \cdot \overline{A'P} = \overline{A'K}$ = area $CRAB$ since, by hypothesis, the area under the graph of f on an interval $[r, s]$ is $F(s) - F(r)$. And, area $CRAB < \overline{AR} \cdot \overline{AB}$ since f is strictly increasing. Therefore, $\overline{A'P} < \overline{AR}$, which means that the point P on the line $\overline{KQ}$ is to the right of the point I on the graph of F.

Similarly, $\overline{AB} = \frac{\overline{P'R'}}{\overline{KR'}}$ by similar triangles. $\overline{P'R'} = \overline{AB} \cdot \overline{KR'} <$ area $C'TAB = \overline{I'R'}$.

Therefore the point P' on the line $\overline{KQ'}$ (the extension of the line $\overline{KQ}$) is strictly below the point I' on the graph of F.

Students may be tempted to conclude that we have shown that

$$\frac{d}{dx} \int_0^x f(u)\,du = f(x)$$

from which, of course, the Fundamental Theorem is a short hop. However, what has actually been shown is purely geometric. There is no analytic content. There is no computational algorithm for determining the area under f, or the slope of the tangent line to the graph of F. It was Newton's genius to create, on this geometric skeleton, the muscle of computational methods which, axiomatized and extended, we call calculus.

References

1. Boyer, C. B., *The History of the Calculus and its Conceptual Development*, Dover, 1959.
2. Struik, D. J., ed., *A Source Book in Mathematics*, Princeton University Press, 1986.

The Point-slope Formula leads to the Fundamental Theorem of Calculus

Anthony J. Macula

The point-slope formula for a line should be familiar to all beginning calculus students. After showing how to compute the average value of a step function, I will show how this simple formula, together with the idea of approximating continuous functions by piecewise linear functions and step functions, leads naturally to the fundamental theorem of calculus. To indicate how easily these ideas can be introduced in the classroom, I present them as a series of "discovery exercises," with commentary. This approach has much in common with an article by Gilbert Strang [Sums and differences vs. integrals and derivatives, *College Mathematics Journal* 21 (1990) 20–27], but where Strang uses velocity and distance, I emphasize slope and area.

Exercise 1. Consider the continuous piecewise linear function F in Figure 1, for which the slopes of the linear segments are shown in Figure 2. If $F(1) = 17.6$, what is $F(4)$?

Solution. Many students would apply the point-slope formula three times to obtain $F(2) = 19.85$, $F(3) = 20.55$, and finally $F(4) = 22.15$. They are impressed when I show them that a simpler way is to compute the average of the local slopes: $\overline{m} = (2.25 + 0.7 + 1.6)/3 = 1.51\overline{6}$, and then use this single average slope in the point-slope formula to find $F(4) = F(1) + \overline{m}(4 - 1) = 22.15$.

Is it true for any continuous piecewise linear function F on $[a, b]$, that

$$F(b) - F(a) = \overline{m}(b - a) \tag{1}$$

where $\overline{m}$ is the average slope of F on $[a, b]$? Exercise 2 is designed to remind students that they know very well how to compute weighted averages of data values, and Exercise 3 demonstrates that (1) holds if the average slope $\overline{m}$ is taken to be a weighted average of the local slopes, i.e., the slopes of the linear segments of F.

The Point-slope Formula leads to the Fundamental Theorem of Calculus

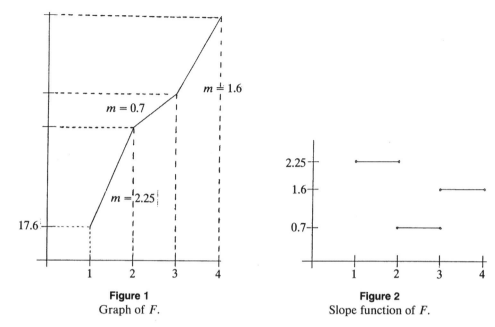

Figure 1
Graph of F.

Figure 2
Slope function of F.

Exercise 2. Suppose in Calculus I there were three quizzes plus midterm and final examinations, which carried point totals 30, 40, 80, 120, and 130 respectively, for a total of 400 possible points for the term. The scores on your papers, however, were expressed as percentages: 82%, 72%, 88%, 93%, and 95%. If the professor requires a score of 90% or better for an A, rounding to the nearest percent, would you qualify for an A?

Solution.

$$\text{Term average} = \frac{\text{Points scored}}{\text{Total points}}$$

$$= \frac{(0.82)(30) + (0.72)(40) + (0.88)(80) + (0.93)(120) + (0.95)(130)}{400} = 0.897,$$

which rounds to 90%. Just under the wire!

Exercise 3. Given the continuous piecewise linear function F in Figure 3, with $F(1) = 5.8$ and with local slopes shown in Figure 4, find the average slope $\overline{m}$ and verify that the point-average slope equation (1) gives the correct value for $F(5)$.

Solution. The average slope of F is the weighted average of the local slopes, using as weights the lengths of the subintervals on which F has constant slope:

$$\overline{m} = \frac{(4.0)(0.3) + (1.0)(0.4) + (-1.0)(1.2) + (-2.5)(0.8) + (3.0)(1.3)}{4} = 0.575.$$

The value for $F(5)$ found by using equation (1) can be laboriously verified by using the point-slope formula to successively calculate $F(1.3), F(1.7), \ldots, F(5)$.

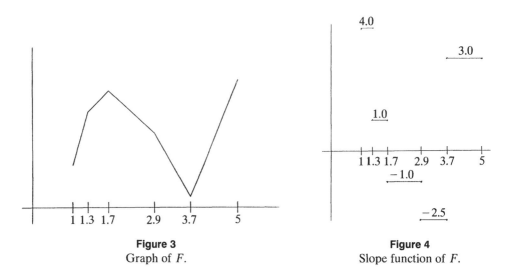

Figure 3
Graph of F.

Figure 4
Slope function of F.

Students are pleased with the effort saved by computing a single average slope in exercises like this one. And most have no trouble discovering the geometric interpretation of the average slope, given in the next exercise.

Exercise 4. Connecting the horizontal "steps" in Figures 2 and 4 to the x-axis by vertical lines, we get a collection of rectangular blocks, shown in Figures 5 and 6. Draw the horizontal lines $y = \overline{m}$ on these graphs, using the values of $\overline{m}$ found in Exercises 1 and 3. Show that in each case *the product $\overline{m}(b - a)$ gives the area between the graph of the step function and the x-axis, if area below the x-axis is counted as negative.*

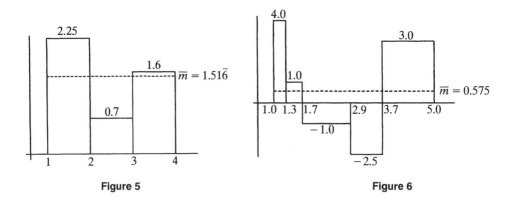

Figure 5

Figure 6

Solution. For the last step, all that is needed is the observation that the numerator in the fraction defining the weighted average $\overline{m}$ is the area A above the x-axis minus the area B below this axis. Since the denominator of this fraction defining $\overline{m}$ is $b - a$, it follows that $\overline{m}(b - a) = A - B$.

When it is pointed out that the step function that gives the local slopes is the *derivative* of the piecewise linear function F, one sees that equation (1) can be expressed in the language of calculus: *If F is any continuous piecewise linear function on [a, b] and F' = f, then*

$$F(b) - F(a) = (\bar{f})(b-a) \qquad (2)$$

where $\bar{f}$ is the average value of the step function f on $[a, b]$.

Now it is natural to turn the process around:

Exercise 5. Consider the step function f defined by

$$f(x) = \begin{cases} 4 & \text{if } -1 < x < 0 \\ -2 & \text{if } 0 < x < 0.5 \\ 3 & \text{if } 0.5 < x < 1.3 \\ -1 & \text{if } 1.3 < x < 2.3 \\ 2 & \text{if } 2.3 < x < 4. \end{cases}$$

(i) Calculate $\bar{f}$ and verify that $(\bar{f})(b-a)$ gives the area between the x-axis and the graph of f, counting areas below the axis as negative.

(ii) If F is the continuous piecewise linear antiderivative of $f(x)$ on $[-1, 4]$ with $F(-1) = 3$, use equation (2) to find $F(4)$.

(iii) Find a formula for $F(x)$ on $[-1, 4]$ and verify that the value you found for $F(4)$ is correct.

Solution. As in Exercise 3 we find

$$\bar{f} = \frac{4(1) + (-2)(0.5) + 3(0.8) + (-1)(1) + 2(1.7)}{5} = 1.56.$$

Thus the point-average slope formula (2) gives $F(4) = 3 + (1.56)(4 + 1) = 10.8$. Integrating f "step by step" using the initial value $F(-1) = 3$, or just using the point-slope formula repeatedly, yields

$$F(x) = \begin{cases} 3 + (4)(x+1) & \text{if } -1 \le x \le 0 \\ 7 + (-2)(x-0) & \text{if } 0 \le x \le 0.5 \\ 6 + (3)(x-0.5) & \text{if } 0.5 \le x \le 1.3 \\ 8.4 + (-1)(x-1.3) & \text{if } 1.3 \le x \le 2.3 \\ 7.4 + (2)(x-2.3) & \text{if } 2.3 \le x \le 4, \end{cases}$$

which allows us to evaluate $F(4)$ the hard way.

The idea of approximating the graph of a continuous function by a piecewise linear graph is familiar to students who have used computer graphics software such as *Derive*, *Maple*, or *Mathematica*. And students who have used a graphing calculator like the TI Model 81 or 85 should recognize that the graph of any piecewise continuous function can be approximated by a step function, since graphs with a low aspect ratio on a graphing calculator are usually just step functions, as in Figure 7. After a discussion of these two ways to approximate functions, it is natural to ask: Does a piecewise continuous function $f(x)$ on $[a, b]$ have an average value $\bar{f}$ such that the product $(\bar{f})(b-a)$ gives the area between the x-axis and the graph of f, counting areas below the axis as negative, as in

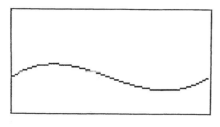
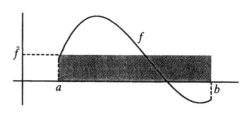

Figure 7

Figure 8

TI-85 graph of $f(x) = (x-2)^3 - x + 3$ on $[1, 3]$.

Figure 8? The requirement is that $(\bar{f})(b-a) = \int_a^b f(x)\,dx$, so the answer is YES:

$$\bar{f} = \frac{1}{b-a} \int_a^b f(x)\,dx.$$

Now we see that the fundamental theorem of calculus is just a generalization of the point-average slope formula (2), from step functions to arbitrary piecewise continuous functions: *If f is a piecewise continuous function on $[a,b]$ and F is a continuous antiderivative of f on $[a,b]$ then*

$$F(b) - F(a) = (\bar{f})(b-a) = \int_a^b f(x)\,dx. \tag{3}$$

Besides the pedagogical benefit of tying the fundamental theorem of calculus to ideas already familiar to students, this approach has two other merits. First, it introduces the fundamental theorem of calculus for piecewise continuous functions, rather than the more limited case of continuous functions. Exercises with step functions and their continuous piecewise linear antiderivatives, such as Exercise 5, can introduce the idea in a simple setting. Second, it is important for students to learn to picture the derivative as an instantaneous rate of change, a *local average velocity*, not just as the local slope of a graph. Students all know that $\Delta x = v \Delta t$ if the velocity is constant; thus the generalization $\Delta x = \bar{v} \Delta t = \int_{t_0}^{t_1} v(t)\,dt$ for a variable velocity function $v(t)$ is very natural, and this is just our formula (3) in a different setting. This approach to the fundamental theorem of calculus provides an interpretation of integration as *transforming a varying local average $f(x)$ on $[a,b]$ into a global average over this interval*:

$$\bar{f} = \frac{\int_a^b f(x)\,dx}{b-a} = \frac{F(b) - F(a)}{b-a}.$$

This point of view can be helpful in understanding other applications of integrals.

A Generalization of the Mean Value Theorem for Integrals

M. Sayrafiezadeh

Let $f(x)$ be a continuous function on $[a, b]$. The Mean Value Theorem for Integrals asserts that there is a point c in (a, b) such that $\int_a^b f(x)\,dx = f(c)(b - a)$. Unlike the proof of the Mean Value Theorem for derivatives, the proof of the Mean Value Theorem for Integrals typically does not use Rolle's Theorem. In this note, we use Rolle's Theorem to introduce a generalization of the Mean Value Theorem for Integrals. Our generalization involves two functions instead of one, and has a very clear geometric explanation.

Theorem. If $f(x)$ and $g(x)$ are continuous functions on $[a, b]$, then there is a value c in (a, b) such that

$$\int_a^c f(t)\,dt + \int_c^b g(t)\,dt = f(c)(b - c) + g(c)(c - a).$$

Proof. Let $h(x)$ be the function defined on $[a, b]$ as

$$h(x) = (x - b)\int_a^x f(t)\,dt + (x - a)\int_x^b g(t)\,dt.$$

Since $f(x)$ and $g(x)$ are both continuous on $[a, b]$, the function $h(x)$ is continuous on $[a, b]$ and differentiable on (a, b). Furthermore, $h(a) = 0$ and $h(b) = 0$. So, by Rolle's Theorem, there is a value c in (a, b) such that $h'(c) = 0$. By the Second Fundamental Theorem of Calculus,

$$h'(x) = (x - b)f(x) + \int_a^x f(t)\,dt - (x - a)g(x) + \int_x^b g(t)\,dt,$$

so from $h'(c) = 0$, we have

$$\int_a^c f(t)\,dt + \int_c^b g(t)\,dt = f(c)(b - c) + g(c)(c - a).$$

The geometric interpretation of this theorem (Figure 1) is that the sum of the area under f's graph on $[a, c]$ and the area under g's graph on $[c, b]$ equals the sum of the areas of two rectangles, one with base $[c, b]$ and height $f(c)$ and the other with base $[a, c]$ and height $g(c)$. ∎

Letting $g(x) = f(x)$ and $g(x) = 0$, we get the following respective corollaries.

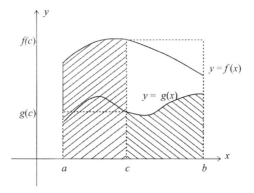

Figure 1.

Corollary 1 (The Mean Value Theorem for Integrals). *If $f(x)$ is continuous on $[a, b]$, then there is a value c in (a, b) such that*

$$\int_a^b f(t)\,dt = f(c)(b - a).$$

Corollary 2. *If $f(x)$ is continuous on $[a, b]$, then there is a value c in (a, b) such that*

$$\int_a^c f(t)\,dt = f(c)(b - c).$$

The interpretation of Corollary 2 is that there is a value c in the interval (a, b) such that the area under f's graph on the subinterval (a, c) equals to the area of the rectangle with height $f(c)$ on (c, b). See Figure 2.

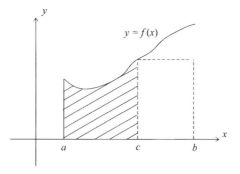

Figure 2.

Proof Without Words: Look Ma, No Substitution!

Marc Chamberland

$$\int_a^1 \sqrt{1-x^2}\,dx = \frac{\cos^{-1} a}{2} - \frac{a\sqrt{1-a^2}}{2}, \quad a \in [-1, 1].$$

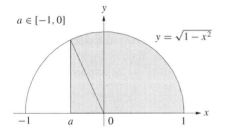
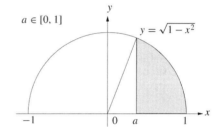

$$\int_a^1 \sqrt{1-x^2}\,dx = \frac{\cos^{-1} a}{2} + \frac{(-a)\sqrt{1-a^2}}{2} \qquad \int_a^1 \sqrt{1-x^2}\,dx = \frac{\cos^{-1} a}{2} - \frac{a\sqrt{1-a^2}}{2}$$

Integration by Parts

V. N. Murty

Introduction. Every student taking a course in calculus in his freshman or sophomore year learns the technique of evaluating an integral by the method known as "*integration by parts*" as embodied in the following well-known rule:

$$\int f(x) \cdot g'(x)\, dx = f(x) \cdot g(x) - \int f'(x) \cdot g(x)\, dx.$$

The above rule is also given in the form

$$\int u\, dv = uv - \int v\, du.$$

Students often find this rule puzzling and have trouble in picking the right u and v. I have adopted the following procedure, which may not be new, but which appears to have succeeded well. I will illustrate the procedure with five examples.

Example 1.

$$\text{Evaluate } \int x^2 \sin x\, dx.$$

The student should factor the integrand into a product of two functions, one whose successive derivatives eventually become zero, and the other whose successive anti-derivatives are easily obtained. In our example, $f(x) = x^2$ can be differentiated successively and the third derivative is zero. The function, $g(x) = \sin x$, when integrated successively, gives $-\cos x$, $-\sin x$, and $\cos x$, respectively. These successive derivatives of $f(x)$ and anti-derivatives of $g(x)$ are written in the following format:

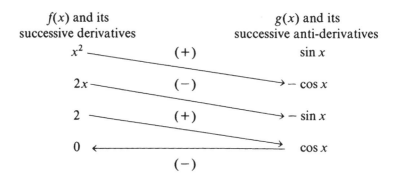

Integration by Parts

Then he is asked to draw arrows as indicated above. Products of functions connecting these arrows are formed and the value of the required integral is

$$\int x^2 \sin x \, dx = (+)(x^2)(-\cos x) - (2x)(-\sin x)$$
$$+ (2)(+1)(\cos x) - \int 0 \cdot \cos x \, dx.$$

The signs for these products are always $+, -, +, -$, etc. The products with the appropriate signs are summed except the product with the horizontal arrow which is integrated, and this clearly is zero. Thus,

$$\int x^2 \sin x \, dx = -x^2 \cos x + 2x \sin x + 2 \cos x + C.$$

The above method can always be used as long as the integrand can be factored as $f(x)g(x)$, where f can easily be differentiated successively and goes to zero and g can easily be integrated successively. Thus, the procedure is workable for $x^n \sin ax$, $x^n \cos ax$, or $x^n e^{ax}$.

Example 2.

$$\int e^{2x} \sin x \, dx.$$

Here successive derivatives of either e^{2x} or $\sin x$ would never become zero. Both are easy to integrate or differentiate successively and the procedure described above is modified slightly.

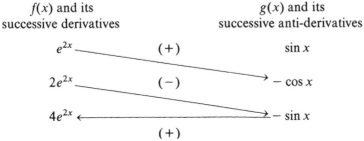

Successive differentiation and integration of $f(x)$ and $g(x)$ is carried out until we reach a stage where the product of the derivative and anti-derivative at that stage is a constant multiple of the function whose integral we wish to evaluate. Proceeding as before, we obtain:

$$\int e^{2x} \cdot \sin x \, dx = -e^{2x} \cos x + 2e^{2x} \sin x - 4 \int e^{2x} \sin x \, dx$$

or

$$\int e^{2x} \sin x \, dx = -\frac{1}{5} e^{2x} \cos x + \frac{2}{5} e^{2x} \sin x + C.$$

This procedure is useful for integrating functions like $e^{ax} \sin bx$, or $e^{ax} \cos bx$.

Example 3.

$$\text{Evaluate } \int x^2 \ln x \, dx.$$

Although the third derivative of x^2 is zero, the procedure used in the first example is not fruitful, since successive integrals of $\ln x$ are hard to obtain. We can, of course, easily find the successive derivatives of $\ln x$, and integrate x^2 successively, but in this case we will never reach a stage where the product of the derivative and the anti-derivative at that stage is a constant multiple of $f(x)g(x)$. In such situations, we stop at the stage where we see that the product of the derivative and the integral at that stage is easily integrable.

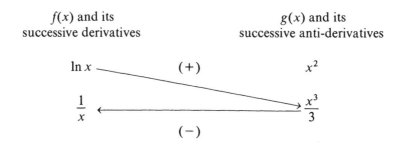

We see that the product of the first derivative of $\ln x$ and the anti-derivative of x^2 is $x^2/3$, which can easily be integrated. So we stop and follow our general rule:

$$\int x^2 \ln x \, dx = \frac{1}{3} x^3 \ln x - \frac{1}{3} \int x^2 \, dx$$

$$= \frac{1}{3} x^3 \ln x - \frac{1}{9} x^3 + C.$$

A word of caution. When the successive derivatives of $f(x)$ do not show a zero at any stage, and the product of the derivative of f and the integral of g at some stage becomes exactly equal to $f(x)g(x)$ with a plus sign, this method does not give the value of the integral.

Example 4.

$$\int \sin^2 x \, dx.$$

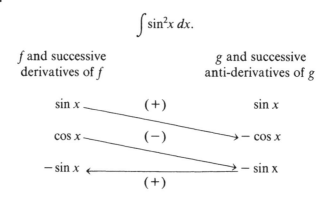

Integrationi by Parts

We stopped with the second derivative since at this stage the product is $\sin^2 x$. Hence,

$$\int \sin^2 x\, dx = -(\sin x)\cos x + \cos x \sin x + \int \sin^2 x\, dx$$

$$= 0 + \int \sin^2 x\, dx.$$

This is true, but useless, for our purpose.

$\int \sin^2 x\, dx$ can be evaluated using integration by parts if we proceed as follows:

$$\int \sin^2 x\, dx = \int (1 - \cos^2 x)\, dx = x - \int \cos^2 x\, dx$$

$$= x - \int \cos x \cdot \cos x\, dx$$

$$= x - \left[\sin x \cos x - \int \sin x (-\sin x)\, dx\right]$$

$$= x - \sin x \cos x - \int \sin^2 x\, dx,$$

which yields

$$\int \sin^2 x\, dx = \frac{1}{2}[x - \sin x \cos x] + C.$$

Integrals of the type $\int \sin ax \cos bx\, dx$, $\int \sin ax \sin bx\, dx$, and $\int \cos ax \cos bx\, dx$ can be evaluated using this procedure when $a \neq b$ and the student need not remember any trigonometric identities in evaluating these integrals involving products of sines and cosines.

Example 5.

Evaluate $\int \sin 5x \cos 3x\, dx = I$.

$\sin 5x$	$(+)$	$\cos 3x$
$5 \cos 5x$	$(-)$	$+\frac{1}{3}\sin 3x$
$-25 \sin 5x$	$(+)$	$-\frac{1}{9}\cos 3x$

$$I = \tfrac{1}{3}\sin 3x \sin 5x + \tfrac{5}{9}\cos 3x \cos 5x + \tfrac{25}{9}I$$

$$I = -\tfrac{3}{16}\sin 3x \sin 5x - \tfrac{5}{16}\cos 3x \cos 5x + C.$$

Tabular Integration by Parts

David Horowitz

Only a few contemporary calculus textbooks provide even a cursory presentation of tabular integration by parts [see for example, G. B. Thomas and R. L. Finney, *Calculus and Analytic Geometry*, Addison-Wesley, Reading, MA, 1988]. This is unfortunate because tabular integration by parts is not only a valuable tool for finding integrals but can also be applied to more advanced topics including the derivations of some important theorems in analysis.

The technique of tabular integration allows one to perform successive integrations by parts on integrals of the form

$$\int F(t)G(t)\,dt \qquad (1)$$

without becoming bogged down in tedious algebraic details [V. N. Murty, Integration by parts, *Two-Year College Mathematics Journal* 11 (1980) 90–94]. There are several ways to illustrate this method, one of which is diagrammed in Table 1. (We assume throughout that F and G are "smooth" enough to allow repeated differentation and integration, respectively.)

Table 1

Column #1	Column #2
$+F$	G
$-F^{(1)}$	$G^{(-1)}$
$+F^{(2)}$	$G^{(-2)}$
$-F^{(3)}$	$G^{(-3)}$
$\vdots$	$\vdots$
$(-1)^n F^{(n)}$	$G^{(-n)}$
$(-1)^{n+1} F^{(n+1)}$	$G^{(-n-1)}$

Tabular Integration by Parts

In column #1 list $F(t)$ and its successive derivatives. To each of the entries in this column, alternately adjoin plus and minus signs. In column #2 list $G(t)$ and its successive antiderivatives. (The notation $G^{(-n)}$ denotes the nth antiderivative of G. Do not include an additive constant of integration when finding each antiderivative.) Form successive terms by multiplying each entry in column #1 by the entry in column #2 that lies just *below* it. The integral (1) is equal to the sum of these terms. If $F(x)$ is a polynomial, then there will be only a finite number of terms to sum. Otherwise the process may be truncated at any level by forming a remainder term defined as the integral of the product of the entry in column #1 and the entry in column #2 that lies directly *across* from it. Symbolically,

$$\int F(t)G(t)\,dt = FG^{(-1)} - F^{(1)}G^{(-2)} + F^{(2)}G^{(-3)} - \cdots$$
$$+ (-1)^n F^{(n)} G^{(-n-1)} + (-1)^{n+1} \int F^{(n+1)}(t) G^{(-n-1)}(t)\,dt$$
$$= \sum_{k=0}^{n} (-1)^k F^{(k)} G^{(-k-1)} + (-1)^{n+1} \int F^{(n+1)}(t) G^{(-n-1)}(t)\,dt. \quad (2)$$

A proof follows from continued application of the formula for integration by parts [K. W. Folley, Integration by parts, *American Mathematical Monthly* 54 (1947) 542–543].

The technique of tabular integration by parts makes an appearance in the hit motion picture *Stand and Deliver* in which mathematics instructor Jaime Escalante of James A. Garfield High School in Los Angeles (portrayed by actor Edward James Olmos) works the following example.

Example. $\int x^2 \sin x\,dx$

column #1	column #2
$+x^2$	$\sin x$
$-2x$	$-\cos x$
$+2$	$-\sin x$
0	$\cos x$

Answer: $-x^2 \cos x + 2x \sin x + 2\cos x + C$

The following are some areas where this elegant technique of integration can be applied.

Miscellaneous Indefinite Integrals. Most calculus textbooks would treat integrals of the form

$$\int \frac{P(t)}{(at+b)^r}\,dt,$$

where $P(t)$ is a polynomial, by time-consuming algebraic substitutions or tedious partial fraction decompositions. However, tabular integration by parts works particularly well on such integrals (especially if r is not a positive integer).

Example.

$$\int \frac{12t^2 + 36}{\sqrt[5]{3t+2}} \, dt$$

column #1	column #2
$+\ 12t^2 + 36$	$(3t+2)^{-1/5}$
$-\ 24t$	$\dfrac{5}{12}(3t+2)^{4/5}$
$+\ 24$	$\dfrac{25}{324}(3t+2)^{9/5}$
0	$\dfrac{125}{13608}(3t+2)^{14/5}$

Answer: $(5t^2 + 15)(3t+2)^{4/5} - \dfrac{50t}{27}(3t+2)^{9/5} + \dfrac{125}{567}(3t+2)^{14/5} + C$

Laplace Transforms. Computations involving the Laplace transform

$$L\{f(t)\} = \int_0^\infty e^{-st} f(t) \, dt \qquad (3)$$

often require several integrations by parts because of the form of the integrand in (3). Tabular integration by parts streamlines these integrations and also makes proofs of operational properties more elegant and accessible. (Many introductory differential equations textbooks omit formal proofs of these properties because of the lengthy detail involved in their derivations.) The following example uses this technique to establish the fundamental formula for the Laplace transform of the nth derivative of a function.

Theorem. *Let n be a positive integer and suppose that $f(t)$ is a function such that $f^{(n)}(t)$ is piecewise continuous on the interval $t \geq 0$. Furthermore suppose that there exist constants A, b, and M such that*

$$|f^{(k)}(t)| \leq Ae^{bt} \text{ if } t \geq M$$

for all $k = 0, 1, 2, \ldots, n-1$. Then if $s > b$,

$$L\{f^{(n)}(t)\} = -f^{(n-1)}(0) - sf^{(n-2)}(0) - \cdots - s^{n-1}f(0) + s^n L\{f(t)\}. \qquad (4)$$

Proof.

	column #1	column #2
$+$	e^{-st}	$f^{(n)}(t)$
$-$	$-se^{-st}$	$f^{(n-1)}(t)$
$+$	$s^2 e^{-st}$	$f^{(n-2)}(t)$
	$\vdots$	$\vdots$
	$(-1)^{n-1}(-1)^{n-1}s^{n-1}e^{-st}$	$f^{(1)}(t)$
	$(-1)^n(-1)^n s^n e^{-st}$	$f(t)$

$$L\{f^{(n)}(t)\} = \left[e^{-st}f^{(n-1)}(t) + se^{-st}f^{(n-2)}(t) + \cdots + s^{n-1}e^{-st}f(t)\right]_{t=0}^{t=\infty}$$
$$+ \int_0^\infty s^n e^{-st} f(t)\, dt$$
$$= -f^{(n-1)}(0) - sf^{(n-2)}(0) - \cdots - s^{n-1}f(0) + s^n L\{f(t)\}.$$

Taylor's Formula. Tabular integration by parts provides a straightforward proof of Taylor's formula with integral remainder term.

Theorem. *Suppose $f(t)$ has $n+1$ continuous derivatives throughout an interval containing a. If x is any number in the interval then*

$$f(x) = f(a) + f^{(1)}(a)(x-a) + \frac{f^{(2)}(a)}{2!}(x-a)^2 + \cdots + \frac{f^{(n)}(a)}{n!}(x-a)^n$$
$$+ \int_a^x \frac{f^{(n+1)}(t)}{n!}(x-t)^n\, dt. \tag{5}$$

Proof.

	column #1	column #2
$+$	$-f^{(1)}(t)$	-1
$-$	$-f^{(2)}(t)$	$(x-t)$
$+$	$-f^{(3)}(t)$	$-\dfrac{(x-t)^2}{2!}$
	$\vdots$	$\vdots$
	$(-1)^{n+1} - f^{(n)}(t)$	$(-1)^n \dfrac{(x-t)^{n-1}}{(n-1)!}$
	$(-1)^{n+2} - f^{(n+1)}(t)$	$(-1)^{n+1} \dfrac{(x-t)^n}{n!}$

$$\int_a^x [-f^{(1)}(t)][-1]\, dt$$
$$= \left[-f^{(1)}(t)(x-t) - \frac{f^{(2)}(t)}{2!}(x-t)^2 - \cdots - \frac{f^{(n)}(t)}{n!}(x-t)^n\right]_{t=a}^{t=x}$$
$$+ \int_a^x \frac{f^{(n+1)}(t)}{n!}(x-t)^n\, dt$$

Equation (5) follows immediately.

Residue Theorem for Meromorphic Functions. Tabular integration by parts also applies to complex line integrals and can be used to prove the following form of the residue theorem.

Theorem. *Suppose $f(z)$ is analytic in $\mathscr{D} = \{z : 0 < |z - z_0| < R\}$ and has a pole of order m at z_0. Then if $0 < r < R$*

$$\oint_{|z-z_0|=r} f(z)\,dz = \frac{2\pi i}{(m-1)!} \lim_{z \to z_0} \left[\frac{d^{m-1}}{dz^{m-1}} (z-z_0)^m f(z) \right]. \tag{6}$$

Proof.

column #1	column #2
$+\ (z-z_0)^m f(z)$	$(z-z_0)^{-m}$
$-\dfrac{d}{dz}(z-z_0)^m f(z)$	$-\dfrac{(z-z_0)^{-m+1}}{m-1}$
$+\dfrac{d^2}{dz^2}(z-z_0)^m f(z)$	$\dfrac{(z-z_0)^{-m+2}}{(m-1)(m-2)}$
$\vdots$	$\vdots$
$(-1)^{m-1}\dfrac{d^{m-1}}{dz^{m-1}}(z-z_0)^m f(z)$	$(-1)^{m-1}\dfrac{(z-z_0)^{-1}}{(m-1)!}$

Thus,

$$\oint_{|z-z_0|=r} (z-z_0)^m f(z)(z-z_0)^{-m}\,dz$$

$$= \left[-\sum_{k=0}^{m-2} \frac{(m-k-2)!\,\dfrac{d^k}{dz^k}(z-z_0)^m f(z)}{(m-1)!(z-z_0)^{m-k-1}} \right] \oint_{|z-z_0|=r}$$

$$+ \frac{1}{(m-1)!} \oint_{|z-z_0|=r} \frac{\dfrac{d^{m-1}}{dz^{m-1}}(z-z_0)^m f(z)}{z-z_0}\,dz, \tag{7}$$

where the summation term in (7) must be evaluated for the closed circle $|z - z_0| = r$. However, each of the functions in column #1 has a removable singularity at z_0 and is therefore single-valued in $\mathscr{D}$. Furthermore, each of the functions in column #2 is also single-valued in $\mathscr{D}$. Thus, the summation term in (7) must vanish. The integral term on the right side of (7) is evaluated by the Cauchy integral formula and the result (6) follows directly. (The right-hand side of (6) can be recognized as the formula for the residue of $f(z)$ at the pole z_0.) See [Ruel V. Churchill and James W. Brown, *Complex Variables and Applications*, McGraw-Hill, New York, 1984].

More on Tabular Integration by Parts

Leonard Gillman

This note comments on the engaging article by David Horowitz on "Tabular Integration by Parts" [*College Mathematics Journal* 21 (1990) 307–311]. The method is based on iterating the diagram

$u \longrightarrow v'$ The integral of this product

           equals this product

$u' \longrightarrow v$ minus the integral of this product,

and can be ended at any stage. Table 1 shows how the procedure handles the integral $\int x^2 \sin x \, dx$. A preliminary column lists alternating plus and minus signs, starting with plus. The integrand is written as a product of two factors, which head

Table 1

$$\int x^2 \sin x \, dx = -x^2 \cos x + 2x \sin x + 2 \cos x + C$$

	Column 1	Column 2
+	x^2	$\sin x$
−	$2x$	$-\cos x$
+	2	$-\sin x$
−	0	$\cos x$

columns 1 and 2. Column 1 lists successive *derivatives* of the head entry, and column 2 successive *antiderivatives* of *its* head entry. The value of the integral is the sum of the indicated diagonal products, plus the integral of the product along the last row, all taken with the indicated signs. Note that if you reach a 0 in Column 1, as in the illustration, then the integral across that row provides the constant of integration.

As Horowitz shows, the method is a blockbuster, as can be seen from the way it knocks off an integral like

$$\int \frac{12t^2 + 36}{\sqrt[5]{3t+2}}\, dt,$$

not to mention Taylor's formula and the other examples he presents.

The purpose of this note is to call attention to a variant scheme, suggested in [Leonard Gillman and Robert H. McDowell, *Calculus*, W. W. Norton; 1st edition (1973) 328; 2nd edition (1978) 409], where it was used to integrate $\sec^3 x$ (but was not developed further). In the second edition this occurred *before* the section on integration by parts; the spirit was "mess around and see what happens:" What do we differentiate in order to get $\sec^3 x$? Clearly it will arise when we differentiate $\sec x \tan x$; but so will another term that we don't want. Well, let's differentiate $\sec x \tan x$ anyway and then see if we can get rid of the unwanted term.

Table 2 shows the details of this scheme for $\int x^2 \sin x\, dx$, the problem treated above. The sum in the 2nd column is $x^2 \sin x$, the integrand, and the sum in the first column is its integral, so the answer is easy to read off. (One could build in a constant of integration by putting C in the first column and 0 opposite it in the second.) The scheme is really self-explanatory, but here are some comments. We write the integrand as uv' (here $u = x^2$, $v' = \sin x$), then enter uv in column 1 and its derivative in column 2; this guarantees that the first term in column 2 will be the original integrand. Our task is then to cancel out the second term; we take it as our new uv' and continue the procedure.

Table 2

$\int x^2 \sin x\, dx = -x^2 \cos x + 2x \sin x + 2 \cos x + C$

uv	$uv' + u'v$
$-x^2 \cos x$	$x^2 \sin x - 2x \cos x$
$+2x \sin x$	$+2x \cos x + 2 \sin x$
$+2 \cos x$	$-2 \sin x$

This more contemplative scheme seems more informative than the other: you can see the mechanism, the work is very easy to check, and the final answer is very easy to read off. A disadvantage is that there is more writing to do, in fact about twice as much.

A mild point shows up in evaluating integrals such as $\int e^x \sin x\, dx$, where ordinarily one integrates by parts until the original integrand reappears on the other side of the equation, then transposes and solves. In the variant scheme the integrand comes back on the same side of the equation, so that what is taking place remains somewhat more transparent; see Table 3. (In the second step, we

More on Tabular Integration by Parts

Table 3

$$\int e^x \sin x \, dx = \tfrac{1}{2}e^x(\sin x - \cos x) + C$$

uv	$uv' + u'v$
$-e^x \cos x$	$e^x \sin x - e^x \cos x$
$+e^x \sin x$	$+e^x \cos x$
	$+e^x \sin x$

write the $e^x \cos x$ first because it is the uv' term.)

In the original method one still has to keep alert. A case in point is an integrand of the form $f(\ln x)$. Suppose we start off with

$$+ \quad f(\ln x) \quad\quad 1$$
$$- \quad \frac{f'(\ln x)}{x} \quad\quad x$$

Continuing the algorithm at this stage is unlikely to lead to anything except algebraic complication. Instead, we leave the product $xf(\ln x)$ untouched but rewrite the second line in the equivalent form

$$- \quad f'(\ln x) \quad\quad 1$$

and continue from there. Table 4 shows the work for $\int (\ln x)^2 \, dx$. The two lines marked * are identified, as are the two marked **. In this example, the variant scheme seems to be smoother (Table 5).

Table 4

$$\int (\ln x)^2 \, dx = x[(\ln x)^2 - 2\ln x + 2] + C$$

	Column 1	Column 2	
+	$(\ln x)^2$	1	
−	$\dfrac{2\ln x}{x}$	x	*
−	$2\ln x$	1	*
+	$\dfrac{2}{x}$	x	**
+	2	1	**
−	0	x	

Table 5

$$\int (\ln x)^2 \, dx = x[(\ln x)^2 - 2\ln x + 2] + C$$

uv	$uv' + u'v$
$(\ln x)^2(x)$	$(\ln x)^2 + 2\ln x$
$-(2\ln x)(x)$	$-2\ln x - 2$
$+2x$	$+2$

Another textbook favorite is $\int \sin \ln x \, dx$. The original method runs into the same problem as the preceding example; but the variant, Table 6, is smooth and in fact is isomorphic to Table 3.

Table 6

$$\int \sin \ln x \, dx = \tfrac{1}{2}x(\sin \ln x - \cos \ln x) + C$$

uv	$uv' + u'v$
$(\sin \ln x)(x)$	$\sin \ln x + \cos \ln x$
$-(\cos \ln x)(x)$	$-\cos \ln x$
	$+\sin \ln x$

A Quotient Rule Integration by Parts Formula

Jennifer Switkes

In a recent calculus course, I introduced the technique of Integration by Parts as an integration rule corresponding to the Product Rule for differentiation. I showed my students the standard derivation of the Integration by Parts formula as presented in [1]:
By the Product Rule, if $f(x)$ and $g(x)$ are differentiable functions, then

$$\frac{d}{dx}[f(x)g(x)] = f(x)g'(x) + g(x)f'(x).$$

Integrating on both sides of this equation,

$$\int [f(x)g'(x) + g(x)f'(x)]\,dx = f(x)g(x),$$

which may be rearranged to obtain

$$\int f(x)g'(x)\,dx = f(x)g(x) - \int g(x)f'(x)\,dx.$$

Letting $U = f(x)$ and $V = g(x)$ and observing that $dU = f'(x)\,dx$ and $dV = g'(x)\,dx$, we obtain the familiar Integration by Parts formula

$$\int U\,dV = UV - \int V\,dU. \tag{1}$$

My student Victor asked if we could do a similar thing with the Quotient Rule. While the other students thought this was a crazy idea, I was intrigued. Below, I derive a Quotient Rule Integration by Parts formula, apply the resulting integration formula to an example, and discuss reasons why this formula does not appear in calculus texts.
By the Quotient Rule, if $f(x)$ and $g(x)$ are differentiable functions, then

$$\frac{d}{dx}\left[\frac{f(x)}{g(x)}\right] = \frac{g(x)f'(x) - f(x)g'(x)}{[g(x)]^2}.$$

Integrating both sides of this equation, we get

$$\frac{f(x)}{g(x)} = \int \frac{g(x)f'(x) - f(x)g'(x)}{[g(x)]^2}\,dx.$$

That is,

$$\frac{f(x)}{g(x)} = \int \frac{f'(x)}{g(x)} dx - \int \frac{f(x)g'(x)}{[g(x)]^2} dx,$$

which may be rearranged to obtain

$$\int \frac{f'(x)}{g(x)} dx = \frac{f(x)}{g(x)} + \int \frac{f(x)g'(x)}{[g(x)]^2} dx.$$

Letting $u = g(x)$ and $v = f(x)$ and observing that $du = g'(x)\,dx$ and $dv = f'(x)\,dx$, we obtain a Quotient Rule Integration by Parts formula:

$$\int \frac{dv}{u} = \frac{v}{u} + \int \frac{v}{u^2} du. \qquad (2)$$

As an application of the Quotient Rule Integration by Parts formula, consider the integral

$$\int \frac{\sin(x^{-1/2})}{x^2} dx.$$

Let

$$u = x^{1/2}, \qquad dv = \frac{\sin(x^{-1/2})}{x^{3/2}} dx,$$

$$du = \frac{1}{2} x^{-1/2} dx, \qquad v = 2\cos(x^{-1/2}).$$

Then

$$\int \frac{\sin(x^{-1/2})}{x^2} dx = \frac{2\cos(x^{-1/2})}{x^{1/2}} + \int \frac{2\cos(x^{-1/2})}{x} \cdot \frac{1}{2} x^{-1/2} dx$$

$$= \frac{2\cos(x^{-1/2})}{x^{1/2}} - 2\int \cos(x^{-1/2}) \cdot \left(-\frac{x^{-3/2}}{2}\right) dx$$

$$= \frac{2\cos(x^{-1/2})}{x^{1/2}} - 2\sin(x^{-1/2}) + C,$$

which may be easily verified as correct.

Why do we not find the Quotient Rule Integration by Parts formula in calculus texts?

First, the Quotient Rule Integration by Parts formula (2) results from applying the standard Integration by Parts formula (1) to the integral

$$\int \frac{dv}{u}$$

with

$$U = \frac{1}{u}, \qquad dV = dv,$$
$$dU = -\frac{1}{u^2} du, \qquad V = v,$$

to obtain

$$\int \frac{dv}{u} = \int U \, dV$$
$$= UV - \int V \, dU$$
$$= \frac{1}{u} \cdot v - \int v \cdot \left(-\frac{1}{u^2}\right) du$$
$$= \frac{v}{u} + \int \frac{v}{u^2} du$$

as before.

Secondly, there is the potential only for slight technical advantage in choosing formula (2) over formula (1). An identical integral will need to be computed whether we use (1) or (2). The only difference in the required differentiation and integration occurs in the computation of du versus dU. In our example, for instance, we differentiated $u = x^{1/2}$ rather than $U = x^{-1/2}$.

Acknowledgment. The author wishes to thank Cal Poly Pomona student Victor Reynolds for his insightful idea.

Reference

1. J. Stewart, *Calculus: Early Transcendentals*, 4th ed., Brooks/Cole, 1999.

Partial Fraction Decomposition by Division

Sidney H. Kung

In this note, we present a method for the partial fraction decomposition of two algebraic functions: (i) $f(x)/(ax+b)^t$ and (ii) $f(x)/(px^2+qx+r)^t$, where $f(x)$ is a polynomial of degree n, t is a positive integer, and px^2+qx+r is an irreducible ($q^2 < 4pr$) quadratic polynomial. Our algorithm is relatively simple in comparison with those given elsewhere [1, 2, 3, 4, 5, 6, 7, 8]. The essence of the method is to use repeated division to re-express the numerator polynomial in powers of the normalized denominator. Then upon further divisions, we obtain a sum of partial fractions in the form $A_i/(ax+b)^i$ or $(B_j x + C_j)/(px^2+qx+r)^j$ for the original function.

For (i), we let $c = b/a$, and express $f(x)$ as follows:

$$f(x) = A_n(x+c)^n + A_{n-1}(x+c)^{n-1} + \cdots + A_2(x+c)^2 + A_1(x+c) + A_0$$

$$= (x+c)[A_n(x+c)^{n-1} + A_{n-1}(x+c)^{n-2} + \cdots + A_2(x+c) + A_1] + A_0,$$

where each A_i is a real coefficient to be determined. Then the remainder after we divide $f(x)$ by $x+c$ gives the value of A_0. The quotient is

$$q_1(x) = (x+c)[A_n(x+c)^{n-2} + A_{n-1}(x+c)^{n-3} + \cdots + A_3(x+c) + A_2] + A_1.$$

If we now divide $q_1(x)$ by $x+c$, we see that the next remainder is A_1 and the quotient is

$$q_2(x) = (x+c)[A_n(x+c)^{n-3} + A_{n-1}(x+c)^{n-4} + \cdots + A_3] + A_2.$$

Continuing to divide in this manner $n-1$ times, we get the quotient $q_{n-1}(x) = A_n(x+c) + A_{n-1}$. Finally, dividing $q_{n-1}(x)$ by $x+c$, we obtain the last two coefficients, A_{n-1} and A_n. Thus, it follows that

$$\frac{f(x)}{(ax+b)^t} = \frac{1}{a^t}\left[\frac{A_n}{(x+c)^{t-n}} + \frac{A_{n-1}}{(x+c)^{t-n+1}} + \cdots + \frac{A_1}{(x+c)^{t-1}} + \frac{A_0}{(x+c)^t}\right]. \quad (1)$$

For example, to find the partial fraction decomposition of $(x^4 + 2x^3 - x^2 + 5)/(2x-1)^5$, we use $c = -1/2$ and perform synthetic division to obtain A_0 through A_n.

		1	2		-1		0		5	
1/2)			1/2		5/4		1/8		1/16	
		1	5/2		1/4		1/8	|	81/16	$\Leftarrow A_0$
			1/2		3/2		7/8			
		1	3		7/4	|	1	$\Leftarrow A_1$		
			1/2		7/4					
		1	7/2	|	7/2	$\Leftarrow A_2$				
			1/2							
$A_4 \Rightarrow$	1	4	$\Leftarrow A_3$							

Partial Fraction Decomposition by Division

Substituting the coefficients into (1), we have

$$\frac{x^4 + 2x^3 - x^2 + 5}{(2x-1)^5} = \frac{1}{2^5}\left[\frac{1}{(x-1/2)} + \frac{4}{(x-1/2)^2} + \frac{7/2}{(x-1/2)^3} + \frac{1}{(x-1/2)^4}\right.$$

$$\left. + \frac{81/16}{(x-1/2)^5}\right]$$

$$= \frac{1/16}{2x-1} + \frac{1/2}{(2x-1)^2} + \frac{7/8}{(2x-1)^3}$$

$$+ \frac{1/2}{(2x-1)^4} + \frac{81/16}{(2x-1)^5}.$$

For (ii), we let $u = q/p$, $v = r/p$, and express $f(x)$ in the following form:

$$f(x) = B_{(n-1)/2}(x^2 + ux + v)^{(n-1)/2} + B_{(n-3)/2}(x^2 + ux + v)^{(n-3)/2} + \cdots$$

$$+ B_1(x^2 + ux + v) + B_0,$$

where each coefficient B_k, $k = 0, 1, \ldots, (n-1)/2$, is a linear function of x, and where we assume that $n \leq 2t - 1$. In this case, dividing $f(x)$ and each successive quotient by $x^2 + ux + v$ as described above, we obtain

$$\frac{f(x)}{(px^2 + qx + r)^t} = \frac{1}{p^t}\left[\frac{B_{(n-1)/2}}{(x^2 + ux + v)^{t-(n-1)/2}}\right. \tag{2}$$

$$\left. + \frac{B_{(n-3)/2}}{(x^2 + ux + v)^{t-(n-3)/2}} + \cdots + \frac{B_0}{(x^2 + ux + v)^t}\right].$$

For instance, take the rational function $(x^5 - 4x^4 + 3x^2 - 2)/(x^2 - x + 2)^3$. Then $u = -1$ and $v = 2$. Since $(n-1)/2 = 2$, we let

$$x^5 - 4x^4 + 3x^2 - 2 = (Mx + N)(x^2 - x + 2)^2 + (Kx + L)(x^2 - x + 2) + Ix + J.$$

Since most students are not familiar with the synthetic division technique when the divisor is a quadratic polynomial, long division can be used in place of the following computation to find the coefficients I, J, K, L, M, and N.

```
    1   -4    0    3    0   -2  |  1   -3   -5    4
    1   -1    2                  |  1   -1    2
   ─────────────────
        -3   -2    3
        -3    3    6
       ─────────────
             -5    9    0
             -5    5  -10
            ──────────────
                   4   10   -2
                   4   -4    8
                  ──────────────
                       14  -10
                        I    J
```

$$\begin{array}{ccccccc}
 & & & & & M & N \\
1 & -3 & -5 & 4 & | & 1 & -2 \\
1 & -1 & 2 & & | & 1 & -1 \quad 2 \\
\hline
-2 & -7 & 4 & & & & \\
-2 & 2 & -4 & & & & \\
\hline
-9 & 8 & & & & & \\
K & L & & & & &
\end{array}$$

Substituting the coefficients in (2) (note that $t - (n-1)/2 = 1$) gives

$$\frac{x^5 - 4x^4 + 3x^2 - 2}{(x^2 - x + 2)^3} = \frac{x - 2}{x^2 - x + 2} + \frac{-9x + 8}{(x^2 - x + 2)^2} + \frac{14x - 10}{(x^2 - x + 2)^3}.$$

Note also that $x^2 - x + 2 = (x - 1/2)^2 + 4/7$. On the right hand side of the above expression, replacing the coefficients -2, 8, and -10 in the numerators by $-2 + 1/2M$, $8 + 1/2M$, and $-10 + 1/2M$, respectively, we get

$$\frac{x^5 - 4x^4 + 3x - 2}{((x - 1/2)^2 + 7/4)^3} = \frac{(x - 1/2) - 3/2}{(x - 1/2)^2 + 7/4} + \frac{-9(x - 1/2) + 7/4}{((x - 1/2)^2 + 7/4)^2}$$
$$+ \frac{14(x - 1/2)^2 - 3}{((x - 1/2)^2 + 7/4)^3}.$$

This last expression is an easily antidifferentiable form.

References

1. S. Burgstahler, An alternative for certain partial fractions, *College Math. J.* **15** (1984) 57–58.
2. X.-C. Huang, A short cut to partial fractions, *College Math. J.* **22** (1991) 413–415.
3. P. T. Joshi, Efficient techniques for partial fractions, *College Math. J.* **14** (1983) 110–118.
4. J. E. Nymann, An alternative for partial fractions (part of the time), *College Math. J.* **14** (1983) 60–61.
5. P. Schultz, An algebraic approach to partial fractions, *College Math. J.* **14** (1983) 346–348.
6. M. R. Spiegel, Partial fractions with repeated linear or quadratic factors, *Amer. Math. Monthly* **57** (1950) 180–181.
7. T. N. Subramaniam and D. E. G. Malm, How to integrate rational functions, *Amer. Math. Monthly* **99** (1992) 762–772.
8. J. Wiener, An algebraic approach to partial fractions, *College Math. J.* **17** (1986) 71–72.

Partial Fractions by Substitution

David A. Rose

The standard method for finding the partial fraction decomposition for a rational function involves solving a system of linear equations. In this note, we present a quick method for finding the partial fraction decomposition of a rational function in the special case when the denominator is a power of a single linear or irreducible quadratic factor, that is, the denominator is either $(ax+b)^k$ or $(ax^2+bx+c)^k$ with $4ac > b^2$. For example, we note that substituting $t+2$ for x and then expanding the numerator transforms

$$\frac{x^2+4x-3}{(x-2)^3} \quad \text{to} \quad \frac{t^2+8t+9}{t^3}.$$

Since this last expression splits into

$$\frac{1}{t} + \frac{8}{t^2} + \frac{9}{t^3},$$

it follows that our original function has

$$\frac{1}{x-2} + \frac{8}{(x-2)^2} + \frac{9}{(x-2)^3}$$

as its partial fraction decomposition. We observe that the numbers 9, 8, and 1 in the numerators of the decomposition could also have been obtained as the remainders by successive division of x^2+4x-3 by $x-2$. This method was considered by Kung [4] in this journal. Our substitution-expansion method avoids such repeated division as well as the usual systems of equations. (For other methods, see for example [1, 2, 3, 5, 6, 7].) It also works equally well on improper fractions, eliminating the need for the initial polynomial division. For our discussion of the general problem of this type, we assume that the denominator is monic (that is, $a=1$), and consider a rational function

$$R(x) = \frac{N(x)}{D(x)}$$

in the linear and irreducible quadratic cases separately.

The linear case, $D(x) = (x + b)^k$. Let $x = t - b$. Then

$$\frac{N(x)}{(x+b)^k} = \frac{G(t)}{t^k},$$

say, and this immediately yields the desired decomposition. The coefficients of G are the coefficients in the numerators of the partial fractions and can be obtained by binomial expansion or as Taylor polynomial coefficients. We will use binomial expansion, although Brenke [1] used Taylor expansion coefficients in a more general case than ours (he required only that the denominator of the fraction have no irreducible quadratic factors).

To illustrate our method, we decompose the function in Kung's first example [4],

$$\frac{x^4 + 2x^3 - x^2 + 5}{(2x - 1)^5}.$$

After factoring out the 2 and letting $x = t + \frac{1}{2}$, straightforward algebra converts this to

$$\frac{1}{32}\left(\frac{1}{t} + \frac{4}{t^2} + \frac{\frac{7}{2}}{t^3} + \frac{1}{t^4} + \frac{\frac{81}{16}}{t^5}\right),$$

so our decomposition is

$$\frac{\frac{1}{16}}{2x - 1} + \frac{\frac{1}{2}}{(2x - 1)^2} + \frac{\frac{7}{8}}{(2x - 1)^3} + \frac{\frac{1}{2}}{(2x - 1)^4} + \frac{\frac{81}{16}}{(2x - 1)^5}.$$

To see how the substitution method works for an improper fraction with the same denominator type, consider

$$R(x) = \frac{2x^5 - x^3 + x - 4}{(x + 2)^3}.$$

Taking $t = x + 2$, we get

$$2t^2 - 20t + 79 - \frac{154}{t} + \frac{149}{t^2} - \frac{62}{t^3},$$

which gives us the decomposition

$$R(x) = 2x^2 - 12x + 47 - \frac{154}{x + 2} + \frac{149}{(x + 2)^2} - \frac{62}{(x + 2)^3}.$$

The irreducible quadratic case, $D(x) = (x^2 + bx + c)^k$. We first complete the square in order to express $D(x)$ in the form $[(x + p)^2 + q]^k$. Now make two substitutions, first $t = x + p$ as before, and then $s = t^2 + q$. We illustrate in the following example:

$$R(x) = \frac{4x^5 - 17x^4 + 45x^3 - 58x^2 + 48x - 8}{(x^2 - 2x + 3)^3}.$$

Then after completing the square, we get

$$\frac{4x^5 - 17x^4 + 45x^3 - 58x^2 + 48x - 8}{[(x-1)^2 + 2]^3}.$$

Setting $t = x - 1$ and simplifying, we get

$$\frac{4t^5 + 3t^4 + 17t^3 + 15t^2 + 19t + 14}{(t^2 + 2)^3}.$$

Letting $s = t^2 + 2$ eventually results in

$$\frac{4t + 3}{s} + \frac{t + 3}{s^2} + \frac{t - 4}{s^3},$$

which gives us the decomposition

$$R(x) = \frac{4x - 1}{x^2 - 2x + 3} + \frac{x + 2}{(x^2 - 2x + 3)^2} + \frac{x - 5}{(x^2 - 2x + 3)^3}.$$

We mentioned earlier Brenke's method using Taylor expansion coefficients which applies to rational functions whose denominators have more than one prime factor. Unfortunately, when irreducible quadratic factors are present, the method requires first, complex linear factorization of quadratic factors, and then, that partial fractions be recombined at the end of the process to recover fractions in real polynomial form.

References

1. W. C. Brenke, On the resolution of a fraction into partial fractions, *Amer. Math. Monthly* **36** (1929) 319–323.
2. X.-C. Huang, A short cut to partial fractions, *College Math. J.* **22** (1991) 413–415.
3. P. T. Joshi, Efficient techniques for partial fractions, *Two-Year College Math. J.* **14** (1983) 110–118.
4. S. H. Kung, Partial fraction decomposition by division, *College Math. J.* **37** (2006) 132–134.
5. P. Schultz, An algebraic approach to partial fractions, *Two-Year College Math. J.* **14** (1983) 346–348.
6. M. R. Spiegel, Partial fractions with repeated linear or quadratic factors, *Amer. Math. Monthly* **57** (1950) 180–181.
7. J. Wiener, An algebraic approach to partial fractions, *College Math. J.* **17** (1986) 71–72.

Proof Without Words: A Partial Fraction Decomposition

Steven J. Kifowit

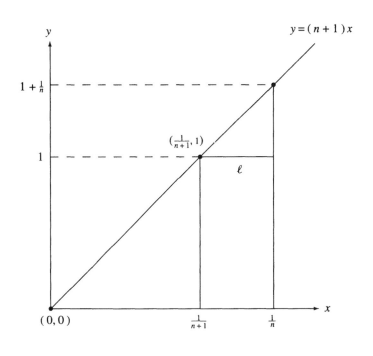

$$\ell = \frac{1}{n} - \frac{1}{n+1} \quad \text{and} \quad \frac{\frac{1}{n+1}}{1} = \frac{\ell}{\frac{1}{n}}$$

$$\Downarrow$$

$$\frac{1}{n} - \frac{1}{n+1} = \frac{1}{n}\left(\frac{1}{n+1}\right)$$

Four Crotchets on Elementary Integration

Leroy F. Meyers

All of the results below are well known to too few people.

1. Integral of Exponential Times Polynomial. Most students, when confronted with

$$\int_0^2 e^{3x}(x^3 + 6x^2 + 11x + 6)\, dx,$$

write the integral as the sum of four integrals and evaluate them separately, using integration by parts six times altogether. They haven't learned that a polynomial is a single function, so that only three successive integrations by parts are needed. However, explicit integration by parts can be avoided by use of a single formula, which is more useful than many of the integration formulas customarily memorized by students.

Let P be a polynomial, and m a nonzero constant. Then

$$\int e^{mx}P(x)\, dx = \frac{e^{mx}}{m}\left(P(x) - \frac{P'(x)}{m} + \frac{P''(x)}{m^2} - \frac{P'''(x)}{m^3} + \cdots\right) + c.$$

(This is essentially exercise 15 on p. 225 of Courant [2].) The proof is by repeated use of the recurrence formula

$$\int e^{mx}P(x)\, dx = \frac{e^{mx}}{m}P(x) - \int e^{mx}\frac{P'(x)}{m}\, dx,$$

obtained by a single integration by parts.

Similar but slightly more complicated formulas may be obtained in the same way for $\int (\sin(mx))P(x)\, dx$ and $\int (\cos(mx))P(x)\, dx$.

Many students haven't learned how to use integration by parts with definite integrals. Thus, $\int_0^2 e^x x\, dx$ is often "evaluated" as

$$e^x x - \int_0^2 e^x\, dx = e^x x - e^2 + 1,$$

which incorrectly depends on x. A better formulation is

$$\int_a^b u(x)v'(x)\,dx = [u(x)v(x)]_{x=a}^b - \int_a^b u'(x)v(x)\,dx,$$

or

$$\int_a^b u(x)v'(x)\,dx = \left[u(x)v(x) - \int u'(x)v(x)\,dx\right]_{x=a}^b.$$

The second form is useful when the separate terms in the first form are undefined, as often occurs in improper integrals.

2. Using limits of integration. The notation $[F(x)]_{x=a}^b$ for $F(b) - F(a)$ avoids the ambiguities associated with the commonly used $F(x)|_a^b$ in two ways: (1) it specifies the letter to be substituted **for**, here x; and (2) it specifies the scope of the formula to be substituted **into**, here $F(x)$. Without these specifications, it is uncertain what $x - 2y + xy^2|_3^5$ means. The notation $[F(x)]_{x=a}^b$ can be used more generally to mean $\lim_{x \nearrow b} F(x) - \lim_{x \searrow a} F(x)$, where the limits are taken from inside the interval (a,b), as with improper integrals. Use of slanted arrows for one-sided limits avoids the confusion caused by nonnumbers like $-(2^+)$ and $(-2)^+$.

3. Integration by substitution. Many students evaluate

$$\int_0^4 12(2x-3)^5\,dx$$

by first multiplying out. (They don't know the binomial theorem.) More enterprising students make the substitution $u := 2x - 3$, evaluate an indefinite integral, and then substitute back before using the limits of integration:

$$\int 12(2x-3)^5\,dx = \int 6u^5\,du = u^6 + c = (2x-3)^6 + c.$$

However, it is usually simpler to find the appropriate limits of integration in terms of u (when t goes from 0 to 4, then u goes from -3 to 5) instead of substituting back. But many students do not change the limits of integration properly, and so write

$$u^6|_0^4 = 4096 \text{ instead of } [u^6]_{u=-3}^5 = 14896.$$

(Note the use of subscript u as a reminder that the limits of integration are now for u.)

4. Differential equations with initial conditions. To solve a first-order differential equation with initial condition, the procedure followed by nearly all textbooks is to find a general solution of the differential equation by means of indefinite integra-

Four Crotchets on Elementary Integration

tion, and then to find the constant of integrity by substitution. This interrupts the sequence of steps (equivalences) in the solution. It is better to use **definite** integration, as in the following example.

Express x in terms of t, where

$$dx/dt = 3x + 6 \text{ for all real } t, \text{ and } x = -5 \text{ when } t = 2.$$

The differential equation is both separable and linear. In general, such equations are best treated as linear equations (see [1], pp. 34–35, 39–40, etc.), since doing so avoids the error-prone process of getting rid of logarithms. After rewriting the equation in standard linear form and multiplying by the integrating factor $e^{\int(-3)dt}$ (choosing one antiderivative), we obtain $e^{-3t}\,dx/dt - 3e^{-3t}x = 6e^{-3t}$ for all real t, with initial condition. Now we integrate with respect to t from 2 to t, using the corresponding limits -5 and x for x, and obtain

$$[e^{-3t}x]_{(t,x)=(2,-5)}^{(t,x)} = [-2e^{-3t}]_{t=2}^{t} \quad \text{for all real } t$$

$$\Leftrightarrow e^{-3t}x + 5e^{-6} = -2e^{-3t} + 2e^{-6} \quad \text{for all real } t$$

$$\Leftrightarrow x = -2 - 3e^{3t-6} \quad \text{for all real } t.$$

In the conventional method, after integrating indefinitely and solving for x in terms of t, we obtain

$$x = -2 + ce^{3t} \quad \text{for all } t.$$

Use of the initial condition then requires "unsolving" the equation for c.

If, however, we solve the equation as a separable equation, we obtain

$$\frac{dx}{x+2} = 3\,dt \quad \text{with initial conditions}$$

$$\Leftrightarrow [\ln(x+2)]_{x=-5}^{x} = [3t]_{t=2}^{t} \quad \text{for all } t$$

$$\Leftrightarrow \ln(x+2) - \ln(-3) = 3t - 6 \quad \text{for all } t\,(??!!)$$

$$\Leftrightarrow x + 2 = -3e^{3t-6} \quad \text{for all } t\, (\text{tricky!}).$$

The antiderivative for the left side (without constant) is often written using absolute values as

$$\ln|x+2|,$$

which, however, does not indicate that, because of the initial condition, $x + 2$ must have the same sign as -3. In general, without absolute values,

$$\int_a^b \frac{1}{x}\, dx = \ln\frac{b}{a}, \quad \textbf{provided that } ab > 0.$$

References

1. Ralph Palmer Agnew, *Differential Equations*, McGraw-Hill, New York and London, 1942.
2. R. Courant, *Differential and Integral Calculus*, Vol. 1, 2nd ed., translated by E. J. McShane, Blackie and Son, London and Glasgow, 1937.

An Application of Geography to Mathematics: History of the Integral of the Secant

V. Frederick Rickey and Philip M. Tuchinsky

Every student of the integral calculus has done battle with the formula

$$\int \sec\theta \, d\theta = \ln|\sec\theta + \tan\theta| + c. \tag{1}$$

This formula can be checked by differentiation or "derived" by using the substitution $u = \sec\theta + \tan\theta$, but these ad hoc methods do not make the formula any more understandable. Experience has taught us that this troublesome integral can be motivated by presenting its history. Perhaps our title seems twisted, but the tale to follow will show that this integral should be presented not as an application of mathematics to geography, but rather as an application of geography to mathematics.

The secant integral arose from cartography and navigation, and its evaluation was a central question of mid-seventeenth century mathematics. The first formula, discovered in 1645 before the work of Newton and Leibniz, was

$$\int \sec\theta \, d\theta = \ln\left|\tan\left(\frac{\theta}{2} + \frac{\pi}{4}\right)\right| + c, \tag{2}$$

which is a trigonometric variant of (1). This was discovered, not through any mathematician's cleverness, but by a serendipitous historical accident when mathematicians and cartographers sought to understand the Mercator map projection. To see how this happened, we must first discuss sailing and early maps so that we can explain why Mercator invented his famous map projection.

From the time of Ptolemy (c. 150 A.D.) maps were drawn on rectangular grids with one degree of latitude equal in length to one degree of longitude. When restricted to a small area, like the Mediterranean, they were accurate enough for sailors. But in the age of exploration, the Atlantic presented vast distances and higher latitudes, and so the navigational errors due to using the "plain charts" became apparent.

The magnetic compass was in widespread use after the thirteenth century, so directions were conveniently given by distance and compass bearing. Lines of fixed compass direction were called **rhumb** lines by sailors, and in 1624 Willebrord Snell dubbed them **loxodromes**. To plan a journey one laid a straightedge on a map between origin and destination, then read off the compass bearing to follow. But rhumb lines are spirals on the globe and curves on a plain chart —facts sailors had difficulty understanding. They needed a chart where the loxodromes were represented as straight lines.

It was Gerardus Mercator (1512–1594) who solved this problem by designing a map where the lines of latitude were more widely spaced when located further from the equator. On his famous world map of 1569 ([1], p. 46), Mercator wrote:

> In making this representation of the world we had...to spread on a plane the surface of the sphere in such a way that the positions of places shall correspond on all sides with each other both in so far as true direction and distance are concerned and as concerns correct longitudes and latitudes... . With this intention we have had to employ a new proportion and a new arrangement of the meridians with reference to the parallels. ... It is for these reasons that we have progressively increased the degrees of latitude towards each pole in proportion to the lengthening of the parallels with reference to the equator.

Mercator wished to map the sphere onto the plane so that both angles and distances are preserved, but he realized this was impossible. He opted for a conformal map (one which preserves angles) because, as we shall see, it guaranteed that loxodromes would appear on the map as straight lines.

Unfortunately, Mercator did not explain how he "progressively increased" the distances between parallels of latitude. Thomas Harriot (c. 1560–1621) gave a mathematical explanation in the late 1580's, but neither published his results nor influenced later work (see [6], [11]-[15]). In his *Certaine Errors in Navigation*... [22] of 1599, Edward Wright (1561–1615) finally gave a mathematical method for constructing an accurate Mercator map. The Mercator map has its meridians of longitude placed vertically and spaced equally. The parallels of latitude are horizontal and unequally spaced. Wright's great achievement was to show that the parallel at latitude θ should be stretched by a factor of $\sec\theta$ when drawn on the map. Let us see why.

FIGURE 1 represents a wedge of the earth, where AB is on the equator, C is the center of the earth, and T is the north pole. The parallel at latitude θ is a circle, with center P, that includes arc MN between the meridians AT and BT. Thus BC and NP are parallel and so angle $PNC = \theta$. The "triangles" ABC and MNP are similar figures, so

$$\frac{AB}{MN} = \frac{BC}{NP} = \frac{NC}{NP} = \sec\theta,$$

or $AB = MN\sec\theta$. Thus when MN is placed on the map it must be stretched horizontally by a factor $\sec\theta$. (This argument is not the one used by Wright [22]. His argument is two dimensional and shows that $BC = NP\sec\theta$.)

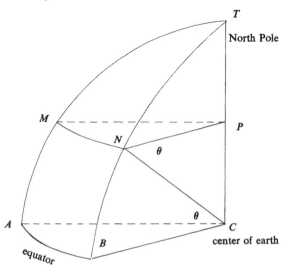

FIGURE 1.

Suppose we can construct a map where angles are preserved, i.e., where the globe-to-map function is conformal. Then a loxodrome, which makes the same angle with each meridian, will appear on this map as a curve which cuts all the map's meridians (a family of parallel straight lines) at the same angle. Since a curve that cuts a family of parallel straight lines at a fixed angle is a straight line, loxodromes on the globe will appear straight on the map. Conversely, if loxodromes are mapped to straight lines, the globe-to-map function must be conformal.

In order for angles to be preserved, the map must be stretched not only horizontally, but also vertically, by $\sec\theta$; this, however, requires an argument by infinitesimals. Let $D(\theta)$ be the distance *on the map* from the equator to the parallel of latitude θ, and let dD be the infinitesimal change in D resulting from an infinitesimal change $d\theta$ in θ. If we stretch vertically by $\sec\theta$, i.e., if

$$dD = \sec\theta\, d\theta$$

then an infinitesimal region on the globe becomes a similar region on the map, and so angles are preserved. Conversely, if the map is to be conformal the vertical multiplier must be $\sec\theta$.

Finally, "by perpetuall addition of the Secantes," to quote Wright, we see that the distance on the map from the equator to the parallel at latitude θ is

$$D(\theta) = \int_0^\theta \sec\theta\, d\theta.$$

Of course Wright did not express himself as we have here. He said ([2], pp. 312–313):

> the parts of the meridian at euery poynt of latitude must needs increase with the same proportion wherewith the Secantes or hypotenusae of the arke, intercepted betweene those pointes of latitude and the aequinoctiall [equator] do increase. ... For...by perpetuall addition of the Secantes answerable to the latitudes of each point or parallel vnto the summe compounded of all former secantes,...we may make a table which shall shew the sections and points of latitude in the meridians of the nautical planisphaere: by which sections, the parallels are to be drawne.

Wright published a table of "meridional parts" which was obtained by taking $d\theta = 1'$ and then computing the Riemann sums for latitudes below 75°. Thus the methods of constructing Mercator's "true chart" became available to cartographers.

Wright also offered an interesting physical model. Consider a cylinder tangent to the earth's equator and imagine the earth to "swal [swell] like a bladder." Then identify points on the earth with the points on the cylinder that they come into contact with. Finally unroll the cylinder; it will be a Mercator map. This model has often been misinterpreted as the cylindrical projection (where a light source at the earth's center projects the unswollen sphere onto its tangent cylinder), but this projection is not conformal.

We have established half of our result, namely that the distance on the map from the equator to the parallel at latitude θ is given by the integral of the secant. It remains to show that it is also given by $\ln|\tan(\frac{\theta}{2} + \frac{\pi}{4})|$.

In 1614 John Napier (1550–1617) published his work on logarithms. Wright's authorized English translation, *A Description of the Admirable Table of Logarithms*, was published in 1616. This contained a table of logarithms of sines, something much needed by astronomers. In 1620 Edmund Gunter (1581–1626) published a table of common logarithms of tangents in his *Canon triangulorum*. In the next twenty years numerous tables of logarithmic tangents were published and so were widely available. (Not even a table of secants was available in Mercator's day.)

In the 1640's Henry Bond (c. 1600–1678), who advertised himself as a "teacher of navigation, survey and other parts of the mathematics," compared Wright's table of meridional parts with a log-tan table and discovered a close agreement. This serendipitous accident led him to conjecture that $D(\theta) = \ln|\tan(\frac{\theta}{2} + \frac{\pi}{4})|$. He published this conjecture in 1645 in Norwood's *Epitome of Navigation*. Mainly through the correspondence of John Collins this conjecture became widely known. In fact, it became one of the outstanding open problems of the mid-seventeenth century, and was attempted by such eminent mathematicians as Collins, N. Mercator (no relation), Wilson, Oughtred and John Wallis. It is interesting to note that young Newton was aware of it in 1665 [18], [21].

The "Learned and Industrious *Nicolaus Mercator*" in the very first volume of the *Philosophical Transactions* of the Royal Society of London was "willing to lay a *Wager* against any one or more persons that have a mind to engage... *Whether the Artificial* [logarithmic] *Tangent-line be the true Meridian-line*, yea or no?" ([9], pp. 217–218). Nicolaus Mercator is not, as the story is often told, wagering that he knows more about logarithms than his contemporaries; rather, he is offering a prize for the solution of an open problem.

The first to prove the conjecture was, to quote Edmund Halley, "the excellent Mr. *James Gregory* in his *Exercitationes Geometricae*, published *Anno* 1668, which he did, not without a long train of Consequences and Complication of Proportions, whereby the evidence of the Demonstration is in a great measure lost, and the Reader wearied before he attain it" ([7], p. 203). Judging by Turnbull's modern elucidation [19] of Gregory's proof, one would have to agree with Halley. At any rate, Gregory's proof could not be presented to today's calculus students, and so we omit it here.

Isaac Barrow (1630–1677) in his *Geometrical Lectures* (Lect. XII, App. I) gave the first "intelligible" proof of the result, but it was couched in the geometric idiom of the day. It is especially noteworthy in that it is the earliest use of partial fractions in integration. Thus we reproduce it here in modern garb:

$$\int \sec\theta\, d\theta = \int \frac{1}{\cos\theta}\, d\theta$$

$$= \int \frac{\cos\theta}{\cos^2\theta}\, d\theta$$

$$= \int \frac{\cos\theta}{1-\sin^2\theta}\, d\theta$$

$$= \int \frac{\cos\theta}{(1-\sin\theta)(1+\sin\theta)}\, d\theta$$

$$= \frac{1}{2}\int \frac{\cos\theta}{1-\sin\theta} + \frac{\cos\theta}{1+\sin\theta}\, d\theta$$

$$= \frac{1}{2}[-\ln|1-\sin\theta| + \ln|1+\sin\theta|] + c$$

$$= \frac{1}{2}\ln\left|\frac{1+\sin\theta}{1-\sin\theta}\right| + c$$

$$= \frac{1}{2}\ln\left|\frac{1+\sin\theta}{1-\sin\theta}\cdot\frac{1+\sin\theta}{1+\sin\theta}\right| + c$$

$$= \frac{1}{2}\ln\left|\frac{(1+\sin\theta)^2}{1-\sin^2\theta}\right| + c$$

$$= \frac{1}{2}\ln\left|\frac{(1+\sin\theta)^2}{(\cos\theta)^2}\right| + c$$

$$= \ln\left|\frac{1+\sin\theta}{\cos\theta}\right| + c$$

$$= \ln|\sec\theta + \tan\theta| + c.$$

We became interested in this topic after noting one line of historical comment in Spivak's excellent *Calculus* (p. 326). As we ferreted out the details and shared them with our students, we found an ideal soapbox for discussing the nature of mathematics, the process of mathematical discovery, and the role that mathematics plays in the world. We found this so useful in the classroom that we have prepared a more detailed version for our students [17].

References

The following works contain interesting information pertaining to this paper. The best concise source of information about the individuals mentioned in this paper is the excellent *Dictionary of Scientific Biography*, edited by C. C. Gillespie.

[1] Anonymous, Gerard Mercator's Map of the World (1569), Supplement no. 2 to Imago Mundi, 1961.
[2] Florian Cajori, On an integration ante-dating the integral calculus, Bibliotheca Mathematica, 3rd series, 14(1915) 312–319.
[3] H. S. Carslaw, The story of Mercator's map. A chapter in the history of mathematics, Math. Gaz., 12(1924) 1–7.
[4] Georgina Dawson, Edward Wright, mathematician and hydrographer, Amer. Neptune, 37(1977) 174–178.
[5] Jacques Delevsky, L'invention de la projection de Mercator et les enseignements de son histoire, Isis, 34(1942) 110–117.
[6] Frank George, Hariot's meridional parts, J. Inst. Navigation, London, 21(1968) 82–83.
[7] E. Halley, An easie demonstration of the analogy of the logarithmick tangents to the meridian line or sum of secants: with various methods for computing the same to the utmost exactness, Philos. Trans., Roy. Soc. London, 19(1695–97) 202–214.
[8] Johannes Keuning, The history of geographical map projections until 1600, Imago Mundi, 12(1955) 1–24.
[9] Nicolaus Mercator, Certain problems touching some points of navigation, Philos. Trans., Roy. Soc. London, 1(1666) 215–218.
[10] E. J. S. Parsons and W. F. Morris, Edward Wright and his work, Imago Mundi, 3(1939) 61–71.
[11] Jon V. Pepper, Hariot's calculation of the meridional parts as logarithmic tangents, Archive for History of Exact Science, 4(1967) 359–413.
[12] _____, A note on Hariot's method of obtaining meridional parts, J. Inst. Navigation, London, 20(1967) 347–349.
[13] _____, The study of Thomas Hariot's manuscripts, II: Hariot's unpublished papers, History of Science, 6 (1967) 17–40.
[14] _____, Hariot's earlier work on mathematical navigation: theory and practice. With an appendix, 'The early development of the Mercator chart,' in Thomas Hariot: Renaissance Scientist, Clarendon Press, Oxford, 1974, John W. Shirley, editor, pp. 54–90.
[15] D. H. Sadler and Eva G. R. Taylor, The doctrine of nauticall triangles compendious. Part I–Thomas Hariot's manuscript (by Taylor). Part II–Calculating the meridional parts (by Sadler), J. Inst. Navigation, London, 6(1953) 131–147.
[16] Eva G. R. Taylor, The Haven-Finding Art, Hollis and Carter, London, 1971.
[17] P. M. Tuchinsky, Mercator's World Map and the Calculus, Modules and Monographs in Undergraduate Mathematics and its Applications (UMAP) Project, Education Development Center, Newton, Mass., 1978.
[18] H. W. Turnbull, editor, The Correspondence of Isaac Newton, Cambridge Univ. Press, 1959–1960, vol. 1, pp. 13–16, and vol. 2, pp. 99–100.
[19] H. W. Turnbull, James Gregory Tercentenary Memorial Volume, G. Bell & Sons, London, 1939, pp. 463–464.
[20] D. W. Waters, The Art of Navigation in England in Elizabethan and Early Stuart Times, Yale Univ. Press, New Haven, 1958.
[21] D. T. Whiteside, editor, The Mathematical Papers of Isaac Newton, vol. 1, Cambridge Univ. Press, 1967, pp. 466–467, 473–475.
[22] Edward Wright, Certaine Errors in Navigation, Arising either of the ordinaire erroneous making or vsing of the sea Chart, Compasse, Crosse staffe, and Tables of declination of the Sunne, and fixed Starres detected and corrected, Valentine Sims, London, 1599. Available on microfilm as part of Early English Books 1475–1640, reels 539 and 1018 (these two copies from 1599 have slightly different title pages). The preface and table of meridional parts have been reproduced as "Origin of meridional parts," International Hydrographic Review, 8(1931) 84–97.

How to Avoid the Inverse Secant (and Even the Secant Itself)

S. A. Fulling

An irritating technicality that must be handled somehow in a standard calculus course is the definition of the inverse secant function at negative arguments. There is no particularly natural choice of principal branch from among the infinitely many candidates, and the derivatives of different branches can differ in sign (see the figure below). Some textbook authors adopt the convention

$$\pi \leq \sec^{-1} x < \frac{3\pi}{2} \quad \text{for } x \leq -1, \qquad \frac{d}{dx} \sec^{-1} x = \frac{1}{x\sqrt{x^2 - 1}}, \tag{1}$$

while others prefer

$$\frac{\pi}{2} < \sec^{-1} x \leq \pi \quad \text{for } x \leq -1, \qquad \frac{d}{dx} \sec^{-1} x = \frac{1}{|x|\sqrt{x^2 - 1}}. \tag{2}$$

Then authors and lecturers have a responsibility to warn students about the existence of the other convention.

One might have expected that after more than a decade of calculus reform, the secant function and its inverse would have been de-emphasized to the vanishing point, along with its even less useful siblings, cosecant and cotangent, and their inverses. The persistence of $\sec^{-1}$ presumably stems from the perceived need to provide a formula (see (17)) for the indefinite integral

$$\int \frac{dx}{x\sqrt{x^2 - 1}} \tag{3}$$

by inverting whichever of the differentiation formulas (1) and (2) one adopts. More generally, it is argued that the secant and tangent functions unavoidably arise when algebraic functions are integrated by trigonometric substitutions, so students should have at least a nodding acquaintance with the variety of integrals involving them.

There is, however, an alternative approach to integrals involving $\sqrt{x^2 - 1}$ or $\sqrt{x^2 + 1}$, which has unaccountably fallen out of favor in recent decades: *hyperbolic substitution.*

The hyperbolic sine and cosine are defined by

$$\cosh u \equiv \tfrac{1}{2}(e^u + e^{-u}), \qquad \sinh u \equiv \tfrac{1}{2}(e^u - e^{-u}). \qquad (4)$$

Their usefulness as replacements for trigonometric functions in integration by substitution stems from the identities

$$\cosh^2 u - \sinh^2 u = 1, \qquad \frac{d}{du}\cosh u = \sinh u, \qquad \frac{d}{du}\sinh u = \cosh u. \qquad (5)$$

The hyperbolic functions (4) are important in their own right, constituting the natural basis of solutions of the differential equation

$$\frac{d^2 y}{du^2} = y$$

satisfying unit Dirichlet and Neumann initial data at $u = 0$; in this role they are indispensable in upper-division courses in applied mathematics. In a linear algebra course, (4) provides a beautiful example of a change of basis, of undeniable practical importance, in a real vector space with no intrinsic inner product or geometrical interpretation. Most important in the present context, hyperbolic substitutions are much simpler and nicer than trigonometric substitutions of the tangent and secant varieties. For one thing, the inverse hyperbolic functions can be expressed in terms of more elementary functions (logarithms and square roots—see (11) and (12)). Furthermore, the branch structure of these inverses is very simple: sinh is bijective, and cosh has a two-branched inverse (just like the square root), with no multiples of $\frac{\pi}{2}$ to be memorized. Nevertheless, many teachers of calculus have a strange dislike for the hyperbolic functions and prefer not to cover them at all. (An independent case in favor of the hyperbolics has been made by Gearhart and Shultz [1].)

In this article I hope to convince the reader that there is nothing that the secant and inverse secant do in the traditional "techniques of integration" chapter that cannot be done better by the hyperbolic sine and cosine and their inverses. *It is time for sec, csc, cot, sec^{-1}, csc^{-1}, and cot^{-1} to be retired from our calculus syllabus, replaced by sinh and cosh. Our students will learn about two elegant and useful transcendental functions while being freed from six complicated and boring ones.*

The rest of the article has two parts. First, we derive several alternative antiderivative formulas, (13), (14), (16), with clear advantages over the traditional formulas (17). Then we'll see that hyperbolic substitution provides an easy way of evaluating or evading (depending on context) the difficult integrals of powers of the secant that crop up in trigonometric substitutions.

The integral formerly known as sec^{-1}

Since the integration problem (3) is the inverse secant's alleged reason for existence, let us see what a hyperbolic substitution does to it. Recall first that (3) makes no sense (in real analysis) unless $|x| \geq 1$. We assume temporarily that $x > 0$, hence $x \geq 1$, and set

$$x = \cosh u \quad (u \geq 0); \qquad dx = \sinh u\, du, \qquad x^2 - 1 = \sinh^2 u. \qquad (6)$$

Therefore,
$$\int \frac{dx}{x\sqrt{x^2-1}} = \int \frac{du}{\cosh u}. \tag{7}$$

(As an aside, note that not much is gained here by introducing the name sech u for the integrand of (7).) To continue we need

$$\int \frac{du}{\cosh u} = 2\tan^{-1}(e^u) + C, \tag{8}$$

which can be either verified by differentiation or "discovered" through these intermediate steps:

$$\int \frac{2\,du}{e^u + e^{-u}} = \int \frac{2e^u\,du}{e^{2u}+1} = \int \frac{2\,dv}{v^2+1} = 2\tan^{-1} v + C.$$

Formula (8) is arguably less recondite than

$$\int \sec\theta\,d\theta = \ln|\sec\theta + \tan\theta| + C, \tag{9}$$

which inevitably plays a leading role in traditional treatments of trigonometric substitution. (And alas, there is seldom time in class to present the interesting history [3] of (9).)

Remark. The function

$$\operatorname{gd} u \equiv 2\tan^{-1}(e^u) - \frac{\pi}{2}, \tag{10}$$

which satisfies the convenient initial condition $\operatorname{gd} 0 = 0$, is called the *Gudermannian* and can be used to express hyperbolic functions as trigonometric functions (of a different variable) and vice versa [2, Secs. 1.48 and 1.49]; for example, if $\theta = \operatorname{gd} u$, then $\sec\theta = \cosh u$, a formula quite pertinent to the equivalence of (14) and (16) below. For more on the history and applications of gd, see Robertson [4].

As previously remarked, one of the great charms of hyperbolic functions (as opposed to trig functions) is that their inverses can be expressed in terms of already familiar functions:

$$\sinh^{-1} x = \ln(x + \sqrt{x^2+1}) \quad \text{for all } x; \tag{11}$$

$$\cosh^{-1} x = \ln(x + \sqrt{x^2-1}) \quad \text{for } x \geq 1, \tag{12}$$

where $\cosh^{-1}$ denotes the positive branch. Note that the right-hand side of (11) is indeed an odd function, though it may not look like one. To prove (11) and (12), simply apply the definitions (4) to their right-hand sides and simplify down to x.

Combining (7), (8), (6), and (12), one arrives at

$$\int \frac{dx}{x\sqrt{x^2-1}} = 2\tan^{-1}(x + \sqrt{x^2-1}) + C, \tag{13}$$

at least for $x \geq 1$. It is now an elementary, though lengthy, exercise in differentiation to verify (13) for $x \leq -1$ as well. However, an alternative approach leads to a neater result. For negative x we can let $x = -\cosh u$ and repeat the previous calculation to obtain

$$\int \frac{dx}{x\sqrt{x^2-1}} = 2\tan^{-1}(|x| + \sqrt{x^2-1}) + C, \qquad (14)$$

which of course agrees with (13) in the positive case. The formula (14) is *even in x*, whereas (13) is not. This is not a contradiction: The constants of integration on the two disconnnected domains, $x \geq 1$ and $x \leq -1$, are independent. At negative x, (13) and (14) with the same C simply differ by a constant (see the figure).

To relate these formulas to the traditional ones based on (1) and (2), we start with a known but nontrivial trigonometric identity,

$$\sec\theta + \tan\theta = \tan\left(\frac{\theta}{2} + \frac{\pi}{4}\right), \qquad (15)$$

whose proof we omit. Now let $|x| = \sec\theta$ with θ in the first quadrant. Then $\tan\theta = \sqrt{x^2-1}$, and (14) becomes, up to the arbitrary constant,

$$\int \frac{dx}{x\sqrt{x^2-1}} = 2\tan^{-1}(\sec\theta + \tan\theta)$$

$$= 2\left(\frac{\theta}{2} + \frac{\pi}{4}\right)$$

$$= \sec^{-1}|x| + \frac{\pi}{2}.$$

In other words,

$$\int \frac{dx}{x\sqrt{x^2-1}} = \sec^{-1}|x| + C \qquad (16)$$

is an alternative formula for the indefinite integral (for either sign of x). Similarly, after appropriate bookkeeping with quadrants, one can relate (13) or (14) at negative x to one's favorite definition of $\sec^{-1}$ there.

It is to be hoped that any authors and lecturers who are still unconvinced of the virtues of hyperbolic substitution will at least adopt (16) in place of either of the traditional formulas,

$$\int \frac{dx}{x\sqrt{x^2-1}} = \sec^{-1}x + C, \qquad \int \frac{dx}{|x|\sqrt{x^2-1}} = \sec^{-1}x + C. \qquad (17)$$

For (16) is analogous to the familiar formula

$$\int \frac{dx}{x} = \ln|x| + C, \qquad (18)$$

it is even in x (unlike either of (17)), and it avoids any reference to the inverse secant function at negative arguments!

Our conclusions so far are summarized and made more precise in the following theorem and figure.

Theorem. *Define*

$$f_1(x) = 2\tan^{-1}\left(|x| + \sqrt{x^2-1}\right),$$

$$f_2(x) = 2\tan^{-1}\left(x + \sqrt{x^2-1}\right),$$

$$f_3(x) = \sec^{-1}|x|.$$

Further, let $f_4(x)$ be the branch of the inverse secant determined by (1), and $f_5(x)$ the branch determined by (2) (with $0 \leq \sec^{-1} x < \pi/2$ for $x \geq 1$). Then

(a) *f_1 and f_3 (and only they) are even functions, and $f_1 = f_3 + \pi/2$.*
(b) *On the interval $[1, \infty)$, all five functions are antiderivatives of $1/(x\sqrt{x^2-1})$, and $f_1 = f_2$, $f_3 = f_4 = f_5$.*
(c) *On the interval $(-\infty, -1]$, the first four functions and $-f_5$ are antiderivatives of $1/(x\sqrt{x^2-1})$, and $f_2 = f_1 - \pi$, $f_4 = f_3 + \pi$, $f_5 = \pi - f_3$.*

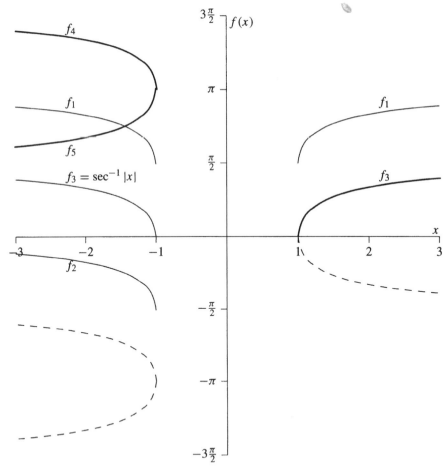

Figure 1. Graphs of the functions f_1–f_5 discussed in the theorem. (Redundant names for the curves on the right are omitted.) Thick curves are the traditional rival branches of the inverse secant. Dashed curves are graphs of other logically possible definitions of $\sec^{-1}$. Thin solid curves are not branches of $\sec^{-1} x$, but nevertheless are useful as antiderivatives of $1/x\sqrt{x^2-1}$.

Integrating powers of the secant

Trigonometric substitution has an unpleasant habit of resulting in integrals like the one in (9), or, more generally, $\int \sec^p \theta \, d\theta$ with p a positive integer, that look at least as hard as the original algebraic integration problem. Making the natural substitution $x = \tan \theta$ ($-\frac{\pi}{2} < \theta < \frac{\pi}{2}$) yields

$$\int \sec^p \theta \, d\theta = \int (x^2 + 1)^{(p-2)/2} \, dx. \tag{19}$$

The right-hand side of (19) is elementary if p is even, but what if p is odd? The advice given to the student by the traditional textbook is, "Make the trigonometric substitution $x = \tan \theta$," which takes us straight back to the left-hand side of (19). Velleman [5] (see also [3]) shows that the substitution $y = \sin \theta$ turns the left-hand side into the integral of a rational function, which can be integrated by partial fractions. Here we investigate what hyperbolic substitution in the right-hand side has to offer.

The appropriate substitution this time is

$$x = \sinh u; \quad dx = \cosh u \, du, \quad \sqrt{x^2 + 1} = \cosh u. \tag{20}$$

It turns (19) into $\int \cosh^{p-1} u \, du$.

Let's concentrate first on the case $p = 1$ (that is, (9)):

$$\int (x^2 + 1)^{-1/2} \, dx = \int du = u + C$$
$$= \sinh^{-1} x + C$$
$$= \ln(x + \sqrt{x^2 + 1}) + C \tag{21}$$

by (11). If our original interest was in integrating the algebraic function, we are done; if we really cared about the secant integral for its own sake, we now use $x = \tan \theta$ to go from (21) to (9) in the quadrants where $\sec \theta > 0$. (How to handle the other quadrants is left to the reader's taste.)

For $p = 3$ we have

$$\int \sqrt{x^2 + 1} \, dx = \int \cosh^2 u \, du. \tag{22}$$

There are two ways to proceed, depending on taste. First, the hyperbolic functions obey identities in close parallel to those for trigonometric functions; one can memorize, look up, or rederive the identity exactly corresponding to the one one would use to evaluate $\int \cos^2 \theta \, d\theta$. On the other hand, it is again a great charm of hyperbolic functions that they can always be eliminated through (4) in favor of the exponential function, which obeys a much simpler and shorter, but equally powerful, list of identities. (Indeed, many students learn, even if their calculus books never tell them, that the best way to recover trig identities is to use the function $e^{i\theta}$ in this same way.) Thus we have

$$\int \cosh^2 u \, du = \frac{1}{4} \int (e^{2u} + e^{-2u} + 2) \, du$$

$$= \frac{1}{4} \left(\frac{1}{2} e^{2u} - \frac{1}{2} e^{-2u} + 2u \right)$$

$$= \frac{1}{4} (\sinh 2u + 2u)$$

$$= \frac{1}{2} (\sinh u \cosh u + u)$$

$$= \frac{1}{2} (x\sqrt{x^2 + 1} + \sinh^{-1} x),$$

hence

$$\int \sqrt{x^2 + 1} \, dx = \frac{1}{2} x \sqrt{x^2 + 1} + \frac{1}{2} \ln(x + \sqrt{x^2 + 1}) + C, \tag{23}$$

and ultimately

$$\int \sec^3 \theta \, d\theta = \frac{1}{2} \tan \theta \sec \theta + \frac{1}{2} \ln |\sec \theta + \tan \theta| + C. \tag{24}$$

Larger odd values of p can in principle be treated in the same way, although, as always in this type of problem, the complexity increases.

Acknowledgments. I thank Philip Yasskin for comments on the manuscript, and a referee for contributing the more heuristic proof of (8).

References

1. W. B. Gearhart and H. S. Shultz, Tugging a barge with hyperbolic functions, *College Math. J.* **34** (2003) 42–49.
2. I. S. Gradshteyn and I. M. Ryzhik, *Table of Integrals, Series, and Products*, Academic Press, 1965.
3. V. F. Rickey and P. M. Tuchinsky, An application of geography to mathematics: History of the integral of the secant, *Math. Mag.* **53** (1980) 162–166.
4. J. S. Robertson, Gudermann and the simple pendulum, *College Math. J.* **28** (1997) 271–276.
5. D. J. Velleman, Partial fractions, binomial coefficients, and the integral of an odd power of sec θ, *Amer. Math. Monthly* **109** (2002) 746–749.

The Integral of $x^{1/2}$, etc.

John H. Mathews

In calculus we derive $\int_0^x t^2\, dt = (x^3/3)$ by a limiting process involving Riemann sums. Known formulas for summing $\sum_{k=1}^n k^m$, where m is a positive integer, are used in computing the limit. Using a similar technique we show that $\int_0^x t^{1/2}\, dt = (2x^{3/2}/3)$. The method can be used to integrate $f(t) = t^{p/q}$, where p and q are positive integers.

For the integrand $f(t) = t^{1/2}$, the partition of $0 \le t \le x$ is chosen to be

$$\left\{ x_k = \frac{k^2 x}{n^2} \right\}_{k=0}^{k=n}$$

(which involves k^2 because $q = 2$), and the corresponding function values are $f(x_k) = (kx^{1/2}/n)$. The definite integral is a limit of Riemann sums:

$$\int_0^x t^{1/2}\, dt = \lim_{n \to \infty} \sum_{k=1}^n f(x_k)\, \Delta x_k$$

$$= \lim_{n \to \infty} \sum_{k=1}^n \frac{kx^{1/2}}{n}\left(\frac{k^2 x}{n^2} - \frac{(k-1)^2 x}{n^2} \right)$$

$$= \lim_{n \to \infty} \sum_{k=1}^n \frac{x^{3/2} k \left(k^2 - (k-1)^2 \right)}{n^3}$$

$$= \lim_{n \to \infty} \sum_{k=1}^n \frac{x^{3/2} k (2k-1)}{n^3}$$

$$= \lim_{n \to \infty} \frac{x^{3/2}}{n^3} \left(\sum_{k=1}^n 2k^2 - \sum_{k=1}^n k \right)$$

$$= x^{3/2} \lim_{n \to \infty} \frac{1}{n^3} \left(\frac{2n^3 + 3n^2 + n}{3} - \frac{n^2 + n}{2} \right)$$

$$= x^{3/2} \lim_{n \to \infty} \frac{4n^3 + 3n^2 - n}{6n^3}$$

$$= \frac{2x^{3/2}}{3}$$

For illustration purposes, a Riemann sum for $f(t) = t^{1/2}$ over $[0, 4]$ using $n = 8$ subintervals involves the partition $\{0, \frac{1}{25}, \frac{4}{25}, \frac{9}{25}, \frac{16}{25}, 1, \frac{36}{25}, \frac{49}{25}, \frac{64}{25}, \frac{81}{25}, 4\}$ and is shown in Figure 1.

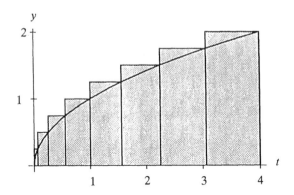

Figure 1
A Riemann sum with $n = 8$ for $f(t) = t^{1/2}$ over $[0, 4]$.

As an example of extending the method, consider $\int_0^x t^{4/3} \, dt = (3x^{7/3}/7)$. For $f(t) = t^{4/3}$, the partition of $0 \leq t \leq x$ is chosen to be

$$\left\{ x_k = \frac{k^3 x}{n^3} \right\}_{k=0}^{k=n}$$

(which involves k^3 because $q = 3$). The corresponding function values are $f(x_k) = (k^4 x^{4/3}/n^4)$. The integral is a limit of Riemann sums:

$$\int_0^x t^{4/3} \, dt = \lim_{n \to \infty} \sum_{k=1}^{n} f(x_k) \, \Delta x_k$$

$$= \lim_{n \to \infty} \sum_{k=1}^{n} \frac{k^4 x^{4/3}}{n^4} \left(\frac{k^3 x}{n^3} - \frac{(k-1)^3 x}{n^3} \right)$$

$$= \lim_{n \to \infty} \sum_{k=1}^{n} \frac{x^{7/3} k^4 (k^3 - (k-1)^3)}{n^7}$$

$$= \lim_{n \to \infty} \sum_{k=1}^{n} \frac{x^{7/3} k^4 (3k^2 - 3k + 1)}{n^7}$$

$$= \lim_{n \to \infty} \frac{x^{7/3}}{n^7} \left(\sum_{k=1}^{n} 3k^6 - \sum_{k=1}^{n} 3k^5 + \sum_{k=1}^{n} k^4 \right)$$

$$= x^{7/3} \lim_{n \to \infty} \frac{1}{n^7} \left(\frac{n - 7n^3 + 21n^5 + 21n^6 + 6n^7}{14} - \frac{2n^6 + 6n^5 + 5n^4 - n^2}{4} \right.$$

$$\left. + \frac{6n^5 + 15n^4 + 10n^3 - n}{30} \right)$$

$$= x^{7/3} \lim_{n \to \infty} \frac{16n + 105n^2 - 70n^3 - 315n^4 + 84n^5 + 420n^6 + 180n^7}{420n^7}$$

$$= \frac{3x^{7/3}}{7}$$

A Direct Proof of the Integral Formula for Arctangent

Arnold J. Insel

In this capsule, we give a direct proof that the Arctangent is an integral of $1/(1 + x^2)$. It then becomes possible to use the Arctangent to determine the tangent and the other trigonometric functions. Here (Figure 1) for any real number a, we define Arctan a as the angle θ (in radians) determined by angle OPR, where θ is taken as negative if $a < 0$.

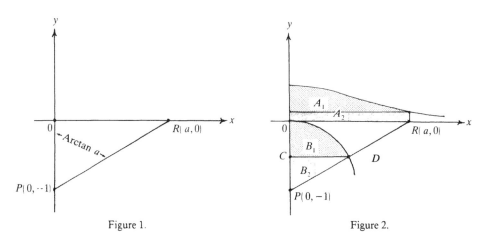

Figure 1.   Figure 2.

In what follows, we fix a number $a > 0$. This will determine two regions, as shown in Figure 2. The region above the x-axis is bounded by the graph of $y = \dfrac{1}{2(1 + x^2)}$ and the x-axis, where $0 \leq x \leq a$. Therefore, the total area of this region is

$$\int_0^a \frac{dx}{2(1 + x^2)}.$$

The region below the x-axis is a sector of a circle having center $(0, -1)$ and radius 1. The sides of the sector are determined by the y-axis and the line connecting $(0, -1)$ to $(a, 0)$. Thus, these sides determine the angle with value Arctan a. Since the area of a sector of a circle of radius r and angle θ (in radians) is $r^2\theta/2$, the total area of this shaded region is $\frac{1}{2}$ Arctan a.

A Direct Proof of the Integral Formula for Arctangent

We shall show that these two shaded regions have equal areas. From this, it follows that the Arctangent can be represented as an integral of the function $y = 1/(1 + x^2)$.

First, consider the region above the x-axis (Figure 2). This region is divided into two subregions, A_1 and A_2. The rectangle A_2 has area $\mathcal{A}(A_2) = \dfrac{a}{2(1+a^2)}$.

The shaded sector below the x-axis is also divided into two subregions, B_1 and B_2. Since triangle CPD is similar to triangle OPR, the legs PC and CD of triangle CPD have lengths $1/\sqrt{1+a^2}$ and $a/\sqrt{1+a^2}$, respectively. Thus, B_2 has area $\mathcal{A}(B_2) = \dfrac{a}{2(1+a^2)}$. In particular, $\mathcal{A}(A_2) = \mathcal{A}(B_2)$.

It remains to be shown that $\mathcal{A}(A_1) = \mathcal{A}(B_1)$. First, solve the equation $y = 1/2(1 + x^2)$ for x to obtain $x = \sqrt{1-2y}/\sqrt{2y}$. Then integrate this along the y-axis to obtain

$$\mathcal{A}(A_1) = \int_{1/2(1+a^2)}^{1/2} \frac{\sqrt{1-2y}}{\sqrt{2y}}\, dy.$$

Likewise, the circular boundary of B_1 can be represented as the graph of $x = \sqrt{1-(y+1)^2}$, where $\dfrac{1}{\sqrt{1+a^2}} - 1 \le y \le 0$. Therefore,

$$\mathcal{A}(B_1) = \int_{\frac{1}{\sqrt{1+a^2}}-1}^{0} \sqrt{1-(y+1)^2}\, dy.$$

Finally, we show that the integral for $\mathcal{A}(A_1)$ can be transformed into the integral for $\mathcal{A}(B_1)$ by means of the substitution $t = \sqrt{2y} - 1$. Indeed, $dt = \dfrac{dy}{\sqrt{2y}}$ and $(t+1)^2 = 2y$. Therefore,

$$\mathcal{A}(A_1) = \int_{1/2(1+a^2)}^{1/2} \frac{\sqrt{1-2y}}{\sqrt{2y}}\, dy$$
$$= \int_{(1/\sqrt{1+a^2})-1}^{0} \sqrt{1-(t+1)^2}\, dt$$
$$= \mathcal{A}(B_1).$$

We have therefore shown that $\mathcal{A}(A_1) + \mathcal{A}(A_2) = \mathcal{A}(B_1) + \mathcal{A}(B_2)$. Thus,

$$\int_0^a \frac{dx}{2(1+x^2)} = \frac{1}{2}\,\text{Arctan}\, a$$

or

$$\int_0^a \frac{dx}{(1+x^2)} = \text{Arctan}\, a \qquad (*)$$

for $a > 0$.

A simple symmetry argument establishes the validity of $(*)$ for $a < 0$. Equation $(*)$ is clearly valid for $a = 0$. Thus, $(*)$ is valid for all real values a.

We outline a method for obtaining the derivatives of the trigonometric functions from (∗). First, apply the fundamental theorem of calculus to obtain the derivative of the Arctangent. The function $f(x) = \tan x \ (-\pi/2 < x < \pi/2)$ is the inverse of the Arctangent, and its derivative $f'(x) = \sec^2 x$ can be obtained from the inverse function theorem. Since the tangent function is a repetition of f on all intervals of the form $((n - \frac{1}{2})\pi, (n + \frac{1}{2})\pi)$, we have

$$\frac{d}{dx} \tan x = \sec^2 x.$$

Next, use the tangent function to represent the secant, and differentiate to obtain the usual formula for the derivative of the secant. For the derivatives of the sine and cosine, observe that $\cos x = 1/(\sec x)$ and $\sin x = \tan x \cos x$ for $x \neq (n + \frac{1}{2})\pi$. Differentiate to obtain the usual formulas with this restriction which can be removed by use of the identities

$$\sin x = \cos\left(x - \frac{\pi}{2}\right) \text{ and } \cos x = \sin\left(x + \frac{\pi}{2}\right).$$

Finally, the derivatives of the cotangent and cosecant can be obtained from the derivatives of the sine and cosine in the usual way.

Riemman Sums and the Eponential Function

Sheldon P. Gordon

Every standard calculus textbook contains the derivations for the definite integral of x and x^2 using Riemann sums based on the known results for the sums of the first k integers and the first k squares of the integers. Now that some of the reform calculus projects have moved the exponential function "up front" because of its importance, it is nice to have a comparable derivation for its definite integral. In the process, we can reemphasize some important ideas.

Let $[a, b]$ be any interval and consider a uniform partition of the interval into n subdivisions of size $h = \Delta x = (b - a)/n$. Therefore,

$$\int_a^b e^x \, dx \approx \sum_{k=0}^{n-1} e^{a+kh} h$$

$$= he^a \sum_{k=0}^{n-1} (e^h)^k$$

$$= he^a \frac{(1 - e^{nh})}{1 - e^h}$$

$$= \frac{he^a(e^{b-a} - 1)}{e^h - 1},$$

using the sum of the first n terms of a geometric progression and the fact that $b = a + nh$. Therefore, in the limit as $n \to \infty$ and hence $h \to 0$, we obtain

$$\int_a^b e^x \, dx = \lim_{h \to 0} \frac{(e^b - e^a)}{(e^h - 1)/h}.$$

Recognizing that the limit as h goes to 0 of the term in the denominator is precisely the definition of the derivative of e^x at $x = 0$, we immediately conclude that

$$\int_a^b e^x \, dx = e^b - e^a.$$

Acknowledgment. This work was supported by National Science Foundation grants #USE-89-53923 for the Harvard Calculus Reform project and USE-91-50440 for the PreCalculus/Math Modeling project.

Proof Without Words Under the Magic Curve

Füsan Akman

There is no limit to the limits you can demonstrate under the magic curve $y = 1/x$. So you have dutifully defined the natural logarithm as an "area",

$$\ln t = \int_1^t \frac{1}{x} dx$$

for $t > 0$, and introduced e as the positive number at which $\ln e = 1$. Before proving that $\ln ab = \ln a + \ln b$ by change of variables, please count to $\ln 485165195.4$ and read the following.

Would your students buy the fact that any two upper (or lower) "Riemann boxes" under the magic curve with equal endpoint ratios have the same area?

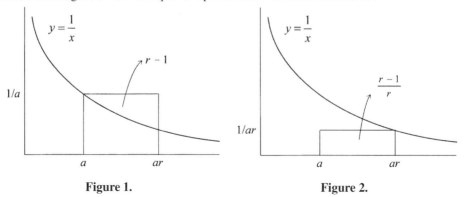

Figure 1.   **Figure 2.**

Once they are hooked, show that this is true for actual areas under the curve.

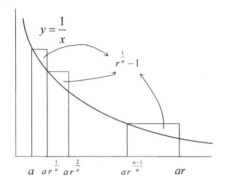

Figure 3.

The length of the longest subinterval in Figure 3, namely $\Delta x_n = ar^{\frac{n-1}{n}}(r^{1/n} - 1)$, goes to zero as n goes to infinity! The limiting area is independent of a. This, of course, leads to

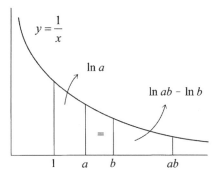

Figure 4.

$\ln ab = \ln a + \ln b$ because of Figure 4 (where $b \geq a > 1$, but the regions may overlap too). Then $\ln a = \ln(b \cdot (a/b))$ does the trick for the companion identity. As for the "down with the power!" rule, use the equal-area property again, with endpoints $1, a, a^2, \ldots, a^p$,

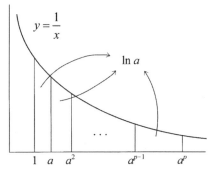

Figure 5.

to show that $\ln a^p = p \ln a$ (p a positive integer) and use the fact that $q \ln(a^{1/q}) = \ln a$ (q a positive integer) to conclude that $\ln(a^{p/q}) = (p/q) \ln a$ for a positive rational number p/q. In case of a negative power, change a to $1/a$ and make the power positive.

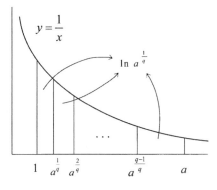

Figure 6.

Now comes the dreaded part where you prove that

$$\lim_{n\to\infty}\left(1+\frac{1}{n}\right)^n = e,$$

probably by using "logarithmic limits. That is your prerogative, of course, but why not continue with the theme of equal areas?

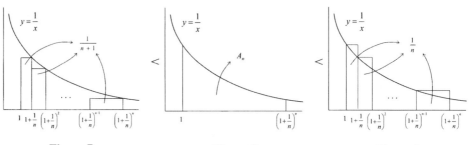

Figure 7.   **Figure 8.**   **Figure 9.**

It should not be too hard to convince our students that

$$\frac{n}{n+1} < A_n < 1 \quad \Rightarrow \quad A_n \to 1.$$

If the sandwiched-in area under $y = 1/x$ approaches 1, what do you think happens to $x = (1+\frac{1}{n})^n$?

Why does the graph of the natural logarithm look the way it does? We can account for the x-intercept, the increase, and the negative concavity easily enough, but why do the ends go off to plus or minus infinity? Each of us has a little classroom strategy to deal with this problem, possible involving some inequalities, but please step under the magic curve once again and consider the following pictures:

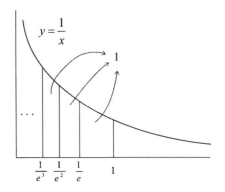

Figure 10.

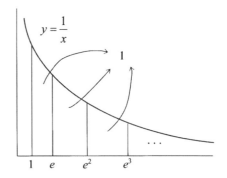

Figure 11.

Figure 10 shows that $\lim_{x \to 0^+} \ln x = -\infty$, and Figure 11 that $\lim_{x \to \infty} \ln x = \infty$.

Note that you can recycle these two pictures in Calculus II to "prove"

$$\int_0^1 \frac{1}{x} dx = \infty \quad \text{and} \quad \int_1^\infty \frac{1}{x} dx = \infty.$$

Speaking of Calculus II, the telescoping series

$$\sum_{k=1}^\infty \frac{1}{k(k+1)} = 1, \quad \text{with} \quad \sum_{k=1}^n \frac{1}{k(k+1)} = 1 - \frac{1}{n+1},$$

also makes an appearance under the curve, though not through equal areas (see Figure 12).

Figure 13 (with $a > 1$) shows my all-time favorite bonus question on the subject of area.

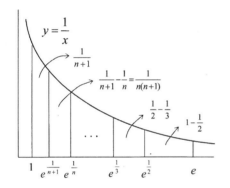

Figure 12.

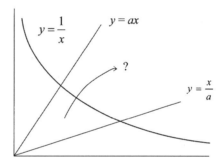

Figure 13.

The answer is $\ln a$, generalized from Figure 14 (with $a \geq b > 0$).

I think somebody (with tenure) ought to define the logarithm using Figure 13 and prove all properties of logarithms based on that, promising to ask for the proofs on the next test.

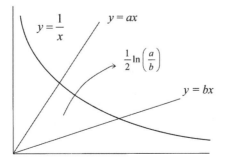

Figure 14.

Acknowledgments. I owe many thanks to Bogdan Mihaila and to Olcay Akman for technical assistance. I thank Prashant Sansgiry for pointing out two related, but different, Proofs Without Words that I include in the references. I don't claim to have thought of the proofs in this article before everybody else, but it certainly felt like that at the time.

References

1. J. Ely, A visual proof that ln(ab) = ln a + ln b, *College Mathematics Journal* **27** (1996) 304.
2. A. H. Stein and D. McGavran, Proof of a common limit, *College Mathematics Journal* **29** (1998) 147.

Mathematics Without Words: Integrating the Natural Logarithm

Roger Nelsen

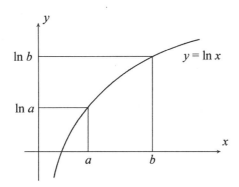

$$\int_a^b \ln x\, dx = b\ln b - a\ln a - \int_{\ln a}^{\ln b} e^y\, dy = x\ln x\Big|_a^b - (b-a) = (x\ln x - x)\Big|_a^b$$

Integrals of Products of Sine and Cosine with Different Arguments

Sherrie J. Nicol

$$\cos(ax)\cos(bx) = \tfrac{1}{2}[\cos((a+b)x) + \cos((a-b)x)],$$
$$\sin(ax)\cos(bx) = \tfrac{1}{2}[\sin((a+b)x) + \sin((a-b)x)],$$
$$\sin(ax)\sin(bx) = \tfrac{1}{2}[\cos((a-b)x) - \cos((a+b)x)]$$

to evaluate integrals of the form

$$\int \cos(ax)\cos(bx)\,dx,$$

$$\int \sin(ax)\cos(bx)\,dx,$$

$$\int \sin(ax)\sin(bx)\,dx.$$

Most students balk in anticipation of more formulas to memorize.

These integrals are typically found in the section of a text dealing with integrating powers of trigonometric functions, which follows the section on integration by parts. I contend that these integrals should be done by repeated (iterated) integration by parts, just as integrals of the form $\int e^{kx}\cos(ax)\,dx$. Although not so easy as using the above identities, integration by parts is not difficult. For example consider the integral

$$I = \int \sin(2x)\cos(3x)\,dx.$$

Let $u = \sin(2x)$ and $dv = \cos(3x)\,dx$. Then $du = 2\cos(2x)\,dx$, and $v = \tfrac{1}{3}\sin(3x)$. Thus

$$I = \frac{1}{3}\sin(2x)\sin(3x) - \frac{2}{3}\int \cos(2x)\sin(3x)\,dx.$$

Now let $p = \cos(2x)$ and $dq = \sin(3x)\,dx$. Then $dp = -2\sin(2x)\,dx$, and $q = -\tfrac{1}{3}\cos(3x)$ yielding

$$I = \tfrac{1}{3}\sin(2x)\sin(3x) - \tfrac{2}{3}(-\tfrac{1}{3}\cos(2x)\cos(3x) - \tfrac{2}{3}I),$$
$$I = \tfrac{1}{3}\sin(2x)\sin(3x) + \tfrac{2}{9}\cos(2x)\cos(3x) + \tfrac{4}{9}I,$$
$$\tfrac{5}{9}I = \tfrac{1}{3}\sin(2x)\sin(3x) + \tfrac{2}{9}\cos(2x)\cos(3x) + C.$$

Finally,

$$\int \sin(2x)\cos(3x)\,dx = \tfrac{3}{5}\sin(2x)\sin(3x) + \tfrac{2}{5}\cos(2x)\cos(3x) + C.$$

For the student who has been taught tabular integration by parts the calculation runs as follows:

u	dv	
$\sin(2x)$	$\cos(3x)$	$I = \tfrac{1}{3}\sin(2x)\sin(3x) + \tfrac{2}{9}\cos(2x)\cos(3x) + \tfrac{4}{9}I,$
$2\cos(2x)$	$\tfrac{1}{3}\sin(3x)$	$\tfrac{5}{9}I = \tfrac{1}{3}\sin(2x)\sin(3x) + \tfrac{2}{9}\cos(2x)\cos(3x) + C$
$-4\sin(2x)$	$-\tfrac{1}{9}\cos(3x)$	$I = \tfrac{3}{5}\sin(2x)\sin(3x) + \tfrac{2}{5}\cos(2x)\cos(3x) + C.$

The integral is evaluated without the use of trigonometric identities and, as I prefer, in terms of the arguments of the trigonometric functions found in the original problem. As Grant [Moments on a rose petal, *CMJ* (1990) 225–227] mentions, when the result is in terms of the original arguments, checking an integral by differentiation is a viable option, even for the more complex integrals $\int \sin\theta \sin^n(m\theta)\,d\theta$ and $\int \cos\theta \sin^n(m\theta)\,d\theta$ which Grant tackles using integration by parts. (Incidentally, checking the example above and a few others by differentiation may prompt some to notice the forms that appear as antiderivatives and thereby to sense the possibility of yet another method: undetermined coefficients.)

Moments on a Rose Petal

Douglass L. Grant

For several years, I have been insisting that my first-year calculus students check every nontrivial integration by differentiation. This teaching strategy, based on the pious hope that students will acquire the habit of finding and eliminating readily detectable errors, is occasionally at cross-purposes with the integration strategies commonly suggested in textbooks.

The best examples are integrals of the form $\int f(mx)g(nx)\,dx$, where f, g are the sine or cosine functions and m, n are numbers, neither of which is an integer multiple of the other. Most texts (34 of 44 surveyed) recommend the use of the "prosthaphaeresis" rule [V. E. Thoren, Prosthaphaeresis revisited, *Historia Mathematica* 15 (1988) 32–39]:

$$\sin A \sin B = [\cos(A - B) - \cos(A + B)]/2$$
$$\cos A \cos B = [\cos(A + B) + \cos(A - B)]/2 \qquad (1)$$
$$\sin A \cos B = [\sin(A + B) + \sin(A - B)]/2.$$

While conceptually simple if one can remember the rule, the technique yields forms of the integral that are extremely difficult to check. Just watch a student's eyes glaze when confronted with maneuvers like

$$\cos x = \cos(5x - 4x) = \cos 5x \cos 4x + \sin 5x \sin 4x.$$

The minority report [M. R. Embry, J. F. Schell, and J. P. Thomas, *Calculus and Linear Algebra: An Integrated Approach*, Saunders, Philadelphia, 1972 and A. Shenk, *Calculus and Analytic Geometry*, Scott, Foresman, Glenview, IL, 1978] advocates the use of integration by parts. This technique is somewhat longer but spares the student memorization of what are by now virtually single-purpose formulas and simplifies checking considerably by preserving the original arguments in the trigonometric functions. Omitting constants and assuming $m^2 - n^2 \neq 0$, the integrals obtained are

$$\int \sin mx \sin nx\,dx = (m \cos mx \sin nx - n \sin mx \cos nx)/(n^2 - m^2)$$

$$\int \cos mx \cos nx\,dx = (n \cos mx \sin nx - m \sin mx \cos nx)/(n^2 - m^2) \qquad (2)$$

$$\int \cos mx \sin nx\,dx = (n \cos mx \cos nx + m \sin mx \sin nx)/(m^2 - n^2).$$

Moments on a Rose Petal

This recommendation underwent the severest test to date when, in the dying seconds of a lecture to a second-year intermediate calculus class, I challenged the students to compute the centroid of one "petal" of the 3-leaved rose $r = \sin 3\theta$. Several students successfully formulated

$$M_x = \iint y\, dA = \int_0^{\pi/3} \int_0^{\sin 3\theta} r^2 \sin\theta\, dr\, d\theta = (1/3) \int_0^{\pi/3} \sin\theta \sin^3 3\theta\, d\theta,$$

and M_y analogously, but none was able to find the required indefinite integral.

Integration by parts being initially unappealing, prosthaphaeresis was given two chances. Three applications of (1) gave the remarkable formulation

$$\sin\theta \sin^3 3\theta = \frac{1}{8}(3\cos 2\theta - 3\cos 4\theta - \cos 8\theta + \cos 10\theta),$$

but a check by differentiation would be daunting indeed. Writing

$$\sin\theta \sin^3 3\theta = (\sin\theta \sin 3\theta)\sin^2 3\theta$$
$$= \frac{1}{2}[\cos 2\theta - \cos 4\theta](1 - \cos 6\theta)/2$$

seems more promising, but merely leads more quickly to the same identity. The reader is invited to explore the delights of substituting $\sin 3\theta = 3\sin\theta - 4\sin^3\theta$, expanding the cube, and hacking through the resulting jungle of double angle formulas.

Head unbowed though sufficiently bloody to keep Macbeth from his banquet, I tried the time-honored technique of generalizing the problem. Let n be any positive integer no smaller than 2, and consider $I = \int \sin\theta \sin^n m\theta\, d\theta$. While it is feasible to integrate the factor $\sin^3 3\theta$ from the original example, it is entirely unfeasible to integrate the more general term $\sin^n m\theta$ so we are drawn inexorably to choose $u = \sin^n m\theta$, $dv = \sin\theta\, d\theta$.

Two applications of the parts formula and one of $\cos^2 m\theta = 1 - \sin^2 m\theta$ yields the following reduction formula:

$$(1 - m^2 n^2)I = -\cos\theta \sin^n m\theta + mn \sin^{n-1} m\theta \cos m\theta \sin\theta$$
$$- m^2 n(n-1) \int \sin^{n-2} m\theta \sin\theta\, d\theta. \tag{3}$$

Applying (3) to our original integral, we obtain, with the aid of (2):

$$-80I = G(\theta)$$
$$= -\cos\theta \sin^3 3\theta + 9\sin^2 3\theta \cos 3\theta \sin\theta - \frac{27}{4}(\cos\theta \sin 3\theta - 3\sin\theta \cos 3\theta).$$

(The reader is invited to check by differentiation.)

The moment M_x of the petal with respect to the horizontal axis is then $-[G(\pi/3) - G(0)]/240 = 27\sqrt{3}/640$. Hence,

$\bar{y} = M_x/(\text{Area}) = 81\sqrt{3}/160\pi$, and by symmetry, $\bar{x} = \bar{y}\cot(\pi/6) = 243/160\pi$.

Similarly, one can obtain the analogous reduction formula for $J = \int \cos\theta \sin^n m\theta\, d\theta$, which is given by

$$(1 - m^2n^2)J = \sin\theta \sin^n m\theta + mn \sin^{n-1} m\theta \cos m\theta \cos\theta$$
$$+ m^2n(n-1)\int \cos\theta \sin^{n-2} m\theta\, d\theta.$$

This formula also assists in computing second moments. For instance, returning to the example of $r = \sin 3\theta$, we find

$$I_x = \int\int y^2\, dA = \int_0^{\pi/3}\int_0^{\sin 3\theta} r^3 \sin^2\theta\, dr\, d\theta$$

$$= (1/4)\int_0^{\pi/3} \sin^2\theta \sin^4 3\theta\, d\theta$$

$$= (1/8)\int_0^{\pi/3} (1 - \cos 2\theta)\sin^4 3\theta\, d\theta$$

$$= (1/8)\int_0^{\pi/3} \sin^4 3\theta\, d\theta - (1/16)\int_0^{2\pi/3} \cos\alpha \sin^4(3\alpha/2)\, d\alpha,$$

where $\alpha = 2\theta$.

Note: Shenk's fourth edition, published by Scott, Foresman in 1988, has fallen from grace by switching to the prosthaphaeresis method for products of sines and cosines with dissimilar arguments.

A Calculation of $\int_0^\infty e^{-x^2}dx$

Alberto L. Delgado

If you are looking for a way to tie together different strands of material your students see in their second semester of a standard three-semester calculus course, the result in the title will serve you very well. Of course, the result is

$$\int_0^\infty e^{-x^2}\,dx = \frac{\sqrt{\pi}}{2}.$$

Students find it surprising (where does the π come from?) and meaningful (when they learn its relationship to the bell curve). You can also use this significant improper integral as an entry into a discussion of approximations of definite integrals over finite regions.

There are numerous ways to evaluate the integral in the title. In just a few minutes in the library, I located two evaluations based on Wallis' infinite product expansion of π, [8] and [5]; a calculation using contour integration in the complex plane, [1]; a reduction to Euler's integral of the first kind, [2]; two evaluations using differentiation under the integral sign, [7] and [9]; and a calculation based on solids of revolution, [3]. You can find the standard computation based on a double integral in polar coordinates in almost any statistics or calculus book. The integral is commonly associated with Gauss, although he credits Laplace with its discovery and publication in 1805. Euler was working with similar integrals thirty years earlier, but he seems to have missed its exact formulation. By 1813, the result appears to have been well known. See [6] for a more detailed history.

Few of the techniques for evaluating this integral are easily accessible to a begining student. Here is a technique that students can appreciate. Although the method may not be new (the main idea is already in [3] and [4]) our variation of it is sufficiently elementary that it can be presented to students.

Let I denote the value of the integral of interest. A comparison with e^{-x} confirms that I is finite. Let's start by finding the volume V of the solid of revolution made by revolving the graph of e^{-x^2} for $x \in [0, \infty)$ about the y-axis (Figure 1).

Using the usual technique of nested cylindrical shells, we obtain

$$V = \int_0^\infty 2\pi x e^{-x^2}\,dx = \pi.$$

Now let's compute V a second time. This time take cross sections parallel to the x-axis (Figure 2).

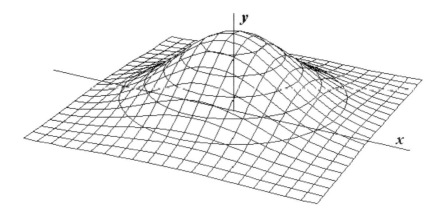

Figure 1. Graph of e^{-x^2} rotated about y-axis.

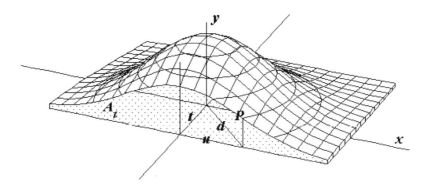

Figure 2. Cross section at distance t from xy-plane.

Denoting the area of the cross section at distance t from the x-axis by A_t, we get $V = 2\int_0^\infty A_t\, dt$. In order to compute A_t, first find the equation for the graph of the cross section. Consider the point P on the cross section. Since the surface is radially symmetric, we can find P's height by rotating back to the value $d = \sqrt{u^2 + t^2}$ on the x-axis. Thus, P's height is $e^{-(\sqrt{u^2+t^2})^2}$, and

$$A_t = 2\int_0^\infty e^{-(u^2+t^2)}\, du = 2\int_0^\infty e^{-u^2}e^{-t^2}\, du = 2e^{-t^2}\int_0^\infty e^{-u^2}\, du = 2e^{-t^2}I.$$

Therefore,

$$V = 2\int_0^\infty A_t\, dt = 2\int_0^\infty 2e^{-t^2}I\, dt = 4I\int_0^\infty e^{-t^2}\, dt = 4I^2.$$

Equating the two expressions for V, we obtain $\pi = 4I^2$ and the result follows.

References

1. Darrell Desbrow, On evaluating $\int_{-\infty}^{\infty} e^{ax(x-2b)}\,dx$ by contour integration round a parallelogram, *American Mathematical Monthly* **105** (1998) 726–731.
2. Édouard Goursat, *A course in mathematical analysis, vol. I*, Ginn & Co., 1904.
3. C. P. Nicholas, Another look at the probability integral, *American Mathematical Monthly* **64** (1957) 739–741.
4. C. P. Nicholas and R. C. Yates, The probability integral, *American Mathematical Monthly* **57** (1950) 412–413.
5. T. J. Stieljes, *Oevres Complètes, vol. II*, Wolters-Noordhoff, 1918.
6. S. C. van Veen, Historische bizjonderheden over $\int_0^{\infty} e^{-x^2}\,dx$, *Mathematica B* **12** (1943–1944), 1–4.
7. Robert Weinstock, Elementary evaluations of $\int_0^{\infty} e^{-x^2}\,dx$, $\int_0^{\infty} \cos(x^2)\,dx$, and $\int_0^{\infty} \sin(x^2)\,dx$, *American Mathematical Monthly* **97** (1990) 39–42.
8. Robert M. Young, *Excursions in Calculus*, Mathematical Association of America, 1992.
9. ———, On the evaluation of certain improper integrals, *Mathematical Gazette* **75** (1991) 88–89.

Calculus in the Operating Room

Pearl Toy and Stan Wagon

Here is a realistic, and potentially important, application of a familiar topic from calculus. Imagine a hospital patient about to undergo surgery. Suppose he has 5 liters (L) of blood in his body, 40% of which consists of red blood cells (this percentage is called the *hematocrit*), and, during the surgery, he will bleed 2½ liters of blood. This is a realistic estimate for certain types of hip replacement, for example. His blood volume is maintained at 5 L by controlled injection of saline solution (no blood cells), which we assume to mix instantaneously with his blood. This means that the blood lost through bleeding becomes less rich in red cells as the operation progresses.

Question 1. What is the patient's volume of red blood cells at the end of the operation?

Some of the lost blood can be recovered, washed, and returned to the patient after the operation; but there is some loss and washing is an expensive procedure. Suppose that, before the operation, some blood is removed from the patient and replaced with saline solution. This blood will be returned to the patient afterward. This procedure, called acute normovolemic hemodilution (ANH), will decrease the loss of red blood cells during the operation. However, during the transfusion the patient's total blood volume is maintained at 5 L; as with the bleeding during surgery, this affects the rate of red blood cell removal.

Question 2. If it is known that the patient's hematocrit can go as low as 20%, but no lower, how much blood should be replaced in the ANH procedure just described?

Both of these questions can be answered by a simple exponential decay model, once one makes the observation that the rate of blood cell loss during the operation is proportional to the amount of red cells present. For the numbers given, and with time measured as a fraction of the length of the operation, the proportionality constant is ½, since 2.5 of the 5 L are lost.

Now both questions can be answered. If $f(t)$ is the volume of red blood cells remaining at time t, then $f(t) = f(0)e^{-t/2}$. For Question 1, $f(1) = 2000/\sqrt{e} = 1213$ milliliters.

For Question 2 we need to know what value of $f(0)$ will cause $f(0)/\sqrt{e}$ to be 1000 ml (20% of 5 L); this is $f(0) = 1000\sqrt{e}$. In order to figure out how much should be removed by ANH, we reverse the technique just discussed and solve $2000e^{-k/5000} = 1000\sqrt{e}$ to get $k = 966$ ml. This leaves $1000\sqrt{e}$, or 1649 ml of red blood cells in the patient for a hematocrit of 33%, which will become 20% after surgery.

Note that without the transfusion the red blood cell loss is $2000-1213 = 787$ ml. With the transfusion the patient starts with 1649 ml of red blood cells and ends up with 1000; a loss of 649 ml. There is a net savings of 138 ml of red blood cells. This savings may in fact not be large enough to justify the procedure; it must be balanced with overall expense, risks associated with ANH, and the risk of an adverse reaction to blood the patient may have to receive from a blood bank after the operation (see M. E. Brecher and M. Rosenfeld, Mathematical and computer modeling of acute normovolemic hemodilution, *Transfusion* 1994 (34), 176-179). But it is noteworthy that a simple freshman-calculus model applies to the basic situation.

Physical Demonstrations in the Calculus Classroom

Tom Farmer and Fred Gass

The tremendous success of mathematical modeling is an article of faith among scientists. Indeed, in his well-known paper "The Unreasonable Effectiveness of Mathematics in the Natural Sciences," Eugene P. Wigner expresses a feeling akin to awe at the ability of modern mathematical science to predict as well as to describe empirical events. As calculus instructors wishing to convey a sense of this remarkable interplay between our subject and the "real world," we began looking for ways to elicit more student interest and involvement in our treatment of applications.

The idea we eventually decided to pursue occurred the day when one of us used a glass of water with a pencil standing in it to illustrate Snell's law after deriving it from Fermat's principle. The novelty of this simple physical example linked to calculus drew a favorable response in class and suggested that more ambitious demonstrations might be even more satisfying. The main purpose was to make discussions more memorable rather than to pursue an application area in greater depth; consequently we looked for activities that would involve minimal equipment and class time while inviting hands-on student participation.

The most effective demonstration we have used in calculus classes involves what is known as Torricelli's law, which concerns the rate at which a fluid drains out of a hole in a container. In our experiment, the container is a cylinder with vertical axis and in this case Torricelli's law states that the time rate of change of volume V of water in the draining container is proportional to the square root of the water's depth h. Since the volume of a cylinder is proportional to height, the law reduces to $h'(t) = k\sqrt{h}$.

We have used this demonstration at the point in the Calculus I course when antiderivatives have been introduced and one can solve separable differential equations. The class period begins with a discussion of the question: if water is draining out of a vertical cylindrical tank, will the volume of water in the tank decrease at a constant rate or will the rate vary with time? The class surely will agree that we expect the rate to decrease with time because experience tells us that the stream of water coming out of the hole will be greatest at first when the depth is great and will be reduced to a dribble when the depth is close to zero. Thus, we expect that the linear model $h'(t) = k$ will not very accurately describe this experiment. We intend to compare the linear model with the model given by Torricelli's law.

The following information is then presented to the students:

Problem. A small hole is drilled in the side of a cylindrical container and the height of the water level (above the hole) goes from 10 cm down to 3 cm in 68 seconds. Estimate the height at intermediate times.

The linear model:

$$\frac{dh}{dt} = k, \; h(0) = 10, \; h(68) = 3$$

can be seen to have approximate solution $h(t) = -0.103t + 10$.

The Torricelli model:

$$\frac{dh}{dt} = k\sqrt{h}, \; h(0) = 10, \; h(68) = 3$$

has approximate solution $h(t) = 0.00044t^2 + (-0.133)t + 10$.

The table below gives the values of h predicted by each of these models for various intermediate times. We will fill in the column of observed values when we perform the experiment.

time t	linear h	Torricelli h	observed h
0	10.0	10.0	
10	9.0	8.7	
20	7.9	7.5	
30	6.9	6.4	
40	5.9	5.4	
50	4.8	4.4	
60	3.8	3.6	
68	3.0	3.0	

The equipment for this demonstration is easily obtained at little or no cost. We prepare a two-liter clear plastic soft drink bottle, whose midsection is essentially cylindrical, by drilling a clean 4 millimeter hole near the bottom of the cylindrical part. We also attach to the bottle a strip of masking tape with centimeters marked on it and zero corresponding to the top of the hole. We bring this bottle to class along with another bottle of water (as a water source), a laundry bucket (as a water sink) and a board to span the bucket. We can probably always rely on a student to have a digital watch to keep time but, to be safe, we bring such a watch. In advance of class time, we run through the process several times to establish the time required to drain the cylinder; with our apparatus it consistently took 68 seconds for the water depth above the hole to drop from 10 centimeters to 3 centimeters. Given this, the problem is to predict what the depth will be at times 10, 20, 30, 40, 50, and 60 seconds.

A vital part of the demonstration is to have students participate. We have had no difficulty in getting three volunteers to play all the roles. The timekeeper helps out initially by pouring the water into the leaky bottle while the bottle keeper covers the hole with a finger. Meanwhile, the recorder copies the table of predicted values on the blackboard. As the experiment progresses, the time keeper calls out ...8, 9, **10**, ..., 18, 19, **20**, ... and at the appropriate times the bottle keeper estimates out loud the depth reading which can be done with accuracy within a tenth of a centimeter. This person should be cautioned not to look at the predicted values in order to avoid being influenced by them. As the depths are called out the recorder records them on the blackboard. Our experience in trying this in three different classes was that the experimental data were virtually identical to the Torricelli predictions, making for an impressive and memorable demonstration.

A second demonstration we have used involves less apparatus but one item that must be made in advance: a thin "plate" shaped like the interior of a parabola from the vertex back to a line perpendicular to the axis. The basic idea is to determine from first principles, via integral calculus, just where the balance point ought to be. The chance for people to test the result on a physical model gives a "payoff" to the example.

We are still experimenting with ideas along the lines described above. Such topics as Newton's law of cooling, period of a pendulum, differentials for inverse square laws, and spark tapes for velocity/acceleration seem promising, if one can find the right mix of calculus and physical interaction in the classroom.

Acknowledgment. We would like to acknowledge the assistance of our colleague from physics, Paul Scholten.

Who Needs the Sine Anyway?

Carlos C. Huerta

The straw has broken the camel's back. In the course of one week, more than ten of my students asked why mathematicians couldn't be like physicists. At first I did not know what they were referring to, but upon investigation I found it to be the simple pendulum. It seems that during that week one of the problems that the physics instructor did on the board was the derivation of the equation of motion for the simple pendulum. Part of that famous derivation includes the step where to simplify the equation, the instructor says that for small x, say $x \ll \pi/2$, we can replace $\sin x$ with x. Thinking how kind physics is to the student, my students wanted to know why mathematicians couldn't do the same thing.

The next day in class I decided I had had enough. Why should I fight physics? What's good for the goose is good for the gander, (temporarily ignoring who's the goose and who's the gander), no? I decided to rework problems involving $\sin x$ by replacing it with x, and to do some error analysis to see how bad the damage was.

I considered first $D_x \sin x$. Normally I would write $D_x \sin x = \cos x$, but now I wrote, to the cheers of the students, $D_x \sin x \approx D_x x = 1$ for $x \ll \pi/2$. At $x = 0$ the derivative for the sine is exact with the physicist's substitution. The problems develop as we move away from zero. At $x = \pi/8$ we get $D_x \sin x|_{x=\pi/8} = 0.923795$ which is only about 8% off. At $x = \pi/4$ things got a little worse, but not bad enough for my students to give up. We saw that $D_x \sin x|_{x=\pi/4} = 0.707107$ which was about 30% off the mark.

The next thing we tried was integration. We considered the following: $\int_0^1 \sin x\, dx \approx \int_0^1 x\, dx = 0.5$. The actual answer is 0.459698 which means that our substitution is only a little less that 9% off. My students quickly pointed out that the error radically reduces the closer we get to zero. For instance, $\int_0^{0.25} \sin x\, dx = 0.031088$ where $\int_0^{0.25} x\, dx = 0.03125$. This gives us an error of only a little more than 0.05%. I saw then that unless I did something drastic I would never be able to use the sine in class again.

As a last resort I went to differential equations. I considered the equation $y' = \sin y$ vs. $y' = y$. The solution of the first one is $y = 2\tan^{-1}(k \cdot e^x)$, and the solution of the second equation is $y = ke^x$, where in both cases k is a constant. Finally I thought I had a winner here. Those two functions look as different as night and day. But alas, every calculus class has one, a student with a Hewlett Packard 28S-Advanced Scientific Calculator. She plotted both curves and came up with the following graphs:

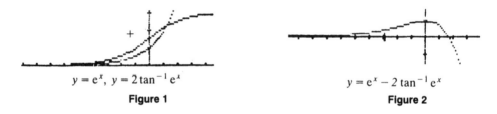

$y = e^x,\ y = 2\tan^{-1} e^x$

Figure 1

$y = e^x - 2\tan^{-1} e^x$

Figure 2

In Figure 1 we see the plots of $y = e^x$ and $y = 2\tan^{-1} e^x$, the solutions of our equations with $k = 1$. In Figure 2 we see the difference of the solutions plotted. As my student quickly pointed out to me, the physicist's substitution has small error asymptotic to zero for large negative numbers. The error slowly increases and reaches a maximum at $x = 0$, then the error goes to zero at $x \approx 1.169231$. Afterwards the error gets dramatically bigger in absolute value. So as long as $x < 1.169231$ she thought that she could live with the error in return for the ease in solving the problem.

I was defeated. Now I no longer use the sine when I teach calculus. If it works for the physicist it must work for the mathematician. I wonder, with what can I replace the cosine and tangent?

Finding Bounds for Definite Integrals

W. Vance Underhill

Students in elementary calculus are often dismayed to learn that not every function has an antiderivative, and consequently not every definite integral can be evaluated by the Fundamental Theorem. Although most textbooks discuss such things as Simpson's Rule and the Trapezoid Rule, these methods are usually long and tedious to apply. In many cases, reasonably good bounds for definite integrals can be obtained with little effort by the use of well-known theorems. The fact that techniques for doing this have never been discussed in one place is the motivation for this note.

Except for very specialized and esoteric results, the following three theorems provide methods for obtaining such bounds.

Theorem A. *If f, g, and h are integrable and satisfy $g(x) \leq f(x) \leq h(x)$ on the interval $[a,b]$, then*

$$\int_a^b g(x)\,dx \leq \int_a^b f(x)\,dx \leq \int_a^b h(x)\,dx.$$

Theorem B. *On the interval $[a,b]$, suppose that f and g are integrable, g never changes sign, and $m \leq f(x) \leq M$. Then*

$$m\int_a^b g(x)\,dx \leq \int_a^b f(x)g(x)\,dx \leq M\int_a^b g(x)\,dx.$$

Theorem C. *If f and g are integrable on $[a,b]$, then*

$$\int_a^b f(x)g(x)\,dx \leq \sqrt{\int_a^b f^2(x)\,dx}\ \sqrt{\int_a^b g^2(x)\,dx}\ .$$

Example 1. Find bounds for $\int_1^2 \frac{x\,dx}{\sqrt{x^3+8}}$. Using Theorem B, we choose $f(x) = \frac{1}{\sqrt{x^3+8}}$ and $g(x) = x$. Then $\frac{1}{4} \leq f(x) \leq \frac{1}{3}$ for $x \in [1,2]$, and $\int_1^2 g(x)\,dx = \frac{3}{2}$. Therefore,

$$.375 \leq \int_1^2 \frac{x\,dx}{\sqrt{x^3+8}} \leq .500. \tag{1}$$

If we use Theorem C, it is natural to choose $f(x) = \frac{x}{\sqrt{x^3+8}}$ and $g(x) = 1$. It follows that

$$\int_1^2 \frac{x\,dx}{\sqrt{x^3+8}} \leq \sqrt{\int_1^2 \frac{x^2\,dx}{x^3+8}} \sqrt{\int_1^2 1^2\,dx} = \sqrt{\frac{2}{3}\ln\frac{4}{3}} < .438, \tag{2}$$

a considerable improvement over the upper bound obtained in (1). Since $x^2 \leq x^3 \leq x^4$ on $[1,2]$, Theorem A yields

$$\int_1^2 \frac{x\,dx}{\sqrt{x^4+8}} \leq \int_1^2 \frac{x\,dx}{\sqrt{x^3+8}} \leq \int_1^2 \frac{x\,dx}{\sqrt{x^2+8}}.$$

The integral on the right equals $2\sqrt{3} - 3$, while the one on the left equals $\frac{1}{2}\ln\left(\frac{2+\sqrt{6}}{2}\right)$. Thus,

$$.399 < \int_1^2 \frac{x\,dx}{\sqrt{x^3+8}} < .465. \tag{3}$$

Combining (1), (2), and (3), we obtain

$$.399 < \int_1^2 \frac{x\,dx}{\sqrt{x^3+8}} < .438. \tag{4}$$

Note that in each of the three theorems, we were not forced into the choices actually made. Other possibilities exist, resulting in different bounds.

In using Theorem B, one interprets a given integrand as the product of two functions f and g, where g is of one sign and can easily be integrated. The calculation of m and M is usually straightforward. One's inclination, of course, is to let g be something easy to integrate. There is sometimes more than one reasonable "decomposition," as the following example shows.

Example 2. For the integral $\int_0^1 x^2 e^{-x^2}\,dx$, we can let $f(x) = e^{-x^2}$ and $g(x) = x^2$. Then $\frac{1}{e} \leq f(x) \leq 1$ on $[0,1]$ and $\int_0^1 g(x)\,dx = \frac{1}{3}$. Hence,

$$\frac{1}{3e} \leq \int_0^1 x^2 e^{-x^2}\,dx \leq \frac{1}{3}.$$

Finding Bounds for Definite Integrals

Other choices of f and g exist, however. Since xe^{-x^2} can easily be integrated, let $g(x) = xe^{-x^2}$ and $f(x) = x$. Then $0 \leq f(x) \leq 1$ on $[0, 1]$ and $\int_0^1 g(x)\,dx = (e-1)/2e$ yield

$$0 \leq \int_0^1 x^2 e^{-x^2}\,dx \leq \frac{e-1}{2e} < .316. \tag{5}$$

What if $g(x) = x$ and $f(x) = xe^{-x^2}$? Then $\int_0^1 g(x)\,dx = \frac{1}{2}$, and f has its minimum value $m = 0$ and maximum value $M = 1/\sqrt{2e}$. Consequently,

$$0 \leq \int_0^1 x^2 e^{-x^2}\,dx \leq \frac{1}{\sqrt{8e}}. \tag{6}$$

A quick check with the calculator shows that the upper bound in (6) is substantially better than in (4) and (5). Combining the best of all cases,

$$.122 < \frac{1}{3e} \leq \int_0^1 x^2 e^{-x^2}\,dx \leq \frac{1}{\sqrt{8e}} < .215. \tag{7}$$

If Theorem A is used on this example, we might reason that $x^2 \leq x$ on $[0, 1]$, and this leads to the inequality

$$\int_0^1 x^2 e^{-x}\,dx \leq \int_0^1 x^2 e^{-x^2}\,dx.$$

The integral on the left equals $2 - (5/e) > .160$, giving us a better lower bound than in (7). Consequently,

$$.160 < \int_0^1 x^2 e^{-x^2}\,dx < .215.$$

Example 3. Find bounds for $\int_0^{\pi/3} x\,dx/\cos x$. Using Theorem B, the obvious decompositions do not yield particularly good results. Suppose, however, that $f(x) = x/\sin x$ (with $f(0)$ taken to be 1) and $g(x) = \sin x/\cos x$. Then f has $m = 1$ and $M = 2\pi/3\sqrt{3}$, and so

$$.693 < \ln 2 \leq \int_0^{\pi/3} \frac{x\,dx}{\cos x} \leq \frac{2\pi}{3\sqrt{3}} \ln 2 < .839. \tag{8}$$

Another estimate can also be obtained from Theorem A. Since $1 - x^2/2 \leq \cos x \leq 1$,

$$\int_0^{\pi/3} x\,dx \leq \int_0^{\pi/3} \frac{x\,dx}{\cos x} \leq \int_0^{\pi/3} \frac{x\,dx}{1 - \frac{1}{2}x^2}.$$

The lower bound here (.548) is worse than that in (8), but the value of the right-hand integral is $\ln\left(\frac{18}{18 - \pi^2}\right)$. Hence,

$$.693 < \int_0^{\pi/3} \frac{x\,dx}{\cos x} < .795.$$

Students find these methods a welcome change of pace from the routine numerical techniques. They enjoy the challenge to improve on bounds already obtained, and they gain valuable experience in working with inequalities, the heart of analysis.

Estimating Definite Integrals

Norton Starr

Many definite integrals arising in practice can be difficult or impossible to evaluate in finite terms. Series expansions and numerical integration are two standard ways to deal with the situation. Another approach, primitive but often very effective, yields cruder estimates by replacing a nasty integrand with nice functions that majorize or are majorized by it. With luck and skill, the bounds achieved suffice for the task at hand. I was introduced to this method as a grad student instructor over forty years ago, when I had the good fortune to learn some innovative teaching methods from Arthur Mattuck. His supplementary notes for MIT's calculus course included a section on the estimation of definite integrals by an approach barely covered in texts back then. Many first year calculus texts of that era touched on the method in connection with comparison tests for improper integrals, but they seldom did anything with proper integrals.

The situation has improved somewhat in recent years, with prominent texts at least mentioning the basic idea within the chapter introducing the definite integral. Sometimes this is labeled the "Domination Rule" or "Comparison Property". An informal survey shows that most such books offer very few, if any, exercises in the method, usually relatively trivial ones. The texts by Edwards & Penney [1] and Stewart [4] are exceptional in providing more than a token selection of such problems. Unfortunately, their exercises are duplicated in the early transcendental versions of these two texts, thus making no use of the broader array of available functions.

Here is a sketch of the way I develop this form of estimation in my intermediate calculus course. All integrals are understood to be over a closed, bounded interval $[a, b]$ and all functions assumed to be (Riemann) integrable. I start with the primitive observation that if f is nonnegative and integrable on $[a, b]$ then $\int_a^b f \geq 0$. (This affords an opportunity to remind students of a definition of $\int_a^b f$ and some of its implications.) Next, using the linearity of the integral and the fact that sums and differences of integrable functions are integrable, I infer that for integrable f and g, if $f \leq g$ on $[a, b]$ then $\int_a^b f \leq \int_a^b g$. This last inequality is the key tool. As with any tool, its effectiveness depends on the skill with which it's wielded. For a given function f or g, the trick is to dream up an appropriate comparison function that leads to a useful bound.

Several fundamental inequalities facilitate this effort:

1) $\sin x < x$ for all positive x. (This is easily illustrated with graphics devices, but asking students why it's true can lead to a review of basic methods from a first course in calculus.)

2) $\ln x < x$ and $\ln x < \sqrt{x}$ for all $x > 0$. (Again, easily illustrated and an occasion for pointing to the relevance of earlier methods.)
3) $x^c < x^d$ if $c > d$ and $0 < x < 1$. (As shown below, this simple observation can turn messy integrals into trivial ones.)
4) If f and g are continuous on $[a, b]$ and $f \leq g$ yet $f(c) < g(c)$ for some c in $[a, b]$, then $\int_a^b f < \int_a^b g$. (This can serve as an occasion to review continuity.)
5) If f is integrable on $[a, b]$ then so is $|f|$, and $|\int_a^b f| \leq \int_a^b |f|$. (This continuous analog of the triangle inequality is an easy way to eliminate pesky terms of the form $(-1)^n$.)

Examples.

i) Which of $\int_1^6 \arctan(\sin x)\, dx$, $\int_1^6 \arctan \sqrt{x}\, dx$, $\int_1^6 \arctan x\, dx$ has the greatest value? Which has the least value? Why? This simply relies on the monotone increasing nature of the inverse tangent function, together with the obvious relations among $\sin x$, $\sqrt{x}$, and x over the interval $[1, 6]$.

ii) Similarly, comparing $\int_0^{1/2} \cos x^2\, dx$ with $\int_0^{1/2} \cos \sqrt{x}\, dx$ amounts to combining the inequality $x^2 < \sqrt{x}$ on $(0, 1)$ with the observation that cosine is decreasing on this interval, so $\cos(x^2) > \cos(\sqrt{x})$. Thus the first integral is larger.

iii) Verify that

$$\int_{10}^{15} \frac{t^3}{t^6 + t^2 - 1}\, dt < .003.$$

On $[10, 15]$, $t^2 - 1 > 0$, so deleting the $t^2 - 1$ in the denominator increases the integrand to t^{-3}. Integrating this from 10 to 15 gives as an upper bound:

$$-\frac{1}{2}\left(\frac{1}{15^2} - \frac{1}{10^2}\right) = \frac{1}{360},$$

which is $< .003$.

iv) Verify that

$$\frac{3}{4} < \int_0^1 \frac{1}{1+t^4}\, dt < \frac{9}{10}.$$

This can be delicate. The lower bound is readily obtained by observing that $1 + t^4 < 1 + t^2$ on $(0, 1)$ and using the inverse tangent. The second inequality appeared on a final examination at our college over thirty years ago, and caused considerable consternation. One route would have been via partial fractions, but our aim was to avoid such extensive computation. Various upper bounds for $1/1 + t^4$ were unsuccessfully proposed among those teaching the course (they would have to grade the test and explain things to curious or unhappy students). Finally my colleague David Armacost suggested showing, in-

Estimating Definite Integrals

stead, that one *minus* the integral is greater than one tenth. This is relatively simple:

$$1 - \int_0^1 \frac{1}{1+t^4} dt = \int_0^1 \frac{t^4}{1+t^4} dt > \int_0^1 \frac{t^4}{1+1} dt = \frac{1}{10}.$$

v) One of the standard examples in calculus involves computing the volume and surface area of the solid of revolution obtained by rotating the graph of $y = 1/x$ about the x-axis for $x \geq 1$. Known as *Gabriel's Horn*, this paradoxical object has an easily computed finite volume, while the surface area is infinite. The area computation,

$$\lim_{A \to \infty} \int_1^A 2\pi \frac{1}{x} \sqrt{1 + \frac{1}{x^4}} \, dx,$$

is a good exercise in techniques of integration. However, to show the unboundedness of the surface area it's sufficient simply to observe that the integrand is larger than $2\pi \frac{1}{x}$ and that the resulting $2\pi \ln A$ blows up when $A \to \infty$. (Cf. "The Paradox of Gabriel's Horn" in the Varberg, Purcell, Rigdon text [5, pp. 418, 419].)

vi) As a final application, we note that without appeal to the theorem allowing termwise integrability of series, it's possible to represent $\arctan(x)$ and $\ln(1+x)$ as power series. Starting with the fundamental finite geometric sum formula:

$$\frac{1}{1-x} = 1 + x + x^2 + x^3 + \cdots + x^n + \frac{x^{n+1}}{1-x} = \left(\sum_{k=0}^{n} x^k \right) + \text{error term},$$

substitute $x = -t^2$ and $-t$, respectively. Integrating from 0 to x yields $\arctan(x)$ and $\ln(1+x)$ as finite power series plus integrals of the error term, respectively. It's not difficult to show that as $n \to \infty$, the integral of the error term disappears for $-1 \leq x \leq 1$ in the case of $\arctan(x)$ and for $-1 < x \leq 1$ in the case of $\ln(1+x)$, yielding the familiar power series for these functions.

For instance, in the case of $\ln(1+x)$, let $\varepsilon \in (0, 1)$ and suppose $-1 + \varepsilon \leq x \leq 0$. The error term integral is

$$\left| \int_0^x \frac{(-t)^{n+1}}{1+t} dt \right| = \left| \int_0^{|x|} \frac{v^{n+1}}{1-v}(-1) dv \right| \leq \int_0^{|x|} \left| \frac{v^{n+1}}{1-v} \right| dv = \int_0^{|x|} \frac{v^{n+1}}{1-v} dv$$

$$\leq \int_0^{|x|} \frac{v^{n+1}}{1-|x|} dv = \frac{|x|^{n+2}}{(n+2)(1-|x|)}.$$

This shrinks to zero as $n \to \infty$, indeed uniformly for $x \in [-1+\varepsilon, 0]$.

The corresponding result when $x \in [0, 1]$ is similar but easier, because there is no need to avoid an endpoint at which the integrand blows up. For the arctan function it's also easy to show that, for $-1 \leq x \leq 1$, the error integral goes to zero as $n \to \infty$. This latter analysis has appeared in George Thomas's calculus texts at least since 1960 (see [2, pp. 689–690].)

The basic approach used in this sort of work may remind readers of the comparison test for infinite series. However, the series method tends to be more simple in classroom practice, for two reasons. First, the traditional repertory of comparison series is relatively limited (p-series and geometric series, with perhaps also factorial series). Second, in the study of a series the primary goal is to find out whether it converges, and for this the limit comparison test obviates the need to confirm an inequality. When the task turns to bounding or estimating the sum of a convergent series or to assessing the rate of growth of a divergent series, the approaches become more similar.

The above examples are elementary, yet they suggest the flexibility needed in approximations as well as the significant simplification provided by changing our point of view. Students find these problems difficult precisely because they cannot be solved by rote-learned algorithms. As Richard Feynman remarked when discussing skill in making approximations, "This is very difficult to teach because it's an art." [3, p. 16]

This topic reinforces the view that calculus is the study of the behavior of functions. More importantly, it shows mathematics as an art, as an endeavor probing relationships among elements in an infinite universe.

Acknowledgment. I am grateful to the referee for alerting me to the presence of this method in contemporary calculus texts.

References

1. C. H. Edwards and D. E. Penney, *Calculus*, 6th ed., Prentice Hall, 2002.
2. R. L. Finney, M. D. Weir, and F. R. Giordano, *Thomas' Calculus*, 10th ed., Addison-Wesley Longman, 2003.
3. D. and J. Goodstein, " 'The Uncreative Scientist'—Feynman's *Other Lost Lecture*", *Caltech News*, **31**:4 (1997) 12–17.
4. J. Stewart, *Calculus*, 5th ed., Thomson Learning, 2003.
5. D. Varberg, E. J. Purcell, and S. E. Rigdon, *Calculus*, 8th ed., Prentice Hall, 2000.

Added in proof: It has come to the author's attention that there was an earlier Capsule in this *Journal* illustrating the repeated use of the basic comparison property to obtain improved estimates of definite integrals. See W. V. Underhill, Finding bounds for definite integrals, *The College Mathematics Journal* **15** (1984) 426–429.

Proof Without Words: The Trapezoidal Rule (for Increasing Functions)

Jesús Urías

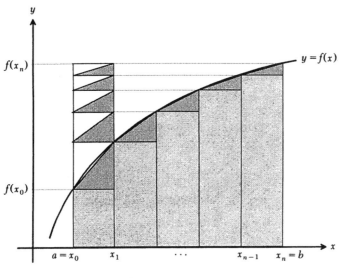

$$\int_a^b f(x)\,dx \cong \sum_{i=0}^{n-1} f(x_i)\frac{b-a}{n} + \frac{1}{2}[f(x_n)-f(x_0)]\frac{b-a}{n}$$

Behold! The Midpoint Rule is Better Than the Trapezoidal Rule for Concave Functions

Frank Buck

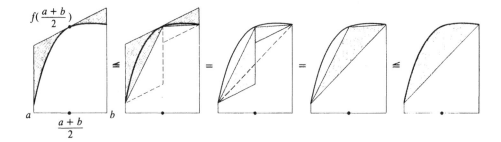

An Elementary Proof of Error Estimates for the Trapezoidal Rule

D. Cruz-Uribe & C.J. Neugebauer

Essentially every calculus textbook contains the trapezoidal rule for estimating definite integrals; this rule can be stated precisely as follows:

If f is continuous, then for each integer $n > 0$ the integral of f on $[a, b]$ is approximated by

$$T_n(f) = \frac{b-a}{2n}\bigl(f(x_0) + 2f(x_1) + 2f(x_2) + \cdots + 2f(x_{n-1}) + f(x_n)\bigr),$$

where $x_i = a + i(b-a)/n$, $0 \leq i \leq n$. Further, if f is twice differentiable, f'' is continuous, and $f''(t) \leq M$ for $t \in [a, b]$, then

$$E_n^T(f) = \left| T_n(f) - \int_a^b f(t)\,dt \right| \leq \frac{(b-a)^3}{12n^2} M. \qquad (1)$$

(See, for instance, Stewart [7].) There are two problems, however, with this result. First, calculus books generally omit the proof, and instead refer the reader to an advanced text on numerical analysis. In such books the trapezoidal rule is usually derived as a corollary to a more general result for Newton-Cotes quadrature methods, and the proof, depending on polynomial approximation, is generally not accessible to calculus students. (See, for example, Ralston [5].)

Second, the error estimate given by (1) is not applicable to such well-behaved functions as $x^{3/2}$ and $x^{1/2}$ on $[0, 1]$, since neither has a bounded second derivative. In other words, you can use the trapezoidal rule to approximate their integrals, but for a given n you have no idea, *a priori*, how good the approximation is.

In this note we give an elementary proof of inequality (1). The key idea is to use integration by parts "backwards." The argument is straightforward and should be readily understood by students in second-semester calculus. This approach is not new and goes back to Von Mises [8] and Peano [4]. (See also Ghizzetti and Ossicini [3].) However, our exact proof is either new or long forgotten—see Cruz-Uribe and Neugebauer [2] for a survey of the literature.

Our proof has two other advantages. First, we can omit the assumption that f'' is continuous, and replace it with the weaker assumption that it is (Riemann) integrable. Second, we can adapt our proof to give estimates for functions that do not have bounded second derivatives, such as $x^{3/2}$ and $x^{1/2}$.

Proving the inequality. The first step is to find an expression for the error on each interval $[x_{i-1}, x_i]$. If we divide the integral of f on $[a, b]$ into the sum of integrals on each of these subintervals, we have

$$E_n^T(f) = \left| \sum_{i=1}^{n} \left(\frac{b-a}{2n}(f(x_{i-1}) + f(x_i)) - \int_{x_{i-1}}^{x_i} f(t)\,dt \right) \right|.$$

For each i, $0 \leq i \leq n$, define

$$L_i = \frac{b-a}{2n}(f(x_{i-1}) + f(x_i)) - \int_{x_{i-1}}^{x_i} f(t)\,dt. \quad (2)$$

Let c_i denote the center of the interval $[x_{i-1}, x_i]$: $c_i = (x_{i-1} + x_i)/2$. Then

$$x_i - c_i = c_i - x_{i-1} = \frac{b-a}{2n},$$

and if we apply integration by parts "backwards," we see that we can express the error in terms of the first derivative:

$$L_i = \int_{x_{i-1}}^{x_i} (t - c_i) f'(t)\,dt. \quad (3)$$

(Given (3), it is easy to apply integration by parts to show that (2) holds; it is more subtle to argue in the other direction.) If we integrate this result by parts, we can express the error in terms of the second derivative:

$$L_i = \frac{1}{2} \int_{x_{i-1}}^{x_i} \left(\left(\frac{b-a}{2n}\right)^2 - (t - c_i)^2 \right) f''(t)\,dt. \quad (4)$$

Therefore, if $|f''(t)| \leq M$ for $t \in [a, b]$, we have that

$$E_n^T(f) \leq \sum_{i=1}^{n} |L_i|$$

$$\leq \frac{1}{2} \sum_{i=1}^{n} \int_{x_{i-1}}^{x_i} \left| \left(\frac{b-a}{2n}\right)^2 - (t - c_i)^2 \right| |f''(t)|\,dt$$

$$\leq \frac{M}{2} \sum_{i=1}^{n} \int_{x_{i-1}}^{x_i} \left(\left(\frac{b-a}{2n}\right)^2 - (t - c_i)^2 \right) dt.$$

For each i, a straightforward calculation shows that

$$\int_{x_{i-1}}^{x_i} \left(\left(\frac{b-a}{2n}\right)^2 - (t - c_i)^2 \right) dt = \frac{1}{6} \left(\frac{b-a}{n}\right)^3.$$

Hence,

An Elementary Proof of Error Estimates

$$E_n^T(f) \le \frac{M}{2}\sum_{i=1}^n \frac{1}{6}\left(\frac{b-a}{n}\right)^3 = \frac{(b-a)^3}{12n^2}M,$$

which is precisely what we wanted to prove.

What if the second derivative is not bounded? The heart of the proof of (1) is using (4) to estimate the error. However, a very similar argument works if we start with (3). Suppose $|f'(t)| \le N$ for $t \in [a, b]$. Then

$$|L_i| \le N \int_{x_{i-1}}^{x_i} |t - c_i|\, dt = \frac{N}{4}\left(\frac{b-a}{n}\right)^2,$$

which yields

$$E_n^T(f) \le \frac{(b-a)^2}{4n}N.$$

For example, if $f(x) = x^{3/2}$ on $[0, 1]$ then $|f'(x)| \le 3/2$, so

$$E_n^T(f) \le \frac{3}{8n}.$$

We can get a better estimate if the integrals of f'' and $|f''|$ exist as improper integrals, as is the case for $f(x) = x^{3/2}$. (Alternatively, we could use a more general definition of the integral, such as the Lebesgue integral or the Henstock-Kurzweiler integral. See [1, 6].) For in this case we can still apply integration by parts to derive (4). Since for each i,

$$0 \le \left(\frac{b-a}{2n}\right)^2 - (t-c_i)^2 \le \left(\frac{b-a}{2n}\right)^2,$$

it follows that

$$|L_i| \le \frac{1}{2}\left(\frac{b-a}{2n}\right)^2 \int_{x_{i-1}}^{x_i} |f''(t)|\, dt.$$

Hence,

$$E_n^T(f) \le \frac{(b-a)^2}{8n^2}\int_a^b |f''(t)|\, dt.$$

Again, if we let $f(x) = x^{3/2}$ on $[0, 1]$, this yields the estimate

$$E_n^T(f) \le \frac{3}{16n^2}.$$

Given a function such as $x^{1/2}$ on $[0, 1]$, whose first derivative is integrable as an improper integral, but whose second derivative is not, we can use the same argument to derive similar estimates from (3):

$$E_n^T(f) \le \frac{b-a}{2n} \int_a^b |f'(t)|\,dt.$$

Thus, if $f(x) = x^{1/2}$ on $[0, 1]$,

$$E_n^T(f) \le \frac{1}{2n}.$$

Further extensions.   We prove these and related results for the trapezoidal rule in [2]. We also show that all of these error estimates are sharp: for each n we construct a function f such that equality holds.

We also use the same techniques to prove error estimates for Simpson's rule in terms of the first and second derivative. The classical error estimate for Simpson's rule involve the fourth derivative [5, 7], and it is an open problem to extend our ideas to give an elementary proof of this result.

REFERENCES

1. G. Bartle, *A modern theory of integration*, Graduate Studies in Mathematics, 32, American Mathematical Society, Providence, 2001.
2. D. Cruz-Uribe, SFO and C. J. Neugebauer, Sharp error bounds for the trapezoidal rule and Simpson's rule, *J. Inequal. Pure Appl. Math.* **3** (2002), Issue 4, Article 49. Available at jipam.vu.edu.au/v3n4/index.html.
3. A. Ghizzetti and A. Ossicini, *Quadrature formulae*, Academic Press, New York, 1970.
4. G. Peano, Resto nelle formule di quadratura espresso con un integrale definito, *Atti Accad. naz. Lincei, Rend., Cl. sci. fis. mat. nat.* (5) **22-I** (1913), 562–9.
5. A. Ralston, *A first course in numerical analysis*, McGraw-Hill, New York, 1965.
6. W. Rudin, *Real and Complex Analysis,* Third Ed., McGraw-Hill, New York, 1987.
7. J. Stewart, *Calculus*, 4th Ed., Brooks/Cole, Pacific Grove, 1999.
8. R. von Mises, Über allgemeine quadraturformeln, *J. Reine Angew. Math.* **174** (1935), 56–67; reprinted in *Selected Papers of Richard von Mises*, Vol. 1, 559–574, American Mathematical Society, Providence, 1963.

Pictures Suggest How to Improve Elementary Numeric Integration

Keith Kendig

In a recent introductory numerical methods course, Maple helped students discover some substantial improvements to the trapezoidal and Simpson's methods. Students had learned to do numerical integration using rectangles and trapezoids, and were just starting to use Simpson's formula.

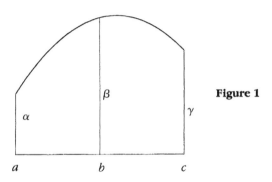

Figure 1

In Figure 1, the base is $(c - a)$, and $\frac{1}{6}(\alpha + 4\beta + \gamma)$ is an average altitude. The formula comes from passing a parabolic arc passing through three points; this arc follows a function's graph more closely than the line segment used in the rectangular or trapezoidal methods. After the class wrote a short computer program using Simpson's formula for approximating integrals, we tested it on $\int_0^2 x^3 \, dx$. Students found that the program gave the exact answer of 4, regardless of the number of subdivisions. This struck a number of them as remarkable since Simpson's formula had been developed to be exact only for quadratics, not cubics.

Why does this happen? The graph of $y = x^3$ passes through $(0, 0)$, $(1, 1)$ and $(2, 8)$, and the approximating parabola through these points is $y = 3x^2 - 2x$. As can be seen in Figure 2, the parabola dips below the cubic from $x = 0$ to 1, and rises above it from $x = 1$ to 2. Since Simpson's formula gives the exact result, the area of the vertically shaded region must be the same as the area of the horizontally shaded region. Students found that an analogous thing happens for simultaneous plots over other intervals and other cubics; through exploration, they stumbled upon one of the great free lunches in mathematics: although Simpson's formula is designed only to integrate quadratics exactly, it in fact exactly integrates every cubic.

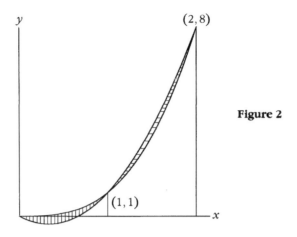

Figure 2

Maple made the proof of this easy. By direct integration, the area over $[a, c]$ and under the cubic

$$f(x) = Ax^3 + Bx^2 + Cx + D$$

is

$$\frac{A}{4}(c^4 - a^4) + \frac{B}{3}(c^3 - a^3) + \frac{C}{2}(c^2 - a^2) + D(c - a).$$

Factoring out the base $(c - a)$ leaves

$$\frac{A}{4}(a^3 + a^2c + ac^2 + c^3) + \frac{B}{3}(a^2 + ac + c^2) + \frac{C}{2}(a + c) + D,$$

which turns out to be the Simpson average altitude $\frac{1}{6}[f(a) + 4f((a + c)/2) + f(c)]$. Maple easily proves this: applying the **simplify** command to their difference gives 0. The error in Simpson-approximating any analytic function therefore comes from terms of degree ≥ 4; the extra, "free" level of accuracy has made this method a perennial favorite.

This suggested taking another look at the rectangular and trapezoidal methods. Just as parabolic arcs are degree two approximations to a curve, the tops of rectangles and trapezoids are degree zero and one approximations. Do these likewise exactly integrate polynomials through one higher degree? For mid-point rectangles, the answer is yes: the curve is a line, and the errors to the right and to the left of the mid-point are the oppositely-signed areas of two congruent triangles, which cancel.

For trapezoids, the answer is clearly no, since the trapezoidal method always over estimates the area under any curve that is concave up, and under-estimates it for any curve that is concave down. We finally decided to see if the trapezoidal method could be improved.

Let us consider the parabola $y = x^2$ between $a = -1$ and $c = 1$, shown in Figure 3. The approximating trapezoid is a rectangle with upper vertices $(-1, 1)$ and $(1, 1)$. Its area is 2, far larger than the area under the parabola, which is only $\frac{2}{3}$.

Figure 3, however, suggests an idea: move the altitudes closer together until the shaded areas cancel out. When does this occur?

In Figure 3, the base of the trapezoid is 2. The parabola intersects the top of the trapezoid at (d, d^2), so the altitude is d^2, giving an area of $2d^2$. If we want this to be equal to the area under the parabola, $\frac{2}{3}$, we take $d = \pm 1/\sqrt{3}$. Thus, moving the bases of the altitudes so they're $\pm 1/\sqrt{3}$ from the center exactly integrates the quadratic in this case.

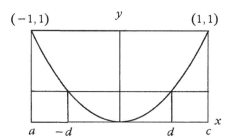

Figure 3

Remarkably, this works more generally. The exact area under $g(x) = Ax^2 + Bx + C$ from $x = a$ to $x = c$ is the same as that of the trapezoid whose base is $(c - a)$ and whose top is determined by altitudes based at points pulled in by a factor $1/\sqrt{3}$ towards the center $b = ((a+c)/2)$. This puts the altitude bases in our improved method at

$$\left(\frac{a+c}{2}\right) \pm \frac{1}{\sqrt{3}}\left(\frac{a-c}{2}\right)$$

rather than at the standard endpoints

$$\left(\frac{a+c}{2}\right) \pm 1\left(\frac{a-c}{2}\right)$$

(that is, a and c).

We can prove our claim much as we did before: first, direct integration shows that the area under the parabola is

$$\frac{A}{3}(c^3 - a^3) + \frac{B}{2}(c^2 - a^2) + C(c - a),$$

and this factors into the base $(c - a)$ times

$$\frac{A}{3}(a^2 + ac + c^2) + \frac{B}{2}(a + c) + C.$$

This last is the same as the average altitude of the trapezoid, which is

$$\frac{1}{2}\left[g\left(\frac{a+c}{2}+\frac{1}{\sqrt{3}}\left(\frac{a-c}{2}\right)\right)+g\left(\frac{a+c}{2}-\frac{1}{\sqrt{3}}\left(\frac{a-c}{2}\right)\right)\right].$$

Maple again easily verifies this—the **simplify** command shows that their difference is 0.

Amazingly, this gives us yet another free lunch: the improved trapezoid method is *so* much better that it actually integrates *all cubics* exactly! To see why, it's enough to show that it does this for $x^3 + g(x)$, with g as above. But since our new trapezoidal method integrates g exactly, we need only check that it does the same for x^3, giving $\frac{1}{4}(c^4 - a^4)$. Factoring out the trapezoid's base $(c - a)$ from this leaves

$$\tfrac{1}{4}(a^3 + a^2c + ac^2 + c^3);$$

this in turn is the average height

$$\frac{1}{2}\left[\left(\frac{a+c}{2}+\frac{1}{\sqrt{3}}\frac{a-c}{2}\right)^3+\left(\frac{a+c}{2}-\frac{1}{\sqrt{3}}\frac{a-c}{2}\right)^3\right],$$

as Maple's **simplify** once again shows.

How do these various integration methods compare in accuracy? We know that for most problems, using midpoint rectangles is the least accurate, using trapezoids is next, and Simpson's formula is best. *However, the improved trapezoid method turns out to be even better than Simpson's formula!* It somehow doesn't seem right that a first-order method should be better than a second-order one. One attentive student asked, "Since the trapezoid method can be improved so it exactly integrates polynomials through degree three, can Simpson's method be improved so it does the same through degree four?" This question turned out to be the right one, and the approach used before of symmetrically moving the altitude base points made this look plausible, at least for $y = x^4$ (see Figure 4). The altitudes intersect the graph of $y = x^4$ in two points, and as these two points approach each other, the parabola passing through them and $(0, 0)$ flattens out, creating areas above and below $y = x^4$. Of course we'd like to get that particular parabola where the error-areas above and below $y = x^4$ cancel out. Let the parabola be $y = d^2x^2$. Over the interval $[-1, 1]$, the area under $y = d^2x^2$ is $2d^2/3$, and the area under $y = x^4$ is $\frac{2}{5}$. If these areas are equal, then $d = \sqrt{\frac{3}{5}}$. We might guess that, more generally, the outer altitudes in

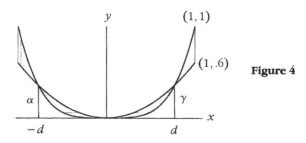

Figure 4

an improved Simpson formula ought to be based at

$$\left(\frac{a+c}{2}\right) \pm \sqrt{\frac{3}{5}}\left(\frac{a-c}{2}\right).$$

With these altitudes, what would the improved Simpson's formula be? The argument is a nice classroom review of the one used to get the usual formula: substituting the coordinates of any single point into the second-order equation $y = Ax^2 + Bx + C$ gives one linear equation in the variables A, B, C, so choosing three distinct points on this parabola gives a system of three linear equations. Solving for $A, B,$ and C then expresses these coefficients in terms of the three chosen points, and one can substitute these into $\int_a^c (Ax^2 + Bx + C)\, dx$ to get the area as Base $\times$ (Average altitude). If the three points lie on the vertical lines through the two endpoints and the midpoint of $[a, c]$, we get the usual expression $(1\alpha + 4\beta + 1\gamma)/(1 + 4 + 1)$ for average altitude. However, if the vertical lines through the endpoints are symmetrically moved in by our factor of $\sqrt{\frac{3}{5}}$ (we continue to call these altitudes $\alpha, \beta,$ and γ), then the average altitude is

$$\frac{5\alpha + 8\beta + 5\gamma}{5 + 8 + 5}.$$

These average altitudes are always numerically identical. Example: if we take the parabola $y = x^2$ over $[-1, 1]$, then the altitudes taken at $x = -1, 0, 1$ are $\alpha = 1, \beta = 0, \gamma = 1$, and the usual Simpson's formula gives an average altitude of $((1 \cdot 1 + 4 \cdot 0 + 1 \cdot 1)/6) = \frac{1}{3}$. If, however, the altitudes are chosen at $x = -\sqrt{\frac{3}{5}}, 0, \sqrt{\frac{3}{5}}$, then $\alpha = \frac{3}{5}, \beta = 0, \gamma = \frac{3}{5}$, and we get $((5 \cdot \frac{3}{5} + 8 \cdot 0 + 5 \cdot \frac{3}{5})/18 = \frac{1}{3}$. The improved approximation to $\int_a^c f(x)\, dx$ is then $(c - a)((5\alpha + 8\beta + 5\gamma)/18)$, where

$$\alpha = f\left(\left(\frac{a+c}{2}\right) - \sqrt{\frac{3}{5}}\left(\frac{a-c}{2}\right)\right); \beta = f\left(\frac{a+c}{2}\right); \gamma = f\left(\left(\frac{a+c}{2}\right) + \sqrt{\frac{3}{5}}\left(\frac{a-c}{2}\right)\right).$$

It turns out that this improvement not only integrates all polynomials through degree 4, it even does it through degree 5!. (Once again, let Maple show that the appropriate difference is 0.)

How good are these improvements? For most differentiable functions, the improved version of Simpson's method will add at least half again the number of decimal places accuracy obtained from the usual Simpson version. Compared to the usual trapezoid method, improved Simpson's at least *triples* the number of accurate decimal places.

It turned out that these were not new discoveries. We'd unexpectedly stumbled upon two old results: the improved trapezoid and Simpson methods are actually cases $n = 2$ and 3 of Gaussian n-point quadrature. (See [2], for example.) The zeros of the Legendre polynomial of degree n lead to an n-point quadrature formula that exactly integrates polynomials through degree $2n - 1$. Our factors $\sqrt{\frac{1}{3}}$ and $\sqrt{\frac{3}{5}}$ are zeros of Legendre polynomials of order 2 and 3; correspondingly, our formulas work through degree 3 and 5. Tabulated solutions for $2 \leq n \leq 20$ appear in [1] and, through $n = 200$, in [3]. Of course, this approach does not fit into a typical beginning course.

Our experience showed that substantial improvements to the trapezoidal and Simpson's methods can be successfully introduced into a course for beginners, using little more than pictures and Maple's **simplify** command.

References

1. M. Abramowitz and I. A. Stegun, *Handbook of Mathematical Functions*, Applied Series No. 55, National Bureau of Standards, 1964.
2. Richard L. Burden and J. Douglas Faires, *Numerical Analysis*, 5th ed., PWS-Kent Publishing Company, 1993.
3. Carl H. Love, *Abscissas and Weights for Gaussian Quadrature*, National Bureau of Standards, Monograph 98, 1966.

Part 4. Polynomial Approximations and Series

The Geometric Series in Calculus

George E. Andrews

1. INTRODUCTION. One of the fairly easily established facts from high school algebra is the Finite Geometric Series:

$$1 + r + r^2 + \cdots + r^n = \sum_{j=0}^{n} r^j = \frac{1 - r^{n+1}}{1 - r}. \tag{1.1}$$

This fact is made convincingly clear to all concerned by direct multiplication

$$
\begin{array}{r}
r^n + r^{n-1} + \cdots + r^2 + r + 1 \\
\times \quad\quad\quad\quad\quad\quad\quad -r + 1 \\
\hline
r^n + r^{n-1} + \cdots + r^2 + r + 1 \\
-r^{n+1} - r^n - r^{n-1} - + \cdots + - r^2 - r \\
\hline
-r^{n+1} \quad\quad\quad\quad\quad\quad\quad\quad + 1
\end{array}
$$

Unfortunately this elementary result is often skipped in algebra and is often first mentioned when infinite series arise in the second semester of calculus.

The object here is to show that the Geometric Series can play a very useful role in simplifying some important but complex topics in calculus. Most of the ideas in this note can be found in only slightly different guise sprinkled throughout Otto Toeplitz's charming book [6], which, unfortunately, is out of print.

2. THE DERIVATIVE OF x^n. Showing students that

$$\frac{d}{dx} x^n = nx^{n-1}$$

usually poses a dilemma. From the standard definition of a derivative, we see that

$$\frac{d}{dx} x^n = \lim_{h \to 0} \frac{(x+h)^n - x^n}{h}.$$

How should we proceed?

One approach [2, p. 162] is to use the Binomial Theorem without saying much save for a few examples. Not exactly convincing!

Or should we prove the Binomial Theorem at this point? Probably not!

Perhaps we could prove a Weak Binomial Theorem:
$$(x+h)^n = x^n + nhx^{n-1} + h^2(\cdots).$$
Again, we have a distraction, at best.

Let us bring the Finite Geometric Series to the rescue. The standard definition of the derivative, viz.
$$f'(x) = \lim_{h \to 0} \frac{f(x+h) - f(x)}{h}, \tag{2.1}$$
is easily seen (both algebraically and geometrically) to be equivalent to
$$f'(x) = \lim_{y \to x} \frac{f(y) - f(x)}{y - x}, \tag{2.2}$$
and if $x \neq 0$ and we look at the ratio of y to x, we find a third equivalent formulation ($y = qx$):
$$f'(x) = \lim_{q \to 1} \frac{f(qx) - f(x)}{qx - x}, \quad x \neq 0. \tag{2.3}$$

We may now use this third definition of $f'(x)$ to determine the derivative of x^n:
$$\begin{aligned}
\lim_{q \to 1} \frac{(xq)^n - x^n}{qx - x} &= x^{n-1} \lim_{q \to 1} \frac{q^n - 1}{q - 1} \\
&= x^{n-1} \lim_{q \to 1} (1 + q + q^2 + \cdots + q^{n-1}) \quad \text{(by (1.1))} \\
&= nx^{n-1}.
\end{aligned} \tag{2.4}$$

While this is valid only if $x \neq 0$, the original definition (2.1) easily treats $x = 0$. The derivative of $x^{n/m}$ can be handled in the same manner by a simple change of the variable q.

3. INTEGRALS AND THE FUNDAMENTAL THEOREM OF CALCULUS. We often hope to say compelling things about Riemann sums when we define
$$\int_a^b f(x)\, dx = \lim_{|P| \to 0} \sum_{i=1}^n f(x_i^*)(x_i - x_{i-1}). \tag{3.1}$$
However, when we try to compute examples of simple integrals with a uniform partition P of $[a, b]$, we can wind up with expressions such as
$$\begin{aligned}
\int_0^1 x^2\, dx &= \lim_{n \to \infty} \sum_{i=1}^n \left(\frac{i}{n}\right)^2 \frac{1}{n} \\
&= \lim_{n \to \infty} \frac{1}{n^3} \sum_{i=1}^n i^2.
\end{aligned}$$

The problem is now analogous to our problem for taking the derivative of x^n. We must either pull
$$\sum_{i=1}^n i^2 = \frac{1}{6} n(n+1)(2n+1)$$
out of a hat, or else spend a substantial amount of time motivating and proving it.

Again the Finite Geometric Series can come to our rescue. As an alternative to the Riemann sum, we can examine a geometric dissection of our interval (see Figure 1).

The area of the rectangles indicated is

$$\begin{aligned}A_q(X) &= f(X)(X - qX) + f(Xq)(Xq - Xq^2) \\ &\quad + f(Xq^2)(Xq^2 - Xq^3) + \cdots \\ &= \sum_{i=0}^{\infty} f(Xq^i)(Xq^i - Xq^{i+1}).\end{aligned} \qquad (3.2)$$

As $q \to 1^-$, it is visually convincing that $A_q(X)$ converges to the area under the curve, and (probably in an appendix) an actual proof that this definition is equivalent to the standard Riemann sum definition is no more difficult than any other portion of the rigorous treatment of Riemann sums.

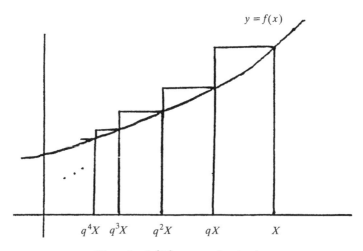

Figure 1. $A_q(X)$ = area of rectangles

In any event, now it is possible to integrate not just x^2, but, indeed, any positive integral power of x. First we note that the Finite Geometric Series directly leads to the Infinite Geometric Series. If $|r| < 1$, then

$$\sum_{i=0}^{\infty} r^i = \lim_{n \to \infty} \sum_{i=0}^{n} r^i = \lim_{n \to \infty} \frac{1 - r^{n+1}}{1 - r} = \frac{1}{1 - r}. \qquad (3.3)$$

The subtleties of infinite series in general need not be introduced here because we have the explicit formula for the partial sums.

Hence

$$\begin{aligned}\int_0^X x^n \, dx &= \lim_{q \to 1^-} A_q(X) \\ &= \lim_{q \to 1^-} \sum_{i=0}^{\infty} (Xq^i)^n (Xq^i - Xq^{i+1})\end{aligned}$$

$$= X^{n+1} \lim_{q \to 1^-} (1-q) \sum_{i=0}^{\infty} q^{i(n+1)}$$

$$= X^{n+1} \lim_{q \to 1^-} \frac{(1-q)}{1-q^{n+1}} \quad \text{(by (3.3))}$$

$$= X^{n+1} \lim_{q \to 1^-} \frac{1}{1+q+q^2+\cdots+q^n} \quad \text{(by (1.1))}$$

$$= \frac{X^{n+1}}{n+1}.$$

(3.4)

Again a simple change of the variable q allows the integration of $x^{n/m}$.

In addition to performing this integration of x^n, the shape of the Fundamental Theorem of Calculus is now much more transparent. From (3.2) (or Figure 1), we see that

$$A_q(X) = f(X)(X - qX) + A_q(Xq).$$

Hence

$$\frac{A_q(Xq) - A_q(X)}{qX - X} = f(X), \tag{3.5}$$

and by recalling (2.3) we see that (3.5) clearly and convincingly suggests the Fundamental Theorem of Calculus.

Although a fully rigorous proof of the Fundamental Theorem can be effected from (3.5), one probably does not really want to do so in a first calculus course.

4. POWER SERIES. I won't dwell on the use of the Infinite Geometric Series in proving the Root Test and Ratio Test. This is well-known and practised in almost all calculus books.

I remark only that when one finally arrives at infinite series, the Infinite Geometric Series is an old and trusted friend rather than something that first arises as the case $p = -1$ of the binomial series [4, p. 605]:

$$(1+x)^p = 1 + px + \frac{p(p-1)}{2}x^2 + \frac{p(p-1)(p-2)}{3}x^3 + \cdots$$

5. EXERCISES ON THE GEOMETRIC SERIES. Given the great utility of the Geometric Series, any exercise that makes it more familiar will be useful. There are countless "plug and chug" type exercises. We close with three more "modern" exercises.

1. ([1, p. 4], Don Cohen): Observe the following dissection of a unit square

Show how this illustrates an instance of the Infinite Geometric Series.

2. The following is from W. Edwards Deming's *The New Economics* [3, p. 136].

The secret for reduction in time of development is to put more effort into the early stages, and to study the interaction between stages. Each stage should have the benefit of more effort than the next stage.

We content ourselves here to adopt a constant ratio of cost from one stage to the next. Specifically, let the cost of any stage be $1 - x$ times the cost of the preceding stage. Then if K be the cost of the opening stage (the 0-th stage, concepts and proposals), then the cost of the n-th stage would be

$$K_n = K(1 - x)^n.$$

The total cost through the n-th stage would be

$$T_n = K\left[1 + (1 - x) + (1 - x)^2 + (1 - x)^3 + \cdots + (1 - x)^n\right].$$

We note that the series in the brackets is merely $1/x$ expanded in powers of $1 - x$. This is easily seen by writing $x = 1 - (1 - x)$. This series will converge if $0 < x \leq 1$, which satisfies our requirements. Further,

$$T_n = K\left\{\left[1 + (1 - x) + (1 - x)^2 + (1 - x)^3 + \cdots \text{ to infinity}\right] - \frac{(1 - x)^{n+1}}{x}\right\}$$

$$= \frac{K}{x}\left[1 - (1 - x)^{n+1}\right].$$

Assignment. Rewrite Deming's argument so that the role of the Finite Geometric Series is **clear**. Why did Deming use the Infinite Geometric Series?

3. A problem attributed by R. Raimi to a Professor Sleator at the University of Michigan in 1941: Two trees are one mile apart. A drib (it is not necessary that you know what a drib is) flies from one tree to the other and back, making the first trip at 10 miles per hour, the return at 20 miles per hour, the next at 40 and so on, each trip at twice the speed of the preceding. When will the drib be in both trees at the same time? Do not spend time wondering or arguing about the drib, but solve the problem.

ACKNOWLEDGMENTS. Richard Askey, Robert Sachs, and Ken Stolarsky drew my attention to most of the ideas described here. Askey has used the material from Sections 2 and 3 in calculus courses. The ideas date back to antiquity going from Archimedes to Fermat to E. Heine, J. Thomae, and F. H. Jackson. The reader interested in an entire calculus book that includes these ideas should consult [6]. The material in this talk was presented at the Fifth Annual Conference on the Teaching of Calculus (Baltimore, June 22, 1996). Partially supported by National Science Foundation Grant DMS-9501101.

REFERENCES

1. D. Cohen, *Calculus By and For Young People*, Don Cohen the Mathman, Champaign, 1988.
2. R. Decker and D. Varberg, *Calculus*, Prentice-Hall, Upper Saddle River, 1996.
3. W. Edwards Deming, *The New Economics*, MIT CAES, Cambridge, 1994.
4. D. Hughes-Hallett et al., *Calculus*, John Wiley, New York, 1994.
5. R. Raimi, *Private communication*, 1996.

6. O. Toeplitz, *Die Entwicklung der Infinitesimalrechnung*, Springer, Berlin, 1949 [transl: The Calculus, A Genetic Approach, U. of Chicago Press, Chicago, 1963].
7. F. Wattenberg, *Calculus in a Real and Complex World*, PWS Publishing, Boston, 1995.

A Visual Approach to Geometric Series

Beata Randrianantoanina

Students introduced to the notion of series frequently consider it counter-intuitive (if not outright illogical) that one can add up an infinite number of terms and get as a result a finite sum instead of infinity, even if the terms are very small. Thus, many visual arguments have been developed to explain summing of infinite series, see, for example, the collections [2] and [3]. In particular, one can find there reprints of very elegant illustrations by Warren Page [4] and Elizabeth M. Markham [1] that illuminate the formula for the sum of a geometric series. This note describes a visual demonstration of this formula that students seem to enjoy. The demonstration is performed in the classroom without the use of any pictures. I use a "scientific model" of an object that consists of an infinite number of parts—a Calculus book. (All Calculus books I have ever used have a number of pages that approximates infinity quite well.)

The demonstration starts by considering all pages of the book as one unit. Next, I divide the pages into 2 equal parts and drop $\frac{1}{2}$ to one side—this is the beginning of my series. I still have $\frac{1}{2}$ of the book in my hand. I split it into 2 equal parts, drop $\frac{1}{4}$ of the book to the side and keep $\frac{1}{4}$ in my hand. (See Figure 1, which shows the book as viewed from one end.) I continue this process several times until I have only one page left in my hand (which visualizes zero), and on the pile are lying:

$$\frac{1}{2} + \frac{1}{4} + \frac{1}{8} + \frac{1}{16} + \cdots$$

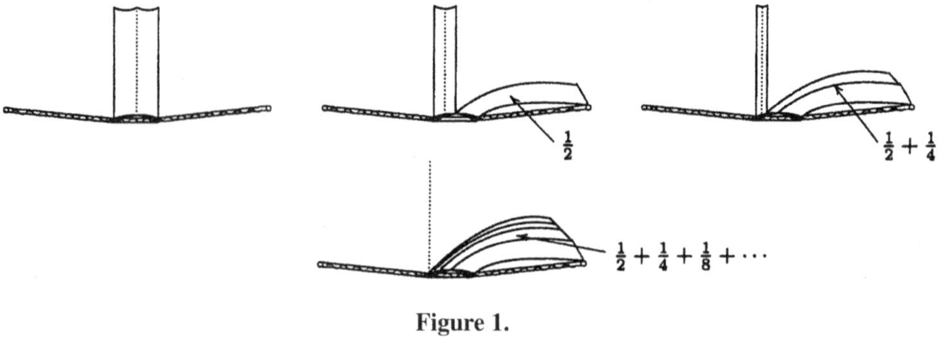

Figure 1.

A Visual Approach to Geometric Series

Based on their observation of the book, most students conclude readily that this means the above sum equals 1.

This demonstration is not limited to geometric series with ratio equal to $\frac{1}{2}$. For example, let $r = \frac{1}{3}$. Then, as illustrated in Figure 2, I split the book into 3 equal pieces, drop $\frac{1}{3}$ to the pile that we "keep", drop $\frac{1}{3}$ to the pile on the other side that we "trash," and hold $\frac{1}{3}$ in my hand "to work with." Continue splitting the part in my hands, putting equal parts on piles on both sides and holding in my hand an equal part.

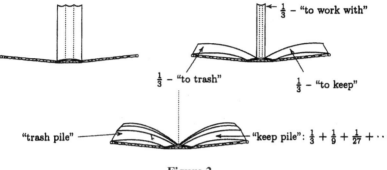

Figure 2.

Again, eventually I am left with only one page in my hands, and there are two piles on each side of equal size. Thus the students conclude that

$$\frac{1}{3} + \frac{1}{9} + \frac{1}{27} + \cdots = \frac{1}{2}.$$

Students really enjoy this activity, so I do it once more for, say, $r = \frac{1}{5}$, as in Figure 3.

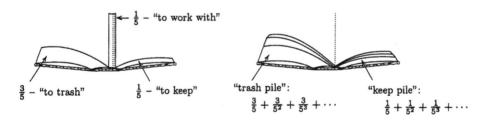

Figure 3.

We put $\frac{1}{5}$ on the "keep pile", hold $\frac{1}{5}$ in our hands "to work with", and put the remainder: $1 - \frac{1}{5} - \frac{1}{5} = \frac{3}{5}$, on the "trash pile". We continue until all pages are distributed and this time the "trash pile" is 3 times as big as the "keep pile". Together they add up to 1 book, so

$$\frac{1}{5} + \frac{1}{5^2} + \frac{1}{5^3} + \cdots = \frac{1}{4}.$$

This is where I usually stop this activity in my classroom and proceed to an algebraic derivation of the formula for the sum of a geometric series, which usually gives my students a, so well-known, "overwhelmed" look on their faces. But, in fact, the above "book demonstration" can be used to "prove" the general formula. Let the ratio of the geometric series be any positive $r < \frac{1}{2}$. Then, as sketched in Figure 4, we put the r portion of the book on the "keep pile", hold another r portion in our hands "to work with", and put the remaining $(1 - 2r)$ on the "trash pile". As before, we repeat this procedure until nothing is left in our hand.

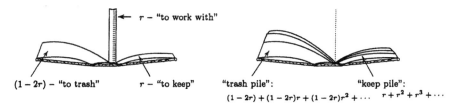

Figure 4.

We end up with two piles of the following sizes:

$$\text{"keep pile"} = r + r^2 + r^3 + \cdots = s,$$
$$\text{"trash pile"} = \frac{(1 - 2r)}{r} \cdot \text{"keep pile"} = \frac{1 - 2r}{r} \cdot s.$$

Since the sum of the "keep pile" and the "trash pile" is equal to 1, we get

$$s + \frac{1 - 2r}{r} \cdot s = 1, \qquad s = \frac{r}{1 - r}, \qquad \sum_{k=1}^{\infty} r^k = \frac{r}{1 - r}.$$

There is also another way to perform this computation. Note that the second step of the procedure of separating the pages of the book is identical to the first step, the only difference being that in the first step we hold in our hand the whole book, i.e. one unit, and in the second step we hold the portion of the book of size r. Thus, portions of the book sent to each pile in the second step need to be multiplied by r. Thus we get

$$\text{"keep pile"} = r + r \cdot r + r^2 \cdot r + \cdots$$
$$\text{"trash pile"} = (1 - 2r) + (1 - 2r)r + (1 - 2r)r \cdot r + \cdots$$
$$= (1 - 2r)[1 + r + r^2 + \cdots],$$

and together they add up to 1. Thus

$$[r + r^2 + r^3 + \cdots] + (1 - 2r)[1 + r + r^2 + \cdots] = 1,$$
$$[1 + [r + r^2 + r^3 + \cdots]] + (1 - 2r)[1 + r + r^2 + \cdots] = 1 + 1,$$
$$(1 + 1 - 2r)[1 + r + r^2 + \cdots] = 2,$$
$$[1 + r + r^2 + \cdots] = \frac{2}{2 - 2r}.$$

A Visual Approach to Geometric Series

So

$$\sum_{k=0}^{\infty} r^k = \frac{1}{1-r}.$$

Note that these two alternative arguments give us the formulas for the sum of a geometric series beginning with a term raised to the power zero or to the power one, according to our preference.

A natural question that arises here is whether this method is restricted to $r < \frac{1}{2}$. The answer is "no." It can be adjusted to work for any positive number $r < 1$ as follows. We agree that our book represents n units instead of 1 unit, where n is a natural number chosen so that $r < \frac{n}{n+1}$. We proceed in the same way as before, we put r on the "keep pile," keep nr in our hand "to work with," and put the remaining $(n - nr - r) = (n - (n+1)r)$ on the "trash pile" (see Figure 5).

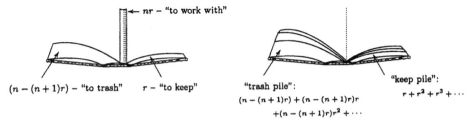

Figure 5.

As before, the second step is identical to the first one, with all quantities multiplied by r. Thus, we get

$$\text{"keep pile"} = r + r^2 + r^3 + \cdots = s,$$
$$\text{"trash pile"} = \frac{n - (n+1)r}{r} \cdot \text{"keep pile"} = \frac{n - (n+1)r}{r} \cdot s,$$

and since the sum of the two piles is now n, we conclude that

$$s + \frac{n - (n+1)r}{r} \cdot s = n, \qquad s = \frac{nr}{n - nr} = \frac{r}{1-r} = \sum_{k=1}^{\infty} r^k.$$

As before, we can also compute

$$\text{"trash pile"} = \big(n - (n+1)r\big) + \big(n - (n+1)r\big)r + \big(n - (n+1)r\big)r^2 + \cdots$$
$$= \big(n - (n+1)r\big)[1 + r + r^2 + \cdots],$$

and thus

$$[r + r^2 + r^3 + \cdots] + \big(n - (n+1)r\big)[1 + r + r^2 + \cdots] = n,$$
$$\big[1 + [r + r^2 + r^3 + \cdots]\big] + \big(n - (n+1)r\big)[1 + r + r^2 + \cdots] = n + 1,$$
$$\big(n + 1 - (n+1)r\big)[1 + r + r^2 + r^3 + \cdots] = n + 1,$$
$$[1 + r + r^2 + r^3 + \cdots] = \frac{n+1}{(n+1) - (n+1)r} = \frac{1}{1-r}.$$

So
$$\sum_{k=0}^{\infty} r^k = \frac{1}{1-r}.$$

A question that remains is whether this procedure can be adjusted to work for negative values of r. Since I don't know the answer, I will leave it to interested readers.

Acknowledgments. I wish to thank Professors Fred Gass and Tom Farmer for their helpful comments.

References

1. Elizabeth M. Markham, Geometric series, *Mathematics Magazine* **66** (1993) p. 242.
2. Roger B. Nelsen, *Proofs Without Words: Exercises in visual thinking*, Classroom Resourse Materials, The Mathematical Association of America, Washington, 1993.
3. ———, *Proofs Without Words II: More Exercises in Visual Thinking*, Classroom Resourse Materials, The Mathematical Association of America, Washington, 2000.
4. Warren Page, Geometric sums, *Mathematics Magazine* **54** (1981) p. 201.

The Telescoping Series in Perspective

Marc Frantz

The telescoping series

$$\sum_{k=1}^{\infty} \frac{1}{k(k+1)}$$

gets its name because the sum of the first n terms collapses:

$$\sum_{k=1}^{n} \frac{1}{k(k+1)} = \sum_{k=1}^{n} \left(\frac{1}{k} - \frac{1}{k+1}\right) = 1 - \frac{1}{n+1}.$$

We conclude, letting $n \to \infty$, that the series converges to 1. The telescoping series is more than just an algebraic curiosity! In fact, we see examples of it almost every day. One is shown in FIGURE 1.

FIGURE 1 shows that the apparent horizontal separations a_n of the telephone poles satisfy (assuming infinitely many poles) $\sum_{n=1}^{\infty} a_n = S$. This series turns out, in fact, to

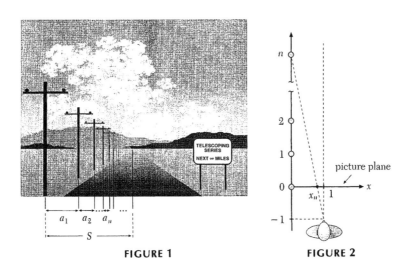

FIGURE 1   **FIGURE 2**

be a version of the telescoping series. For simplicity, we'll consider the special case shown in FIGURE 2. The idea behind a perspective drawing (FIGURE 2) is that a viewer, seen from above, stands in front of a "picture plane" that in this case is perpendicular to the ground and contains the x-axis. The viewer uses only one eye, and thus is idealized as a single point. As light rays travel in straight lines from objects in the real world to the viewer's eye, they pierce the picture plane, leaving behind appropriately colored dots that, taken together, depict the scene.

For convenience, we take as our unit of measure the (uniform) separation between the telephone poles, and we locate the picture plane and the viewer as shown in FIGURE 2. For $n \geq 1$, the use of similar triangles shows that the x-coordinate x_n of the image of the nth pole satisfies

$$\frac{x_n}{n} = \frac{1-x_n}{1} \Rightarrow x_n = \frac{n}{n+1}.$$

Thus, for $n \geq 1$, the nth gap between x_{n-1} and x_n has width a_n, where

$$a_n = x_n - x_{n-1} = \frac{1}{n(n+1)}.$$

Moreover, comparing FIGURE 2 to FIGURE 1 shows that $S = 1$: the row of telephone poles appears to vanish precisely when the viewer looks straight ahead.

Proof Without Words

Richard Hammack and David Lyons

Theorem. *An alternating series $a_1 - a_2 + a_3 - a_4 + a_5 - a_6 + a_7 - a_8 + \cdots$ converges to a sum S if $a_1 \geq a_2 \geq a_3 \geq a_4 \geq \cdots \geq 0$ and $a_n \to 0$. Moreover, if $s_n = a_1 - a_2 + a_3 - \cdots \pm a_n$ is the nth partial sum then $s_{2n} < S < s_{2n+1}$.*

Proof.

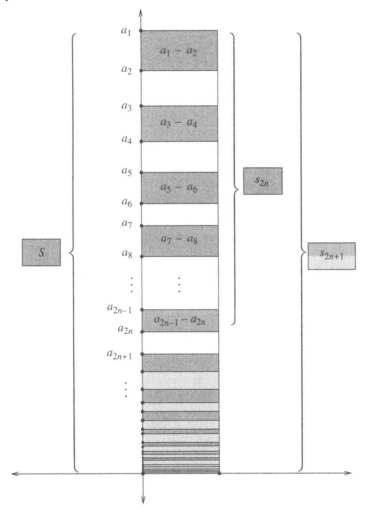

The Bernoullis and the Harmonic Series

William Dunham

Any introduction to the topic of infinite series soon must address that first great counterexample of a divergent series whose general term goes to zero—the harmonic series $\sum_{k=1}^{\infty} 1/k$. Modern texts employ a standard argument, traceable back to the great 14th Century Frenchman Nicole Oresme (see [3], p. 92), which establishes divergence by grouping the partial sums:

$$1 + \frac{1}{2} > \frac{1}{2} + \frac{1}{2} = \frac{2}{2}$$

$$1 + \frac{1}{2} + \left(\frac{1}{3} + \frac{1}{4}\right) > \frac{2}{2} + \left(\frac{1}{4} + \frac{1}{4}\right) = \frac{3}{2}$$

$$1 + \frac{1}{2} + \frac{1}{3} + \frac{1}{4} + \left(\frac{1}{5} + \frac{1}{6} + \frac{1}{7} + \frac{1}{8}\right) > \frac{3}{2} + \left(\frac{1}{8} + \frac{1}{8} + \frac{1}{8} + \frac{1}{8}\right) = \frac{4}{2},$$

and in general

$$1 + \frac{1}{2} + \frac{1}{3} + \cdots + \frac{1}{2^n} > \frac{n+1}{2},$$

from which it follows that the partial sums grow arbitrarily large as n goes to infinity.

It is possible that seasoned mathematicians tend to forget how surprising this phenomenon appears to the uninitiated student—that, by adding ever more negligible terms, we nonetheless reach a sum greater than any preassigned quantity. Historian of mathematics Morris Kline ([5], p. 443) reminds us that this feature of the harmonic series seemed troubling, if not pathological, when first discovered.

So unusual a series could not help but attract the interest of the preeminent mathematical family of the 17th Century, the Bernoullis. Indeed, in his 1689 treatise "Tractatus de Seriebus Infinitis," Jakob Bernoulli provided an entirely different, yet equally ingenious proof of the divergence of the harmonic series. In "Tractatus," which is now most readily found as an appendix to his posthumous 1713 masterpiece *Ars Conjectandi*, Jakob generously attributed the proof to his brother ("Id primus deprehendit Frater"), the reference being to his full-time sibling and part-time rival Johann. While this "Bernoullian" argument is sketched in such mathematics history texts as Kline ([5], p. 444) and Struik ([6], p. 321), it is little enough known to warrant a quick reexamination.

Courtesy of the Lilly Library, Indiana University, Bloomington, IN

The proof rested, quite unexpectedly, upon the *convergent* series
$$\frac{1}{2} + \frac{1}{6} + \frac{1}{12} + \frac{1}{20} + \cdots = \sum_{k=1}^{\infty} \frac{1}{k(k+1)}.$$
The modern reader can easily establish, via mathematical induction, that
$$\sum_{k=1}^{n} \frac{1}{k(k+1)} = \frac{n}{n+1},$$
and then let n go to infinity to conclude that
$$\sum_{k=1}^{\infty} \frac{1}{k(k+1)} = 1.$$

Jakob Bernoulli, however, approached the problem quite differently. In Section XV of *Tractatus*, he considered the infinite series
$$N = \frac{a}{c} + \frac{a}{2c} + \frac{a}{3c} + \frac{a}{4c} + \cdots,$$
then introduced
$$P = N - \frac{a}{c} = \frac{a}{2c} + \frac{a}{3c} + \frac{a}{4c} + \frac{a}{5c} + \cdots,$$
and subtracted termwise to get
$$\begin{aligned}\frac{a}{c} = N - P &= \left(\frac{a}{c} - \frac{a}{2c}\right) + \left(\frac{a}{2c} - \frac{a}{3c}\right) + \left(\frac{a}{3c} - \frac{a}{4c}\right) + \cdots \\ &= \frac{a}{2c} + \frac{a}{6c} + \frac{a}{12c} + \frac{a}{20c} + \cdots.\end{aligned} \quad (1)$$

Thus, for $a = c$, he concluded that

$$\frac{1}{2} + \frac{1}{6} + \frac{1}{12} + \frac{1}{20} + \cdots = \frac{1}{1} = 1. \tag{2}$$

Unfortunately, Bernoulli's "proof" required the subtraction of two divergent series, N and P. To his credit, Bernoulli recognized the inherent dangers in his argument, and he advised that this procedure must not be used without caution ("non sine cautela"). To illustrate his point, he applied the previous reasoning to the series

$$S = \frac{2a}{c} + \frac{3a}{2c} + \frac{4a}{3c} + \cdots$$

and

$$T = S - \frac{2a}{c} = \frac{3a}{2c} + \frac{4a}{3c} + \frac{5a}{4c} + \cdots.$$

Upon subtracting termwise, he got

$$\frac{2a}{c} = S - T = \frac{a}{2c} + \frac{a}{6c} + \frac{a}{12c} + \frac{a}{20c} + \cdots, \tag{3}$$

which provided a clear contradiction to (1).

Bernoulli analyzed and resolved this contradiction as follows: the derivation of (1) was valid since the "last" term of series N is zero (that is, $\lim_{k \to \infty} a/(kc) = 0$), whereas the parallel derivation of (3) was invalid since the "last" term of series S is non-zero (because $\lim_{k \to \infty}(k+1)a/(kc) = a/c \neq 0$). In modern terms, he had correctly recognized that, regardless of the convergence or divergence of the series $\sum_{k=1}^{\infty} x_k$, the new series $\sum_{k=1}^{\infty}(x_k - x_{k+1})$ converges to x_1 *provided* $\lim_{k \to \infty} x_k = 0$. Thus, he not only explained the need for "caution" in his earlier discussion but also exhibited a fairly penetrating insight, by the standards of his day, into the general convergence/divergence issue.

Having thus established (2) to his satisfaction, Jakob addressed the harmonic series itself. Using his brother's analysis of the harmonic series, he proclaimed in Section XVI of *Tractatus*:

XVI. *Summa feriei infinitæ harmonicè progreffionalium*, $\frac{1}{1} + \frac{1}{2} + \frac{1}{3} + \frac{1}{4} + \frac{1}{5}$ &c. *eft infinita.*

He began the argument that "the sum of the infinite harmonic series

$$\frac{1}{1} + \frac{1}{2} + \frac{1}{3} + \frac{1}{4} + \frac{1}{5} \text{ etc.}$$

is infinite" by introducing

$$A = \frac{1}{2} + \frac{1}{3} + \frac{1}{4} + \frac{1}{5} + \frac{1}{6} + \frac{1}{7} + \cdots,$$

which "transformed into fractions whose numerators are 1, 2, 3, 4 etc" becomes

$$\frac{1}{2} + \frac{2}{6} + \frac{3}{12} + \frac{4}{20} + \frac{5}{30} + \frac{6}{42} + \cdots.$$

Using (2), Jakob next evaluated:

$$C = \frac{1}{2} + \frac{1}{6} + \frac{1}{12} + \frac{1}{20} + \cdots = 1$$

$$D = \phantom{\frac{1}{2} + {}} \frac{1}{6} + \frac{1}{12} + \frac{1}{20} + \cdots = C - \frac{1}{2} = 1 - \frac{1}{2} = \frac{1}{2}$$

$$E = \phantom{\frac{1}{2} + \frac{1}{6} + {}} \frac{1}{12} + \frac{1}{20} + \cdots = D - \frac{1}{6} = \frac{1}{2} - \frac{1}{6} = \frac{1}{3}$$

$$F = \phantom{\frac{1}{2} + \frac{1}{6} + \frac{1}{12} + {}} \frac{1}{20} + \cdots = E - \frac{1}{12} = \frac{1}{3} - \frac{1}{12} = \frac{1}{4}$$

$$\vdots \qquad \vdots \qquad \vdots \qquad \vdots$$

By adding this array columnwise, and again implicitly assuming that termwise addition of infinite series is permissible, he arrived at

$$C + D + E + F + \cdots = \frac{1}{2} + \left(\frac{1}{6} + \frac{1}{6}\right) + \left(\frac{1}{12} + \frac{1}{12} + \frac{1}{12}\right) + \cdots$$

$$= \frac{1}{2} + \frac{2}{6} + \frac{3}{12} + \frac{4}{20} + \cdots$$

$$= A.$$

On the other hand, upon separately summing the terms forming the extreme left and the extreme right of the arrayed equations above, he got

$$C + D + E + F + \cdots = 1 + \frac{1}{2} + \frac{1}{3} + \frac{1}{4} + \cdots = 1 + A.$$

Hence, $A = 1 + A$. In Jakob's words, "The whole" equals "the part"—that is, the harmonic series $1 + A$ equals its part A—which is impossible for a finite quantity. From this, he concluded that $1 + A$ is infinite.

Jakob Bernoulli was certainly convinced of the importance of his brother's deduction and emphasized its salient point when he wrote:

> The sum of an infinite series whose final term vanishes perhaps is finite, perhaps infinite.

Obviously, this proof features a naive treatment both of series manipulation and of the nature of "infinity." In addition, it attacks infinite series "holistically" as single entities, without recourse to the modern idea of partial sums. Before getting overly critical of its distinctly 17th-century flavor, however, we must acknowledge that Bernoulli devised this proof a century and a half before the appearance of a truly rigorous theory of series. Further, we can not deny the simplicity and cleverness of his reasoning nor the fact that, if bolstered by the necessary supports of modern analysis, it can serve as a suitable alternative to the standard proof.

Indeed, this argument provides us with an example of the history of mathematics at its best—paying homage to the past yet adding a note of freshness and ingenuity to the modern classroom. Perhaps, in contemplating this work, some of today's students might even come to share a bit of the enthusiasm and wonder that moved Jakob Bernoulli to close his *Tractatus* with the verse [7]

> XVI. *Summa seriei infinitæ harmonicè progressionalium*, $\frac{1}{1} + \frac{1}{2} + \frac{1}{3} + \frac{1}{4} + \frac{1}{5}$ &c. *est infinita.*
>
> Id primus deprehendit Frater: inventa namque per præced. summa seriei $\frac{1}{2} + \frac{1}{6} + \frac{1}{12} + \frac{1}{20} + \frac{1}{30}$, &c. visurus porrò, quid emergeret ex ista serie, $\frac{2}{2} + \frac{3}{6} + \frac{4}{12} + \frac{4}{20} + \frac{5}{30}$, &c. si resolveretur methodo Prop. XIV. collegit propositionis veritatem ex absurditate manifesta, quæ sequeretur, si summa seriei harmonicæ finita statueretur. Animadvertit enim,
>
> Seriem A, $\frac{1}{2} + \frac{1}{3} + \frac{1}{4} + \frac{1}{5} + \frac{1}{6} + \frac{1}{7}$, &c. ∞ (fractionibus singulis in alias, quarum numeratores sunt 1, 2, 3, 4, &c. transmutatis)
>
> seriei B, $\frac{1}{2} + \frac{2}{6} + \frac{3}{12} + \frac{4}{20} + \frac{5}{30} + \frac{6}{42}$, &c. ∞ C+D+E+F, &c.
>
> C. $\frac{1}{2} + \frac{1}{6} + \frac{1}{12} + \frac{1}{20} + \frac{1}{30} + \frac{1}{42}$, &c. ∞ per præc. $\frac{1}{1}$
> D.. $+ \frac{1}{6} + \frac{1}{12} + \frac{1}{20} + \frac{1}{30} + \frac{1}{42}$, &c. ∞ C $- \frac{1}{2}$ ∞ $\frac{1}{2}$
> E... $+ \frac{1}{12} + \frac{1}{20} + \frac{1}{30} + \frac{1}{42}$, &c. ∞ D $- \frac{1}{6}$ ∞ $\frac{1}{3}$
> F..... $+ \frac{1}{20} + \frac{1}{30} + \frac{1}{42}$, &c. ∞ E $- \frac{1}{12}$ ∞ $\frac{1}{4}$
> &c. ∞ &c.
> ∞ G; unde sequitur, se-
>
> riem G ∞ A, totum parti, si summa finita esset. Ego

So the soul of immensity dwells in minutia.
And in narrowest limits no limits inhere.
What joy to discern the minute in infinity!
The vast to perceive in the small, what divinity!

Remark. Jakob Bernoulli, eager to examine other infinite series, soon turned his attention in section XVII of *Tractatus* to

$$1 + \frac{1}{4} + \frac{1}{9} + \frac{1}{16} + \cdots = \sum_{k=1}^{\infty} \frac{1}{k^2}, \tag{4}$$

the evaluation of which "is more difficult than one would expect" ("difficilior est quam quis expectaverit"), an observation that turned out to be quite an understatement. He correctly established the convergence of (4) by comparing it termwise with the greater, yet convergent series

$$1 + \frac{1}{3} + \frac{1}{6} + \frac{1}{10} + \cdots$$
$$= 2\left(\frac{1}{2} + \frac{1}{6} + \frac{1}{12} + \frac{1}{20} + \cdots \right) = 2(1) = 2.$$

But evaluating the sum in (4) was too much for Jakob, who noted rather plaintively

> If anyone finds and communicates to us that which up to now has eluded our efforts, great will be our gratitude.

The evaluation of (4), of course, resisted the attempts of another generation of mathematicians until 1734, when the incomparable Leonhard Euler devised an enormously clever argument to show that it summed to $\pi^2/6$. This result, which Jakob Bernoulli unfortunately did not live to see, surely ranks among the most unexpected and peculiar in all of mathematics. For the original proof, see ([4], pp. 83–85). A modern outline of Euler's reasoning can be found in ([2], pp. 486–487).

REFERENCES

1. Jakob Bernoulli, *Ars Conjectandi*, Basel, 1713.
2. Carl B. Boyer, *A History of Mathematics*, Princeton University Press, 1985.
3. C. H. Edwards, *The Historical Development of the Calculus*, Springer-Verlag, New York, 1979.
4. Leonhard Euler, *Opera Omnia* (1), Vol. 14 (C. Boehm and G. Faber, editors), Leipzig, 1925.
5. Morris Kline, *Mathematical Thought from Ancient to Modern Times*, Oxford University Press, New York, 1972.
6. D. J. Struik (editor), *A Source Book in Mathematics (1200–1800)*, Harvard University Press, 1969.
7. Translated from the Latin by Helen M. Walker, as noted in David E. Smith's *A Source Book in Mathematics*, Dover, New York, 1959, p. 271.

On Rearrangements of the Alternating Harmonic Series

Fon Brown, L. O. Cannon, Joe Elich, and David G. Wright

The two series most familiar to beginning calculus students are the Harmonic Series (usually a student's first example of a divergent series whose terms approach zero) and the Alternating Harmonic Series (the first conditionally convergent series). When Taylor series are studied, it is shown that the Alternating Harmonic Series (abbreviated AHS) actually converges to ln 2.

Because of its familiarity, the AHS is a reasonable candidate for illustrating how conditionally convergent series may be rearranged to change their sums. For example, we may replace each odd term x of the AHS by $(2x - x)$ and get a pattern in which one positive term is followed by two negative terms. If we then multiply each term of the new series by $1/2$, we get a rearrangement of the AHS which converges to half of the original sum. Thus,

$$\ln 2 = 1 - \tfrac{1}{2} + \tfrac{1}{3} - \tfrac{1}{4} + \tfrac{1}{5} - \tfrac{1}{6} + \cdots$$
$$= 2 - 1 - \tfrac{1}{2} + \tfrac{2}{3} - \tfrac{1}{3} - \tfrac{1}{4} + \tfrac{2}{5} - \tfrac{1}{5} - \tfrac{1}{6} + \cdots,$$

and the rearranged AHS satisfies

$$(\tfrac{1}{2})\ln 2 = 1 - \tfrac{1}{2} - \tfrac{1}{4} + \tfrac{1}{3} - \tfrac{1}{6} - \tfrac{1}{8} + \tfrac{1}{5} - \tfrac{1}{10} - \cdots.$$

This is an example of a *regular rearrangement*, in which there is a regular pattern consisting of a fixed number of positive terms taken in order, followed by a fixed number of negative terms taken in order. We use $A(m, n)$ to denote such an ordered rearrangement consisting of *m positive terms* followed by *n negative terms*. Thus, the example above shows that $A(1,2)$ converges to $(1/2)\ln 2$. A simple argument [see Arthur B. Simon's *Calculus with Analytic Geometry*, Scott, Foresman and Co., Illinois (1982), 514] shows that $A(2, 1)$ converges to $(3/2)\ln 2$.

Our investigation was initially prompted by a classroom problem posed by Edwin E. Moise [*Introductory Problem Courses in Analysis and Topology*, Springer-Verlag, New York (1982), 45]:

Find a rearrangement of the AHS that converges to zero.

The standard argument to show the existence of a rearrangement of the AHS which converges to a given limit L is to observe that because the Harmonic Series diverges, it is possible to add enough consecutive positive terms, $1 + 1/3 + \cdots$ to get a partial sum larger than L. Then enough consecutive negative terms are added to make the partial sum smaller than L, and the process is continued. For $L = 0$, a little work with a hand calculator showed some unanticipated regularity. To get a rearrangement of the AHS to converge to zero, it appeared that we needed *one* positive term and then *four* negative terms. The next positive term, $1/3$, made the sum positive, and four more negative terms were needed to get a negative partial sum. The same regularity continued as far as we were able to check by hand, so the class was led to search for a proof that this rearrangement $A(1,4)$ does, in fact, converge to zero. (A complicated inductive proof was discovered based on showing that the partial sums S_m are positive if 5 does not divide m, and S_m are negative if 5 divides m.)

Having discovered that the regular rearrangement $A(1,4)$ converges to zero, students began to ask questions about the convergence of $A(m,n)$ in general. We used a microcomputer to generate data on partial sums for a number of regular rearrangements. A few partial sums are given:

$$A(1,1) \sim .6907 \qquad A(1,2) \sim .3453$$
$$A(2,2) \sim .6919 \qquad A(2,4) \sim .3459$$
$$A(3,3) \sim .6923 \qquad A(3,4) \sim .5489$$
$$A(4,4) \sim .6925 \qquad A(2,1) \sim 1.0372.$$

Students were encouraged to formulate their own conjectures from the partial sums. For example, the data suggest that $A(2,2)$, $A(3,3)$ and $A(4,4)$ all converge to $\ln 2$. Similarly, it appears that $A(1,2)$ and $A(2,4)$ have the same limit. Ultimately we were led to the following:

Theorem. *Every regular rearrangement of the AHS converges. In particular, $A(m,n)$ converges to* $\ln 2 + (1/2)\ln(m/n)$.

Our proof of this theorem is based on the well-known fact [see, for example, Thomas and Finney's *Calculus and Analytic Geometry*, sixth edition, Addison-Wesley, Massachusetts (1984), p. 640, problem 45] that the difference $H_N - \ln N$ between H_N (the Nth partial sum of the Harmonic Series) and $\ln N$ approaches a constant γ (Euler's Constant) as N becomes large.

In working with partial sums for a rearrangement $A(m,n)$, it is most natural to consider sums of N terms, where N is divisible by $m + n$. For convenience of notation, O_N and E_N will, respectively, denote the sum of the first N odd terms and the sum of the first N even terms of the AHS. Thus, $O_N + E_N = H_{2N}$ and $2E_N = H_N$.

Let S_N be the Nth partial sum of $A(m,n)$, where $N = (m+n)k$. Collecting positive and negative terms together, we have:

$$S_{(m+n)k} = O_{mk} - E_{nk} = O_{mk} + E_{mk} - E_{mk} - E_{nk}$$
$$= H_{2mk} - \left(\frac{1}{2}\right)H_{mk} - \left(\frac{1}{2}\right)H_{nk}$$
$$= (H_{2mk} - \ln 2mk) - \left(\frac{1}{2}\right)(H_{mk} - \ln mk) - \left(\frac{1}{2}\right)(H_{nk} - \ln nk)$$

$$+ \ln 2mk - \left(\frac{1}{2}\right)\ln mk - \left(\frac{1}{2}\right)\ln nk$$

$$= (H_{2mk} - \ln 2mk) - \left(\frac{1}{2}\right)(H_{mk} - \ln mk) - \left(\frac{1}{2}\right)(H_{nk} - \ln nk)$$

$$+ \ln 2 + \left(\frac{1}{2}\right)\ln\left(\frac{m}{n}\right).$$

Therefore,

$$\lim_{k \to \infty} S_{(m+n)k} = \gamma - \left(\frac{1}{2}\right)\gamma - \left(\frac{1}{2}\right)\gamma + \ln 2 + \frac{1}{2}\ln\left(\frac{m}{n}\right)$$

$$= \ln 2 + \frac{1}{2}\ln\left(\frac{m}{n}\right).$$

For each fixed $r \in \{1, 2, \ldots, m+n-1\}$, we have $S_{(m+n)k+r} = S_{(m+n)k} + \{r \text{ terms of } A(m,n)\}$. Since the terms of $A(m,n)$ approach zero, $\lim_{k \to \infty} S_{(m+n)k+r} = \ln 2 + (\frac{1}{2})\ln(m/n)$ for each r. Therefore, it is an easy matter to verify that $\lim_{n \to \infty} S_N = \ln 2 + (\frac{1}{2})\ln(m/n)$, even if N is not divisible by $m+n$.

There are some additional observations which may be pertinent here. As was pointed out earlier, there is a standard argument to show the existence of a rearrangement of the AHS which will converge to any given real number L. However, given an arbitrary real number L, there is not necessarily a *regular* rearrangement which will converge to L. From our theorem, we can prove the following:

Corollary. *There is a regular rearrangement of the AHS which converges to L if and only if e^{2L} is a rational number.*

If $A(m,n)$ converges to L, then $L = \ln 2 + (1/2)\ln(m/n)$ and $e^{2L} = 4m/n$ is rational. Suppose, conversely, that $e^{2L} = p/q$ is rational. Then $L = (1/2)\ln(p/q)$. If we choose integers m, n satisfying $m/n = p/4q$, then $A(m,n)$ converges to $\ln 2 + (1/2)\ln(p/4q) = L$.

As we had previously observed, there is not a unique rearrangement of the AHS that converges to a particular limit. More interestingly, we were led to consider regular rearrangements by following the standard argument for rearranging the AHS to converge to zero, but the standard rearrangement procedure does not necessarily lead to a regular rearrangement. Our Theorem shows that $A(4,1)$ converges to $\ln 4$; but if we had used the standard procedure to get a rearrangement converging to $\ln 4$, we would choose only *three* positive terms before the first negative term.

Editor's Note: For related discussions of this theme, see "Rearranging the Alternating Harmonic Series" by C. C. Cowen, K. R. Davidson, and R. P. Kaufman [Amer. Math. Monthly, 87 (December 1980) 817–819], "Sum-Preserving Rearrangements of Infinite Series" by Paul Schaefer [Amer. Math. Monthly, 88 (January 1981) 33–40], and "Rearranging Terms in Alternating Series" by Richard Beigel [Math. Mag., 54 (November 1981) 244–246].

An Improved Remainder Estimate for Use with the Integral Test

Roger Nelsen

Nearly every modern calculus text contains the following result in the chapter on infinite series: If $\sum_{i=0}^{\infty} f(i)$ converges to S by the integral test, and $S_n = \sum_{i=1}^{n} f(i)$ denotes the nth partial sum of the series, then the "remainder" $R_n = S - S_n = \sum_{i=n+1}^{\infty} f(i)$ satisfies

$$\int_{n+1}^{\infty} f(x)\,dx \leq R_n \leq \int_{n}^{\infty} f(x)\,dx. \tag{1}$$

The hypotheses for the integral test require that f be continuous, positive, and decreasing on $[1, \infty)$. In [1], R. K. Morley showed that if, as is often the case, f is also *convex* (concave up) on $[1, \infty)$, then the "traditional" estimate (1) can be improved to

$$\int_{n}^{\infty} f(x)\,dx - \frac{1}{2}f(n) \leq R_n \leq \int_{n}^{\infty} f(x)\,dx - \frac{1}{2}f(n+1). \tag{2}$$

The purpose of this Capsule is to note that, under the same hypotheses, these estimates can be further sharpened to

$$\frac{1}{2}f(n+1) + \int_{n+1}^{\infty} f(x)\,dx \leq R_n \leq \int_{n+1/2}^{\infty} f(x)\,dx. \tag{3}$$

The proof of (3) follows directly from the observation that for convex functions, the midpoint rule underestimates integrals while the trapezoidal rule overestimates integrals. (This observation holds for convergent improper integrals as well as for proper integrals.) The right-hand inequality in (3) follows from the fact that $R_n = \sum_{i=n+1}^{\infty} f(i)$ is a midpoint rule estimate for $\int_{n+1/2}^{\infty} f(x)\,dx$. Similarly, the left-hand inequality in (3) follows from the fact that $\sum_{i=n+1}^{\infty} f(i) - \frac{1}{2}f(n+1)$ is a trapezoidal rule estimate for $\int_{n+1}^{\infty} f(x)\,dx$.

In actual practice, when we use (1)–(3) in estimating S, we replace R_n by $S - S_n$ and solve for S. Thus, these inequalities become

$$S_n + \int_{n+1}^{\infty} f(x)\,dx \le S \le S_n + \int_n^{\infty} f(x)\,dx, \tag{4}$$

$$S_n + \int_n^{\infty} f(x)\,dx - \frac{1}{2}f(n) \le S \le S_n + \int_n^{\infty} f(x)\,dx - \frac{1}{2}f(n+1) \tag{5}$$

$$S_n + \frac{1}{2}f(n+1) + \int_{n+1}^{\infty} f(x)\,dx \le S \le S_n + \int_{n+1/2}^{\infty} f(x)\,dx. \tag{6}$$

For example, consider estimating the sum S of $\sum_{i=1}^{\infty} i^{-4}$ using $S_5 \cong 1.080352$ (all calculations have been rounded to six places). Since $f(x) = x^{-4}$ is convex on $[1, \infty)$, inequalities (4)–(6) yield the following intervals:

Table 1. Using S_5 to approximate $\sum_{i=1}^{\infty} i^{-4}$

Method	Interval
(1)	$1.081895 \le S \le 1.083019$
(2)	$1.082219 \le S \le 1.082633$
(3)	$1.082281 \le S \le 1.082355$

Of course, the sum S is actually equal to $\pi^4/90 \cong 1.082323$.

In this example, the third interval is less than 7% as wide as the first and less than 18% as wide as the second (.000074 versus .001123 and .000414, respectively), a substantial improvement in precision at virtually no "cost" in additional computation. In general, the width of the interval for R_n in (1) is given by $\int_n^{n+1} f(x)\,dx$, the width in (2) is $\frac{1}{2}[f(n) - f(n+1)]$, while the width of the interval in (3) is $\int_{n+1/2}^{n+1} f(x)\,dx - \frac{1}{2}f(n+1)$. The improvement in precision is illustrated in Figure 1, where the areas of the shaded regions are numerically equal to the interval widths in each case.

The geometric interpretation of the interval width in Figure 1(c) leads to an efficient method to determine the number n of terms to use in (3) to obtain an approximation to any desired precision ε. The area of the shaded region in Figure 1(c) is no larger than the area of the triangle with the same vertices, $\frac{1}{4}[f(n+1/2) - f(n+1)]$. If we wish the interval width to be less than ε, then n must satisfy $f(n+1/2) - f(n+1) < 4\varepsilon$. When f is differentiable on $(1, \infty)$, we can apply the mean value theorem to f on $(n+1/2, n+1)$ to obtain $f(n+1/2) - f(n+1) = f'(c)(-1/2)$ for some c in $(n+1/2, n+1)$. But f is convex, so f' is increasing. Thus, $-f'$ is decreasing, and so $-f'(c) \le -f'(n+1/2)$. Hence, $f(n+1/2) - f(n+1) < 4\varepsilon$ whenever

$$-f'(n+1/2) < 8\varepsilon. \tag{7}$$

Returning to our example, let us find the number n of terms necessary to approximate $\sum_{i=1}^{\infty} i^{-4}$ correctly to six decimal places using inequalities (6). Solving (7) with

An Improved Remainder Estimate for Use with the Integral Test

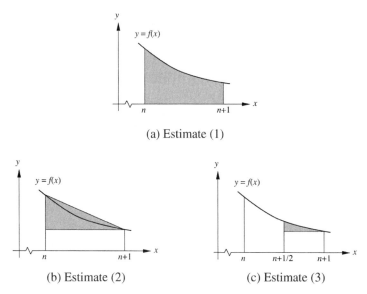

(a) Estimate (1)

(b) Estimate (2)

(c) Estimate (3)

Figure 1.

$f(x) = x^{-4}$ and $\varepsilon = .5 \times 10^{-6}$ yields $n \geq 16$. Using $S_{16} \cong 1.08224917$ (to 8 places) and (3) yields $1.08232300 \leq S \leq 1.08232337$; hence $S \cong 1.082323$, correct to 6 places. To obtain the same accuracy with (2) requires $n \geq 21$, and with (1) requires $n \geq 38$.

Reference

1. R. K. Morley, The remainder in computing by series, *American Mathematical Monthly* **57** (1950), 550–551.

A Differentiation Test for Absolute Convergence

Yaser S. Abu-Mostafa

In this note, we describe a new test which provides a necessary and sufficient condition for absolute convergence of infinite series. The test is based solely on differentiation and is very easy to apply. It also provides a pictorial illustration for absolute convergence and divergence.

The discovery was made a few years ago when I was asked to give a lecture on infinite series to my classmates. The subject was new to us and some basic ideas were not quite appreciated at that early stage. I tried to give an informal or pictorial illustration for any concept that sounded abstract. When I mentioned that an infinite series would converge absolutely if its *far away* terms became *small enough*, I had to explain to the students (and to myself as it turned out) *how* far and *how* small.

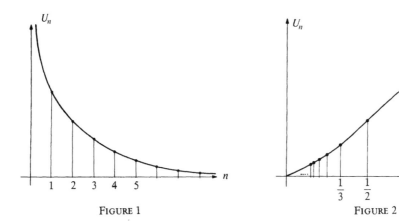

FIGURE 1   FIGURE 2

Plotting the series term by term with the summation index running along the x-axis was not very successful (FIGURE 1). I never seemed to get far enough and the terms looked quite small even with the harmonic series which, I had recently learned, was divergent. I had to come up with a picture that would show what a small term is and how it remains small if one multiplies the whole series by 10^{100} and why no matter how far away a point is, it may still not be far enough.

Remembering the duality of zero and infinity, I thought it might be nice to plot the series term by term with the *inverse* of the summation index running along the x-axis (FIGURE 2) so that we see the whole thing crowded near zero. First of all, the curve of any convergent series had to "hit" the origin since the terms go to zero as $n \to \infty$. I felt that the shape of the curve near the origin should also be related to convergence.

A Differentiation Test for Absolute Convergence

When I plotted divergent series like $\Sigma 1/n$ and $\Sigma 1/\sqrt{n}$, I ended up with positive or infinite slopes at the origin, but the convergent series $\Sigma 1/n^2$ had slope zero at the origin (FIGURE 3). The correspondence between the zero slope and the idea of "small terms" was appealing since a zero slope multiplied by 10^{100} is still a zero slope. The fact that the slope of a curve is a limit concept was in accordance with the "far away" idea. After the lecture, I rushed to check whether my particular illustration might generalize. After some scribbling, it gave rise to a valid criterion for absolute convergence which I call the **differentiation test**.

You probably have guessed the mechanism of the test by now. Roughly speaking, you take the infinite series in question, ΣU_n, and construct the function f defined by $f(1/n) = U_n$. First check that $f(0) = 0$. Now differentiate f and check the value of $f'(0)$: if this is also zero, the series is convergent.

Let us state this formally with the proper qualifying conditions:

DIFFERENTIATION TEST. *Let $\Sigma_{n=1}^{\infty} U_n$ be an infinite series with real terms. Let $f(x)$ be any real function such that $f(1/n) = U_n$ for all positive integers n and d^2f/dx^2 exists at $x = 0$. Then $\Sigma_{n=1}^{\infty} U_n$ converges absolutely if $f(0) = f'(0) = 0$ and diverges otherwise.*

Notice that there is a requirement that $f''(0)$ exists for the test to apply. When this requirement is satisfied, the test is guaranteed to determine whether the series is absolutely convergent or divergent. We will say more about relaxing the requirement of existence for $f''(0)$ later on, but first we present some examples to see how the test works.

We start with simple examples in which we know whether or not the series is convergent (after all, we haven't proved anything yet). Consider the two series

$$\sum_{n=1}^{\infty} \sin\frac{1}{n} \quad \text{and} \quad \sum_{n=1}^{\infty} 1 - \cos\frac{1}{n}.$$

There are many ways to verify that the first series is divergent while the second is convergent. Applying the differentiation test, we examine the functions $\sin x$ and $1 - \cos x$, both of which have second derivatives at $x = 0$. The test now tells us that the first series is divergent

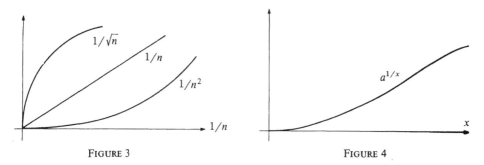

FIGURE 3   FIGURE 4

($f'(0) = 1 \neq 0$) and the second is absolutely convergent ($f(0) = f'(0) = 0$).

Another interesting example is the geometric series $\Sigma_{n=1}^{\infty} a^n$ where $0 < a < 1$. When we substitute x for $1/n$, we get the function $f(x) = a^{1/x} = e^{(\ln a)/x}$ (for $x > 0$ and zero otherwise). Since $\ln a < 0$, $f(x)$ goes to zero very quickly as $x \to 0$ (FIGURE 4). In fact, $f(x)$ has the derivatives of *all* orders at $x = 0$ equal to zero. To see this, differentiate $f(x)$ any number of times and you will always get a (finite) polynomial in $1/x$ multiplied by $f(x)$ itself. Since the exponential is "stronger" than any polynomial, all the derivatives will go to zero as $x \to 0$. This suggests that the geometric series with $0 < a < 1$ is *very* convergent, which is indeed the case.

Now we show some examples where you can determine convergence or divergence right away using the differentiation test while others will require effort to get the result. In fact, for the following examples, using other techniques to determine convergence is practically the same as writing the proof for the differentiation test in the general case.

Consider the infinite series

$$\sum_{n=1}^{\infty} \int_0^{1/n} g(t)\, dt$$

where $g(t)$ is any function that has a derivative at $t=0$. You are required to determine the conditions on g for the series to converge absolutely. How long does it take you to conclude that it will converge absolutely if, and only if, $g(0)=0$? To verify the result, try substituting simple functions like t^2 or e^{-t} for $g(t)$ and carry out the integration, then test the resulting series for convergence or divergence using standard methods.

Now consider:

$$\sum_{n=1}^{\infty} \sinh\left(\tanh\frac{1}{n} - \tan\frac{1}{n} + \sec\frac{1}{n^2} - \cosh\frac{1}{n}\right).$$

Since $\sinh(\tanh x - \tan x + \sec x^2 - \cosh x)$ is analytic and has zero value and zero derivative at $x=0$, the series is absolutely convergent. You can try other compositions of simple functions like these and see that the differentiation test is equivalent to expanding the composite function $f(x)$ in a Taylor series about $x=0$ and checking that the lowest power in the expansion is at least x^2.

If you would like to see other techniques for dealing with these examples, go through the following proof of the differentiation test which depends on such techniques.

Proof. Our proof of the differentiation test depends on L'Hospital's rule, the limit comparison test [1], and the integral test.

Since d^2f/dx^2 is assumed to exist at $x=0$, we are guaranteed (among other things) that $f(x)$ is continuous at $x=0$ and is differentiable in a neighborhood of $x=0$ (we will need the latter to apply L'Hospital's rule). We thus have the following steps relating the conditions on $f(x)$ to the absolute convergence of $\sum_{n=1}^{\infty} U_n = \sum_{n=1}^{\infty} f(1/n)$:

(1) $f(0)=0$ is necessary for any convergence, since $\lim_{n\to\infty} U_n = \lim_{x\to 0} f(x) = f(0)$ and if this is non-zero, the series must diverge.

(2) Suppose that $f(0)$ does equal zero, but $f'(0) = a \neq 0$. Then $\lim_{x\to 0} f(x)/x = \lim_{x\to 0}(f(x)-f(0))/(x-0) = a$. Consequently, $\lim_{n\to\infty} |U_n|/(1/n) = |a| \neq 0$. By the limit comparison test, $\sum_{n=1}^{\infty} U_n$ diverges absolutely since the harmonic series also does.

(3) We have determined that $f(0) = f'(0) = 0$ is necessary for convergence. We now assume that this condition holds and prove sufficiency. Take $0 < u < 1$ and consider the limit

$$\lim_{x\to 0^+} \frac{f(x)}{x^{1+u}} = \lim_{x\to 0^+} \frac{f'(x)}{(1+u)x^u} = \frac{1}{1+u}\lim_{x\to 0^+}\left(\frac{f'(x)-f'(0)}{x-0}\right)x^{1-u} = \frac{f''(0)}{1+u}\lim_{x\to 0^+} x^{1-u} = 0,$$

where the first equality is an application of L'Hospital's rule. Therefore, $\lim_{n\to\infty} |U_n|/(1/n)^{1+u} = 0$ and again by the limit comparison test, $\sum_{n=1}^{\infty} U_n$ must converge absolutely since $\sum_{n=1}^{\infty} 1/n^{1+u}$ converges absolutely by the integral test.

Steps (1), (2), (3) complete the proof.

Perhaps you noticed in part (3) of the proof that the convergence did not depend critically on the existence of $f''(0)$. This is indeed the case and the existence of $f''(0)$ can be replaced by a weaker condition. We note that the condition cannot be completely removed since $\sum_{n=2}^{\infty} 1/n \ln n$, which is absolutely divergent by the integral test, has terms $f(1/n)$ where $f(x) = -x/\ln x$ (for $x > 0$ and zero otherwise); this function $f(x)$ has zero value and zero derivative at $x=0$, but a non-existent second derivative.

The existence of $f''(0)$ in the differentiation test can be replaced, for example, by the existence of $\lim_{x\to 0^+} f'(x)/x^u$ or $d^2|x|^u f(x)/dx^2|_{x=0}$ for some $0 < u < 1$ (both conditions are implied by the existence of $f''(0)$ when $f(0) = f'(0) = 0$). Very minor modification of part (3) of the proof above is needed in these cases. Certain weaker conditions will also work; their discovery is left as a simple exercise.

It is also obvious that only the existence of $f'(0)$ is needed to conclude absolute divergence of a series using the test. Since divergence is seldom good news, I choose to leave the test in its simple symmetric form. Finally, one can apply the test with $f(1/n) = |U_n|$ instead of U_n. This covers complex series as well.

I should like to acknowledge Dr. Brent Smith and one of the referees for their assistance.

References

[1] T. M. Apostol, Mathematical Analysis, Addison-Wesley, 1974, pp. 183–193.
[2] A. E. Taylor & W. R. Mann, Advanced Calculus, John Wiley, 1972, pp. 598–630.

Math Bite: Equality of Limits in Ratio and Root Tests

Prem N. Bajaj

Relations among various tests for convergence of series arose in the note [1] and its corrigendum. For the convergence of series $\sum_{n=0}^{\infty} a_n$ with positive terms, two well-known tests are as follows:

D'ALEMBERT'S RATIO TEST. *Suppose that* $\lim_{n \to \infty} \frac{a_{n+1}}{a_n} = L$. *Then* Σa_n *converges if* $L < 1$ *and diverges if* $L > 1$.

CAUCHY'S ROOT TEST. *Suppose that* $\lim_{n \to \infty} a_n^{1/n} = M$. *Then* Σa_n *converges if* $M < 1$ *and diverges if* $M > 1$.

We show—using the tests themselves—that if the limits L and M exist, they must be equal. To this end, suppose that $L < M$. (The argument for the case $M < L$ is similar.) Choose a real number k such that $L < k < M$.

Now consider the series Σb_n, where $b_n = a_n / k^n$. Then we have

$$\lim_{n \to \infty} \frac{b_{n+1}}{b_n} = \frac{L}{k} < 1, \quad \text{but} \quad \lim_{n \to \infty} b_n^{1/n} = \frac{M}{k} > 1.$$

The first limit implies that Σb_n converges; the second, that Σb_n diverges. This is a contradiction.

REFERENCES

1. D. Cruz-Uribe, SFO, The relation between the root and ratio tests, this MAGAZINE 70 (1997), 214–215; corrigendum, this MAGAZINE 70 (1997), 310–311.

Another Proof of the Formula $\sum_0^\infty (1/n!)$

Norman Schaumberger

The standard derivation of the formula $e = \sum_0^\infty (1/n!)$ using infinite series is well known. A proof using the binomial theorem and $\lim_{n \to \infty}(1 + 1/n)^n = e$ can be presented earlier in a course in calculus. However, this proof is not particularly simple. [See N. D. Kazarinoff, *Analytic Inequalities*, Holt, Rinehart, and Winston, NY, 1961, pp. 40–41.] In this note we use the mean value theorem to give a simple proof of $e = \sum_0^\infty (1/n!)$ that can be presented as soon as the exponential function has been discussed.

If $1 \geq x > 0$, the mean value theorem yields

$$\frac{e^x - 1}{x} = e^c \quad \text{for some } c \in (0, x).$$

Hence,

$$e \geq e^x > \frac{e^x - 1}{x} > 1$$

or

$$1 + ex > e^x > 1 + x.$$

It follows that

$$\int_0^x (1 + ey)\, dy > \int_0^x e^y\, dy > \int_0^x (1 + y)\, dy.$$

That is,

$$1 + x + \frac{ex^2}{2} > e^x > 1 + x + \frac{x^2}{2}.$$

Thus

$$\int_0^x \left(1+y+\frac{ey^2}{2}\right) dy > \int_0^x e^y \, dy > \int_0^x \left(1+y+\frac{y^2}{2}\right) dy$$

or

$$1+x+\frac{x^2}{2!}+\frac{ex^3}{3!} > e^x > 1+x+\frac{x^2}{2!}+\frac{x^3}{3!}.$$

Continuing in this manner, we get

$$1+x+\frac{x^2}{2!}+\frac{x^3}{3!}+\cdots+\frac{ex^{n+1}}{(n+1)!} > e^x > 1+x+\frac{x^2}{2!}+\frac{x^3}{3!}+\cdots+\frac{x^{n+1}}{(n+1)!}.$$

Putting $x=1$ gives

$$e - \frac{1}{(n+1)!} > 1+1+\frac{1}{2!}+\frac{1}{3!}+\cdots+\frac{1}{n!} > e - \frac{e}{(n+1)!}.$$

As $n \to \infty$, the left and right sides tend to e and we obtain the desired result.

Taylor Polynomial Approximations in Polar Coordinates

Sheldon P. Gordon

One of the major features of many of the calculus reform projects currently being conducted, including the Harvard Calculus Project, is a renewed emphasis on Taylor polynomial approximations [1]. Graphics software and graphing calculators allow students to explore the concept of such approximations visually and so come to a much deeper understanding of the process of approximation of one function by another.

We will consider a natural extension of these ideas and attempt to approximate a curve given in polar coordinates using a Taylor polynomial. Suppose we begin naively by considering a polar function $r = f(\theta)$ in the neighborhood of a point $P(r_0, \alpha)$ and define the linear approximation to the function as the tangent line at that point. As is shown in most calculus texts, the slope of the tangent line to a polar curve is given by

$$m = \frac{dy}{dx} = \frac{f'(\theta)\sin\theta + f(\theta)\cos\theta}{f'(\theta)\cos\theta - f(\theta)\sin\theta}.$$

Therefore, the equation of the tangent line to the polar curve at the point P is given in polar coordinates by

$$r \sin\theta - r_0 \sin\alpha = m(r \cos\theta - r_0 \cos\alpha)$$

so that

$$r = \frac{r_0(\sin\alpha - m\cos\alpha)}{\sin\theta - m\cos\theta}.$$

We invite those readers who feel somewhat masochistic to continue this process to determine the polar equation for the approximating parabola and the corresponding higher degree polynomials. You will see that this naive approach is not particularly fruitful. Rather, it is necessary to re-interpret the idea of a Taylor polynomial approximation to reflect the circular symmetries that naturally arise in equations given in polar coordinates.

First, the constant approximation to $f(\theta)$, $r = f(\alpha) = r_0$, is a circle of radius r_0 centered at the pole. Next, the first-order approximation,

$$r = f(\alpha) + f'(\alpha)(\theta - \alpha)$$

is a translation and rotation of the standard Archimedean spiral $r = a\theta$. This "linear" approximation bends along with the original curve and so should be a reasonably accurate approximation to the curve for points near $\theta = \alpha$. In a similar way, the quadratic approximation,

$$r = f(\alpha) + f'(\alpha)(\theta - \alpha) + f''(\alpha)(\theta - \alpha)^2/2$$

is also a counter-clockwise spiral. Notice that the sign of $f''(\alpha)$ does not affect the counter-clockwise unrolling of the spiral as θ increases. However, the constant and "linear" terms may affect the overall spiral behavior when θ is close to α in the sense that the spiral may not unroll in a monotonically increasing manner.

Clearly, if we continue this process, the Taylor polynomial approximation in polar coordinates is a linear combination of terms of the form $(\theta - \alpha)^n$. Each such term is a spiral opening in the counter-clockwise direction. However, since a Taylor polynomial is "centered" about $\theta = \alpha$, it may be helpful to think of the approximation as consisting of two separate spirals, both initiating at the point where $\theta = \alpha$ and where one spiral opens counter-clockwise for $\theta > \alpha$ and increasing and the other opens clockwise for $\theta < \alpha$ and decreasing toward $-\infty$.

Before examining how the Taylor polynomial approximations apply, we digress to consider the properties of such *polynomial spirals*. We immediately observe that the larger n is, the more rapidly the corresponding spiral unrolls for $\theta - \alpha > 1$ and the less rapidly for $\theta - \alpha < 1$. Further, if the coefficient of such a term is negative or if the values of $(\theta - \alpha)^n$ are negative, then the spiral is reflected back through the pole, but will still open in the counter-clockwise direction. See Figure 1.

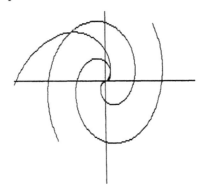

Figure 1
Graphs of $r = \theta$, $r = \theta^2$ and $r = -\theta$.

Suppose now, for simplicity, that we consider a polar polynomial of degree n in θ,

$$P(\theta) = a_0\theta^n + a_1\theta^{n-1} + a_2\theta^{n-2} + \cdots + a_n.$$

In the large, the polar graph of such a polynomial will be a counter-clockwise spiral. On a more local basis, various types of behavior are possible. The following observations are evident after a little thought:

1. The real roots of $P(\theta)$ correspond to angles at which the curve passes through the pole.

2. As θ runs through an interval between successive real roots $\theta_1 < \theta_2$ of $P(\theta)$, the point $(\theta, P(\theta))$ moves around a closed loop of the polar graph of P, leaving the

pole tangent to the line $\theta = \pm \theta_1$ and returning tangent to the line $\theta = \pm \theta_2$.

3. If θ_0 is a real root of P with even multiplicity, the polar graph of P has a cusp at the pole tangent to the line $\theta = \pm \theta_0$. Note that the sign of $P(\theta)$ does not change as θ passes through the value θ_0.

4. Each of the relative maxima and minima of the polynomial corresponds to a point where the distance from the pole to the spiral is a relative maximum or minimum.

5. Each of the points of inflection of the polynomial corresponds to a point where the spiral has the same curvature as a "linear" spiral, $r = a\theta$, for a given value of θ. However, it is much harder to distinguish such points geometrically in the polar plane than in the rectangular plane.

We next see how these properties of polynomial spirals provide the type of agreement we would like to achieve between a polar curve and the corresponding Taylor polynomial approximation in polar coordinates. The agreement can best be appreciated visually. In Figures 2a, 2b and 2c, we show the successive spiral polynomial approximations $r = \theta$, $r = \theta - \theta^3/3!$ and $r = \theta - \theta^3/3! + \theta^5/5!$ to the circle $r = \sin \theta$. By the fifth-degree term, the approximation cannot be distinguished, by eye, from the circle between $\theta = -\pi/2$ and $\pi/2$.

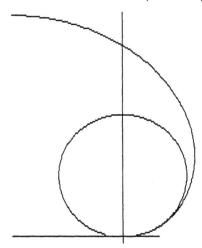

Figure 2a
Graph of $r = \theta$ (outer curve) as approximation to $r = \sin \theta$ on $\theta \in [0, \pi]$.

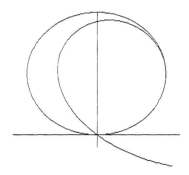

Figure 2b
Graph of $r = \theta - \theta^3/3!$ (inner curve) as approximation to $r = \sin \theta$ on $\theta \in [0, \pi]$.

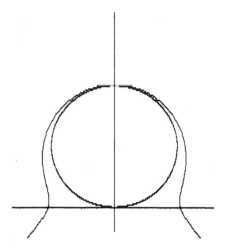

Figure 2c
Graph of $r = \theta - \theta^3/3! + \theta^5/5!$ as approximation to $r = \sin \theta$ on $\theta \in [0, \pi]$.

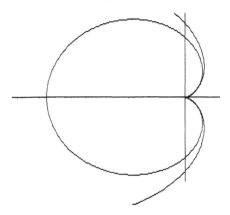

Figure 3a
Graph of $r = \tfrac{1}{2}\theta^2$ (outer curve) as approximation to $r = 1 - \cos \theta$ (inner curve) on $\theta \in [-\pi, \pi]$.

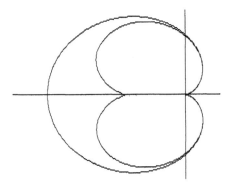

Figure 3b
Graph of $r = \theta^2/2 - \theta^4/4!$ (inner curve) as approximation to $r = 1 - \cos \theta$ (outer curve) on $\theta \in [-\pi, \pi]$.

Taylor Polynomial Approximations in Polar Coordinates

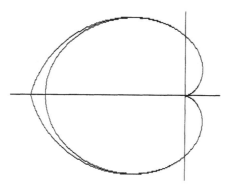

Figure 3c
Graph of $r = \theta^2/2 - \theta^4/4! + \theta^6/6!$ (outer curve) as approximation to $r = 1 - \cos\theta$ (inner curve) on $\theta \in [-\pi, \pi]$.

In Figures 3a, 3b and 3c, we show the successive approximations of degree $n = 2, 4, 6$, respectively, to the cardioid $r = 1 - \cos\theta$. Notice how the resulting polynomials all have double roots at $\theta = 0$ since the leading term in each case is quadratic in θ and so each polynomial accurately approximates the cusp in the cardioid. Notice also that successive approximations are alternatively outside and inside the cardioid. Further, each successive approximating curve is clearly a better fit to the cardioid.

In Figures 4a, 4b and 4c, we show three approximations to the four-leaf rose, $r = \sin 2\theta$, corresponding to Taylor polynomials of degree $n = 1, 3$, and 9 respectively. (In these graphs, we use different viewing "windows" to optimize the view; one unfortunate consequence is that the dimensions of the rose appear somewhat different in each case.) Notice how, as the degree of the approximating polynomial increases and it has more real roots, successively more loops are produced to approximate the petals of the rose curve. By the last case with $n = 9$, the shape of the approximation mirrors the rose rather well and, in fact, it is virtually impossible to distinguish by eye the upper pair of petals of the rose from the upper pair of loops of the approximation.

The above discussions can be presented effectively in the classroom using a computer and projector or graphing calculators if each student has one. Alternatively, they can be assigned to students to investigate on their own using what

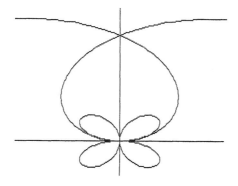

Figure 4a
Graph of $r = 2\theta$ as approximation to $r = \sin 2\theta$ on $\theta \in [-\pi, \pi]$.

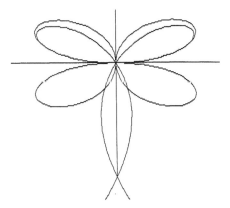

Figure 4b
Graph of $r = 2\theta - (2\theta)^3/3!$ as approximation to $r = \sin 2\theta$ on $\theta \in [-\pi, \pi]$.

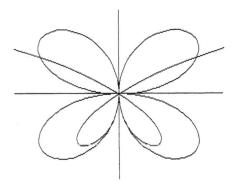

Figure 4c
Graph of ninth degree Taylor polynomial approximation to $r = \sin 2\theta$ on $\theta \in [-\pi, \pi]$.

graphing technology is available. It is a particularly nice exercise because it extends the coverage of two important themes in calculus, Taylor approximations and polar coordinate graphs, by showing the interplay between them. That reinforcement helps students better understand both topics. Equally importantly, too many topics covered in calculus appear, to the students, to be totally unrelated since they appear in different chapters and there is no cross-reference between them. An exercise of this type serves to break down one of the walls between two important ideas and helps the students see some of the fascinating connections between them.

Acknowledgment. The author gratefully acknowledges the support provided by the National Science Foundation under grants #USE-89-53923 and #USE-91-50440. The views expressed here are those of the author and do not necessarily reflect those of the Foundation, the Harvard Calculus Reform Project or the Math Modeling/PreCalculus Reform Project under which the work was conducted.

Reference

1. Sheldon P. Gordon, Introducing Taylor polynomial approximations in introductory calculus, *PRIMUS*, 1 (1991) 305–313.

Proof Without Words: The Taylor Polynomials of $\sin\theta$

John Quintanilla

I.

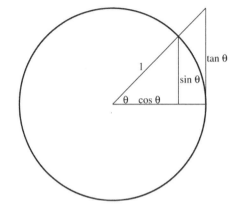

$$\frac{1}{2}\sin\theta\cos\theta < \frac{1}{2}\theta < \frac{1}{2}\frac{\sin\theta}{\cos\theta}$$

$$\cos\theta < \frac{\theta}{\sin\theta} < \frac{1}{\cos\theta}$$
$$\downarrow \qquad\qquad \downarrow$$
$$1 \qquad\qquad 1$$

$$\sin\theta \approx \theta \qquad (\theta \approx 0)$$

II(a).

$$T(\theta) = \frac{1}{2}\left[2\sin\left(\frac{\theta}{2}\right)\right]\left[\cos\left(\frac{\theta}{2}\right)\right] = \frac{\sin\theta}{2}$$

$$A(\theta) = \frac{1}{2}\left[2\sin\left(\frac{\theta}{2}\right)\right]\left[1-\cos\left(\frac{\theta}{2}\right)\right] = \sin\left(\frac{\theta}{2}\right)\left[2\sin^2\left(\frac{\theta}{4}\right)\right]$$

$$= 2\sin\left(\frac{\theta}{2}\right)\sin^2\left(\frac{\theta}{4}\right)$$

$$\approx 2\left(\frac{\theta}{2}\right)\left(\frac{\theta}{4}\right)^2 = \frac{\theta^3}{16}$$

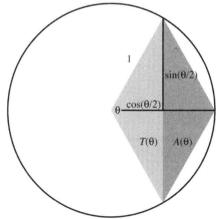

II(b).

$$\frac{\theta}{2} = T(\theta) + A(\theta) + 2A(\theta/2) + 4A(\theta/4) + \cdots$$

$$\frac{\theta}{2} \approx \frac{\sin\theta}{2} + \frac{\theta^3}{16} + 2\left[\frac{(\theta/2)^3}{16}\right] + 4\left[\frac{(\theta/4)^3}{16}\right] + \cdots$$

$$\frac{\theta}{2} \approx \frac{\sin\theta}{2} + \frac{\theta^3}{12}$$

$$\theta - \frac{\theta^3}{6} \approx \sin\theta$$

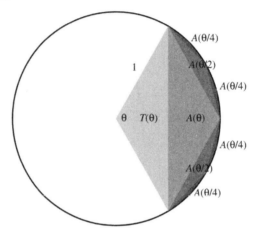

III(a).

$$A(\theta) = 2\sin\left(\frac{\theta}{2}\right)\sin^2\left(\frac{\theta}{4}\right) \approx 2\left[\frac{\theta}{2} - \frac{(\theta/2)^3}{6}\right]\left[\frac{\theta}{4} - \frac{(\theta/4)^3}{6}\right]^2 \approx \frac{\theta^3}{16} - \frac{\theta^5}{256}$$

III(b).

$$\frac{\theta}{2} = T(\theta) + A(\theta) + 2A(\theta/2) + 4A(\theta/4) + \cdots$$

$$\frac{\theta}{2} \approx \frac{\sin\theta}{2} + \left[\frac{\theta^3}{16} - \frac{\theta^5}{256}\right] + 2\left[\frac{(\theta/2)^3}{16} - \frac{(\theta/2)^5}{256}\right]$$

$$+ 4\left[\frac{(\theta/4)^3}{16} - \frac{(\theta/4)^5}{256}\right] + \cdots$$

$$\frac{\theta}{2} \approx \frac{\sin\theta}{2} + \frac{\theta^3}{12} - \frac{\theta^5}{240}$$

$$\theta - \frac{\theta^3}{6} + \frac{\theta^5}{120} \approx \sin\theta$$

ETC.

Appendix I:
Topic Outline for AP Calculus Courses

The topic outline for an Advanced Placement Program calculus course is given below. The topic outline is intended to indicate the scope of the course, but it is not necessarily the order in which the topics need to be taught. Topics marked with an asterisk (*) are on the topic outline for the Calculus BC course, but not on the topic outline for the Calculus AB course. Although the exams are based on the topics listed here, many teachers enrich their courses with additional topics. [Reprinted by permission of the College Board.]

I. Functions, Graphs, and Limits

Analysis of graphs

With the aid of technology, graphs of functions are often easy to produce. The emphasis is on the interplay between the geometric and analytic information and on the use of calculus both to predict and to explain the observed local and global behavior of a function.

Limits of functions (including one-sided limits)
- An intuitive understanding of the limiting process
- Calculating limits using algebra
- Estimating limits from graphs or tables of data

Asymptotic and unbounded behavior
- Understanding asymptotes in terms of graphical behavior
- Describing asymptotic behavior in terms of limits involving infinity
- Comparing relative magnitudes of functions and their rates of change (for example, contrasting exponential growth, polynomial growth, and logarithmic growth)

Continuity as a property of functions
- An intuitive understanding of continuity. (The function values can be made as close as desired by taking sufficiently close values of the domain.)
- Understanding continuity in terms of limits
- Geometric understanding of graphs of continuous functions (Intermediate Value Theorem and Extreme Value Theorem)

* **Parametric, polar, and vector functions** The analysis of planar curves includes those given in parametric form, polar form, and vector form.

II. Derivatives

Concept of the derivative
- Derivative presented graphically, numerically, and analytically
- Derivative interpreted as an instantaneous rate of change
- Derivative defined as the limit of the difference quotient
- Relationship between differentiability and continuity

Derivative at a point
- Slope of a curve at a point. Examples are emphasized, including points at which there are vertical tangents and points at which there are no tangents.
- Tangent line to a curve at a point and local linear approximation
- Instantaneous rate of change as the limit of average rate of change
- Approximate rate of change from graphs and tables of values

Derivative as a function
- Corresponding characteristics of graphs of f and f'
- Relationship between the increasing and decreasing behavior of f and the sign of f'
- The Mean Value Theorem and its geometric interpretation
- Equations involving derivatives. Verbal descriptions are translated into equations involving derivatives and vice versa.

Second derivatives
- Corresponding characteristics of the graphs of f, f', and f''
- Relationship between the concavity of f and the sign of f''
- Points of inflection as places where concavity changes

Applications of derivatives
- Analysis of curves, including the notions of monotonicity and concavity
* Analysis of planar curves given in parametric form, polar form, and vector form, including velocity and acceleration
- Optimization, both absolute (global) and relative (local) extrema
- Modeling rates of change, including related rates problems
- Use of implicit differentiation to find the derivative of an inverse function Interpretation of the derivative as a rate of change in varied applied contexts, including velocity, speed, and acceleration
- Geometric interpretation of differential equations via slope fields and the relationship between slope fields and solution curves for differential equations
* Numerical solution of differential equations using Euler's method
* L'Hospital's Rule, including its use in determining limits and convergence of improper integrals and series

Computation of derivatives
- Knowledge of derivatives of basic functions, including power, exponential, logarithmic, trigonometric, and inverse trigonometric functions

Appendix I

- Derivatives rules for sums, products, and quotients of functions
- Chain rule and implicit differentiation
* Derivatives of parametric, polar, and vector functions

III. Integrals

Interpretations and properties of definite integrals

- Definite integral as a limit of Riemann sums
- Definite integral of the rate of change of a quantity over an interval interpreted as the change of the quantity over the interval:

$$\int_a^b f'(x)dx = f(b) - f(a)$$

- Basic properties of the definite integrals (examples include additivity and linearity)

Applications of integrals

Appropriate integrals are used in a variety of applications to model physical, biological, or economic situations. Although only a sampling of applications can be included in any specific course, students should be able to adapt their knowledge and techniques to solve other similar application problems. Whatever applications are chosen, the emphasis is on using the method of setting up an approximating Riemann sum and representing its limit as a definite integral. To provide a common foundation, specific applications should include finding the area of a region *(including a region bounded by polar curves), the volume of a solid with known cross sections, the average value of a function, the distance traveled by a particle along a line, *the length of a curve (including a curve given in parametric form), and accumulated change from a rate of change.

Fundamental Theorem of Calculus

- Use of the Fundamental Theorem to evaluate definite integrals
- Use of the Fundamental Theorem to represent a particular antiderivative, and the analytical and graphical analysis of functions so defined

Techniques of antidifferentiation

- Antiderivatives following directly from derivatives of basic functions
- Antiderivatives by substitution of variables (including change of limits for definite integrals), *parts, and *simple partial fractions (non-repeating linear factors only)
* Improper integrals (as limits of definite integrals)

Applications of antidifferentiation

- Finding specific antiderivatives using initial conditions, including applications to motion along a line
- Solving separable differential equations and using them in modeling (including the study of the equation $y' = ky$ and exponential growth)
* Solving logistic differential equations and using them in modeling

Numerical approximations to definite integrals

Use of Riemann sums (using left, right, and midpoint evaluation points) and trapezoidal sums to approximate definite integrals of functions represented algebraically, graphically, and by tables of values

*IV. Polynomial Approximations and Series

*Concept of series

A series is defined as a sequence of partial sums, and convergence is defined in terms of the limit of the sequence of partial sums. Technology can be used to explore convergence and divergence.

*Series of constants

* Motivating examples, including decimal expansion
* Geometric series with applications
* The harmonic series
* Alternating series with error bound
* Terms of series as areas of rectangles and their relationship to improper integrals, including the integral test and its use in testing the convergence of p-series
* The ratio test for convergence and divergence
* Comparing series to test for convergence or divergence

*Taylor series

Taylor polynomial approximation with graphical demonstration of convergence (for example, viewing graphs of various Taylor polynomials of the sine function approximating the sine curve)

* Maclaurin series and the general Taylor series centered at $x = a$
* Maclaurin series for the functions e^x, $\sin x$, $\cos x$, and $1/(1-x)$
* Formal manipulation of Taylor series and shortcuts to computing Taylor series, including substitution, differentiation, antidifferentiation, and the formation of new series from known series
* Functions defined by power series
* Radius and interval of convergence of power series
* Lagrange error bound for Taylor polynomials

Source: Copyright © 2008. The College Board. Reproduced with permission. <http://apcentral.collegeboard.com/>.

Appendix II:
Suggested Uses for the Articles in a First-year Calculus Course

In this appendix we present some suggestions from the Advisory Panel and the editors for possible uses of the articles in calculus courses. The list is necessarily incomplete in that many creative teachers will find other ways to use the articles in their courses. We have limited our suggestions to the following:

- EX: *Classroom examples*. These articles present examples that might be used to introduce a topic, or might supplement textbook examples. Teachers might also want to use these examples to reinforce or review a particular idea.
- PR: *Projects*. These articles present topics that a teacher could assign as enrichment projects to individuals or groups of students, especially in an AP class after the May examination.
- BR: *Background reading*. These articles contain material that may give teachers a deeper insight into a particular topic, or a different perspective than what is normally found in calculus texts. Some of these articles will also provide stimulating reading for students. Individual teachers will be able to make reading assignments based on their classroom dynamics and student interests.

Page	Title	EX	PR	BR
2	Touring the Calculus Gallery			✓
21	Calculus: A Modern Perspective			✓
25	Two Historical Applications of Calculus		✓	✓
36	Ideas of Calculus in Islam and India			✓
48	A Tale of Two CD's			✓
54	The Changing Face of Calculus			✓
59	Things I Have Learned at the AP Calculus Reading			✓
69	Book Review: *Calculus with Analytic Geometry*			✓
74	The All-Purpose Calculus Problem			✓
76	Graphs of Rational Functions	✓		

Page	Title	EX	PR	BR
80	Computer-Aided Delusions	✓		
84	An Overlooked Calculus Question	✓	✓	
86	Why Can't We Just Trust the Table?	✓		
87	A Circular Argument		✓	✓
91	A Geometric Proof	✓		
94	The Changing Concept of Change			✓
106	Derivatives Without Limits	✓		
107	Rethinking Rigor in Calculus			✓
118	Rolle over Lagrange	✓	✓	
122	An Elementary Proof of a Theorem in Calculus	✓		
123	A Simple Auxiliary Function	✓		
124	A Note on the Derivative of a Composite Function	✓		
125	Do Dogs Know Calculus?	✓	✓	
130	Do Dogs Know Related Rates?	✓	✓	
133	Do Dogs Know Bifurcations?	✓	✓	
139	The Lengthening Shadow			✓
149	The Falling Ladder Paradox	✓		
154	Solving the Ladder Problem		✓	✓
175	How Not to Land at Lake Tahoe!	✓	✓	
178	The Best Shape for a Tin Can	✓		
182	To Build a Better Box	✓	✓	
189	The Curious 1/3	✓	✓	
191	Hanging a Bird Feeder: Food for Thought	✓		✓
193	Honey, Where Should We Sit?	✓		✓
200	A Dozen Minima for a Parabola	✓	✓	
203	Maximizing the Area of a Quadrilateral	✓	✓	
205	A Generalization of the Minimum Area Problem		✓	
208	Constrained Optimization and Implicit…	✓	✓	
213	For Every Answer There Are Two Questions	✓	✓	
217	Old Calculus Chestnuts: Roast, or Light a Fire?	✓	✓	
220	Cable-laying and Intuition	✓		
223	Descartes Tangent Lines		✓	
226	Can We Use the First Derivative to…		✓	✓
230	Differentiate Early, Differentiate Often!	✓	✓	
235	A Calculus Exercise For the Sums of Integer Powers		✓	
238	L'Hôpital's Rule Via Integration	✓		✓
240	Indeterminate Forms Revisited			✓
245	The Indeterminate Form 0^0		✓	✓

Appendix II

Page	Title	EX	PR	BR
247	On the Indeterminate Form 0^0	✓	✓	✓
249	Variations on a Theme of Newton			✓
257	A Useful Notation for Rules of Differentiation	✓		✓
259	Wavefronts, Box Diagrams, and the Product Rule	✓		
264	$(x^n)' = nx^{n-1}$: Six Proofs	✓		✓
266	Sines and Cosines of the Times	✓	✓	✓
268	The Spider's Spacewalk Derivation of sin′ and cos′	✓		✓
270	Differentiability of Exponential Functions			✓
276	A Discover-e	✓	✓	
278	An Exponential Rule	✓	✓	✓
279	The Derivative of Arctan$x$	✓	✓	✓
281	The Derivative of the Inverse Sine	✓	✓	
282	Graphs and Derivatives of the Inverse Trig Functions	✓	✓	
283	Logarithmic Differentiation			✓
286	Comparison of Two Elementary Approximation…	✓		✓
294	How Should We Introduce Integration?			✓
297	Evaluating Integrals Using Self-Similarity		✓	
308	Self-integrating Polynomials		✓	
311	Symmetry and Integration	✓	✓	
314	Sums and Differences vs Integrals and Derivatives			✓
322	How Do You Slice the Bread?		✓	
326	Disks and Shells Revisited	✓		✓
328	Disks, Shells, and Integrals of Inverse Functions		✓	✓
331	Characterizing Power Functions by Volumes…	✓		✓
334	Gabriel's Wedding Cake		✓	
338	Can You Paint a Can of Paint?	✓		
341	A Paradoxical Paint Pail	✓		
343	Dipsticks for Cylindrical Storage Tanks		✓	
350	Finding Curves with Computable Arc Length	✓	✓	✓
352	Arc Length and Pythagorean Triples	✓	✓	
354	Maximizing the Arclength in the Cannonball Problem		✓	
356	An Example Demonstrating the FTC	✓	✓	
358	Barrow's Fundamental Theorem			✓
360	The Point-slope Formula leads to the FTC			✓
365	A Generalization of the MVT for Integrals	✓		✓
367	Proof Without Words: Look Ma, No Substitution!	✓		
368	Integration by Parts		✓	

Page	Title	EX	PR	BR
372	Tabular Integration by Parts	✓		
377	More on Tabular Integration by Parts	✓		✓
381	A Quotient Rule Integration by Parts Formula		✓	
384	Partial Fraction Decomposition by Division		✓	
387	Partial Fractions by Substitution		✓	
390	Proof Without Words: A Partial Fraction…	✓		
391	Four Crotchets on Elementary Integration	✓		
395	An Application of Geography to Mathematics		✓	✓
400	How to Avoid the Inverse Secant			✓
407	The Integral of $x^{1/2}$, etc.		✓	✓
410	A Direct Proof of the Integral Formula for Arctan	✓		✓
413	Riemann Sums and the Exponential Function	✓		✓
414	Proofs Without Words Under the Magic Curve	✓	✓	
419	Math Without Words: Integrating the Natural Log	✓		
420	Integrals of Products of Sine and Cosine	✓	✓	
422	Moments on a Rose Petal		✓	
425	A Calcululation of $\int_0^\infty e^{-x^2}\,dx$		✓	
428	Calculus in the Operating Room	✓		
430	Physical Demonstrations in the Calculus Classroom	✓		✓
433	Who Needs the Sine Anyway?	✓		✓
435	Finding Bounds for Definite Integrals		✓	✓
439	Estimating Definite Integrals		✓	
443	Proof Without Words: The Trapezoidal Rule	✓		
444	The Midpoint Rule is Better Than the Trapezoidal…	✓		
445	An Elementary Proof of Error Estimates		✓	✓
449	Pictures Suggest How to Improve…	✓		✓
456	The Geometric Series in Calculus			✓
462	A Visual Approach to Geometric Series	✓		
467	The Telescoping Series in Perspective	✓		
469	Proof Without Words [Alternating series]	✓		
470	The Bernoullis and the Harmonic Series			✓
476	On Rearrangements of the Alternating Harmonic…	✓	✓	
479	An Improved Remainder Estimate for Use with…		✓	
482	A Differentiation Test for Absolute Convergence	✓		
486	Math Bite: Equality of Limits in Ratio and Root Tests	✓		✓
487	Another Proof of the Formula $\sum_0^\infty (1/n!)$	✓	✓	
489	Taylor Polynomial Approximations in Polar…		✓	
495	The Taylor Polynomials of $\sin\theta$	✓		

Author Index

Abu-Mostafa, Yaser S.	482	Dundas, Kay	182
Akman, Füsan	414	Dunham, William	2, 470
Andrews, George E.	456		
Anselone, Philip M.	270	Elich, Joe	476
Austin, Bill	139		
		Farmer, Tom	430
Bajaj, Prem N.	486	Ferdinands, John	350
Barnier, William	223	Fink, A. M.	213
Barry, Don	139	Fleron, Julian F.	334
Barshinger, Richard	175	Ford, Ben	283
Berman, David	139	Frantz, Marc	467
Bilodeau, G. E.	278	Frohliger, John A.	193
Boas, R. P.	240	Fulling, S. A.	400
Brannen, Noah Samuel	283		
Bressoud, David M.	xix, 54, 294	Gallego, Jorge	130
Brown, Fon	476	Gardner, Robert B.	257
Buck, Frank	444	Gass, Fred	430
Byrd, Stan	76	Gatica, Juan A.	213
		Gethner, Robert M.	338
Cannon, L. O.	476	Gillman, Leonard	377
Carlip, Walter	326	Gordon, Russell A.	205
Chamberland, Marc	367	Gordon, Sheldon P.	413, 489
Cibes, Margaret	217	Grabiner, Judith V.	94
Corless, Robert M.	249	Graham, Jeffrey A.	308
Couch, Eugene	84	Grant, Douglass L.	422
Cruz-Uribe, David	445		
		Hahn, Alexander J.	25
Dawson, John W., Jr.	191, 259	Hahn, Brian	193
Dawson, Robert	230	Hall, Leon M.	200
Delgado, Alberto L.	425	Hall, Richard L.	80
DeYoung, Gary W.	208	Hammack, Richard	469
Diamond, Harvey	286	Hartig, Donald	238
Dou, Ze-Li	354	Hendel, Russell Jay	264
Dudley, Underwood	68	Hesterberg, Tim	268
Duemmel, James E.	189	Hill, James Colin	322

Horowitz, David	372		
Huerta, Carlos C.	433	Quintanilla, John	495
Insel, Arnold J.	410	Randrianantoanina, Beata	462
		Raphael, Louise	286
Jantosciak, James	223	Richman, Fred	86
Johnson, Craig	281	Richmond, Bettina	331
		Richmond, Donald E.	122
Kalman, Dan	154	Richmond, Tom	331
Katz, Victor J.	36, 266	Rickey, V. Frederick	395
Kendig, Keith	449	Roe, P. L.	178
Kennedy, Dan	48, 59, 74	Roitberg, Joseph	220
Key, Eric	328	Roitberg, Yael	220
Kifowit, Steven J.	390	Rose, David A.	387
Knisley, Jeff	21	Rotando, Louis M.	245
Korn, Henry	245		
Kouba, Duane	226	Sayrafiezadeh, M.	365
Kung, Sidney H.	384	Sanchez, David A.	343
		Schaumberger, Norman	279, 487
Lee, John W.	270	Schwenk, Allen J.	86
Lipkin, Leonard J.	247	Sedinger, Harry	106
Littleton, Pam	343	Sholten, Paul	149
Lynch, Mark	341	Silverman, Herb	123
Lyons, David	469	Simonson, Andrew	149
		Skala, Helen	276
Macula, Anthony J.	360	Smith, Robert S.	118
Malm, Eric	322	Staples, Susan G.	354
Mathews, John H.	91, 407	Starr, Norton	439
Meyers, Leroy F.	391	Strang, Gilbert	314
Minton, Roland	133	Strichartz, Robert S.	297
Moen, Courtney	352	Switkes, Jennifer	381
Moran, Daniel A.	282		
Murty, V. N.	124, 368	Toy, Pearl	428
		Tuchinsky, Philip M.	395
Nelsen, Roger	311, 419, 479	Tucker, Thomas W.	107
Neugebauer, C. J.	445		
Nicol, Sherrie J.	420	Underhill, W. Vance	435
Nord, Gail	322	Urías, Jesús	443
Nord, John	322		
		Wagon, Stan	428
Palais, Bob	356	Wagner, Jack	358
Pennings, Timothy J.	125, 133	Walters, Terry	76
Perruchet, Pierre	130	Wiener, Joseph	235
Peter, Thomas	203	Wright, David G.	476

About the Editors

Caren L. Diefenderfer (A.B. Dartmouth College, M.A., Ph.D. University of California at Santa Barbara) is professor of mathematics at Hollins University. Caren has been active with the AP Calculus Program for over 20 years and served as Chief Reader for AP Calculus from 2004–2007. She has been a leader with MAA efforts on Quantitative Reasoning and is currently the Chair of the MAA's Special Interest Group for the Teaching of Advanced High School Mathematics (SIGMAA TAHSM).

Roger B. Nelsen (B.A. DePauw University, Ph.D. Duke University) is professor emeritus of mathematics at Lewis & Clark College. Roger has been an AP Calculus Reader for many years and has authored or coauthored four books for the MAA: *Proofs Without Words: Exercises in Visual Thinking* (1993); *Proofs Without Words II: More Exercises in Visual Thinking* (2000); *Math Made Visual: Creating Images for Understanding Mathematics* (with Claudi Alsina, 2006); and *When Less is More: Visualizing Basic Inequalities* (with Claudi Alsina, 2009).